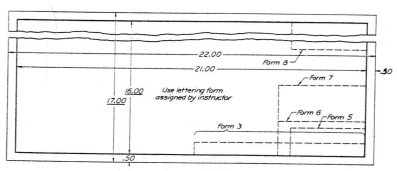

Fig. VI   Size C Sheet (17.00″ × 22.00″)

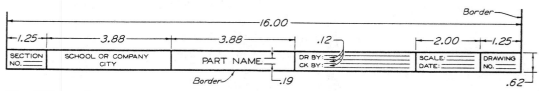

Fig. VII   Form 4. Title Block

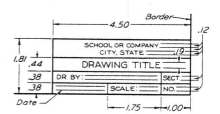

Fig. VIII   Form 5. Title Block

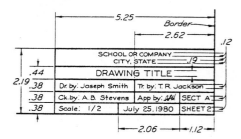

Fig. IX   Form 6. Title Block

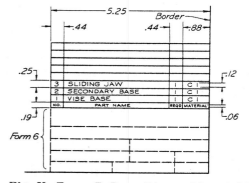

Fig. X   Form 7. Parts List or Material List

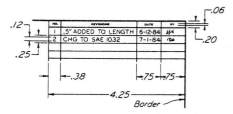

Fig. XI   Form 8. Revision Block

Books by the Authors

ENGINEERING GRAPHICS, 7th ed., by F. E. Giesecke, A. Mitchell, H. C. Spencer, I. L. Hill, R. O. Loving, J. T. Dygdon, J. E. Novak, with S. Lockhart (Prentice Hall, Inc., 2000)

TECHNICAL DRAWING, 11th ed.,by F. E. Giesecke, A. Mitchell, H. C. Spencer, I. L. Hill, J. T. Dygdon, J. E. Novak, with S. Lockhart (Prentice Hall, Inc., 1999)

ENGINEERING DRAWING, PROBLEM SERIES 1, by F. E. Giesecke, A. Mitchell, H. C. Spencer, I. L. Hill, J. T. Dygdon, and J. E. Novak (Prentice Hall, Inc., 1997)

ENGINEERING DRAWING, PROBLEM SERIES 2, by F. E. Giesecke, A. Mitchell, H. C. Spencer, I. L. Hill, R. O. Loving, J. T. Dygdon, and J. E. Novak (Prentice Hall, Inc., 1997)

ENGINEERING DRAWING, PROBLEM SERIES 3, by P. Davis and K. Juneau (Prentice Hall, Inc., 1998)

MODERN GRAPHIC COMMUNICATION, 1st ed., by F. E. Giesecke, A. Mitchell, H. C. Spencer, I. L. Hill, J. T. Dygdon, J. E. Novak, and S. Lockhart (Prentice Hall, Inc., 1998)

ENGINEERING DESIGN COMMUNICATION, 1st ed., by S. Lockhart and C. Johnston (Prentice Hall, Inc., 2000)

A TUTORIAL GUIDE TO AUTOCAD 2000, 1st ed., by S. Lockhart (Prentice Hall, Inc., 2001)

A TUTORIAL GUIDE TO MECHANICAL DESKTOP, 1st ed., by S. Lockhart and C. Johnston (Prentice Hall, Inc., 2001)

A TUTORIAL GUIDE TO PRO/ENGINEER, 1st ed., by S. Lockhart and W. Mark Perkins (Prentice Hall, Inc., 2001)

A TUTORIAL GUIDE TO AUTOCAD R.14, 1st ed., by S. Lockhart (Addison-Wesley, 1998)

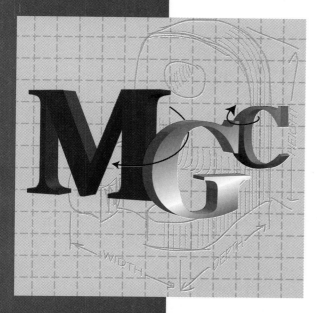

MODERN GRAPHICS
COMMUNICATION

Second Edition

FREDERICK E. GIESECKE

ALVA MITCHELL

HENRY CECIL SPENCER

IVAN LEROY HILL

JOHN THOMAS DYGDON

JAMES E. NOVAK

SHAWNA LOCKHART
Instructor of Mechanical Engineering
University of Montana–Bozeman

PRENTICE HALL
Upper Saddle River, NJ 07458

Library of Congress Cataloging-in-Publication Data

Modern graphics communication 2e / Frederick E. Giesecke . . . [et al.]
    p.   cm.
  Includes bibliographical references and index.
  ISBN 0–13–031724–1   (alk. paper)
  1. Engineering graphics 2. Computer-aided design.   I. Giesecke, Frederick Ernst
T353.M675   2001
604.2—dc21
                                    96–46927
                                       CIP

Acquisitions Editor: *Eric Svendsen*
Editorial/Production Supervision: *Rose Kernan*
Developmental Editor: *Rose Kernan*
Editorial Director: *Marcia Horton*
Managing Editor: *David A. George*
Vice-President of Production and Manufacturing: *David W. Riccardi*
Marketing Manager: *Danny Hoyt*
Manufacturing Buyer: *Pat Brown*
Cover Art: (3D Piston): *Slim Films*
Cover Art: (background sketch): *Michael Ohar*
Cover/Interior Designer: *Tom Nery*
Art Manager: *Gus Vibal*
Creative Director: *Paul Belfanti*
Art Director: *John Boylan*
Associate Editor: *Joe. Russo*
Editorial Assistant: *Kristen Blanco*
Insert Designer: *Joe Sengotta*
Manager of Formatting: *Jim Sullivan*
Composition: *Donna Marie Paukovits*

© 2001 by **PRENTICE-HALL, INC.**
Simon & Schuster/A Viacom Company
Upper Saddle River, NJ 07458

Hands On 2.1 (page 45), Figure 3.22 (page 83), and Hands On 5.7 (page 140) images used with express permission of Image Club Graphics.™ Adobe® and Image Club Graphics™ are trademarks of Adobe Systems Incorporated.

Printed in the United States of America
10 9 8 7 6 5 4 3 2

ISBN 0-13-031724-1

Prentice-Hall International (UK) Limited, *London*
Prentice-Hall of Australia Pty. Limited, *Sydney*
Prentice-Hall Canada, Inc., *Toronto*
Prentice-Hall Hispanoamericana, S.A., *Mexico*
Prentice-Hall of India Private Limited, *New Delhi*
Prentice-Hall of Japan, Inc., *Tokyo*
Simon & Schuster Asia Pte. Ltd., *Singapore*
Editora Prentice-Hall do Brasil, Ltda., *Rio de Janeiro*
Prentice-Hall, Inc. *Upper Saddle River, New Jersey*

BRIEF CONTENTS

CONTENTS

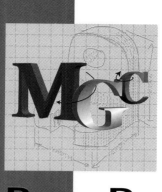

P R E F A C E

■ ABOUT THIS BOOK

We have designed *Modern Graphics Communication* to present succinctly the practices and techniques of sketching, visualization, design and CAD that are important to today's graphics curriculum. Based on the long standing authoritative text on the subject, Giesecke's *Technical Drawing,* this text preserves many of the time tested graphics techniques that remain fundamental to the class, and were so clearly explained in Giesecke's original volume. However, we have rewritten this book to provide a shorter, more lively presentation that covers current course trends, and includes pedagogy appropriate for the modern student. The topics of sketching and visualization skills are this book's primary focus, and provide a solid conceptual basis for the CAD instruction most graphics students receive. We have employed a new format that utilizes integrated on-the-page activities to help students visualize and retain key concepts, unifying art and text into powerful instructional tools easily digested by today's hurried students. Students who complete *Modern Graphics Communication* will leave with a full repertoire of graphical skills they will find invaluable both in education and industry.

■ KEY ELEMENTS

We have designed the second edition of this text based upon the input of first edition users and reviewers. Elements include:

- **NEW**—*Increased coverage of the design process in Chapter 1*—This chapter now includes more coverage of 3-D solid Modeling, and parametric or constraint based modeling. Each chapter also features several student design projects.

- **NEW**—*More Sketching Worksheets*—A selection of new worksheets is included at the end of the book and focus more on sketching techniques.

- **NEW**—*Thoroughly updated material on manufacturing as well as new coverage of geometric tolerancing.*—We are especially thankful to Serope Kalpakjian for help with these sections.

- **NEW**—*Student CD-ROM contains 25 animations keyed to concepts in the text.*—Each concept with an accompanying animation is marked in the text with a 💿 icon.

- *Continued Emphasis on Sketching, Visualization, Design, and CAD*—Instead of extensive material on using instruments, this text highlights a sketching approach to graphics.

- *Exceptional Student Pedagogy*—This text continues to feature student pedagogy designed to integrate art and text into self-contained teaching units. These features include:

STEP BY STEP—These boxes explain essential techniques and processes in detail, using an easy-to-read, visual format. They are vital learning tools that should be given equal importance with the body of the text.

PRACTICAL TIP—These boxes contain useful tips for sketching techniques or CAD procedures.

HANDS-ON—This feature provides an opportunity for the student to practice a technique s/he just learned by trying a quick and easy activity right on the page. These are great confidence builders.

TEAR-OUT WORKSHEETS—Sixteen activities that help students visualize and retain information presented in the text. All are located at the end of the book.

 GRAPHICS SPOTLIGHT—Boxes that highlight up-to-date information on how graphics and CAD are used in the field. They appear in each chapter and include the following topics:

A Day at Ideo U

Unlocking the Power of Solid Modeling

Sketching and Parametric Modeling

From Art to Part

Multiview Projection from a 3D Model

3D Pictorials Aid Designers of Future Electronic Cars

Modeling Irregular Surfaces

High Technology is First Mate in the Race for America's Cup

Digital Polish for Factory Floors

Semiautomatic Dimensioning Using CAD

Geometric Tolerances with AutoCAD 2000

Fastener Libraries

Technical Document Management Systems

 WEB LINKS—Addresses of web sites are included to provide starting points for independent research that apply to the subject matter.

- Each chapter contains an *introductory overview*, *objectives*, *key terms*, a *chapter summary*, and *review questions* to help students organize their reading and review what they have learned.

- *Design projects* and *problems* provide ample context for students to practice solving graphic problems. These are located at the end of each chapter.

- *An eight-page color insert* provides a cost effective means to give students a feel for how color is used in CAD software without obscuring the basic context of the text. It also allows us to keep the cost of the textbook at an affordable price for students.

- *A complete appendix and index* provide readily accessed reference tools that students will find useful during the course and afterward in professional settings.

■ SUPPLEMENTS

 INSTRUCTOR'S MANUAL WITH CD-ROM—INCLUDES SOLUTIONS AND POWERPOINT SLIDES

This text will have its own instructors guide containing solutions to review questions and teaching tips for each chapter. It will also contain a CD holding 1) animations keyed to the text for classroom use and distribution, 2) an extensive set of Powerpoint lectures for instructor use, 3) CAD solutions of most of the text's problems, and 4) selected text images in pdf format. Material for this supplement was prepared by Tom Kane, of Pueblo Community College, Shawna Lockhart, and W. Mark Perkins of the University of Montana–Bozeman.

HTTP://WWW.PRENHALL.COM/GIESECKE

In order to provide instructors and students with the most current information possible, Prentice Hall has created the Giesecke Web site. This expanded site now features information appropriate to the entire Giesecke series of texts. Forty animations (twenty-five of these keyed directly to this text) are viewable on-line or can be downloaded to a local PC or network. A Question and Answer feature lets students answer questions on-line and submit their responses to instructors via E-mail. Case studies focus on how actual companies design actual products. A section on VRML 3D browsing lets students download a 3D browser and explore various web sites. Lastly, an extensive set of web links, including every citation in this text, helps students navigate the World Wide Web.

WORKBOOKS

For those who need extra projects for students, Prentice Hall offers a set of 3 workbooks:

- **Series 1** (ISBN 0–13–658536–1) This book contains traditional workbook problems.

- **Series 2** (ISBN 0–13–658881–6) This book contains traditional problems with an emphasis on engineering concepts.

- **Series 3** (ISBN 0–13–125954–3) A brand new workbook by Paige Davis and Karen Juneau of the Louisiana State University. This book contains modern problems, as well as an extensive, CAD-based project.

■ ACKNOWLEDGMENTS

We would like to thank Frederick E. Giesecke, Alva Mitchell, Henry Cecil Spencer, Ivan Leroy Hill, John Thomas Dygdon, and James E. Novak for their important work in creating the basis for this text. We have valued the Giesecke series for years and were happy to have the chance to work with this excellent material.

We would like to thank the reviewers whose suggestions have guided this revision:

Second Edition Reviewers
Jerrar Andon—University of California-Santa Barbara
Marvin Bollman—Milwaukee School of Engineering
Dr. Ryan Keith Brown—Illinois State University
Karen Coen-Brown—University of Nebraska-Lincoln
Paige R. Davis—Louisiana State University
Kathy Holliday-Darr—Penn State University-Erie
Mary A. Jasper—Mississippi State University
John P. Marchand—Wentworth Institute of Technology
Howard L. Reynolds—Texas Tech University
Emerald Roider—Louisiana State University
Keith Stansbury—Clark College
Scott Tolbert—University of North Dakota

First Edition Reviewers
Abdul B. Sadat—California Polytechnic Institute at Pomona
Carol L. Hoffman—University of Alabama
Jim Hardell—Virginia Tech
Karla D. Kalasz—Washington State University
Tim Hogue—Oklahoma University
Michael Pleck—University of Illinois
Nick DiPirro—SUNY Buffalo
Dan Hiett—Shoreline Community College
Tom Sawasky—Oakland Community College
Dennis Lambert—Georgia Southern University
Hank Metcalf—University of Maine
Julia Jones—University of Washington
Fred Brasfield—Tarrant County Junior College

We would like to express our appreciation to the New Media Group, Madaline Landaeta, Steve Gagliostro, and Erik Unhjem, whose web-site and animation work are an important part of this teaching package.

We would also like to thank the following companies who helped with illustrations, advice, and interest in the project: Bridgeport Machines, Intel, Logitech, Inc., Chartpak, Ritter Manufacturing, and Autodesk.

Shawna Lockhart and Marla Goodman would like to extend special thanks to Gene and Cecelia Goodman, Bob Knebel, Catey Lockhart, Nick Lockhart, Todd Radel, Wren Goodman, and Billy Ray Harvey for their support and assistance.

TEXT AND ILLUSTRATION TEAM

SHAWNA LOCKHART

Shawna Lockhart teaches engineering design graphics at the University of Montana–Bozeman. She is the author of several successful AutoCAD tutorial guides and contributed her expertise to write the CAD At Work segments for Prentice Hall's popular *Technical Drawing, 11th Edition*. Ms. Lockhart has received awards for excellence in teaching and for her community involvement in programs that encourage women and people with disabilities to pursue careers in engineering and technology.

MARLA GOODMAN

Marla Goodman received her BA in Fine Art from Montana State University. As a freelance copywriter and graphic designer she has provided promotional materials and graphic production for a number of businesses and publications. Before starting her design firm, Ms. Goodman supervised art and graphic design departments at a daily newspaper and an internationally marketed manufacturing company. Her design work has received regional and national recognition.

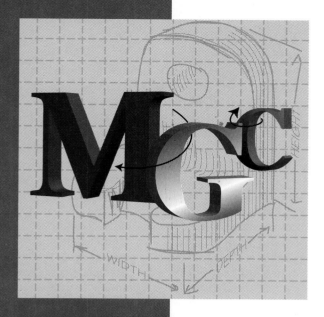

C H A P T E R   1

DESIGN AND GRAPHIC COMMUNICATION

OBJECTIVES

After studying the material in this chapter, you should be able to:

1. Define the different types of design.
2. Define engineering design and examine how the proper objectives and motivation can turn any one into a good designer.
3. Describe the role of the engineer on a design team.
4. Discuss the different sources for design ideas for both the individual designer as well as a design team.
5. Describe the stages of the design process.
6. Discuss solid modeling, parametric modeling, prototyping and rapid prototyping.
7. Use a case study to illustrate each stage of the design process.
8. Explain why standards are important.
9. Identify examples of parallel and perspective projection.
10. Define *plane of projection* and *projectors.*
11. Identify uses of technical graphics.
12. Sketch a diagram showing the steps in the design process.
13. Create examples of each stage of the design process.

OVERVIEW

A new machine, structure, or system must exist in the mind of the engineer or designer before it can become a reality. The design process is an exciting and challenging effort, during which the engineer-designer uses graphics as a means to create, record, analyze, and communicate design concepts or ideas.

Everyone on the engineering and design team needs to be able to communicate quickly and accurately in order to compete in the world market. Like carpenters learning to use the tools of their trade, engineers, designers, and drafters must learn the tools of technical drawing. The design team progresses through five stages in the design process. To be a successful member of the design team, you must understand the process and know your role.

Various types of drawings are required at each stage of the design process. CAD can help the drafter create drawings, but it takes a skilled drafter to know which drawings are required at each design stage. Much of the design process is refinement of existing products. Refinement creates improved products, lowers costs, and increases profit for the company that manufactures the product. Models are an important part of the design process. Some models are created to scale in a model shop. Other models are generated by the computer and printed or displayed in virtual reality. All models allow interaction with the design to further improve the design. Revising a drawing is an important part of the design process. Revisions must be tracked, identified, logged, and saved for future reference. Both paper and electronic storage is an important part of the drafter's responsibility on the design team. Assembly and working drawings show how multiple parts fit together. They describe the end result of creating individual pieces that must fit together to work.

Design concepts are usually communicated to others through freehand sketches or drawings created using computer-aided design or computer-aided drafting (CAD). Preliminary sketches are followed by more exact sketches and drawings as the idea is developed more fully. CAD can help you create drawings, but it takes skill to know which drawings and what level of detail is required at each design stage. While CAD has replaced traditional drafting tools for many design teams, the basic concepts of graphic communication remain the same. Your proficiency in communicating using graphics will be valuable to you and to your eventual employer.

■ INTRODUCTION

In developing any new machine, structure, system, or improvement of an existing system, the concept has to exist in your mind before it can become a reality. In the engineering profession, you are frequently called upon to communicate concepts to others using freehand sketches, as shown in Figure 1.1.

To progress from an idea to a finished project, you need to create idea sketches, calculate stresses, analyze motions, size parts, specify materials and production methods, and make design layouts. You also have to know how to prepare the drawings and specifications that will control the numerous details of product manufacture, assembly, and maintenance. To perform or supervise these many tasks, engineers and technicians use freehand sketches that record and communicate ideas quickly. Ease in freehand sketching and the ability to work with computer drawing techniques requires knowledge of **standards** for graphic communication. Engineers and designers who use a computer for drawing and design work still need to know how to create and interpret drawings.

■ 1.1 "DESIGN" DEFINED

Design is a process, a series of linked steps with stated objectives. It is a way of conceiving and creating new ideas and then communicating those ideas to others in a way that can be easily understood. This is accomplished most efficiently through the use of graphics. Design can be used to reflect personal expressions or to enhance product development. This reflection of personal expression is most often referred to as *aesthetic design* while the enhancement of product development is considered *functional design*. Aesthetics and function can work hand in hand to create a product that is not only appealing to the senses, but fulfills specific product demands. A well-designed automobile is a good example of how aesthetics and function can work together. (See Figure 1.2.)

There are two general types of design: *empirical design*, sometimes referred to as conceptual design, and *scientific design*. In scientific design, use is made of the principles of physics, mathematics, chemistry, mechanics, and other sciences in the new or revised design of devices, structures, or systems intended to function under specific conditions. In empirical design, much use is made of the information in handbooks, which in turn has been learned by experience. Nearly all technical design is a combination of scientific and empirical design. Therefore, a competent designer has both adequate engineering and scientific knowledge and access to the many handbooks related to the field.

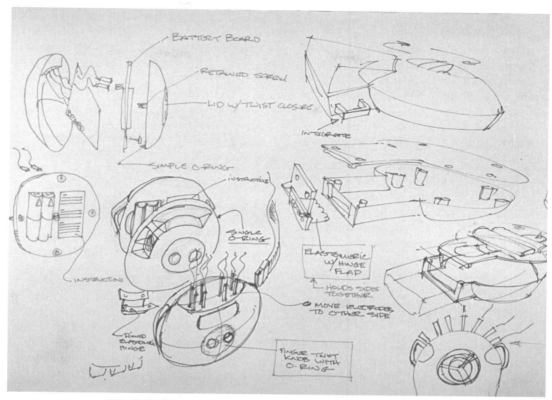

FIGURE 1.1   **An Initial Idea Sketch.** *Courtesy of Ratio Design Lab, Inc.*

■ 1.2 "ENGINEERING DESIGN" DEFINED

Engineering design is also a process. This process is used to solve society's needs, desires, and problems through the application of scientific principles, experience, and creativity. Some people are creative and are naturally gifted at design, but everyone can become a designer if they learn to use the proper tools and techniques involved with the design process. Becoming a designer is much like learning to play a musical instrument; some people are better at it then others, but everyone can learn to play if they learn the steps involved.

Two key elements to any successful design plan is gaining the proper *motivation* and stating the *objectives* to the plan. Design is the single most important activity practiced by engineers. Design separates engineering from the rest of the sciences in that it is the application of scientific principles to create solutions. Your motivation in any design plan should be to create the most efficient solution to any given problem. The objective statement will provide a framework within which any engineering design problem can be addressed in a methodical manner. Proper planning and

scheduling are also key to successful designs plans. Setting a deadline for the completion of each design phase is imperative. We will the discuss the steps in the design process at length in §1.4

FIGURE 1.2   **Aesthetic and functional design combine to give this sports car not only a look of elegance, but of speed. Although pleasing to the eye, this car is the superb product of aerodynamic and mechanical engineering.**
Reprinted with permission from Karl Snitcher.

■ 1.3 DESIGN CONCEPTS— SOURCES FOR NEW IDEAS

INDIVIDUAL CREATIVITY TECHNIQUES   New ideas or design concepts usually begin in the mind of a single individual—the designer. But how does one go about developing new ideas? There is an old saying in the engineering industry:

"Good design is to borrow. Genius design is to steal."

Unlike your other courses of study where plagiarism is considered bad and should be subject to punishment, copying good ideas is highly advisable in design. Students are urged to copy not only from existing products and classmates, but they should study catalogs, manufacturers' patents, and nature.

Look through industry catalogs and handbooks for existing designs. Think of ways in which these existing designs can be used or modified to work in your design plan. Manipulate them through freehand sketches or through the use of computer software (CAD).

STUDY PATENT DRAWINGS (See Figure 1.3)   A patent is issued by the U.S. government granting the holder the "right to exclude others from making, using or selling" a specific product. The patent process was first developed as a way of disclosing technical advances by granting a period of protection for a limited amount of time (a patent is issued for 17 years). The U.S. Patent and Trademark Office has extremely strict regulations as to the presentation of materials (i.e., no freehand sketches are accepted and all patent drawings must be made with drafting instruments or by a process which will make them easily reproducible). You cannot use or copy any existing patents, but they can be a great source of ideas. The U.S. Patent Office, therefore, can be a valuable resource in quest for design ideas. (Patent drawings are discussed in §13.17.)

EXAMINE MANUFACTURED PRODUCTS.   Dismantle them, evaluate them, and study how their parts are designed to work together. This is referred to a *reverse engineering*. Sophisticated reverse engineering involves evaluating a product using a machine called a *coordinate measuring machine* (CMM). (See Figure 1.4.) The machine is a electromechanical devices containing a probe on one end. The probe measures the object and then places all of the pertinent information into a CAD database where it can be manipulated. Although you may not have access to such a complicated machine, examine manufactured products that are available to

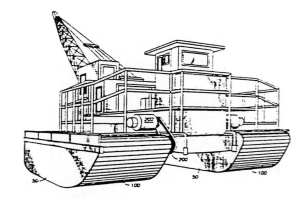

FIGURE 1.3   Pictorial Patent Drawing.

you. Think of ways to improve or change these existing designs. Where could they be improved? What would you do differently?

Study a product that is no longer performing to the manufacturer's existing or upgraded standards, referred to as *functional decomposition*. How could you expand/change the design to guarantee better performance? What could you do to expand the life of the product? How could you make it more efficient, more cost effective, etc.?

STUDYING THE NATURAL WORLD.   Noting how other creatures interact with their surroundings can provide a wealth of information and creativity. Such things as bee hives or spiders' webs are masterpieces of structural design. A hummingbird's wings are aerodynamic wonders. (See Figures 1.5 and 1.6.) Study them and expand on their designs.

Use all computer hardware and software available to you. Animation and CAD programs have become numerous in recent years. Although you may not have the resources to invest in numerous new software programs, be sure to make use of those readily available at your school. Even programs such as Adobe Illustrator® or Photoshop® can help in your search for new ways of viewing and conceiving objects.

   Excellent resources for engineering and design are available on the World Wide Web. Search for terms like design, engineering, technology, or for more specific terms, depending on your interests. The following Web sites are useful for engineering design:

* http://www.yahoo.com/headlines/
  Yahoo's site for the latest technology news and a one-week archive

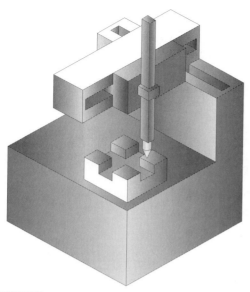

FIGURE 1.6 The wings of a hummingbird are aerodynamic wonders. Study of such designs in nature can be enlightening to any designer. *Courtesy of Photo Researchers, Inc.*

FIGURE 1.4 Coordinate Measuring Machine (CMM). Used to Accurately Measure a Part for Reverse Engineering or Quality Control.

- http://www.techweb.com/
  TechWeb site from CMP media
- http://www.uspto.gov/
  U.S. Patent Office on-line search site

Using the techniques discussed above and listed in Figure 1.7, the designer should have a few ideas as to where to begin. In order to capture, preserve, and develop these ideas, the designer makes liberal use of freehand sketches of views and pictorials. These sketches are revised or redrawn as the concept is developed. All sketches should be preserved for reference and dated as a record of the development of the design.

FIGURE 1.5 Studying designs in nature, such as this spider's web, can provide the designer with new information and ideas. *Courtesy of Peter Arnold, Inc.*

GROUP CREATIVITY TECHNIQUES At some point in the development of the idea, you will probably find it to your advantage to pool your ideas with those of others and begin working in a team effort; such a team may include others familiar with problems of materials, production, marketing, and so on. In industry, the project becomes a team effort long before the product is produced and marketed. Obviously, the design process is not a haphazard operation of an inventor working in a garage or basement, although it might well begin in that manner. Industry could not long survive if its products were determined in a haphazard manner. Hence, nearly all successful companies support a well-organized design effort, and the vitality of the company depends, to a large extent, on the *planned* output of its designers and design teams. Groups play an important role in the creative design process.

The two most commonly used group creativity techniques are brainstorming and storyboarding. *Brainstorming* occurs when a group of individuals come to together to discuss new ideas. A brainstorming session can stimulate, enlighten, and motivate designers to look at their product designs in a different light based upon input from other team members. The most important rule to follow during a brainstorming session is that no criticism of others' work should be tolerated. Criticism only serves to stymie the creative process. A second rule for any good brainstorming session is to come prepared with as many new ideas as possible. This is not the time to be conservative; be open to presenting new ideas.

Storyboarding is a technique often used by designers to graphically illustrate the progression of their designs, as well as the manufacturing process required to create a final product. Storyboards are rough sketches, usually created freehand by the designer. Storyboards are a valuable tool in any brainstorming session as they can be used as a base to be built on creatively by group members during the session. These freehand sketches should be rough so that they can be updated, revised, or modified based on input from team members.

Reintegration of the ideas generated during the brainstorming session into an individual's design is of utmost importance. Once the design team has settled on a specific design, it is imperative that the individual designer incorporate this input into his/her design. (See Figure 1.7.)

Since it is important for you to be able to work effectively with others in a group or team, you must be able to express yourself clearly and concisely. Do not underestimate the importance of your communication skills, your ability to express your ideas verbally (written and spoken), symbolically (equations, formulas, etc.), and *graphically*.

These graphical skills include the ability to present information and ideas clearly and effectively in the form of sketches, drawings, graphs, and so on. This textbook is dedicated to helping you develop your communication skills in graphics.

■ 1.4 THE DESIGN PROCESS

Design is the ability to combine ideas, scientific principles, resources, and often existing products into a solution of a problem. This ability to solve problems in design is the result of an organized and orderly approach to the problem known as the *design process*.

The design process leading to manufacturing, assembly, marketing, service, and the many activities necessary for a successful product is composed of several easily recognized phases. Although many industrial groups may identify them in their own particular way, a convenient procedure for the design of a new or improved product is in five stages as follows:

1. Identification of problem, need, or "customer."
2. Concepts and ideas.
3. Compromise solutions.
4. Models and/or prototypes.
5. Production and/or working drawings.

Ideally, the design moves through the stages as shown in Figure 1.8, but if a particular stage proves

---

INDIVIDUAL CREATIVITY TECHNIQUES INCLUDE:

Studying Industry Catalogs/Handbooks
Examining Manufactured Items
Studying Patent Drawings
Conducting Reverse Engineering
Examining Functional Decomposition
Studying the Natural World
Using Software Products
Utilizing Design, Engineering, and Technology
   Web Sites

GROUP CREATIVITY TECHNIQUES INCLUDE:

Brainstorming
Storyboarding
Reintegration

FIGURE 1.7 Individual and group creativity techniques.

---

unsatisfactory, it may be necessary to return to a previous stage and repeat the procedure as indicated by the dashed-line paths. This repetitive procedure is often referred to as *looping*.

■ 1.5 STAGE 1—IDENTIFICATION OF THE PROBLEM AND THE CUSTOMER

The design activity begins with the recognition of a *problem* and/or the determination of a *need* or want for a product, service, or system and the economic feasibility of fulfilling this need.

The designer not only must identify the problem or need but also the *customer*. Who will be affected or influenced by the design? The creation of any new design and the related design process ultimately should be driven by its end users. Determine if the design should be geared toward a single, very specific user, a specific purchaser or purchasers, a manufacturer or group of manufacturers, or to the general public. A part to be used in the space shuttle, for example, would not need to be designed or manufactured for operation by the general public. It has a limited market and customer base. But a design for a new home gym, which requires the user to complete the final assembly, should take into account a wide range of users and mechanical abilities. It is important that the designer identify the end user before beginning the design process.

It is also important to determine if the product to be designed must meet with any government

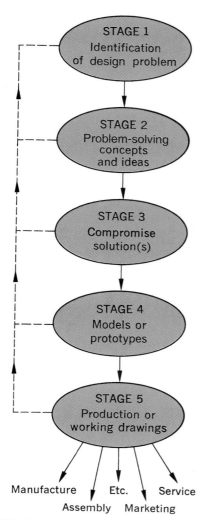

FIGURE 1.8   Stages of the Design Process.

beneficial. If the designer has the requirements broken into these categories before implementing the design, decision making as to what to change and when will become easier when the need arises.

Engineering design problems may range from the simple and inexpensive container opener such as the pull tab (Figure 1.9) commonly used on beverage cans to the more complex problems associated with the needs of air and ground travel, space exploration, environmental control, and so forth. Although the product may be very simple, such as the pull tab on a beverage can, the production tools and dies require considerable engineering and design effort. The airport automated transit system design, Figure 1.10, meets the need of moving people efficiently between the terminal areas. The system is capable of moving 3300 people every 10 minutes.

The Lunar Roving Vehicle, Figure 1.11, is a solution to a need in the space program to explore larger areas of the lunar surface. This vehicle is the end result of a great deal of design work associated with the support systems and the related hardware.

At the problem identification stage, either the designer recognizes that there does exist a need requiring a design solution or, perhaps more often, a directive is received to that effect from management. No attempt is made at this time to set goals or criteria for the solution.

Information concerning the identified problem becomes the basis for a problem proposal, which may be a paragraph or a multipage report presented for formal consideration. A proposal is a plan for action that will be followed to solve the problem. The proposal, if approved, becomes an agreement to follow the plan. In the classroom, the agreement is made between you and your instructor on the identification of the problem and your proposed plan of action.

Following approval of the proposal, further aspects of the problem are explored. Available information related to the problem is collected, and parameters or guidelines for time, cost, function, and so on are defined within which you will work. For example: What is the design expected to do? What is the estimated cost limit? What is the market potential? What can it be sold for? When will the prototype be ready for testing? When must production drawings be ready? When will production begin? When will the product be available on the market?

The parameters of a design problem, including the time schedule, are established at this stage. Nearly all designs represent a compromise, and the amount of time budgeted to a project is no exception.

standards/regulations or adhere to any professional organizations' standards or codes before starting the design process.

Any design process will involve compromises such that all of the designer's original requirements cannot be met. Government standards may limit the use of certain materials, for example, and the design may have to take a different form than that originally conceived by the designer. Materials or manufacturing processes may become too costly, resources may become unavailable, etc. It is important for the designer to approach the design process knowing that compromises may have to be made. Prioritize the design requirements. It is a good idea to break the *design requirements* into four categories: essential, important, desirable, or

FIGURE 1.9   **Pull-Tab Can Opener.** *John Schultz—PAR/NYC.*

FIGURE 1.11   **Lunar Roving Vehicle.** *Courtesy of NASA.*

FIGURE 1.10   **Airport Transit System.** *Courtesy of Westinghouse Electric Corp.*

■ 1.6 STAGE 2—CONCEPTS AND IDEAS

At this stage, many ideas are collected—reasonable and otherwise—for possible solutions to the problem. The ideas are broad and unrestricted to permit the possibility of new and unique solutions. The ideas may be from individuals, or they may come from group or team brainstorming sessions where one suggestion often generates many more ideas from the group. As the ideas are elicited, they are recorded for future consideration and refinement. No attempt is made to evaluate ideas at this stage. All notes and sketches are signed, dated, and retained for possible patent proof.

The larger the collection of ideas, the greater are the chances of finding one or more ideas suitable for further refinement. All sources of ideas, such as technical literature, reports, design and trade journals, patents, and existing products are explored. Ideas can come from such sources as the Greenfield Village Museum in Dearborn, Michigan, the Museum of Science and Industry in Chicago, trade exhibitions, the World Wide Web, large hardware and supply stores, mail order catalogs. Even the user of an existing product is an excellent source, because that person often has suggestions for improvement. The potential user may be helpful with specific reactions to the proposed solution.

No attempt is made to evaluate ideas at this stage. All notes and sketches are signed, dated, and retained for possible patent proof.

■ 1.7 STAGE 3—COMPROMISE SOLUTIONS

Various features of the many conceptual ideas generated in the preceding stages are selected after careful consideration and combined into one or more promising compromise solutions. At this point the best solution is evaluated in detail, and attempts are made to simplify it so that it performs efficiently and is easy to manufacture, repair, and even dispose of when its lifetime is over.

Refined design sketches are often followed by a study of suitable materials and of motion problems that may be involved. What source of power is to be used—manual, electric motor, or what? What type of motion is needed? Is it necessary to translate rotary motion into linear motion or vice versa? Many of these problems are solved graphically using schematic drawings in which various parts are shown in skeleton form. For example, pulleys and gears are represented by circles, an arm by a single line, and a path of motion by centerlines. Certain basic calculations, such as those related to velocity and acceleration, may also be made at this time.

Preliminary studies are followed by a design layout—usually an accurate CAD drawing, showing actual sizes so that proportions and fits can be clearly visualized—or by a clearly dimensioned layout sketch. An example is shown in Figure 1.12. At this time all parts are carefully designed for strength and function. Costs are constantly kept in mind, because no matter how well the device performs, it must sell at a profit; otherwise the time and development costs will have been a loss.

During the layout process, experience provides a sense of proportion, size, and fit that permits noncritical features to be designed by eye or with the aid of empirical data. Stress analysis and detailed computation may be necessary in connection with high speeds, heavy loads, or special requirements or conditions.

Figure 1.13 shows the layout of basic proportions of parts and how they fit together in an assembly drawing. Special attention is given to clearances of moving parts, ease of assembly, and serviceability. Standard parts are used wherever possible, because they are less expensive than custom parts. Most companies maintain some form of an engineering standards manual, which contains much of the empirical data and detailed information that is regarded as "company standard." Materials and costs are carefully considered. Although functional considerations must come first, manufacturing problems must be kept constantly in mind.

■ 1.8 STAGE 4—MODELS AND PROTOTYPES

A model to scale is often constructed to study, analyze, and refine a design. To instruct the model-shop craftsperson in the construction of the prototype or model, dimensioned sketches or three dimensional computer models are required. A full-size working model made to final specifications, except possibly for materials, is known as a prototype. The prototype is tested and modified where necessary, and the results are noted in the revision of the sketches and working drawings. Figure 1.14 shows a prototype of the magnetic levitation train.

If the prototype is unsatisfactory, it may be necessary to return to a previous stage in the design process and repeat the procedures. It must be remembered that time and expenses always limit the duration of this looping. Eventually a decision must be reached for the production model.

SOLID OR 3D MODELING   CAD systems that offer solid modeling usually have options or commands for creating complex solids using primitive shapes, including boxes, prisms, cylinders, spheres, cones, tori, and sometimes wedges and pyramids. Figure 1.15 shows an example of some solids generated using CAD. Many CAD systems also have a primitive command which creates any regular solid. If there is not a specific command to create the solid you want, you usually have the option of creating new solid objects through processes called extrusion and revolution (both generally referred to as sweeping solids).

Extrusion is named for the manufacturing process which forms material by forcing it through a shaped opening. You can also think of extrusion as taking the cross-sectional shape of the object and sweeping it along a path to enclose a solid volume. Features which have a continuous cross section along a straight axis can be created using extrusion. Some CAD software can only extrude shapes along a straight-line path, while others can do straight or curved paths. Most have the option to taper the extruded feature as it is created. An example of an extruded solid model is shown in Figure 1.16.

Revolution is the process of forming a solid by revolving the cross-sectional shape of the object along a circular path to enclose a solid volume. Many objects that cannot be created by extrusion can be created by revolution. A solid object created by revolving a shape by 270° is shown in Figure 1.17.

Complex solid models can be formed by joining primitives and solids formed by extrusion and revolution using *Boolean operators*. Boolean operators are named for 18th century mathematician and logician, Charles Boole. Most CAD programs that allow solid modeling support three Boolean operators: *union* (sometimes called addition), *difference* (sometimes called subtraction), and *intersection*. Venn diagrams are often used to show how sets are joined together using Boolean operators. Figure 1.18 shows Venn diagrams for union, difference, and intersection. Boolean operators can also be used to join solids you create by extruding or revolving other solids or solid primitives.

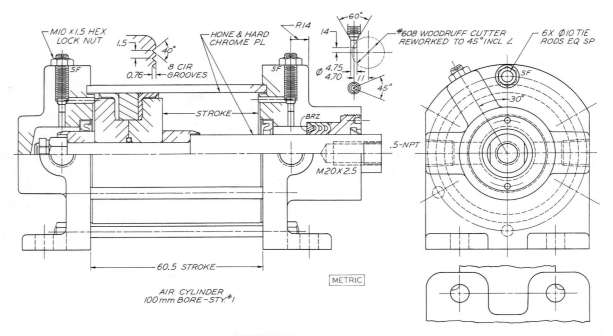

FIGURE 1.12 Design Layout.

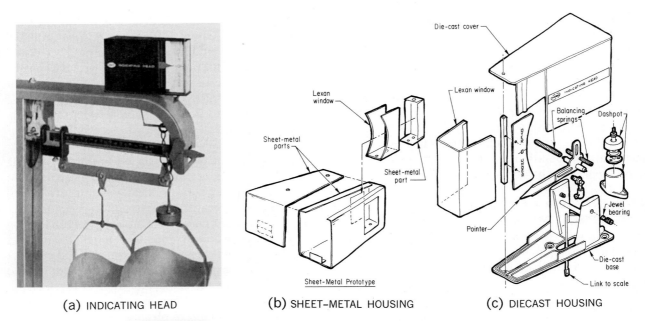

(a) INDICATING HEAD (b) SHEET–METAL HOUSING (c) DIECAST HOUSING

FIGURE 1.13 Improved Design of Indicating Head. *Courtesy of Ohaus Scale Corp. and Machine Design.*

FIGURE 1.14 **A Prototype of the Magnetic Levitation Train Car During a Test Run.**

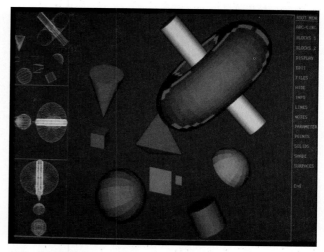

FIGURE 1.15 **Examples of Solids Created with CAD.**
Courtesy of American Small Business Computers.

The union of solid A and solid B forms a single new solid that is their combined volume without any duplication where they have overlapped. Solid A difference solid B is similar to subtracting B from A. The order of the operation does make a difference in the result (unlike union). For A difference B, any volume from solid B that overlaps solid A is eliminated and the result forms a new solid. A intersect B results in a new solid where only the volume common both to solid A and solid B is retained. You can use these primitives, extrusion, revolution, and Boolean operators to create solid models for a variety of objects. Objects that cannot be created in this way are warped surfaces, such as those on the exteriors of automobiles and airplanes.

PARAMETRIC SOLID MODELING   CAD systems, especially those allowing parametric solid modeling (where design parameters control the model geometry), provide many benefits for shortening the design cycle time. (In parametric design, the curve paths are controlled by mathematical functions rather than a set of coordinates.) In this way, parametric solid modeling allows concurrent design, where members of the design team along with members from the company's manufacturing and marketing divisions can work together at the same time to provide a total design solution. In parametric models, constraints and parametric dimensions control the model geometry. As the design changes, so can the constraints and dimensions; the model and drawings update automatically. Three-dimensional models can also be exported to rapid prototyping equipment and to direct manufacturing to allow quick progress from design to product.

RAPID PROTOTYPING   While refining the design ideas, engineers often work concurrently with manufacturing to determine the best ways to make an assemble the necessary parts. After several cycles of refining, analyzing, and synthesizing the best ideas, the final design is ready to go into production. *Rapid prototyping* systems allow parts to quickly be generated directly from 3D models for mockup and testing. Rapid prototyping can also be used in situations were prototypes are still deemed necessary for various reasons (i.e., customer request). Many companies are designing their products with 3D design packages an then feeding the CAD data into separate software programs that not only evaluate the design but also generate a separate set of data, which is then usable in a variety of ways. The new data can be forwarded to a CNC mill and a sample model can then be cut from the data. Data can also be

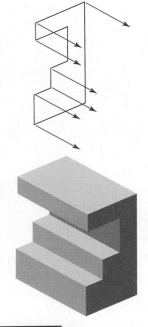

FIGURE 1.16 Extruded Solid.

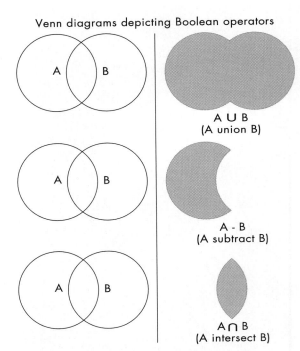

Venn diagrams depicting Boolean operators

A ∪ B
(A union B)

A - B
(A subtract B)

A ∩ B
(A intersect B)

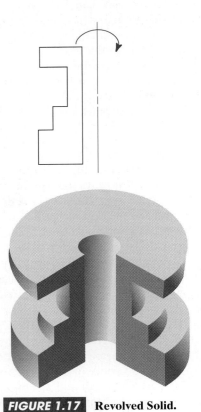

FIGURE 1.17 Revolved Solid.

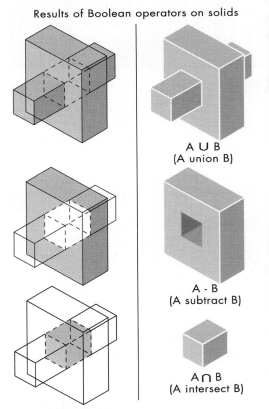

Results of Boolean operators on solids

A ∪ B
(A union B)

A - B
(A subtract B)

A ∩ B
(A intersect B)

FIGURE 1.18 Boolean Operators.

FIGURE 1.19   Solid modeling allows you to quickly get from design to product.   *Courtesy of Solid Concepts, Inc.*

forwarded to rapid prototyping equipment using such technologies as *stereolithography* (SLA), *selective laser sintering* (SLS), *ballistic particle manufacturing* (BPM), and *laminated object manufacturing* (LOM).

SLA builds part from layers of laser-cured photopolymer. The SLS process builds parts layer by layer with a laser from powdered materials such as nylon, polycarbonate, or a composite glass-nylon material. Some rapid prototyping machines build objects by spraying molten particles of a thermoplastic. The LOM process builds layer by layer from rolls of sheet goods similar to paper. (Refer to the Graphics Spotlight on pages 20-21 for a good example of rapid prototyping in action.)

■ 1.9 STAGE 5—PRODUCTION OR WORKING DRAWINGS

To produce or manufacture a product, a final set of production or working drawings is made, checked, and approved.

In industry the approved production design layouts are turned over to the engineering department for the production drawings. The drafter, or detailers, "pick off"

the details from the layouts with the aid of the scale or dividers. The necessary views are drawn for each part to be made, and complete dimensions and notes are added so that the drawings will describe these parts completely. These are called *detail drawings*.

Unaltered standard parts do not require a detail drawing but are shown conventionally on the assembly drawing and listed with specifications in the parts list.

A detail drawing of one of the parts from the design layout of Figure 1.12 is shown in Figure 1.20.

After the parts have been detailed, an *assembly drawing* is made, showing how all the parts go together in the complete product. The assembly may be made by tracing the various details in place directly from the detail drawings, or the assembly may be traced from the original design layout, but if either is done, the value of the assembly for checking purposes will be largely lost.

Finally, in order to protect the manufacturer, a *patent drawing*, which is often a form of assembly, is prepared and filed with the u.s. patent office. patent drawings are line shaded, often lettered in script, and otherwise follow rules of the patent office.

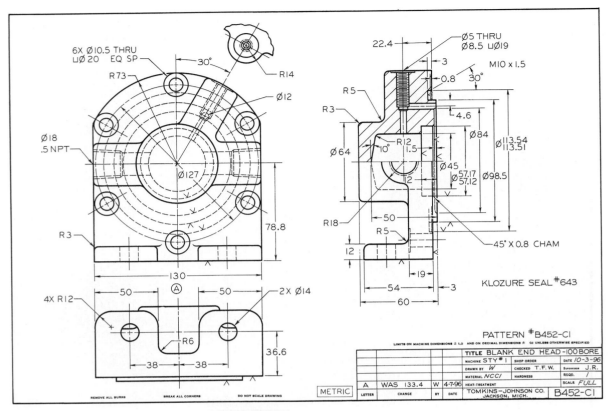

FIGURE 1.20 A Detail Drawing.

■ 1.10 DESIGN OF A NEW PRODUCT

An example of the design and development of a new product is that of IBM's ThinkPad 701C subnotebook computer, shown in Figure 1.21.

STAGE 1 *Problem Identification* When a company wants to determine the feasibility of a new product, it solicits opinions and ideas from many sources, including engineers, designers, drafters, managers, and potential consumers. Price ranges and estimated sales are also carefully explored.

In the case of the ThinkPad, IBM wanted to produce a subnotebook-size computer with a full-size keyboard—that is, a keyboard with the same size keys and the same spacing between keys as in a desktop computer. They also wanted the computer to have the largest available display, but the end product had to be thin, lightweight, and inexpensive enough to be competitive.

STAGE 2 *Concepts and Ideas* In order to fit a full-size keyboard into a subnotebook case, a mechanical engineer working on the project came up with the idea of splitting

the keyboard into two pieces that would interlock when the computer case was open. The two pieces could hang over the sides of the open case, thereby giving a little extra space for the relatively large keyboard. To close the computer case, the two pieces of the keyboard would need to separate and slide inward into new positions.

Once the idea of the split keyboard was accepted, the team moved from rough sketches and mockups to solid modeling.

STAGE 3 *Compromise Solution* Using IBM's CATIA CAD/CAM software and an IBM RISC System/6000 workstation, the development team created a variety of possible keyboard designs for the ThinkPad. This system enabled them to produce virtual prototypes without building and rebuilding a series of actual 3D models. For example, they had to design a system to move the two parts of the keyboard when the computer case was opened and closed. However, the keyboard halves could not simply be moved by the case itself because if the system jammed, forcing the case open would break the keyboard. One feature of the

CATIA software is a system that can identify areas of possible physical interference between parts of a model. Thus, the team could see on a computer screen where solid parts of the new computer might hit each other.

When the ThinkPad case is opened, the TrackWrite keyboard is driven into position by a spring-loaded mechanism that moves the two halves of the keyboard asymmetrically. Closing the computer case moves the two halves of the keyboard back into their storage position by way of an axial cam and resets the spring (see Figures 1.22 and 1.23).

STAGE 4 *Prototypes* Although in the past prototypes referred to actual working models, in the production of the ThinkPad computer, many of the prototypes existed only as 3D computer images. For example, early prototype designs for the keyboard—in the form of CATIA files—were given to the IBM unit that was developing other elements of the computer so that all the parts could be integrated into a functional whole.

STAGE 5 *Production* In the final stages of the design process, completed CATIA models were sent to outside vendors who used them to program numerically control (NC) tools that created the molds for parts of the computer. The same models created by the design process functioned throughout the various stages of the project. (See the Graphics Spotlight for a real-life example of a team using the design process.)

■ 1.11 COMMUNICATING USING GRAPHICS

Although people around the world speak different languages, ***graphic communication*** has existed since the earliest of times. The earliest forms of writing were picture forms, such as the Egyptian hieroglyphics shown in Figure 1.24. Later these forms were simplified and became the abstract symbols used in writing today.

Graphic representation has developed along two distinct lines: artistic and technical. From the beginning of time, artists have used drawings to express aesthetic, philosophic, or other abstract ideas. People learned by looking at sculptures, pictures, and drawings in public places. Everybody could understand pictures, and they were a principal source of information.

Technical drawings that communicate design information, however, convey information differently. From the beginning of recorded history, people have used drawings to represent the design of objects to be built or constructed. Although the earliest drawings no longer exist, we know people could not have designed and built as they did without using fairly accurate drawings.

■ 1.12 EARLY TECHNICAL DRAWING

Perhaps the earliest known technical drawing in existence is the plan view for the design of a fortress drawn by the Chaldean engineer Gudea and engraved on a stone tablet (Figure 1.25). It is remarkable how similar this plan is to those made by modern architects, although it was created thousands of years before paper was invented.

FIGURE 1.24   Egyptian Heiroglyphics.

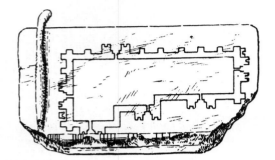

FIGURE 1.25   Plan of a Fortress. This stone tablet is part of a statue now in the Louvre, in Paris, and is classified in the earliest period of Chaldean art, about 4000 B.C. *From Transactions ASCE, May 1891.*

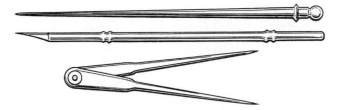

FIGURE 1.26   Roman Stylus, Pen, and Compass. *From Historical Note on Drawing Instruments, published by V & E Manufacturing Co.*

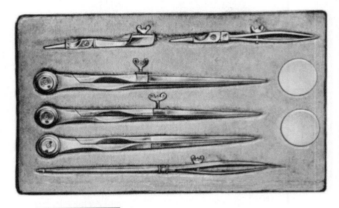

FIGURE 1.27   George Washington's Drawing Instruments. *From Historical Note on Drawing Instruments, published by V & E Manufacturing Co.*

In museums we can see actual specimens of early *drawing instruments*. Early bronze compasses, shown in Figure 1.26, were about the same size as the modern kind. Pens were cut from reeds.

At Mount Vernon are drawing instruments, bearing the date 1749, used by the great civil engineer George Washington. This set is very similar to the conventional drawing instruments used throughout the 20th century. It consists of a divider and compass with pencil and pen attachments, plus a ruling pen with parallel blades (Figure 1.27).

A single view of a part for a design is technically known as a *projection* The theory of projections of objects on imaginary viewing planes apparently was not developed until the early part of the 15th century by the Italian architects Alberti, Brunelleschi, and others. It is well known that Leonardo da Vinci used drawings to record and transmit to others his ideas and designs for mechanical constructions, and many of these drawings are still in existence. Figure 1.28 on page 20 shows Leonardo's drawing of an arsenal. He may even have made mechanical drawings like those used today.

■ 1.13 EARLY DESCRIPTIVE GEOMETRY

Descriptive geometry uses graphics and projections to solve spatial problems. Gaspard Monge (1746–1818) is considered the inventor of descriptive geometry. While he was a professor at the Polytechnic School in France, Monge developed the principles of projection that are now the basis for technical drawing. These principles of descriptive geometry were soon recognized to be of such military importance that Monge was forced to keep them secret. His book, *La Géométrie Descriptive*, published in 1795, is still regarded as the first text on projection drawing. By the early 1800s these ideas were adopted in the United States and taught at universities. They were also used in manufacturing interchangeable parts, particularly in the early arms industry.

■ 1.14 MODERN TECHNICAL DRAWING

In 1876 the blueprint process was introduced at the Philadelphia Centennial Exposition. Up to this time, creating technical graphics was more or less an art, characterized by fine-line drawings made to resemble copperplate engraving, by shade lines, and by water-color washes. These techniques became unnecessary after the introduction of blueprinting, and drawings gradually became less ornate to get better results in reproduction. This was the beginning of modern technical drawing. Technical drawing became a relatively exact method of representation, often making it unnecessary to build a working model before a device could be constructed.

Hands On 1.1
Stages of the design process

Following are descriptions of processes that might take place in the design of a step-in snowboard binding. Read each step and decide at which stage of the design process it would take place. Write the letter of the description in the appropriate space on the design process diagram at right. Stage 1 is done for you.

a) Stress analysis results show that attaching a locking mechanism to existing boots is unsafe. The concept is redesigned as an integrated boot.

b) A need exists for a step-in snowboard binding that allows the user freedom to enter and exit chair lifts and trams while providing the safest possible locking and release mechanisms.

c) 3-D models of the step-in snowboard binding and boot system are exported from CAD for manufacturing.

d) A mockup is made and lab tested. Then a group of snowboarders field test the prototype's performance in various terrains and situations.

e) Inspired by a special locking sytem on his bicycle shoe, the designer sketches an idea for a snowboard boot and binding attachment that will retrofit on existing snowboard boots.

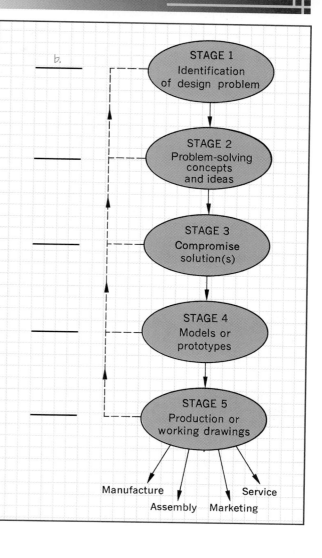

■ 1.15 DRAFTING STANDARDS

Standards for the appearance of technical drawings have been developed to ensure that they are easily interpreted across the nation and around the world. As you learn to create technical graphics, you will adhere to these standards. This will allow you to create drawings which communicate clearly and cannot be misinterpreted by others. In the United States the American National Standards Institute (ANSI), the American Society for Engineering Education (ASEE), the Society of Automotive Engineers (SAE), and the American Society of Mechanical Engineers (ASME) have been the principal organizations involved in developing the standards now in place. As sponsors, they have prepared the *American National Standard Drafting Manual—Y14,* which consists of a number of separate sections that are frequently updated. (See Appendix 1.)

 Refer to the ANSI Web site at http://www.ansi.org for information on ordering up-to-date standards for your specific application.

FIGURE 1.28 **An Arsenal, by Leonardo da Vinci.** *Courtesy of the Bettmann Archive*

■ 1.16 DEFINITIONS

Before you begin creating technical drawings and sketches, a few terms need to be defined:

MECHANICAL DRAWINGS Drawings made with mechanical-drawing instruments or industrial drawings in general, whether or not they are mechanically drawn.

COMPUTER GRAPHICS Graphics created using computer software to draw, analyze, modify, and finalize a variety of graphical solutions.

CAD/CADD Computer-aided design or computer-aided drafting (CAD) and computer-aided design and drafting (CADD). See Figure 1.29. You can use CAD to create a useful database that accurately describes the 3-D geometry of the machine part, structure, or system you are designing. The database can then be used to perform analysis, to directly machine parts, or to create illustrations for catalogs and service manuals.

ENGINEERING DRAWING Technical graphic communication in general, but the term does not clearly include all of the people in diverse fields concerned with technical work and industrial production.

TECHNICAL DRAWING Any drawing used to express technical ideas, or in general, the scope of technical graphic communications. The term has been used by various writers since at least Monge's time and is still widely used.

ENGINEERING GRAPHICS OR ENGINEERING DESIGN GRAPHICS Drawings for technical use, specifically technical drawings representing designs and specifications for physical objects and data relationships as used in engineering and science.

TECHNICAL SKETCHING   A valuable tool for engineers and others engaged in technical work, allowing them to express most technical ideas quickly and effectively without the use of special equipment.

READING DRAWINGS   Interpreting drawings made by others. Also called blueprint reading, although the blueprint process has now been replaced by other processes. Blueprints and other reproductions of a drawing are often just referred to as *prints*. Many companies are moving toward paperless offices that entirely use electronic file transfer, signatures, and storage. Reading a drawing means interpreting its ideas and specifications, whether or not the drawing is a blueprint.

DESCRIPTIVE GEOMETRY   Using technical drawings to solve engineering problems involving spatial relationships.

■ 1.17 BENEFITS OF GRAPHIC SKILLS

Drawings have accompanied and made possible technical advancements throughout history. Today the connection between engineering and science and the ability to visualize and communicate graphically is as vital as ever. Engineers, scientists and technicians need to be proficient in expressing their ideas through technical graphics, using both sketching and CAD. Training in the application of technical drawing is required in virtually every engineering school in the world.

Artistic talent is no longer a prerequisite to learning the fundamentals of technical graphics. To produce technical drawings you need the same aptitudes, abilities, and computer skills used in you other science and engineering courses.

As an engineer, scientist, or technician, you will need to create and interpret graphical representations of engineering structures, designs, and data relationships. You need to understand fundamental principles in order to communicate through technical graphics. You also must be able to do the work with reasonable skill so that others can understand your sketches and design representations. This skill is also beneficial in seeking an entry-level position in an engineering field. When you start out, you'll likely work with or revise CAD drawings and prepare sketches under the direction of an experienced engineer.

In most technical professions the ability to read a drawing is a necessity, whether or not you produce drawings yourself. Technical drawings are found in nearly every engineering textbook, and instructors often require you to supplement calculations with technical

FIGURE 1.29 **DIGITAL Personal Workstation.** *Courtesy of Digital Equipment Corporation.*

sketches, such as *free body diagrams*. So mastering a course in technical drawing using both sketching and CAD will help you not only in your professional work, but also in many courses.

Besides the necessity of producing and interpreting technical graphics, the development of neatness, speed, and accuracy is useful for any engineer, scientist, or technician.

The ability to think in three dimensions is one of the most important skills in any technical profession. Learning to visualize objects in space is one of the benefits of studying technical graphics. Many extraordinarily creative people possess an extraordinary ability to visualize, but anyone can develop it.

■ 1.18 PROJECTIONS

Modern technical graphics uses individual views or projections to communicate the shape of a 3-D object or design on a sheet of paper. You can think of every drawing as involving the spatial relationship of four things:

1. The observer's eye, or the station point
2. The object
3. The *plane of projection*
4. The projectors, also called visual rays or lines of sight.

The rules for Cannonball Run, the final competition at Ideo University, are simple: To win, a team needs to:

1. Build a device that will propel a steel "cannonball" farther than contraptions built by the three opposing teams.
2. Make the ball signal the end of its journey by setting off a buzzer in the designated "cannonball catcher," a round target with the approximate circumference of a coffee cup.
3. Use the provided material, which consists of a bundle of wooden dowels, a chunk of cushiony foam, a roll of electrical tape, a stack of index cards, some paint sticks, a fistful of rubber bands, an extrememly long piece of black string, a mousetrap, and the tubes all this came in.
4. Complete a cannonball launcher within 90 minutes.

Not the simplest of tasks, but then again, no one ever said that learning creativity was easy. In fact, the compeition at Ideo U. consisted of one journalist and 15 engineers from the networking giant Cisco Systems. At the beginning of the competition, they were all skeptical. Creativity couldn't possibliy be taught, let alone at a one-day innovation and design workshop.

IDEO U. DESIGN WORKSHOPS

Ideo, the Silicon Valley firm that's famous for the designing products like Apple's first mouse, a no-mess toothpaste tube for Procter & Gamble's Crest, and, most recently, the Palm V, the silvery, wafer-thin, light-as-a feather addition to 3Com's blockbuster line of hand held organizers.

Founded in 1978 by Standford design professor, David Kelley, Ideo has designed more than 3,000 products for a roster of clients, which include a sizeable portion of the FORTUNE 500. Ideo began offering design workshops for its customers and potential customers after many began clamouring to know exactly how they came up with there designs. Since then, companies ranging from NEC to Kodak to Steelcase have been sending employees to Ideo U.

CAN CREATIVITY BE TAUGHT?

"They say that genius is 99% perspiration and 1% inspiration," says Dennis Boyle, the Ideo principal leading the Cannonball Run Workshop. "Most companies have that 99%. It's the 1% that's really hard, and that's why our clients are asking us to work with their people and not just their product."

To home in on that elusive 1%, Ideo's employees explain the techniques they use to design products and then force participants to put them into practice. That's where Cannonball Run comes in.

BRAINSTORMING

The first step: brainstorming. Earlier that afternoon, Brendan Boyle, head of Ideo's toy-invention studio had explained brainstorming the Ideo way. Ideo takes its rules for brainstorming so seriously that they are printed on a large banner that runs across the top of the classroom's whiteboards They are:

- Defer judgment (otherwise you'll interrupt the flow of ideas);
- Build on the ideas of others (it's far more productive than hogging the glory of your own insights);

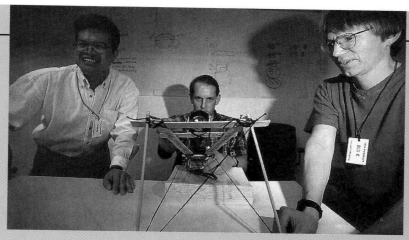

- Stay focused on the topic (no tangents);
- One person at a time (so you don't drown out that quiet, brilliant mumbler in the corner);
- Go for quantity (when Ideo staffers brainstorm, they shoot for 150 ideas in 30 to 45 minutes);
- Encourage wild ideas (to paraphrase Einstein, "If the first idea doesn't sound absurd, then there's no hope for it); and
- Be visual (sketch ideas to help people understand them).

Team 2 stuck to these rules and within a few minutes came up with 15-20 ideas for the launcher.

RAPID PROTOTYPING

To narrow their ideas down, Team 2 used rapid prototyping, another Ideo technique. The idea behind rapid prototyping is that it's easier to discuss a model of something, no matter how primitive, than to talk about a bunch of ideas. "If a picture is worth a thousand words," says Ideo's Steve Vassallo, "a prototype is worth ten thousand."

Rapid prototyping consists of three Rs: Rough, Rapid, and Right. The first two Rs are fairly self-explanatory—make your models rough and make them rapidly. In the early stages perfecting a model is a waste of time. "You learn just as much from a models that's wrong as you do from one that's right," says Vassalo. Even the final R (Right), doesn't mean that your model has to work. Instead it refers to building lots of small models that focus on specific problems. For Team 2's Cannonball Run project, the three Rs worked.

After making a few rough models with the mousetrap, they were certain that they could never fire the cannonball through the air with any precision. That nudged them towards a safer option—building a ramp that would guide the ball from a table top to its target on the floor via two guide rails made out of the wooden dowels. They got busy building their device. Thirty minutes before deadline, they realized that one of the other teams were doing the exact same thing, with one crucial difference: They were taping together the index cards (the most useless looking of all the provided materials) to create a 30-foot long track that would guide the cannonball from their ramp to their cannonball catcher at the far end of the room. What's more they were carefully folding up the sides of each card to make sure the ball didn't get derailed on its way to the destination. Brilliant!

Teams 3 and 4 had gone with different designs that don't look nearly as impressive. So what did Team 2 do? They stole the idea. Suddenly, the competition was reduced to which team, Team 1 or Team 2, would be able to tape the index cards together faster.

USING BRAINSTORMING AND RAPID PROTOTYPING TECHNIQUES IN REAL LIFE

Of course, in real life, you can't always glance across the room and steal your competitor's ideas. But the Cannonball Run project drove home Ideo's philosophy about brainstorming and rapid prototyping methods: They get you to stop dithering and start doing. What you can come up with on the fly won't be nearly as bad as you think—in fact it'll often be better that what you come up with working slowly and deliberately. Besides, coming up with something—anything—is often half the battle.

All 15 participants walked away from the experience with changed ideas of how to approach even the simplest of projects. "I like how it showed that you don't have to spend tons of money to prototype. You can do a lot of trial-and-error modeling on your own without paying a lot of money to go outside just to find something that doesn't work," stated one Team 2 participant. Another teammate added, "When I went to work the next day, I called an emergency brainstorming session and set a goal of 100 ideas in an hour. I thought maybe we'd get 50. We got 103." Would he recommend the workshop to his colleagues? "Oh man, are you kidding me? Absolutely!" But they could be biased. After all, Team 2 won.

Adapted from "Staying Smart: A Day at Innovation U.," by Ed Brown, *Fortune*, April 12, 1999, pp. 163-165.

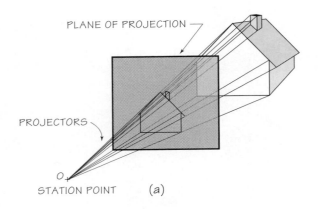

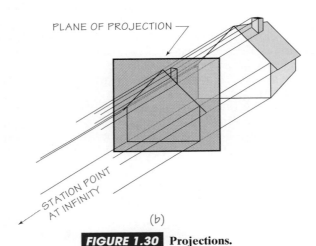

FIGURE 1.30 Projections.

There are two main types of projection—perspective and parallel. Figure 1.30 shows a ***perspective projection***. The projected view of the house is shown on the plane of projection as viewed by the observer. You can think of imaginary rays or projectors beaming from the vertices of the object (the corners of the house in this case) to strike the plane of projection, converging at the observer's eye. The appearance of a perspective projection is similar to the view you would actually see.

In a ***parallel projection***, shown in Figure 1.30, you can think of the projectors or rays as beaming from the house, perpendicular to the plane of projection, so that they would be parallel to each other on to infinity. The result of this is an orthographic (meaning right angle) projection. Orthographic projections can accurately depict the dimensions of the object. Oblique projections are a type of parallel projection where the rays or projectors strike the projection plane at an angle other than 90 degrees.

The main types of projections are further broken down into many subtypes, as shown in Figure 1.31; they will be discussed in later chapters.

■ KEY WORDS

graphic communication	models	boolean operators	computer simulation
technical drawing	prototype	extrusion revolution	working drawing
computer graphics	looping	checking and proofing	rapid prototyping
empirical design	layout sketch	functional design	patent drawings
scientific design	stages of design	aesthetic design	reverse engineering
parametric solid modeling	design requirements	motivation	coordinate measuring machine
design team	models	objectives	
descriptive geometry	virtual reality	creativity techniques	functional decomposition
standards	detail drawing	extruded solid	brainstorming
projection	3D modeling	customer	storyboarding
plane of projection	solid modeling	concepts	reintegration
perspective projection	assembly drawing	refinement	revolved solid
parallel projection	production drawings	parametric solid modeling	venn diagrams

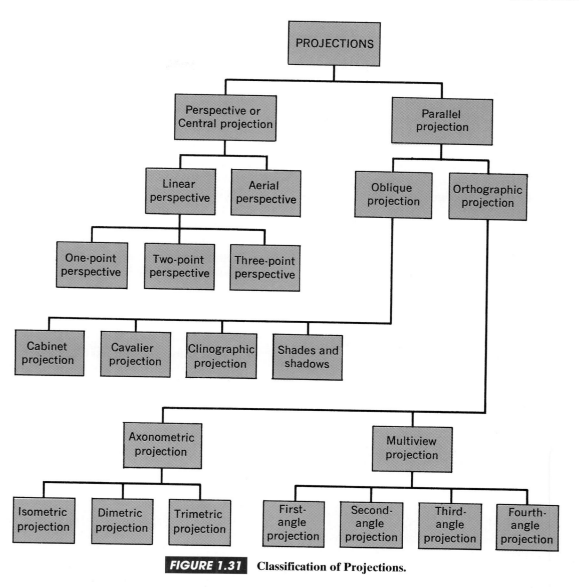

FIGURE 1.31   Classification of Projections.

Classes of Projection	Distance from Observer to Plane of Projection	Direction of Projectors
Perspective	Finite	Radiating from station point
Parallel	Infinite	Parallel to each other
Oblique	Infinite	Parallel to each other and oblique to plane of projection
Orthographic	Infinite	Perpendicular to plane of projection
Axonometric	Infinite	Perpendicular to plane of projection
Multiview	Infinite	Perpendicular to planes of projection

TABLE 1.1   *Classification by Projectors.*

■ CHAPTER SUMMARY

- The members of the engineering design project team must be able to communicate among themselves and with the rest of the project team in order to contribute to the team's success.
- The design team moves through five stages during the design process. Each stage helps the team refine the design until it meets all product requirements.
- Models are important for testing the way parts are assembled. To test the design, the design team uses both scale models created in a model shop or computer-generated models.
- During the design process, all members of the team must understand their specific roles and how they should relate and interact with the rest of the team. Effective teamwork is an essential part of the design process.
- The final drawings created during the design process include assembly drawings, working drawings, design drawings, and patent drawings.
- There are many revisions to drawings during the design process. The drafter must keep track of each version and what changes were made.

- Models are an important way of testing the way parts are assembled. Both scale models created in a model shop and computer-generated virtual reality models are used by the design team to test their design.•
- Graphic communication is the universal language used by every engineering team designing and developing products throughout the world.
- Technical drawing is based on the universal principles of descriptive geometry, developed in the late 18th century in France.
- Every technical drawing is based on standards that prescribe what each symbol, line, and arc means.
- A computer running CAD software is the current tool used to create accurately scaled drawings, but the basic drawing principles are the same ones used for hundreds of years.
- Drawings are based on the projection of an image onto a plane of projection. There are two types of projection: parallel and perspective.
- Successful companies hire skilled people who can add value to their design team. The ability to communicate clearly using verbal, symbolic, and technical graphics is a skill employers require.

■ REVIEW QUESTIONS

1. What is the role of the engineer on the design team?
2. What are the five stages of the design process? Describe each stage.
3. What is the difference between mechanical drawing and sketching?
4. Describe the main difference between parallel projection and perspective projection.
5. When is sketching an appropriate form of graphic communication?
6. Why are standards so important for members of the engineering design team?

7. What is the most important new tool used by drafters?
8. What is a plane of projection?
9. What are projectors and how are they drawn?
10. What are the special requirements of a patent drawing?
11. Name three individual creativity techniques. Name two group creativity techniques.
12. What are the advantages of computer modeling? What are the disadvantages?

■ CHAPTER PROJECTS

If necessary, refer to the appropriate section of the chapter to check your answers.

Proj. 1.1 Define the following terms as they apply to the design process: needs statement, concept, compromise solutions, prototype, layout sketch.

Proj. 1.2 Sketch a flowchart that depicts the design process as you understand it.

Proj. 1.3 The hypothetical "Flies R Us" fishing tackle company has contracted your services to restructure their design process and increase profitability. Read the job descriptions and the design flow chart in Figure 1.32 and decide at what stages of the design process each team member should be involved. Sketch out a flowchart that illustrates the new reporting relationships for these individuals or groups.

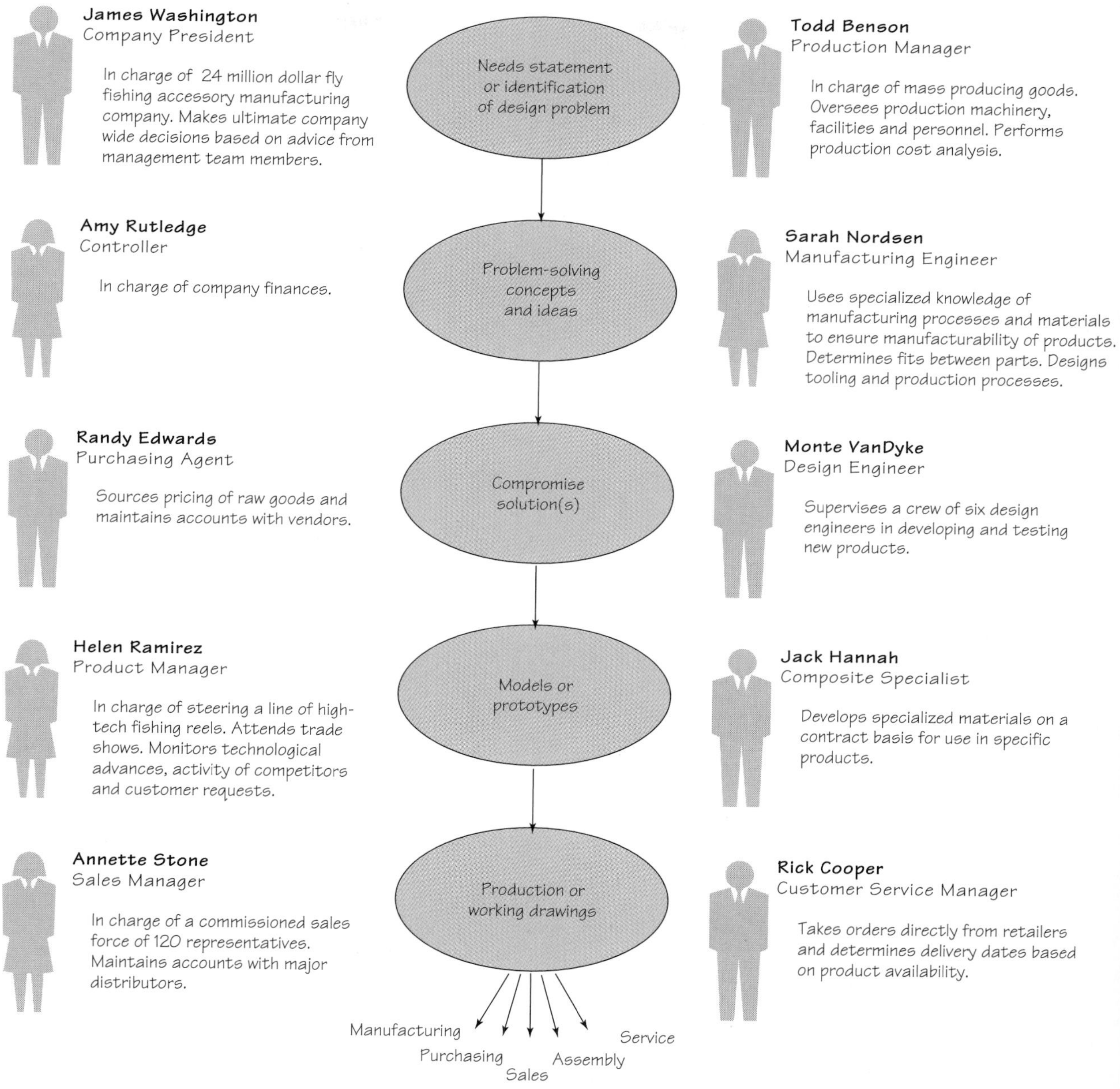

James Washington
Company President

In charge of 24 million dollar fly fishing accessory manufacturing company. Makes ultimate company wide decisions based on advice from management team members.

Amy Rutledge
Controller

In charge of company finances.

Randy Edwards
Purchasing Agent

Sources pricing of raw goods and maintains accounts with vendors.

Helen Ramirez
Product Manager

In charge of steering a line of high-tech fishing reels. Attends trade shows. Monitors technological advances, activity of competitors and customer requests.

Annette Stone
Sales Manager

In charge of a commissioned sales force of 120 representatives. Maintains accounts with major distributors.

Todd Benson
Production Manager

In charge of mass producing goods. Oversees production machinery, facilities and personnel. Performs production cost analysis.

Sarah Nordsen
Manufacturing Engineer

Uses specialized knowledge of manufacturing processes and materials to ensure manufacturability of products. Determines fits between parts. Designs tooling and production processes.

Monte VanDyke
Design Engineer

Supervises a crew of six design engineers in developing and testing new products.

Jack Hannah
Composite Specialist

Develops specialized materials on a contract basis for use in specific products.

Rick Cooper
Customer Service Manager

Takes orders directly from retailers and determines delivery dates based on product availability.

Flow chart: Needs statement or identification of design problem → Problem-solving concepts and ideas → Compromise solution(s) → Models or prototypes → Production or working drawings → Manufacturing, Purchasing, Sales, Assembly, Service

FIGURE 1.32 "Flies R Us" Job Descriptions and Design Flow Chart.

Proj. 1.4 Using Figure 1.32 and your flowchart from Project 1.3 as a guide, answer the following:

What sort of information does Randy Edwards need before he can source raw goods for the manufacture of a new product?

At what point should Todd Benson be involved in the design process, and why?

What might happen if Rick Cooper were out of the communication loop?

Would Annette Stone be likely to see a prototype?

Which team members do you think Monte VanDyke has programmed into the speed dial function of his office phone?

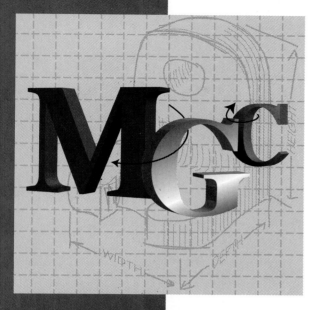

C H A P T E R 2

INTRODUCTION TO CAD

OBJECTIVES

After studying the material in this chapter, you should be able to:

1. List the basic components of a computer-aided drawing workstation.
2. Describe the relationship between computer-aided drawing (CAD) and computer-aided manufacturing (CAM).
3. List the major parts of a computer and describe their function.
4. Describe the purpose of a computer operating system.
5. List several input and output devices.
6. Describe ways in which a computer stores information.
7. Explain the differences between ROM and RAM.
8. Explain the differences between a bit and a byte.
9. Understand issues that affect the choice and the use of a CAD system.
10. Explain features common to most CAD software.

OVERVIEW

The use of electronic computers today in nearly every phase of engineering, science, business, and industry is well known. The computer has altered accounting and manufacturing procedures, as well as engineering concepts. The integration of computers into the manufacturing process from design to prototyping, manufacture, and marketing, is changing the methods used in the education and training of technicians, drafters, designers, and engineers.

Engineering, in particular, is a constantly changing field. As new theories and practices evolve, more powerful tools are developed and perfected to allow the engineer and designer to keep pace with the expanding body of technical knowledge. The computer has become an indispensable and effective tool for design and practical problem solving. New methods for analysis and design, the creation of technical drawings, and the solving of engineering problems, as well as the development of new concepts in automation and robotics, are the result of the influence of the computer on current engineering and industrial practice.

Computers are not new. Charles Babbage, an English mathematician, developed the idea of a mechanical digital computer in the 1830s, and many of the principles used in Babbage's design (Figure 2.1) are the basis of today's computers. The computer has appeared in literature and science fiction to be a mysterious, uncompromising, often sinister, machine, but it is nothing more than a tool. It is capable of data storage, basic logical functions, and mathematical calculations. Computer applications have expanded human capabilities to such an extent that virtually every type of business and industry utilizes a computer directly or indirectly.

■ 2.1 COMPUTER SYSTEMS AND COMPONENTS

Engineers and drafters have used computers for many years to perform the mathematical calculations required in their work. Only recently, however, has the computer been accepted as a valuable tool in the preparation of technical drawings. Traditionally, drawings were made by using drafting instruments and applying ink, or graphite, to paper or film. Revisions and reproductions of these drawings were time consuming and often costly. Now the computer is used to produce, revise, store, and transmit original drawings. This

FIGURE 2.1 A Working Model of Charles Babbage's "Difference Engine" Originally Designed in 1833. *From the New York Public Library Picture Collection.*

method of producing drawings is called *computer-aided design* or *computer-aided drafting* (CAD) and *computer-aided design and drafting* (CADD). Since these and other comparable terms are used synonymously, and since industry and software creators are beginning to standardize, they will be referred to throughout this book simply as CAD.

Other terms, such as *computer-aided manufacturing* (CAM), *computer-integrated manufacturing* (CIM), and computer-assisted engineering (CAE), are often used in conjunction with the term "CAD." The term "CAD/CAM" refers to the integration of computers into the design and production process (see Figure 2.2). The term "CAD/CAE/CAM/CIM" describes the use of computers in the total design and manufacturing process, from design to production, publishing of technical material, marketing, and cost accounting. The single concept that these terms refer to is the use of a computer and software to aid the designer or drafter in the preparation and completion of a task.

Computer graphics is a very broad field. It covers the creation and manipulation of computer-generated images and may include areas in photography, business, cartography, animation, publication, as well as drafting and design.

FIGURE 2.2 CAD/CAM Driven Machine Tool Cutting.
Courtesy of David Sailors.

A complete computer system consists of *hardware* and *software*. The various pieces of physical equipment that comprise a computer system are known as hardware. The programs and instructions that permit the computer system to operate are classified as software. Computer programs are categorized as either *application programs* or *operating systems*. Operating systems, such as DOS, Windows (Figure 2.3), and UNIX, are sets of instructions that control the operation of the computer and peripheral devices as well as the execution of specific programs. This type of program may also provide support for activities and programs (such as input/output [I/O] control, editing, storage assignment, data management, and diagnostics), assign drives for I/O devices, and provide support for standard system commands and networking.

FIGURE 2.3 Screenshot of the Windows 98 Operating System. *Courtesy of Microsoft Corporation.*

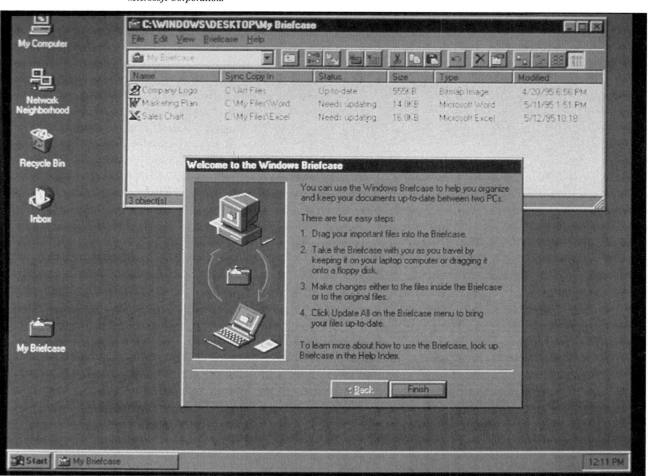

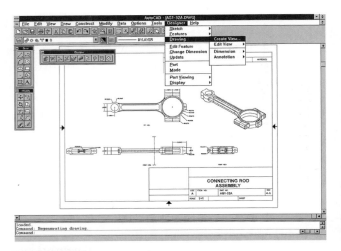

FIGURE 2.4 **AutoCAD is a Widely Used Drafting Application Program.** *This material has been reprinted with the permission from and under the copyright of Autodesk, Inc.*

FIGURE 2.5 **Advanced CAM Technology Used in High-Resolution Color Picture Tube Production.** *Courtesy of Zenith Electronics Corporation and Charlie Westerman.*

Application programs are the link between specific system use and its related tasks—design, drafting, desktop publishing, etc.—and the general operating system program (Figure 2.4).

 The PC Webopaedia is a great place to look up CAD and computer terminology: http://www.pcwebopaedia.com/CAD.htm. You may want to add bookmarks or links to handy reference sites like this one for easy access.

■ 2.2 COMPUTER TYPES

Computers may be classified as one of two distinct types: *analog* or *digital*. An analog computer measures continuously without steps, whereas the digital computer counts by digits, going from one to two, three, and so on, in distinct steps. An electric wall clock with minute and hour hands and the radial speedometer on a car are examples of analog devices. An abacus and a digital watch are examples of digital devices. Digital computers are more widely used than analog computers because they are more flexible and can do a greater variety of jobs.

Analog computers are generally used for mathematical problem solving. This type of computer, which measures continuous physical properties, is often used to monitor and control electronic, hydraulic, or mechanical equipment. Digital computers have extensive applications in business and finance, engineering, numerical control, and computer graphics (Figure 2.5).

Both types of computers have undergone great changes in appearance and in operation. Equipment that once filled the greater part of a large room has now been replaced by machines that occupy small desktop areas. The single most important advancement in computer technology has been the development of the *integrated circuit* (IC). The IC chip has replaced thousands of components on the *printed circuit* (PC) board and made possible the development of microprocessors. The microprocessor is the processing unit of a computer. The difference in size between a PC board with individual components and an IC chip is shown in Figure 2.6. The term "microminiaturization" is applied to advanced integrated circuit chip technology. The evolution of IC chip technology has led to the increased production of low-cost microcomputers. Microcomputers are largely responsible for

FIGURE 2.6 **Size Comparison of a PC Board and an IC Chip.**

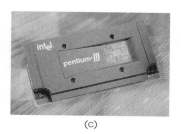

(c)

(a)   (b)

FIGURE 2.7 (a) Computer with Intel Pentium III. *Courtesy of Dell Corporation.* (b) Think pad. *Courtesy of IBM Corporation.* (c) Intel Pentium III chip. *Courtesy of Intel Corporation.*

the increase in use of computer-aided drafting systems in industry. Low-cost microcomputer CAD systems can now be cost justified by industrial users (Figs. 2.7a-c). With the recent release of the Pentium III processor, CAD applications can run faster and more efficiently than ever before (See Figs. 2.7a and c). Thinkpad computers make mobile reference to CAD files not only possible, but commonplace (See Figure 2.7b).

Since CAD systems utilize digital computers, we will restrict our discussion of computer types to digital computers.

■ 2.3 COMPUTER-AIDED DRAFTING

The first demonstration of the computer as a design and drafting tool was given at the Massachusetts Institute of Technology in 1963 by Dr. Ivan Sutherland. His system, called "Sketchpad," used a cathode ray tube and a light pen for graphic input to a computer. An earlier system, called SAGE, was developed in the 1950s for the Air Defense Command and used the light pen for data input. The first commercial computer-aided drafting system was introduced in 1964 by International Business Machines (IBM).

Many changes have taken place since the introduction of the first CAD system. The changes are due to the advent of the microprocessor, more sophisticated software (programs), and new industrial applications. In most cases, the drafter/engineer can create, revise, obtain prints (hard copy), and store drawings with relative ease, utilizing less space. CAD was originally used to aid in creating production drawings. The advent of three-dimensional CAD software made it apparent that a 3D computer model (Figure 2.8) could

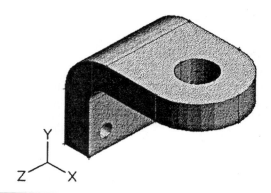

FIGURE 2.8 A CAD Solid Model of the Ball Bracket from a Trailor Hitch Assembly. *From* Machine Design: An Integrated Approach *by Robert Norton, © 1996. Reprinted by permission of Prentice-Hall, Inc., Upper Saddle River, NJ.*

assist not only in the manufacture of the part but also, along with its three-dimensional database, in testing the design with finite element analysis programs (Figure 2.9), in developing technical manuals and other documentation that combine illustrations of the design with text from word processing programs, and in marketing (for which the 3D solid models can be used with a rendering and animation program). Increases in productivity and cost effectiveness are two advantages constantly stressed by CAD advocates. In addition, CAD stations can be linked either directly or through a *local area network* (LAN) to the manufacturing or production equipment, or they can be linked with *numerical control* (NC) equipment to program NC machines automatically in manufacturing operations or in robotics (Figure 2.10).

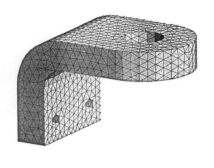

FIGURE 2.9   An FEM Mesh Applied to the Solid Model of the Ball Bracket. *From* Machine Design: An Integrated Approach *by Robert Norton, © 1996. Reprinted by permission of Prentice-Hall, Inc., Upper Saddle River, NJ.*

The primary users of CAD are in mechanical engineering and electronic design, civil engineering, and cartography. The design and layout of printed circuits are a principal application of CAD in the electronics industry, which, prior to 1976, was the largest CAD user. Mechanical engineering has since overtaken electronics and continues to expand its CAD applications and use. Continued expansion in mechanical design applications is expected because the design, analysis, and numerical control capabilities of CAD can be applied to a varied range of products and processes. Cartography, seismic data display, demographic analysis, urban planning, piping layouts, and especially architectural design also show growth in CAD use. A relatively new area in computer graphics is *image processing*, which includes animation, 35-mm slide preparation, photocolor enhancement, and font and character generation (used in television broadcasting and the graphic arts industry).

■ 2.4 CAD SYSTEM CONFIGURATIONS

All computer-aided drafting systems consist of similar hardware components (Figure 2.11), such as input devices, a central processing unit, data storage devices, and output devices. For input devices, the system will have one or more of the following: a keyboard, mouse, trackball, digitizer/graphics tablet, and light pen. For output, The CAD system will include devices such as plotters, printers, and some type of monitor. The system must also have a data storage device, such as a tape drive, a hard (fixed) or soft (floppy) disk drive, or an optical disk drive. Finally, a computer or central processing unit (CPU) is needed to do all the numerical manipulations and to control all the other devices connected to the system.

Frequently, some devices are combined. For example, a terminal or workstation can contain the keyboard, monitor, disk drives, and a CPU all in the same cabinet. Such a combined device is often called a *workstation* (Figure 2.12).

■ 2.5 CENTRAL PROCESSING UNIT

The CPU (Figure 2.13), or computer, receives all data and manages, manipulates, and controls all functions of the CAD system. CAD systems use digital computers. All data must be converted into a binary form or

FIGURE 2.10   Computers Work with NC Machines in Modem Industry. This is an Example of a Computer Numerical Control (CNC) Four-Axis Turning Center.

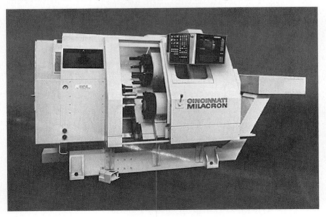

FIGURE 2.11   Complete CAD Systems Need Input Devices, Output Devices, Storage Devices, and a Central Processing Unit. *Courtesy of Hewlett-Packard Company.*

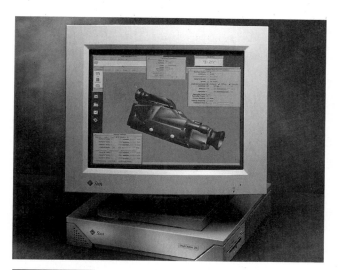

FIGURE 2.12   **A Powerful Computer Workstation.**
Courtesy of Sun Microsystems.

code for the computer to understand and accept. This code is called *binary coded decimal instructions* (BC-DI). This binary code uses a two-digit system, 1 and 0, to transmit all data through the circuits. The number 1 is the "on" signal; the 0 is the "off" signal. A *bit* is a binary digit. Bits are grouped, or organized, into larger instructions. Word length, which is expressed as bits, differs with various computers. The most common word lengths are 8, 16, and 32 bits. The word length

indicates the maximum word size that can be processed, as a unit, during an instruction cycle. Often, computers are categorized by their word length, such as 16-bit or 32-bit computers. The number of bits in the word length indicates the processing power of the computer (the larger the word length, the greater the processing power). A sequential group of adjacent bits in a computer is called a *byte*. The current industry standard is that 8 bits equal 1 byte. A byte represents a character that is operated on as a unit by the CPU. The length of a word on a majority of computer system is currently 4 bytes. This means that each word in any of these storage systems occupies a 32-bit storage location. The memory capacity of a computer is therefore expressed as a number of bytes rather than bits.

Inside the CPU is the brain of the computer—the microprocessor chip (Figure 2.13). IBM-compatible computers have used the 386, 486, Pentium (P5), P6, II, and III chips. MacIntosh computers use the Motorola chips or PowerPC chip. Intel's Pentium III processor along with Apple's PowerPC G3 processor have kept pace with industries' every expanding need for speed and memory. Silicon Graphics and Sun computers use their own chips. The job of the microprocessor is to execute the instructions of the software programs.

The microprocessor chip's speed is rated in *megahertz* (MHz), and its power is rated in *millions of instructions per second* (MIPS). It is the largest chip in the CPU and is mounted on the main circuit board,

FIGURE 2.13   **Inside a CPU Box.**
Courtesy of Apple Computer.

called the *motherboard*. Attached to the motherboard are all the devices inside the CPU: the memory chips, power supply, connection ports (where the external cables are attached), internal modems, video cards, network interface cards, sound cards, and special hard drive controller cards.

Computer memory is stored on small circuit boards called *single inline memory modules* (SIMMs). Most of the SIMMs are for the main memory, called the *random access memory* (RAM). This memory is temporary. When you turn off the computer, the information in the memory is erased. The software programs and user files are stored in RAM when the program is actively running. The more RAM in a computer, the more programs you can run at one time. If you do not have enough memory, you may have a hard time running even one large program.

There is also permanent memory on the motherboard, which is called *read only memory* (ROM). When the computer is turned off, the ROM chips do not forget what is stored in them. The ROM chips contain basic operating system programs, like simple diagnostic programs that check the computer system to make sure all circuits and devices are operational when the computer is turned on. One type of ROM is called *flash* ROM, and it can be reprogrammed. Normal ROM cannot be reprogrammed.

FIGURE 2.14 **A Mother Board with 10 Bus Slots for Adding Capabilities Such as Video Display Cards.** *Courtesy of International Business Machines.*

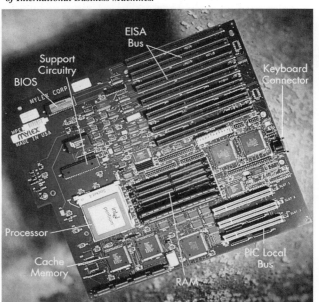

The motherboard contains slots for peripheral devices. (See Figure 2.14.) Internal modems, video display cards, and network interface cards are typical cards that can be plugged into the motherboard. These cards talk to the microprocessor via an electrical pathway called the *bus*. The bus is like an expressway that allows electrical information to be shared between all the devices connected to the motherboard. Just like an expressway, computer busses have speed limits. They are rated in megahertz like the CPU. The faster the bus, the more quickly information can be transported into the computer.

The rear panel of the CPU contains a series of connector ports (plug receptacles). Different types of ports transfer data differently. Each corresponds to a matching cable that connects the CPU to an external peripheral device. Most printers, for example, use a 25-pin D connector to connect to the parallel port. The keyboard and mouse each have their own connection port (usually a small round port). Many of the cards inside the CPU have connection ports attached to the card that protrude through the back of the CPU. For example, the monitor connects to the port on the video display card. The modem telephone line connects to a port on the back of the internal modem. The LAN cable connects to the port on the back of the network interface card.

■ 2.6 DISPLAY DEVICES

Another major reason for the rapid growth in CAD systems is improvements in display devices. These display devices, commonly referred to as monitors (Figure 2.15), utilize a wide variety of imaging principles. Each device has definite characteristics with regard to brightness, clarity, resolution, response time, and color. The purpose of any graphics display is to project an image on a screen. The image that is displayed may be alphanumeric (text symbols, letters, and/or numerals) or graphical (pictorial symbols and/or lines). Users of interactive CAD systems communicate directly or indirectly through graphics terminals. The information requested by the user may be displayed as animated figures, graphs, color-coded diagrams, or simply a series of lines.

Most interactive CAD systems use a raster scan monitor. *Raster scan* devices are similar to conventional television screens. These devices produce an image with a matrix of picture-element dots, called *pixels*, within a grid. Each pixel is either a light or dark image that appears on the screen as a dot. As in conventional

FIGURE 2.15 21″ Monitor for CAD Hi-Res Images.
Courtesy of NEC Technologies, Inc.

television, an electron beam is swept across the entire screen, line by line, top to bottom. This process is called raster scanning. A signal turns on or illuminates a pixel according to a pattern stored in memory. The screen is scanned around 60 times a second to update the image before the phosphor dims.

Most CAD computers use a 17-inch or 21-inch display. These sizes are measured diagonally across the front of the screen. The image on the monitor is generated by the video display card. These cards determine the resolution of the display and the number of colors. Standard *video graphics array* (VGA) resolution is 640 × 480 pixels. The more pixels per inch, the greater the resolution and the easier it is to read details on the monitor. Large-screen monitors can support up to 1600 pixels horizontally and 1200 pixels vertically.

In addition to resolution, video display cards can generate a range of colors. Normal color density is 256 colors. Photographic quality requires a color density of 16.7 million colors. When a monitor has a lot of pixels and many colors on the screen, it takes a lot of processing power and video memory to draw the image on the monitor. The microprocessor chip of the CPU cannot handle this load so the video display card often provides its own processor, called a video accelerator, and its own memory. Professional CAD computers usually have a very fast video accelerator that provides high resolution and a large number of colors. With a fast video accelerator, even the largest, high-resolution monitor can redraw in the blink of an eye. Slower video display cards can take up to a minute to redraw the screen. Most video display cards use a standard 15-pin VGA connector. However, larger monitors may require four separate cables.

Video monitors are rated by the speed with which they can refresh the screen. There are two ratings: horizontal refresh rate and vertical refresh rate. The faster the refresh rate, the easier the monitor is on the eyes. For example, if you are using a monitor with only a 60-Hz vertical refresh rate, you will begin to see a flicker on the screen after using it for several hours. Higher refresh rates reduce this annoying flicker. Quality large-screen monitors can cost as much as the CPU, but they are part of the user interface and can affect the long-term productivity of the operator.

■ 2.7 INPUT DEVICES

A CAD system may use one or a combination of input devices to create images on the display screen. Graphic input devices may be grouped into three categories: (1) *keyboard* and *touch sensitive*, (2) *time-dependent* devices, and (3) *coordinate-dependent* devices.

The keyboard is the universal input device by which data and commands are entered. A typical keyboard consists of alphanumeric character keys for keying in letters, numbers, and common symbols, such as #, &, and %; cursor control keys with words or arrows, printed on them, indicating directional movement of the screen cursor, and special function keys, which are used by some software programs for entering commands with a single keystroke. Many large mainframe-based CAD systems have used a special function keypad, or menu pad, that allows access to a command with a single keystroke (Figure 2.16). Single-stroke command selection was considered so essential for cost effectiveness and ease of use that developers of mini- and microcomputer-based CAD systems included this feature of single and double keystroke command access into their program utilizing the CTRL, ALT, SHIFT, and function keys. Typically, a CAD system will use a keyboard for inputting commands and text, and another input device for cursor control.

A popular input device in use with both large and small CAD systems is a *mouse* (Figure 2.17a-c). A mouse may be of the mechanical type or the optical type. A mechanical mouse uses a roller, or ball, on the underside of the device to detect movement. An optical mouse senses movement and position by bouncing a light off a special reflective surface. These optical mice track most reliably on a mousepad, but also can function on other surfaces, with the exception of glass. Most mice will have from one to three buttons on top

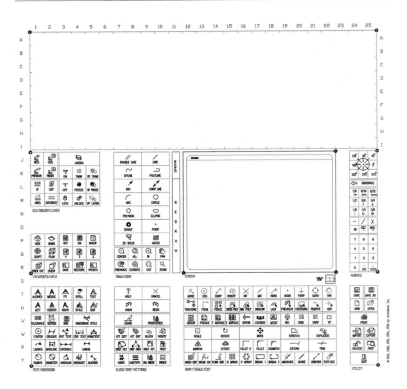

FIGURE 2.16 An AutoCAD Menu Pad. *This material has been reprinted with the permission from and under the copyright of Autodesk, Inc.*

(a) Standard Mouse   (b) Cordless Mouse   (c) Microsoft IntelliEye Optical Mouse

FIGURE 2.17 Mice (a–c). *Courtesy of Logitech, Inc. and Microsoft, Inc.*

of them to select positions or commands. Microsoft's new IntelliEye optical mouse comes with two extra buttons. The advantages of a mouse include ease of use, small required working area, and is relatively low cost. A mouse cannot, however, be used to digitize existing drawings into a CAD format.

Digitizing tablets (Figure 2.18) are another commonly used input device. They can be used to create an original CAD drawing or to convert an existing pen or pencil and paper drawing into a CAD drawing. Digitizing tables range in size from 80×110 to 360×480. Tablets larger than 360×480, called digitizing tables, are used primarily for converting existing drawings to a CAD system format. The *resolution* of digitizing tablets is important. This determines how small a movement the input device can detect (usually expressed in thousandths of an inch) and depends on the number of wires per inch in the tablet's grid system. The working area on a tablet can have areas that are used as menus to pick commands from the CAD system. Attached to the tablet will be either a puck or stylus. A *puck* is a small, hand-held device with a clear plastic extension (or window containing crosshairs) that transfers the location of the puck on the tablet-grid to the relative location on the screen. Single or multiple buttons on top of the puck are used to select points and/or commands. A *stylus* appears to be a ball-point pen with an electronic cable attached to it. The tip of the stylus senses the position on the tablet grid and relays these coordinates to the computer. When the stylus is moved across the tablet, the screen cursor

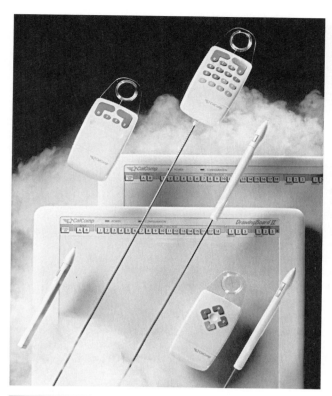

FIGURE 2.18 **Tablets with Picking Devices—Puck and Stylus—Wired and Wireless.** *Courtesy of CalComp Digitizer Division.*

FIGURE 2.19 **Large Format Scanners for Capturing Documents Electronically.** *Courtesy of CalComp Digitizer Division.*

moves correspondingly across the screen. The stylus also contains a pressure-sensitive tip that enables the user to select points or commands by pressing down on the stylus.

Existing paper drawings can also be converted to CAD drawings using a scanner (Figure 2.19) to read the existing drawing. Scanners create raster or bitmapped files, so the scanned images need to be converted to vector (line drawing) formats before they are useful for CAD. There are a number of raster-to-vector conversion programs on the market that help automate this process.

One of the oldest input devices used on CAD systems is the *trackball* (Figure 2.20a). Trackballs were used on many large mainframe-based CAD systems and were often incorporated into the keyboard. Now they are a popular input device on portable computer systems. A trackball consists of a ball nested in a holder or cup, much like the underside of a mechanical mouse, and from one to three buttons for entering coordinate data into the system. Within the holder are

sensors that pick up the movement of the ball. The ball is moved in any direction with the fingers or hand to control cursor movement on the CRT screen. Cursor speed and button functions can be set by the user. Figure 2.20b shows a thumb mouse, similar to a trackball, which is actually a 3-D controller. These controllers allow the user to manipulate objects on x, y, and z axes, as well as control pitch, roll and yaw movements. These 3-D controllers allow the user complete control of graphic objects in six degrees of freedom and are readily used in CAD applications.

A *joystick* (Figure 2.21) is an input device more commonly used with video games today than with CAD systems. This device looks like a small rectangular box with a lever extending vertically form the top surface. This hand-controlled lever is used to manipulate the screen cursor and manually enter coordinate data. Buttons of the device can be used to enter coordinate location or specific commands. This device is inexpensive and requires a very small working area, but it sometimes presents a problem when precise cursor positioning is important.

The *light pen* is the oldest type of CAD input device currently in use (Figure 2.22). It looks much like a ballpoint pen or the stylus on a digitized tablet. A light pen is a hand-held photosensitive device that works only with raster scan or vector refresh monitors and is used to identify displayed elements of a design or to specify a location on the screen where an action is to take place. The pen senses light created by the electron beam as it scans the surface of the CRT. When the

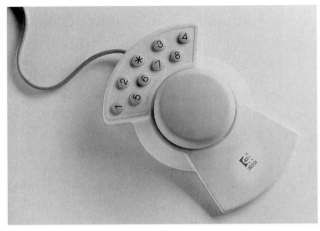

(a)   (b)

FIGURE 2.20   **(a) A Trackball.** *Photo Courtesy of Logitech, Inc.* **(b) This Magellan 3-D Controller Allows the User to Manipulate Graphic Objects with *x, y, z,* Pitch, Roll and Yaw Movement.** *Courtesy of Logitech, Inc.*

FIGURE 2.21   Joystick. *Courtesy of Logitech, Inc.*

FIGURE 2.22   Lightpen. *Courtesy of HEI, Inc.*

pen is held close to or touches the CRT screen, the computer can determine its location and position the cursor under the pen. Because this input device more closely emulates the traditional drafter's pencil or pen than other devices, it quickly gained popularity in the technical drawing field. It is popular for uses in which the user selects buttons or areas from the screen, because it is quick and easy to use.

Touch-panel displays, although not widely used with CAD systems, are another type of input device. These devices allow the user to touch an area on the

screen with a finger to activate commands or functions. This area on the screen contains a pressure-sensitive mesh that picks up the location of the finger and transfers it to the computer, which activates the command associated with that specific location. These devices work well with function or command entry but are inaccurate for CAD coordinate entry.

Bar code readers, although not often used with CAD programs, can offer users a way of labeling, tracking, and storing data and diagrams fro future use. (See Figure 2.23.)

Practical Tips
Organizing your files using Windows Explorer

When using CAD, and computers, in general, it is important to manage and back up the files that you create. A well-planned directory structure can help you keep your files organized so that they are easy to locate and back up. You can think of your hard drive as a file cabinet. A directory, or folder, is similar to a paper file folder. Inside each directory or folder you can store files (like pieces of paper) or you can have folders within other folders to keep your records organized further.

A typical directory structure listed in Windows 95/NT 4.0 is shown in Figure A. The directories or folders have an icon like a folder next to them. Some folders contain program files, such as AutoCAD R14 and Netscape, others contain the operating system files, such as Winnt (the standard directory for Windows NT installation). Others may contain your working files, such as letters you have written to a client. It is not a good idea in general to keep your working files within the directories that contain program files because when you upgrade or reinstall a program it may overwrite your working files. You should usually make a back-up copy of programs files before you install them. You do not generally need to backup the program files stored on your hard drive, because if they become corrupted or lost you can reinstall from your disks. Back-up CAD drawings, letters, spreadsheets, and other working files regularly so that they are not lost.

One good way to organize your files is to create a separate directory where your project files are stored. When you are working in an engineering firm, you will probably have several different projects and clients at a time. You may even have one client for whom you are doing several projects. Within your project directory create subdirectories for each job. Keep all of the files related to that job inside its folder as shown in Figure B.

To make it easier to remember to save your work to the project files, you can use the Taskbar and Start Menu Programs selections to specify a directory for the program to start in, as shown in Figure C. This will become the default directory where you save your files when running that program.

Keeping your files organized and regularly backing up your project directory will save you time in locating files and prevent the loss of important files or information.

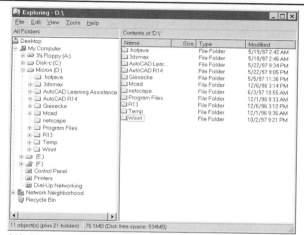

(A) Directory structure listed in WIndows 95/NT 4.0

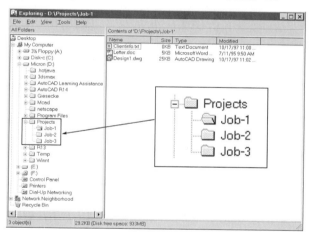

(B) Organizing files by project

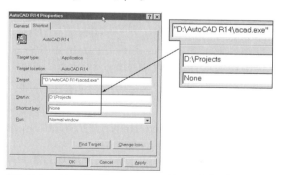

(C) Using the Taskbar to specify a default directory for AutoCAD R14.

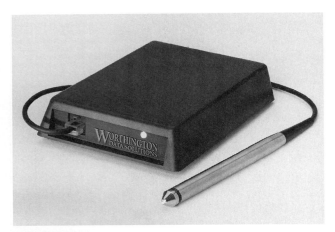

FIGURE 2.23 **Bar code reader.** *Courtesy of Worthington Data Solutions.*

A relatively recent development in data input is *voice recognition* technology. This technology utilizes a combination of specialized integrated circuits and software to recognize spoken words. The system itself must first be "trained" by repeating commands into a microphone. The computer converts the operator's oral commands to digital form and then stores the characteristics of the operator's voice. When the operator gives an oral command, the system will check the sound against the words stored in its memory and then execute the command. A disadvantage of voice recognition systems is that the vocabulary supported by the system is limited. In addition, the memory required for storing complex sound or voice patterns can be extremely large. Access time, or time between the spoken word and command activation, may be several seconds. Finally, if the operator's voice changes in some manner or words have similar voice patterns, the system may not recognize the oral input.

■ 2.8 OUTPUT DEVICES

In most instances, the user of a CAD system will need a record of images that are stored on database files or displayed on the CRT. When an image is placed on paper, film, or other media, it is then referred to as *hard copy*. This hard copy can be produced by one of several types of output devices.

A commonly used device for the reproduction of computerized drawings is the *pen plotter*. Pen plotters may be classified as drum, flatbed, or microgrip. The *drum plotter* utilizes a long, narrow cylinder in combination with a movable pen carriage. The medium to be

drawn on (paper, vellum, or film) is mounted curved rather than flat and conforms to the shape of the cylinder drum. The drum rotates, moving the drawing surface, and provides one axis of movement, while the pen carriage moves the pen parallel to the axis of the cylinder and provides the other axis of movement. The combined movement of drum and pen allows circles, curves, and inclined lines to be drawn. The pen carriage on a drum plotter typically holds more than one pen, so varied line weights or multicolor plots can be drawn. These plotters accept up to E size paper in single sheets or in a roll, which is cut after the drawing is plotted.

Flatbed plotters (Figure 2.24) differ from drum plotters in that the medium to be drawn on is mounted flat and held stationary by electrostatic or vacuum attraction while the pen carriage controls movement in box axes. The area of these plotters may be as small as A size (80 × 110) or larger than E size (340 × 440), and from one to eight pens may be used for varied line weights and multicolor plots.

Microgrip plotters (Figure 2.25) have become one of the most widely used types of output devices. Their popularity is due to their adaptability to all types of computers, their size ranges, their low maintenance requirements, and their relatively inexpensive pricing. Microgrip plotters are similar to drum plotters in that the medium to be drawn on is moved in one axis while the pen moves along the other axis. These plotters get their name from the small rollers that grip the edges of the medium and move it back and forth under the pen carriage. These plotters range from A to E size, with single or multiple pen carriages, and may accept cut sheets or rolls.

FIGURE 2.24 **Flatbed Plotter.** *Courtesy of Houston Instruments, a division of CalComp Canada, Inc., Downsview, Canada.*

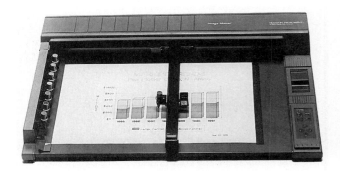

FIGURE 2.25 **Microgrip Plotter.** *Courtesy of Houston Instruments, a division of CalComp, Canada Inc., Downsview, Canada.*

All pen plotters are rated according to specified standards for accuracy, acceleration, repeatability, and speed. Accuracy is the amount of deviation in the geometry the pen plotter is supposed to draw (usually ranging from .0010 to .0050 in.). Acceleration is the rate at which the pen attains plotting speed and is expressed in *G*s (for gravitational force). Pen speed is important because slower speeds usually produce darker lines. The faster the pen attains a constant speed, the more consistent the line work will be. The ability of a plotter to retrace the same drawing over and over again is called repeatability. The deviation of the pen in redrawing the same line is the measure of repeatability and usually varies from .0010 to .0050. Plotter pen speed determines how fast the pen moves across the drawing medium. Most CAD software also allows the operator to adjust the pen speed to achieve maximum line quality and consistency. Slow pen speeds normally produce better quality plots than high pen speeds.

Other factors help determine the quality of a pen-plotted drawing. The variety of pens, inks, and drawing media available allows the operator to coordinate pen, ink, and paper to produce the most desirable hard copy.

Dot matrix or raster plotting is another method by which hard copy can be produced. These devices use a process called rasterization to convert images to a series of dots. The image is transferred optically (or sometimes by a laser) to the surface of the medium on a selenium drum that is electrostatically charged. Sometimes the image may be created by an array of nibs that electrically charge small dots on the medium.

A high-quality raster scan plotter can produce an image so fine and of such quality that it is not obvious how the image was produced unless examined under a magnifying glass.

Electrostatic plotters (Figure 2.26) produce hard copy by placing an electrostatic charge on specially coated paper and having a toner, or ink, adhere to the charged area. Drawing geometry is converted through rasterization into a series of dots. These dots represent the charged area. Resolution of these plotters is determined by the number of dots per inch (dpi), usually ranging from 300 to 600 dpi. This type of plotter produces hard-copy drawings in single color or multicolors much faster than pen plotters, but the cost, power, and environmental requirements are also much greater.

An *ink jet printer/plotter* (Figure 2.27) produces images by depositing droplets of ink on paper. These droplets correspond to the dots created by the rasterization process. This device places a charge on the ink rather than the paper, as in the electrostatic process. Ink jet plotters can produce good quality color-rendered images in addition to standard technical drawings.

Laser technology represents the newest evolution in plotter technology. A *laser printer/plotter* (Figures 2.28 and 2.29) uses a beam of light to create images.

FIGURE 2.26 **Electrostatic Printer.** *Courtesy of Houston Instruments, a division of CalComp, Canada Inc., Downsview, Canada.*

FIGURE 2.27   **Inkjet Printer.** *Photo courtesy of CalComp, Downsview, Ontario.*

FIGURE 2.29   **Full Size Laserjet Printer.** *Photo courtesy of CalComp.*

This device utilizes electrostatic charging and raster scanning to produce a plotted image that is of very high quality.

Dot matrix printers (Figure 2.30) produce images by one of two processes. They can produce images through impact on carbon or ink ribbon, or they may use a heat (thermal) process. Each method uses a number of pins in a specific configuration, such as 5×7, 7×9, or 9×9, set in the printer head. Machine commands control the sequence in which the pins strike the medium to produce an image. Dot matrix printers are normally less expensive than other types, but they are also associated with lower-quality printing.

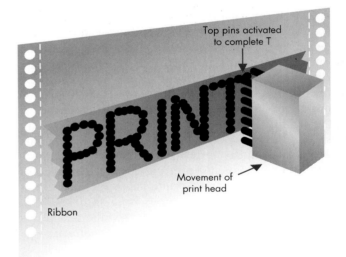

FIGURE 2.30   **Dot Matrix Printer Character Formation—Each Character is Formed in a Matrix as the Print Head Moves Across the Paper.** *Photo from* Computers, *4/E by Long/Long. © 1996. Reprinted by permission of Prentice-Hall, Inc., Upper Saddle River, NJ.*

FIGURE 2.28   **Laserjet Printer.** *Courtesy of Hewlett-Packard Company.*

■ 2.9 DATA STORAGE DEVICES

Since all data kept in RAM will be lost when the computer is turned off, they must be saved, or stored, before the power is off. Data storage devices provide a place to save information permanently for later use. CAD programs, for example, are stored on a disk; when loaded (or activated), portions of the program go into RAM, which is temporary memory. While a drawing is being worked on, all data associated with that drawing are kept in the same temporary memory. Periodically the operator must save that drawing and all the associated data to a storage device before the program is exited or the power is shut off. Otherwise, all accumulated data from that work session will be lost. These storage devices can be considered electronic file cabinets.

Disk drives, optical drives, and magnetic tape are distinct categories of storage devices. Disk storage devices are the most commonly used method of data storage. *Disk drives* may be of the fixed (hard disk) variety, flexible (floppy) variety, or optical type. Disk drives file and read data in random order. This means that the device writes data to any portion of the disk that is empty, and it is able to locate data almost instantly because it has access to the whole disk at once. Disk drives are rated according to their type, access time, capacity, and transfer rate.

The *fixed disk drive*, or hard disk (Figure 2.31) is the most common method of data storage. This type of drive uses an aluminum disk as the medium for storage. These drives may be internal, attached inside the computer case, or external, in a separate case of cabinet. A disk controller or controller card must be installed in the computer to allow the computer and drive to communicate or interface with each other. The storage capacity of these drives will range from 200 MB to several gigabytes. Access time is expressed in milliseconds (ms) and will range from 6 to 80 ms. The lower the number in milliseconds, the faster the access time.

FIGURE 2.32   **3.5 Inch Floppy Disk and Drive.** *Photo from* **Computers,** *4/E by Long/Long. © 1996. Reprinted by permission of Prentice-Hall, Inc., Upper Saddle River, NJ.*

Floppy disk drives (Figure 2.32) derive their name from the removable flexible plastic disks used in this device. The disks used in this drive are typically $3\frac{1}{2}''$ in diameter. The density of a disk refers to the amount of data the disk will hold. Typically, a $3\frac{1}{2}''$ disk, called microdiskette, will hold 1.44 MB of data in double-sided, high-density format. The floppy disk is inexpensive and convenient to use but holds less data and is slower than fixed disk drives.

Zip drives (Figure 2.33) are high-capacity magnetic disk drives, similar to floppy drives, that can store up to 100 MB on a $3\frac{1}{2}''$ disk. The disks used in zip drives are a special high-capacity medium and are not the same as a standard floppy disk. Zip drives are popular because there are external parallel and SCSI transfer models available which can be used to transport large amounts of data from one machine to another. They

FIGURE 2.31   **Fixed or Hard Disk Drive Interior.** *Courtesy of Western Digital Corporation.*

FIGURE 2.33   **Zip Disk and Drive.** *Courtesy of Iomega, Inc.*

FIGURE 2.34 **Jaz Drive and Disk.** *Courtesy of Iomega Corporation.*

can also be used for effective short-term backup (disks have a shelf life of 10 years).

Jaz drives are similar to Zip drives but their disks can hold up to 2 GB of data (Figure 2.34).

Similarly, *superdisks* can hold up to 120 MB of data while $\frac{1}{4}''$ *data cartridges* can hold up to 20 GB of information.

The ever-increasing need for larger storage capacity has spurred the development of new technologies. The newest technology in data storage uses lasers to read and write data and is generally termed optical storage. Optical storage media are capable of holding many gigabytes of data.

CD-ROM drives are a type of optical drive. Previously, CD-ROM disks were usually created from a master, so when you purchased a CD it already had the digital information written to it. Now you can store information on recordable CD-ROM systems, which produce a *write once read many* (WORM) disk. WORM devices allow data to be written to them, but the data become permanent on the disk and cannot be erased. This storage device is especially suited for archival purposes. CD-ROM drives use a laser to read and write data to a chemically coated aluminum disk (Figure 2.35). The data is "burned" into the disk surface by a laser so the information becomes permanent (unlike magnetic storage) and the disk is removable. Recordable CD-ROM systems can store about 650 MB of information on a single disk.

FIGURE 2.35 **CD-Rom and Disk.** *Courtesy of NEC Technologies, Inc.*

Hands On 2.1
Your CAD system's specifications

Will you be using a CAD system?

Fill in the information in the blanks below. You can use your operating system or ask your technical support person.

Monitor resolution _____ x _____ pixels.

Total hard disk space _____

Available hard disk space _____

CPU speed _____ megahertz.

RAM _____ megabytes.

Output device(s) _____

_____

Input device(s) _____

_____

Operating system _____

Software _____

_____

_____

_____

Optical disk drives (Figure 2.36) allow data to be erased and written over. These drives use a laser to change the state of optical magnetic media. These optical magnetic media can be changed again and again (write many read many). Because optical magnetic media are not sensitive to dust, like hard drives, optical disks can be removed from the drive and replaced with a new disk when additional storage capacity is needed.

Magnetic tape storage (Figure 2.37) uses plastic tape coated with magnetic particles. A read/write head in the tape drive charges magnetic particles to store information on the tape. The data being sent are recorded as a series of charges along the tape. Once these particles are charged, they will remain charged until the head writes over them or they are demagnetized. Tape drives file and read data in sequential order. This means they must look through data in the order that the tape is wound or unwound. (This is similar to forwarding or rewinding a videotape to look for a specific scene, or an audiotape to play a particular song.) Gigabytes of data may be stored on some tape systems. These cassettes resemble audiocassettes in both size and appearance. Tape storage is essentially used for backing up data from a hard drive, or for archival purposes (since the tapes can be removed and stored for later use).

■ 2.10 CAD SOFTWARE

CAD software tells the computer how to interact with the flow of data entered by the user through an input device. For example, it lets the computer "use" formulas to solve complictaed questions requiring detailed

FIGURE 2.36   **Optical Disk Cartridge and Optical Disk Drive.** *Courtesy of SyQuest Technology.*

analysis of large amounts of data, such as finding the center of gravity for a truncated cone. It also handles drawing processes, like creating many different models or views of the same object automatically (Figure 2.38). Software helps organize data; it will find previously stored drawing symbols, and it will help create and archive new ones. Software can be used to count, measure, and direct devices to print or plot drawings, create a bill of materials, or exchange files with other programs. CAD software is extremely powerful and has been designed to serve all major branches of engineering. It will be an important tool in most engineering careers.

FIGURE 2.37   **Tape Back-up System.** *Courtesy of Iomega, Inc.*

■ 2.11 COMMON CORE

CAD SOFTWARE   All CAD software generates familiar geometric terminology for creating drawings. But even though the geometry is common and the procedures for construction are similar, every CAD software program will vary in operational procedures typically involving the basic hierarchy of command structure.

Three features are found in all CAD software. You can access these features interactively through basic commands and menu options.

1. Commands for geometry generators (basic geometric construction)
2. Functions to control viewing of drawing geometry
3. Modifiers for changing the drawing or editing variations in the drawing (rotate, mirror, delete, group, etc.).

The commands and menu options may often be selected in two basic ways: by typing or by picking using a tablet (digitizer) or mouse. The Cartesian coordinates may be accessed with the keyboard or the mouse. You may switch between these two input methods at any time to issue commands and select options. The sequence of selection is called the hierarchy of command structure and provides an ease of operation that is the basis for selecting one software program over another.

■ 2.12 CAD CAPABILITY CHECKLIST

CAD software can offer the following characteristics required for creating technical documents (Figure 2.39):

1. Draw construction lines at any convenient spacing through any points, at any angles, and create tangent lines to one or more arcs.
2. Draw any type of line, such as visible, center, hidden, or section.
3. Draw circles and arcs of any size with given data.
4. Perform cross-hatching within specified boundaries.
5. Establish a scale or set a new scale within a drawing for various drawings within a document.
6. Calculate or list pertinent data of graphic construction, such as actual distances or angles.
7. Create a group of geometric figures for editing or copying.
8. Relocate drawing elements to any new position. Correct or change additions in stored documents.
9. Edit all or erase (delete) any part of a line, arc, or any geometric form on a drawing. Correct dimensions.

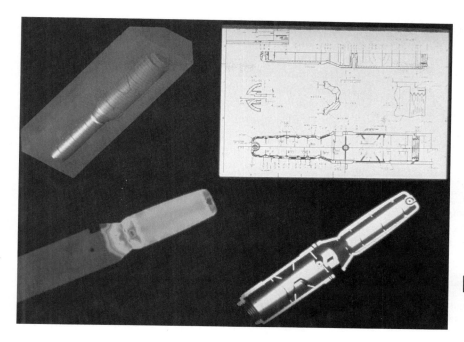

FIGURE 2.38   CAD Software Easily Creates Different Outputs of the Same Object. *Courtesy of SDRC, Milford, OH.*

10. Make a mirrored image or create symmetrical forms.

11. Perform associative or datum unit dimensioning.

12. Label drawings with notes and create title blocks and bills of material.

13. Save the entire drawing or any part for use on other documents and in other formats.

14. Create pictorials from three-view drawings.

15. Create orthographic views from a 3D model.

16. Retrieve and use stored drawings.

However, software programmers are constantly adding new capabilities and options to CAD software. For instance, many new programs offer very sophisticated 3D capabilities, often in a Windows environment (Figure 2.40).

■ 2.13   DRAWING WITH CAD SOFTWARE

The geometry that is created, drawn, or generated with CAD programs is generally referred to as objects, entities, or elements. These geometric entities are individually constructed figures or groups of elements that consist of points, lines, arcs, circles, rectangles, polygons, splines, solids, surfaces, symbols or blocks, cross-hatching, dimensions, and notes. These basic building units are selected from the menu and constructed at specific graphic locations of the monitor screen by the CAD user. Many CAD systems can store standard drawing symbols (called blocks) and use overlays on the digitizer for retrieval.

■ 2.14   USING A CAD SYSTEM

After a CAD system has been installed, the beginning user must become thoroughly familiar with it and learn how to use it effectively. This will require learning some new skills as well as a different vocabulary. The CAD operator will need to learn to create accurate drawings and construct them to various scales. In addition, the operator will need to learn to interact

FIGURE 2.39   CAD Assembly Drawing. *Courtesy of Sputnik Equipment Corporation, Inc.*

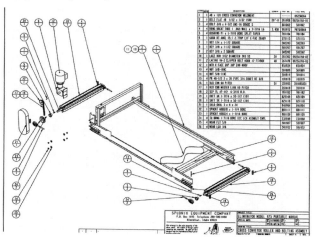

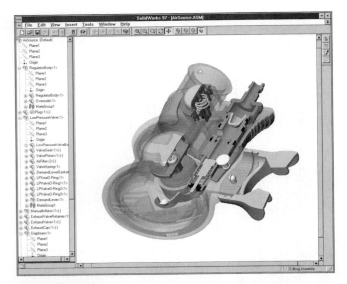

FIGURE 2.40 **Desktop PCs Running the Windows Operating System Can Now Provide Capabilities to Create Parametric Solid Models.**

effectively with the operating system to store and copy drawings and to routinely perform backups so that no data are lost. Most CAD manufacturers offer training programs and tutorials that will make the learning process much easier (Figure 2.41). They will provide instruction and training manuals that give information and details about the operation of the system. These manuals can be used not only during the initial training period but also for reference purposes during later operation of the system.

Most experienced drafters have developed shorter or simpler methods for creating a drawing, such as using overlays or templates. CAD systems also have simplified methods for drawing. Some systems have symbol libraries that contain many of the frequently used symbols, such as electrical relays, switches, transformers, resistors, bolts, nuts, keys, piping, and architectural symbols. These symbols may be in a symbol library, or the symbol may be located on one of the templates in the library of templates. Most CAD systems allow users to customize their symbol libraries. The desired symbols must first be drawn by the user on the CAD system the same as they would appear on a drawing board. This process may initially take as much time as it would manually, but once the image has been entered in the computer database, it need never be drawn again. The symbol can easily be retrieved from the symbol library whenever required.

■ 2.15 SELECTING A CAD SYSTEM

As the number of manufacturers of computer system equipment has increased, there has been a corresponding decrease in the cost of these systems. Some CAD systems may contain only those features necessary to produce simple two-dimensional entities, while others have the capability to create true three-dimensional drawing objects. The manufacturers of some of these systems are eager to promote new or improved functions that may not be found on other systems. Almost daily the computer industry announces amazing advances.

Since the newer CAD systems are generally easier to operate, the words "user friendly" are frequently used by manufacturers to promote their systems. However, there has been a tendency to exaggerate what particular systems can do. A prospective user should therefore view all claims with some skepticism until proved, for they can be misleading and often lead to disappointment. Unfortunately, many firms have purchased a system only to discover that it did not perform as well as expected.

Before purchasing a CAD system, a careful, well-thought-out plan for selecting a system should be developed and followed. This plan can be divided into five phases:

1. Establish the need for a CAD system.
2. Survey and select system features.
3. Request CAD system demonstrations.
4. Review selected systems.
5. Select, purchase, and install a CAD system.

Let us now discuss each of these phases in greater detail.

FIGURE 2.41 **CAD Training Helps the User Learn the Software Quickly.** *Courtesy of Jeff Kaufman and FPG International.*

ESTABLISH THE NEED FOR A CAD SYSTEM The first consideration in the selection process is to determine whether a CAD system is needed. All potential users of the CAD system should be consulted regarding how, when, and where a system would be used and whether it would be cost effective in their particular operations (see Figure 2.42). Never purchase a system simply because it may be considered a first step into the future and your firm wants to project a progressive image. It is important to investigate and evaluate the time- and cost-saving claims of manufacturers by contacting firms that have CAD systems in operation. Contact as many firms as possible, especially those with a wide range of experience. Prepare a brief questionnaire to survey these firms, asking questions regarding costs, training periods, system operation, and so on. Also ask at what point after installation the system became cost effective. Evaluate the responses to these questions and compare them with your specific requirements. This information will assist you in deciding if a CAD system can be beneficial to your firm at present.

Undoubtedly, you will hear many spectacular claims about the productivity of CAD systems. The productivity, however, will depend on the type of engineering operation. For example, among the first companies to introduce and use CAD systems were the electronic industry manufacturers. They found that drafters were constantly redrawing many symbols (resistor, transistor, etc.) and standard hardware parts (nuts, screws, etc.). Although they used plastic templates, common in all drafting departments, a considerable amount of time was spent doing tedious, repetitive drawing tasks. Since much of the time was devoted to tracing or redrawing something that was previously drawn, it was determined that a CAD system would save as much as 50% of the drafting time compared to using traditional methods. It is important to remember that with CAD it is only necessary to draw something once, even though initially it may take as long, or perhaps longer, to create it on the system as on the drawing board. From that point on, however, you need only recall this information from data storage, which a CAD system can do with amazing speed. In this example the benefits of CAD are obvious.

It is difficult to place a direct monetary value on many benefits of using CAD. These benefits include things like shorter design cycle time, improved ability to visualize complex fits between parts, links to direct manufacturing, better ability to reuse existing drawing and designs, and improved analysis early in the design phase.

FIGURE 2.42 **CAD Systems Can Be Expensive, So it is Important to Make Informed Decisions.** *Courtesy of Ron Chapple and FPG International.*

The process of determining the need for a CAD system can be time consuming and frustrating. Nevertheless, speed should be sacrificed to careful deliberation in this phase.

SURVEY AND SELECT SYSTEM FEATURES It is generally agreed that software should be selected before hardware. However, some CAD systems are *turnkey* system—that is, a total system with software and hardware combined and inseparable. Therefore, you should examine all the features of any given system very carefully before being attracted by spectacular hardware. For example, some systems use dual monitors. The chance of being impressed by this feature may overshadow the question of whether one really needs the two displays.

Consider whether the system will be multipurpose or used strictly for CAD. Will other office operations, such as word processing or accounting, be done on this machine? The answer to this question may add or eliminate CAD programs based on their operating system software.

Investigate how well a system will exchange information or interface with other CAD or CAM systems. Many systems do not have this capability. One important consideration is the CAD system's ability to exchange information with other CAD systems and other engineering applications. One standard for such exchange is the initial graphics exchange specification (IGES). There are also a number of other common formats. Make sure that the system you select can export common file formats, particularly to other applications you are planning to use.

It is suggested that you survey the people who will use the system to determine what they think is desirable in a CAD system. From this survey, develop a checklist of hardware and software features that your future system should have.

REQUEST CAD SYSTEM DEMONSTRATIONS　After the checklist has been created and approved by all parties concerned, make arrangements to see the various CAD systems in operation. A list of vendors can be compiled from advertisements in trade journals, magazines, and so on, or from the various directories of computer graphics manufacturers that are published. Contact the vendors to arrange demonstrations. Explain to them exactly what you expect the system to do. If the vendors are completely aware of your requirements, they will be able to give a more realistic presentation. Most of all, be prepared to ask questions. With each succeeding demonstration, your questions will be more effective. Moreover, the answers that you receive will be more meaningful. It is a good idea to select your own project which you would like to have demonstrated. Some CAD systems run their own canned example well but may not be flexible enough to meet your needs. Having your own project also helps you to contrast the systems fairly. Some systems may have a fancy demonstration, while others will not look as good initially. You want to select a system that will perform well for your needs, not look good running a canned demonstration.

Some of the questions that you should ask are as follows:

1. What are the brand names of the equipment used?
2. Are service contracts available? If so, what are the details and the cost of the contract? In most instances, service contracts are very expensive. Are you required to buy the service contract? Doing so may not be cost effective compared to having repairs done as you need them. Can you afford to have some downtime if something goes wrong with your system? If not, you may want a service contract. It is also possible in some situations to obtain service contracts from third parties that specialize in this type of business. The service provided by these firms for preventive and downtime maintenance may be less costly than that offered by the original manufacturer.

3. What is the warranty period, or periods, if parts of the system are furnished by different vendors? Also, is the warranty paid for by the original vendor or is the cost shard by the user and vendor?
4. What types of CAD software are provided, and what operations can be performed?
5. What kind and length of training will be provided to the user staff? How long after installation will it take for the staff to be proficient enough for the system to be cost effective?
6. Does the vendor issue software updates? If so, how often is this done and what is the cost?
7. What is the reputation of the vendor for providing user support in the form of toll-free telephone assistance or other methods of communication? Is the vendor readily available when the user encounters software and/or hardware problems and requires assistance?
8. Who currently uses the particular system? Will the vendor provide a list of current users? Users that have several months' experience with a CAD system are best qualified to answer question 7.

A CAD system analysis worksheet can be helpful when demonstrations are presented. Items to be included on the worksheet may be arranged according to hardware and software specifications. The hardware features that should be listed will generally be included in the following five categories: (1) central processor, (2) input devices, (3) monitors, (4) data storage, and (5) output devices. You may want to create a worksheet similar to the one shown in Figure 2.43 that you can use to track CAD system features during demonstrations. The worksheet may be expanded or modified to reflect particular requirements. You should also develop a similar worksheet listing CAD system software feature requirements.

With the completion of this phase, you will have achieved several important goals. Your knowledge and understanding of CAD systems will have increased, and you will be in a position to identify those system features that you require and can afford. In addition, you will have assembled a list of current users of CAD systems for future reference or exchange of information.

REVIEW SELECTED SYSTEMS　It is important at this stage that all of the collected information be organized and carefully reviewed. A list should be made of only those systems that merit further serious consideration. You may wish to request another demonstration of the particular systems that are on your revised list and to request additional information from current users regarding equipment performance, staff training, vendor support, and so on. Remember, if you select a system made by a reputable and financially stable manufacturer, and purchase it from a dependable retailer, you will sacrifice nothing in terms of service and support. Throughout this chapter we have made every effort to feature or illustrate systems and equipment made by such firms.

 Find Worksheet 2.1 in the tear-out section. This or a similar CAD system evaluation checklist can be helpful during demonstrations. Worksheet items may be arranged according to hardware and software specifications. Hardware features listed will generally fall into five categories: (1) central processor, (2) input devices, (3) monitors, (4) data storage, and (5) output devices. A sample worksheet is included in the tear-out worksheet section of the book, but you may want to create a similar worksheet of your own to track CAD system features during demonstrations. The worksheet may be expanded or modified to reflect particular requirements. You should also develop a similar worksheet listing CAD software requirements.

FINAL SELECTION, PURCHASE, AND INSTALLATION OF A CAD SYSTEM This last phase occurs when the final decision will be made regarding whether or not to purchase a CAD system. The decision invariably will depend on how you plan to use the system and how much you can afford to pay for it. You may want to consider a leasing arrangement, or lease-purchase arrangement. This can be effective for keeping the technology up to date. New technology is available constantly, and the lifetime of computer equipment is generally considered to be about three years. If your system cannot pay for itself in three years, perhaps it is not a good investment. You should plan to upgrade or purchase new equipment and software on a regular cycle. If the decision is made to acquire a system, determine costs, choose a delivery date, and arrange for installation and training.

■2.16 SUMMARY

The information presented in this chapter is intended to familiarize the student with the basic concepts, hardware, peripherals, and systems in CAD. It is not possible (nor was it intended) to present a comparison of CAD programs or all the commands used on CAD systems.

When possible, the instructor should arrange for students to visit nearby engineering and drafting departments that have CAD systems in operation. Those students who wish to obtain additional information on this subject should consult their school or local library.

The CAD system on the personal computer is replacing many drafting instruments, drafting tables, and drafting files. However, like no other drafting tool before, it raises engineering productivity without replacing the basic functions of the designer, engineer, and drafting technician. CAD developers, in their quest to harness computer technologies, have had a profound impact on the high-tech teams as they resolve problems in research, development, design, production, and operation (the five basic engineering functions).

The skills learned "on the board" are related and complementary to those needed by the CAD user. Learning the performance skills needed for creating drawings with CAD tools is time consuming and requires practice and manual dexterity. Both methods of drafting use simple and familiar geometric terminology for structuring the graphic production of technical documents, and both have the same goal—drawings that will meet industry standards.

The basic principles of drafting are common to traditional drafting and computer-aided drafting. The American National Standards Institute (ANSI) has well-established standards for shaping engineering drawings. Knowledge of drafting principles from the alphabet of lines to dimensioning and sectioning procedures, continues to be essential in shaping CAD documents. CAD can help you produce consistent lettering and regulate line work to improve the production of working drawings better than any other tool.

The CAD user is responsible for preparing engineering documents that are an integral part of the total manufacturing process (Figure 2.44). The ability to interact with all forms of technical information increases the significant role of the drafting technician.

FIGURE 2.43   Worksheet for Evaluating CAD System Hardware.

Item	Y/N	Size/Type	Comments	Cost
Central Processor				
memory (MB)				
word size (16/32 bit)				
cache				
speed (MHz/Mips)				
bus type (eisa, vesa, pci)				
expansion/upgrade				
Operating System				
32 bit				
multitasking				
software availability				
Data Input Devices				
mouse				
trackball				
digitizer				
light pen				
thumb wheel				
Display				
monochrome				
color				
screen size				
resolution				
Video Card				
memory				
software support				
dual display support				
Storage, Hard Drive				
type				
access time				
capacity				
expansion				
removable				
Storage, Floppy				
type				
access time				
capacity				
removable				
CD-ROM				
type				
speed				
capacity				
read/write				

FIGURE 2.43   *(cont.)*

Item	Y/N	Size/Type	Comments	Cost
Backup System				
type				
capacity				
speed				
automation				
Output Devices				
type				
provided with system?				
medium				
cost per sheet				
speed				
resolution/accuracy				
color				
Maintenance				

■ KEY WORDS

CAD	megahertz (Mhz)	mouse	menu
CAM	motherboard	digitizing tablet	wire frame
software	SIMM	printer	CD-ROM
operating systems	RAM	plotter	Zip disk
hardware	worm disk	floppy disk	Jaz disk
CPU	port	hard disk drive	superdisks
microprocessor	video display monitor	modem	data cartridges
MIPS			

■ CHAPTER SUMMARY

- Computers have revolutionized the drawing process. New technologies are constantly invented which make this process quicker, more versatile, and more powerful.

- CAD is the tool of choice for engineering design companies. The effective user of this tool requires an understanding of technical drawing fundamentals as well as training on the CAD software program.

- The microprocessor, RAM, and hard disk drive of the computer are essential components of a computer system. The keyboard and mouse are typical input devices. Printers and plotters output the drawing to paper for review and approval. The display monitor shows the drafter what is being drawn and offers command choices.

- CAD software can draw in three dimensions (width, height, and depth), unlike paper drawing which only consists of two dimensions in a single view.

- Different CAD packages have different operational procedures, and different strengths and weaknesses. Three features found in all CAD software are commands for geometry generators, functions to control the viewing of drawing geometry, and modifiers for changing the drawing or editing variations.

- Operating a CAD system typically has required extensive training. Newer CAD systems are becoming more user friendly, but one should not overestimate the claims CAD packages make. It is important to evaluate each package thoroughly and make an informed decision.

GRAPHICS SPOTLIGHT
Unlocking the Power of Solid Modeling

Successful use of solid modeling should increase the size of a company's market, increase its market share, and increase its profit margin. Early CAD systems basically only automated the drafting process, but solid modeling has the potential to affect the entire production process, from preliminary design through engineering and manufacturing. Ancillary functions, such as purchasing and marketing, can also be affected.

Solid modeling can create all the critical information for a product, and a company needs to take advantage of all the information contained in these models. Good implementation of solid modeling is marked by the wide-ranging use of the solid models in downstream applications. This maximizes return because it permits many operations to work from the original solid model rather than re-creating the design for each operation. Solid models should be the basis for virtual prototypes, engineering analyses, machine tool paths, purchase orders, marketing images, etc. Anywhere a design is re-created or its related data is retyped into another computer is a sign that solid modeling may be underutilized.

Although downstream use is an important consideration, there is nothing to be gained by insisting that all design be done with solid-modeling when various operations are just as well served by 2D processes. Another consideration in introducing a new technology is whether it affects existing bottlenecks in the production process. Unless the use of solid modeling (or any other technological innovation) helps to eliminate or decrease the bottlenecks, or constraints, in the overall process, it will not improve productivity or profits.

Introduction of solid modeling into company operations requires careful planning. First, there are many solid modeling systems on the market, and choice of the correct one is of paramount importance. Second, it will probably be necessary to undertake an expensive hardware upgrade because solid modeling requires larger workstations, better graphics, more memory, etc. than simpler programs. Third, everyone who will use the solid model should be given extensive training. Introducing solid modeling in a pilot project is also widely recommended. The gradual implementation of solid modeling on a project-by-project basis has been found to be more successful than a one-step installation throughout the company.

Small and medium-sized companies have an advantage over large companies in using solid modeling for maximum return. Large companies may have the resources and dedication to make it work, but the smaller companies have the most flexibility in terms of organizational structure.

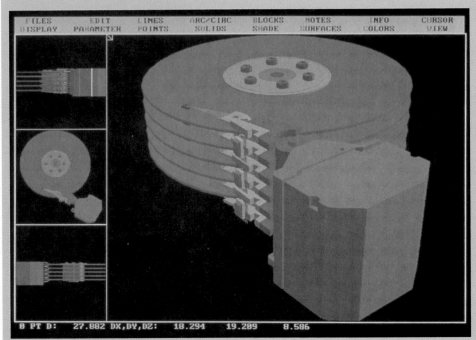

Multiview CAD Solid Model. *Courtesy of CADKEY.*

Adapted from "Unlocking the Power of Solid Modeling" by Caren D. Potter, *Computer Graphics World*, Nov. 1995, Vol. 18, No. 11, p. S3(4). ©PennWell Publishing Company 1995.

FIGURE 2.44 CAD Documents Supply a Multitude of Information which Engineers and Technicians Must Understand and Use. *Courtesy of SDRC, Milford, OH.*

■ REVIEW QUESTIONS

1. What are the basic components of a computer-aided drawing (CAD) system?

2. Discuss the relationship between CAD and CAM in modern design and manufacturing facilities?

3. List the similarities and differences between a mouse and a digitizing pad with puck?

4. What are the main advantages of CAD over traditional drawing methods?

5. What is faster, a computer with a 100 Mhz microprocessor and a 500 MB hard disk, or a 500 Mhz computer with a 100 MB hard disk? Which will store more information?

6. What is the difference between RAM storage and hard disk storage? What computer parts are typically found on the motherboard?

7. What is the difference between plotting and printing?

8. What are the hardware and software specifications of your school's CAD system?

9. What are questions you should ask about any CAD system you consider buying?

■ CAD PROJECTS

When necessary, refer to the appropriate sections of the chapter to check your answers.

Proj. 2.1 Define the following terms: computer system, hardware, software, analog, digital, computer graphics, CAD, CADD, and CAM.

Proj. 2.2 What are the principal components of a computer system? A CAD system? Draw a systems flowchart that illustrates the sequence of operations for each of the systems.

Proj. 2.3 Prepare a list of CAD system hardware components and give examples of each.

Proj. 2.4 Call a hardware company and compare the prices for three graphics monitors of different resolutions and sizes. Which purchase do you recommend? Give your reasoning.

Proj. 2.5 Prepare a list of possible data storage devices and determine total storage capabilities for each.

Proj. 2.6 Determine what would be your best storage device, in terms of value, if your average CAD drawing file size is 1500k and you store 10–20 drawings per week.

Proj. 2.7 Arrange a visit to the computer center at your school, or to a local engineering design office, and prepare a written report on the use of computers in design and drafting at these facilities.

The following problems ask you to use CAD software to solve some typical geometric construction that is similar from one software program to another. All the problems have been prepared on a CAD system. Prepare the required CAD drawing problems, as shown with your CAD system, and produce a hard copy with a printer or plotter for approval.

Proj. 2.8 Prepare a list of modifying (or editing) commands available on the CAD system you will be using.

Proj. 2.9 The unknown distance KA in Figure 2.45 has been determined and the angle measured using CAD. Re-create this problem with your CAD system, changing the 908 angle at H to 758; then determine the angles at K and A and the length of line KA.

Proj. 2.10 Prepare a revised version of the CAD drawing (Figure 2.46) by increasing the radius 0.40 to 0.4375 and changing the slot dimension 1.60 to 1.70.

Proj. 2.11 Prepare a detailed CAD drawing of the Safety Key (Figure 2.47) with the following changes: Correct the right-side view and add the missing dimension 0.40. Examine the placement of dimensions and relocate where necessary. Change 1.12 to 1.25 and add the difference to dimension 4.70.

Proj. 2.12 Using a CAD system, determine the true length of lines AD and CD (Figure 2.48) when the horizontal projection of point A is relocated to a new coordinate reading of (0, 3.125) and the horizontal projection of point D is relocated to a Cartesian coordinate of (1.75, 1.625). Revise the drawing using the "F" notation for the frontal projections instead of the V notation, as shown. What is the new slope of line CD?

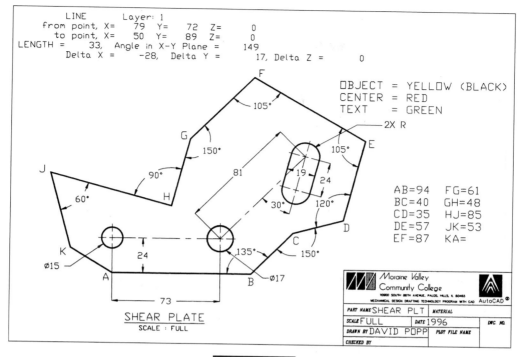

```
          LINE       Layer: 1
     from point, X=    79  Y=   72  Z=      0
       to point, X=    50  Y=   89  Z=      0
LENGTH =      33,  Angle in X-Y Plane =      149
       Delta X =   -28,  Delta Y =       17, Delta Z =       0
```

OBJECT = YELLOW (BLACK)
CENTER = RED
TEXT   = GREEN

2X R

AB=94   FG=61
BC=40   GH=48
CD=35   HJ=85
DE=57   JK=53
EF=87   KA=

Moraine Valley
Community College
10900 SOUTH 88TH AVENUE, PALOS HILLS, IL 60465
MECHANICAL DESIGN DRAFTING TECHNOLOGY PROGRAM WITH CAD   AutoCAD®

PART NAME SHEAR PLT	MATERIAL	
SCALE FULL	DATE 1996	DWG. NO.
DRAWN BY DAVID POPP	PLOT FILE NAME	
CHECKED BY		

SHEAR PLATE
SCALE : FULL

FIGURE 2.45

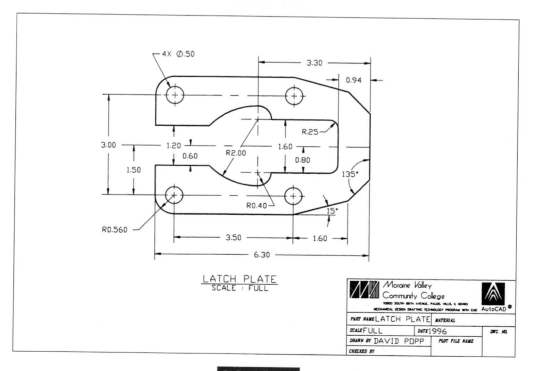

Moraine Valley
Community College
10900 SOUTH 88TH AVENUE, PALOS HILLS, IL 60465
MECHANICAL DESIGN DRAFTING TECHNOLOGY PROGRAM WITH CAD   AutoCAD®

PART NAME LATCH PLATE	MATERIAL	
SCALE FULL	DATE 1996	DWG. NO.
DRAWN BY DAVID POPP	PLOT FILE NAME	
CHECKED BY		

LATCH PLATE
SCALE : FULL

FIGURE 2.46

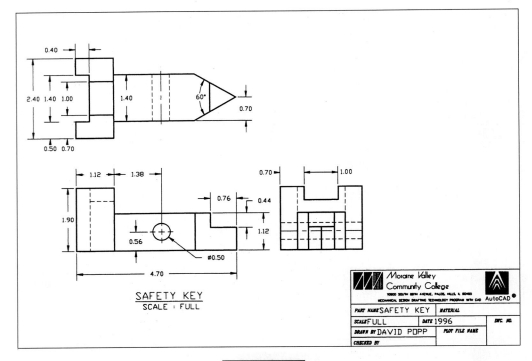

SAFETY KEY
SCALE : FULL

Moraine Valley
Community College
10900 SOUTH 88TH AVENUE, PALOS, HILLS, IL 60465
MECHANICAL DESIGN DRAFTING TECHNOLOGY PROGRAM WITH CAD

AutoCAD®

PART NAME SAFETY KEY	MATERIAL	
SCALE FULL	DATE 1996	DWG. NO.
DRAWN BY DAVID POPP	PLOT FILE NAME	
CHECKED BY		

FIGURE 2.47

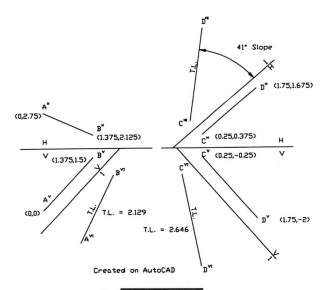

Created on AutoCAD

FIGURE 2.48

C H A P T E R 3

FREEHAND SKETCHING AND LETTERING TECHNIQUES

OBJECTIVES

After studying the material in this chapter, you should be able to:

1. Create freehand sketches using the correct sketching techniques.
2. Sketch parallel, perpendicular, and evenly spaced lines.
3. Sketch a circle and an arc of a given diameter.
4. Use techniques to keep your sketch proportionate.
5. Enlarge an object using grid paper.
6. Sketch various line types.
7. Add lettering to a sketch.

OVERVIEW

Sketching technique is one of the most important skills for engineering visualization. Sketching is a quick way to communicate ideas with other members of the design team. A picture is often worth a thousand words (or 1K words, as it were). Sketching is a time-efficient way to plan out the drawing processes needed to create a complex object. Sketches act like a road map for the completion of a final paper or CAD drawing. When you sketch basic ideas ahead of time, you can often complete a drawing sooner and with fewer errors. Legible hand lettering is used on the sketch to specify important information.

■ 3.1 TECHNICAL SKETCHING

Freehand sketches are a helpful way to organize your thoughts and record ideas. They provide a quick, low-cost way to explore various solutions to a problem so that the best choice can be made. Investing too much time in doing a scaled layout before exploring your options through sketches can be a costly mistake. Sketches are also used to clarify information about changes in design or provide information on repairing existing equipment.

The degree of precision needed in a given sketch depends on its use. Quick sketches to supplement verbal descriptions may be rough and incomplete. Sketches that are supposed to convey important and precise information should be drawn as carefully and accurately as possible.

The term *freehand sketch* does not mean a sloppy drawing. As shown in Figure 3.1, a freehand sketch shows attention to proportion, clarity, and correct line widths.

■ 3.2 SKETCHING MATERIALS

One advantage of freehand sketching is that it requires only pencil, paper, and eraser. Small notebooks or sketch pads are useful in the field (when working at a site) or when an accurate record is needed. Graph paper can be helpful in making neat sketches like the one in Figure 3.2. Paper with 4, 5, 8, or 10 squares per inch is convenient for maintaining correct proportions.

Find a style of pencil that suits your use. Figure 3.3 shows three styles which are all good for preparing sketches. Automatic mechanical pencils (shown as C in the illustration) come in .3-mm, .5-mm, .7-mm, and .9-mm leads that advance automatically and are easy to use. The .5-mm lead is a good general size, or you can use a .7-mm lead for thick lines and .3-mm for thin

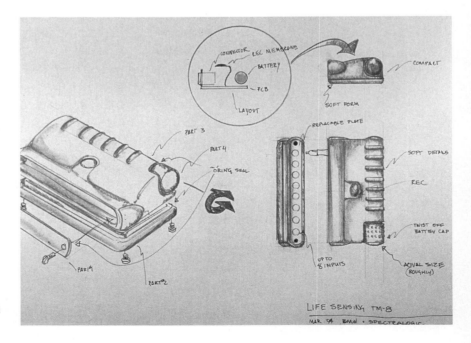

FIGURE 3.1 **Great Ideas Often Start as Freehand Sketches Made on Scratch Paper.** *Courtesy of ANATech, Inc.*

lines. The lead holder shown as part B requires a special sharpener, so it is not usually suitable for work in the field. Plain wooden pencils work great. They are inexpensive and make it easy to produce thick or thin lines by the amount you sharpen them.

A sketch pad of plain paper with a master grid sheet showing through underneath works well as a substitute for grid paper. You can create your own master grid sheets for different sketching purposes using CAD. Specially ruled isometric paper is available for isometric sketching.

Figure 3.4 shows the grades of lead and their uses. Use soft pencils, such as HB or F, for freehand sketching. Soft vinyl erasers are recommended.

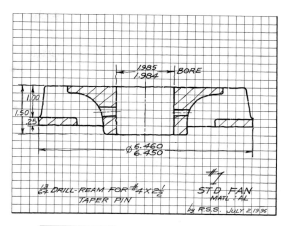

FIGURE 3.2 Sketch on Graph Paper.

FIGURE 3.3 Drawing Pencils.

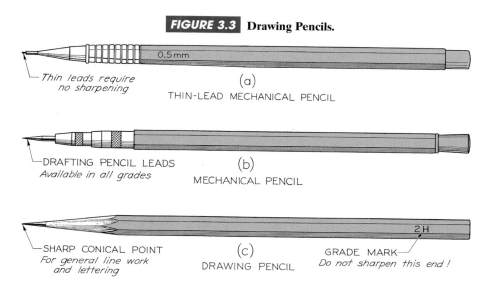

(a)
THIN-LEAD MECHANICAL PENCIL
Thin leads require no sharpening

(b)
MECHANICAL PENCIL
DRAFTING PENCIL LEADS
Available in all grades

(c)
DRAWING PENCIL
SHARP CONICAL POINT
For general line work and lettering
GRADE MARK
Do not sharpen this end!

FIGURE 3.4 Lead Grade Chart.

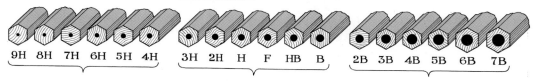

9H 8H 7H 6H 5H 4H   3H 2H H F HB B   2B 3B 4B 5B 6B 7B

Hard	Medium	Soft
The hard leads in this group (left) are used where extreme accuracy is required, as on graphical computations, charts, and diagrams. The softer leads (right) are sometimes used for line work on engineering drawings, but their use is restricted because the lines are apt to be too light.	These grades are general purpose. The softer grades (right) are used for technical sketching, lettering, arrowheads, and other freehand work. The harder leads (left) are used for line work on machine drawings and architectural drawings. H and 2H leads are used on pencil tracings for reproduction.	These leads are too soft to be useful in mechanical drafting. Their use for such work results in smudged, rough lines that are hard to erase, and the lead must be sharpened continually. These grades are used for art work of various kinds, and for full-size details in architectural drawing.

Practical Tips
Freehand lines

- To sketch long lines, mark the ends of the line with a light 6H pencil.

- Draw long sweeps between, keeping your eye on the mark you are moving your pencil toward.

- Make several strokes, improving the accuracy of your line.

- Finally darken in the distinct line with an HB lead.

- If your line looks like this you may be gripping your pencil too tightly or trying too hard to imitate mechanical lines.

- Slight wiggles are okay as long as the line continues on a straight path.

- Occasional gaps are fine and make it easier to draw straight.

One of the most important factors in making good sketches is line quality. For many people, the biggest problem with line quality is not making object lines thick and dark enough. Make object lines black and relatively thick.

Practice with permanent markers

- Engineers are often required to maintain permanent records during the design process in a design notebook. This can help when applying for patents and in showing the steps the design went through. These records are required to be in ink. Practice sketching with felt-tip markers so that you can draw neat permanent sketches such as will be required in your design notebook.

- Disposable engineering markers are available in various widths and may be used freehand or with templates to make neat permanent sketches.

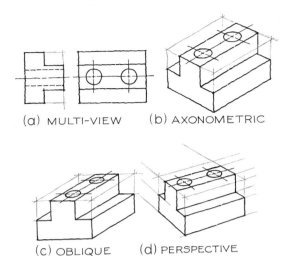

(a) MULTI-VIEW   (b) AXONOMETRIC

(c) OBLIQUE   (d) PERSPECTIVE

FIGURE 3.5 Types of Projection.

■ 3.3 TYPES OF SKETCHES

Technical sketches of 3-D objects are usually one of four standard types of projection, shown in Figure 3.5:

- Multiview projection
- Axonometric (isometric) projection
- Oblique projection
- Perspective sketches

Multiview projection shows one or more necessary views. You will learn multiview projection in Chapter 5. Axonometric, oblique, and perspective sketches are methods of showing the object pictorially in a single view. They will be discussed in Chapter 6.

■ 3.4 SCALE

Sketches are not usually made to a specific scale. Sketch objects in their correct proportions as accurately as possible by eye. Grid paper helps you sketch the correct proportions by providing a ready-made scale (by counting squares). The size of the sketch is up to you, depending on the complexity of the object and the size of the paper available. Sketch small objects oversize to show the details clearly.

■ 3.5 TECHNIQUE OF LINES

The main difference between an instrument drawing and a freehand sketch is in the style or technique of the lines. A good freehand line is not expected to be precisely straight or exactly uniform, as is a CAD or instrument-drawn line. Freehand lines show freedom and variety. Freehand construction lines are very light, rough lines. All other lines should be dark and clean.

Practical Tips
Drawing freehand lines

Line Weights

- Make dimension, extension, and centerlines thin, sharp, and black.

- Make hidden lines medium and black.

- Make visible and cutting plane lines thick and black.

- Make construction lines thick and light.

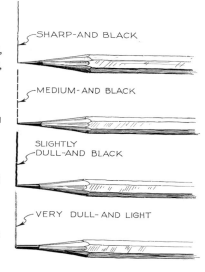

SHARP-AND BLACK

MEDIUM-AND BLACK

SLIGHTLY
DULL-AND BLACK

VERY DULL- AND LIGHT

Horizontal lines

- Hold your pencil naturally, about 1 inch back from the point, and approximately at right angles to the line to be drawn.

- Draw horizontal lines from left to right with a free and easy wrist-and-arm movement.

Drawing vertical lines

- Draw vertical lines downward with finger and wrist movements.

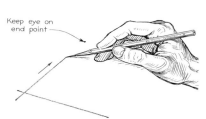

Keep eye on
end point

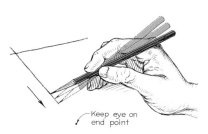

Keep eye on
end point

Two methods for blocking in horizontal and vertical lines

Method A

Hold your pencil firmly, sliding a rigid finger along the edge of the paper to maintain a uniform border.

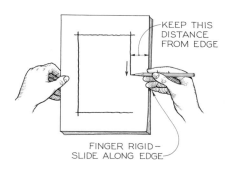

KEEP THIS
DISTANCE
FROM EDGE

FINGER RIGID-
SLIDE ALONG EDGE

Method B

Mark the distance on a piece of paper and use it like a ruler. Hold your pencil as shown and make distance marks by tilting the pencil down to paper. Slide your hand to the next location.

STRIP OF PAPER

Two methods for finding midpoints

Method A

Gauge the half-distance with your pencil. Try the distance on the left and then right, adjusting until it matches on both sides.

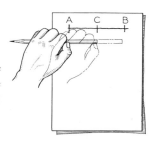

A C B

Method B

Mark the endpoints of the line on a strip of paper. Fold the paper to bring the endpoints together. Use the crease as the midpoint.

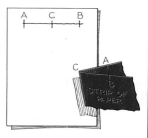

A C B

C A

B
STRIP OF
PAPER

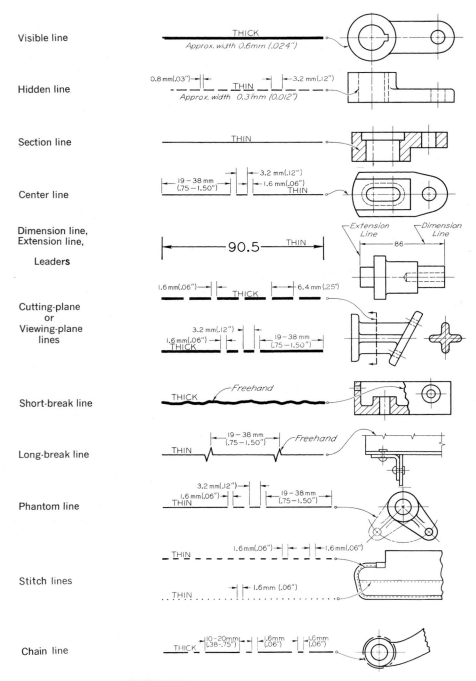

FIGURE 3.6 **Alphabet of Lines (Full Size).**

■ 3.6 STYLES OF LINES

Each line on a technical drawing has a definite meaning. Drawings use two different line widths—thick and thin—and different line styles indicate the meaning of the line. A person reading a drawing depends on line styles to communicate whether a line is visible or hidden, if it represents a center axis, or if its purpose is to convey dimension information. Without making these distinctions, drawings would become a confusing jumble of lines. To make your drawings clear and easy to

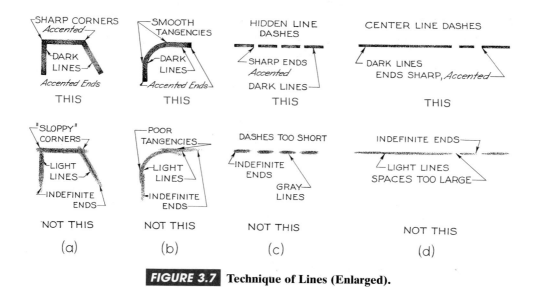

FIGURE 3.7 Technique of Lines (Enlarged).

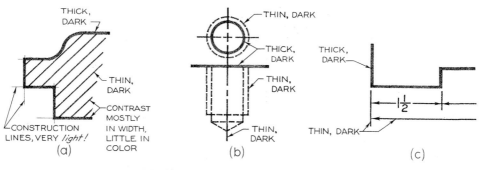

FIGURE 3.8 Contrast of Lines (Enlarged).

read, make the contrast between the two widths of lines distinct. Thick lines such as visible lines and cutting plane lines should be twice as thick as thin lines. Thin lines are used for construction lines, hidden lines, dimension lines, extension lines, center lines, and phantom lines. Figure 3.6 shows the different styles of lines that you will be using. All lines except for construction lines should be sharp and dark. Construction lines should be very light so that they are not visible (or are barely visible) in the completed drawing. Figures 3.7 and 3.8 show examples of technique for sketching using different line patterns.

 Find Worksheet 3.1 in the tear-out section and practice sketching freehand lines. You will probably want to use your own paper for additional practice.

Step by Step 3.1
Sketching circles

Three methods for sketching circles

Method A

1. Sketch lightly the enclosing square and mark the midpoint of each side

2. Draw light arcs connecting the midpoints.

3. Darken in the final circle.

Method B

1. Sketch the two centerlines

2. Add light 45-degree radial lines. Sketch light arcs across the lines at an estimated radius distance from the center.

3. Darken the final circle.

Method C

1. Mark the estimated radius on the edge of a card or scrap of paper and set off from the center as many points as desired.

2. Sketch the final circle through these points.

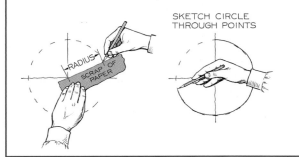

3.7 SKETCHING CIRCLES, ARCS, AND ELLIPSES

Small circles and arcs can be sketched in one or two strokes without any preliminary blocking in. Sketching arcs is similar to sketching circles. In general, it is easier to sketch arcs by holding your pencil on the inside of the curve. In sketching arcs, look closely at the actual geometric constructions and carefully approximate all points of tangency so that the arc touches a line or other entity at the right point. Circle templates also make it easy to sketch accurate circles of various sizes.

If a circle is tipped away from your view, it appears as an ellipse. Figure 3.9 shows a coin viewed so that it appears as an ellipse. You can learn to sketch small ellipses with a free arm movement similar to the way you sketch circles, or you can use ellipse templates to help you easily sketch ellipses. These templates are usually grouped according to the amount a circular shape the would be rotated to form the ellipse. They provide a number of sizes of ellipses on each template, but usually only one or a couple typical rotations.

 Find Worksheet 3.2 in the tear-out section and practice sketching circles and ellipses. Additional practice on white paper, a dry erase board, or a chalkboard will help you create circles and ellipses quickly and accurately.

 After you have gone through the steps for drawing circles, ellipses, and arcs shown in in Step by Steps 3.1, 3.2, and 3.3, find Worksheet 3.3 in the tear-out section. Apply the line and curve techniques you have learned to sketch actual parts.

FIGURE 3.9 Circle Viewed as an Ellipse.

Step by Step 3.2
Three ways to sketch an ellipse

Method A

1. Hold the pencil naturally, resting your weight on your upper forearm, and move the pencil rapidly above the paper in the elliptical path you want.

2. Lower the pencil to draw several light overlapping ellipses.

3. Darken the final ellipse.

Method B

1. Sketch lightly the enclosing rectangle.

2. Mark the midpoints of the sides and sketch light tangent arcs, as shown.

3. Complete the ellipse lightly and darken in the final ellipse.

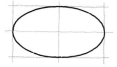

Method C

1. To sketch an ellipse on given axes, lightly sketch in the major and minor axes of the ellipse.

2. Mark the distance along the axes and lightly block in the ellipse.

3. Darken the final ellipse.

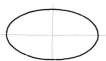

Practical Tip
Using a trammel

The trammel method is a good way to sketch accurate ellipses.

To make a trammel, mark one-half the desired length of the minor axis on the edge of a card or strip of paper (A-B). Using the same starting point, mark one-half the length of the major axis (A-C). The measurements will overlap.

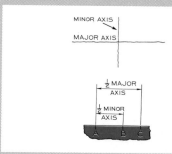

Next, line up the last two trammel points (B and C) on the axes and mark a small dot at the location of the first point (A). Move the trammel to different positions keeping B and C on the axes and mark points at A to define the ellipse.

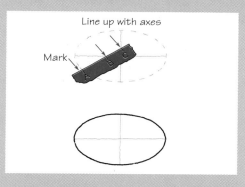

Sketch the final ellipse through the points, as shown.

Step by Step 3.3
Three methods for sketching arcs

Method A

1. Locate the center of the arc and lightly block in perpendicular lines. Mark off the radius distance along the lines.

2. Draw a 45-degree line through the center-point and mark off the radius distance along it.

3. Lightly sketch in the arc as shown. Darken the final arc.

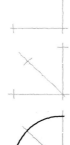

Method B

1. Locate the center of the arc and lightly block in perpendicular lines. Mark off the radius distance along the lines.

2. Mark the radius distance on a strip of paper and use it as a trammel.

3. Lightly sketch in the arc as shown. Darken the final arc.

Method C
Use these steps to draw arcs sketched to points of tangency.

1. Locate the center of the arc and sketch in the lines to which the arc is tangent.

2. Draw perpendiculars from the center to the tangent lines.

3. Draw in the arc tangent to the lines ending at the perpendicular lines.

4. Darken in the arc and then darken the lines from the points of tangency.

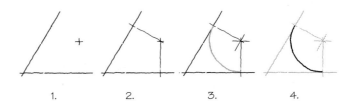

1.   2.   3.   4.

Practical Tips
Freehand circles

This method of drawing freehand circles is particularly quick and easy.

Using your hand like a compass, you can create circles and arcs with surprising accuracy after a few minutes of practice.

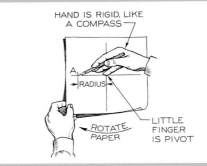

HAND IS RIGID, LIKE A COMPASS

A
RADIUS

ROTATE PAPER

LITTLE FINGER IS PIVOT

1. Place the tip of your little finger or the knuckle joint of your little finger at the center.

2. "Feed" the pencil out to the radius you want as you would do with a compass.

3. Hold this position rigidly and revolve the paper with your free hand.

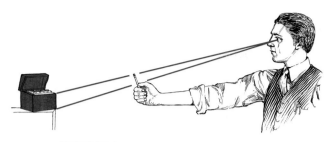

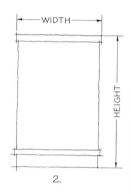

Step by Step 3.4
Sketching a utility cabinet

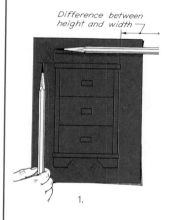

FIGURE 3.10 Estimating Dimensions.

■ 3.8 MAINTAINING PROPORTIONS

The most important rule in freehand sketching is to keep the sketch in proportion. No matter how brilliant the technique or how well-drawn the small details, if the proportions are bad, the sketch will be of little use. To keep your sketch in proportion, first determine the relative proportions of the height to the width and lightly block them in. Then lightly block in the medium-size areas and the small details.

Compare each new estimated distance with already-established distances. One way to estimate distances is to mark an arbitrary unit on the edge of a card or strip of paper. Then see how many units wide and how many units high the object is.

To sketch an object with many curves to a different scale, use the squares method. On the original picture, rule accurate grid lines to form squares of any convenient size. It is best to use a scale and some convenient spacing, such as 1/2 inch or 10 mm. On the new sheet, rule a similar grid, marking the spacing of the lines proportional to the original, but reduced or enlarged as needed. Draw the object's contours in and across the new grid lines to match the original as closely as you can by eye.

 Find Worksheet 3.4 in the tear-out section. On the grid lines provided, you will use the squares method to enlarge the shape of a car and a photo detail of your choice.

In sketching from an actual object, you can compare various distances on the object by using your pencil as a sight, as shown in Figure 3.10. Hold your pencil at arm's length and don't change your position. Compare the length you sighted with the other dimensions of the object to maintain the correct proportions. If the object is small, such as a machine part, you can compare distances by actually placing the pencil against the object. In establishing proportions, use the blocking-in method shown in Step by Step 3.5, especially for irregular shapes.

1. If you are working from a given picture, such as this utility cabinet, first establish the relative width compared to the height. One way is to use the pencil as a measuring stick. In this case, the height is about 1-3/4 times the width.

2. Sketch the enclosing rectangle in the correct proportion. This sketch is to be slightly larger than the given picture.

Difference between height and width

WIDTH

HEIGHT

1.

2.

3. Divide the available drawer space into three parts with the pencil by trial. Hold your pencil about where you think one third will be and then try that measurement. If it is too short or long, adjust the measurement and try again. Sketch light diagonals to locate centers of the drawers and block in drawer handles. Sketch all remaining details.

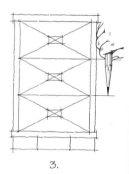

3.

4. Darken all final lines, making them clean, thick, and black.

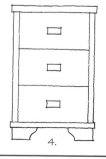

4.

Step by Step 3.5
How to block in an irregular object

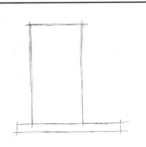 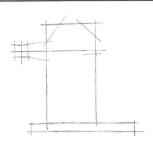 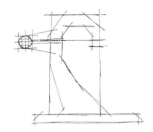 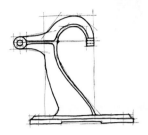

1. Capture the main proportions with simple lines.

2. Block in the general sizes and direction of flow of curved shapes.

3. Lightly block in additional details.

4. Darken the lines of the completed sketch.

Hands On 3.1
Block in the irregular object

Block in the part at right using the steps shown in Step by Step 3.5.

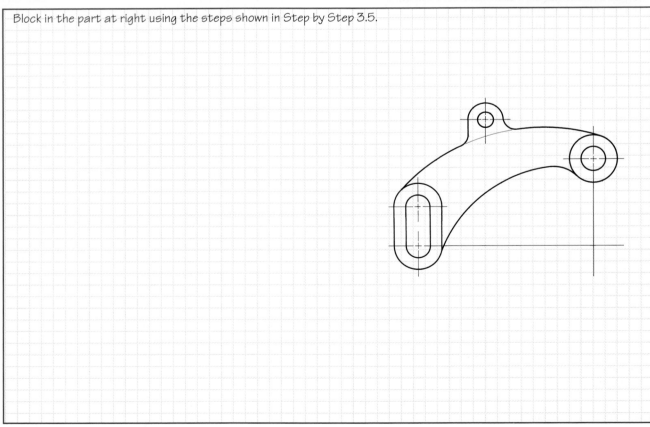

ABCDEFGH
abcdefghij

ABCDEFGHI
abcdefghijklm

Roman *All letters having elementary strokes "accented" or consisting of heavy and light lines, are classified as Roman.*

Italic- All slanting letters are classified as Italics~ These may be further designated as Roman-Italics, Gothic Italics or Text Italic.

FIGURE 3.11 Serif and Sans-Serif Lettering.

LETTERING

Lettered text is often necessary to completely describe an object or to provide detailed specifications. Lettering should be legible, be easy to create, and use styles acceptable for traditional drawing and CAD drawing.

■ 3.9 FREEHAND LETTERING

Most engineering lettering is single-stroke Gothic font. A font is the name for a particular shape of letters. Figure 3.11 shows some common fonts. Most hand-drawn notes are lettered 1/8 inch high and are drawn within light horizontal guidelines. CAD notes are typed from the keyboard and sized according to the plotted size of the drawing. Certain rules applying to the placement of lettering and CAD notes will be discussed in Chapter 9.

■ 3.10 LETTERING STANDARDS

The modern styles of letters were derived from the design of Roman capital letters, whose origins date all the way back to Egyptian hieroglyphics. The term *Roman* refers to any letter that has wide downward strokes, thin connecting strokes, and ends terminating in spurs called serifs. In the late 19th century, the development of technical drawing created a need for a simplified, legible alphabet that could be drawn quickly with an ordinary pen. Single-stroke Gothic sans-serif (meaning without serifs or spurs) letters are used today because they are very legible.

■ 3.11 COMPUTER LETTERING

Lettering is a standard feature available in computer graphics programs. Using CAD software, you can add titles, notes, and dimensioning information to a drawing. Several fonts and a variety of sizes may be selected.

When modifications are required, it is easy to make appropriate lettering changes on the drawing by editing existing text.

CAD drawings typically use a Gothic style of lettering, but often use a Roman style of lettering for titles. When adding lettering to a CAD drawing, a good rule of thumb is not to use more than two fonts within the same drawing. You may want to use one font for the titles and a different one for notes and other text. However, you may have a couple different sizes of lettering in the drawing and perhaps some slanted lettering all using the same font. It is sometimes tempting to use many different fonts in a drawing because of the wide variety available on CAD systems, but drawings that use too many different fonts have been jokingly referred to as having a ransom note style of lettering.

■ 3.12 LETTERING TECHNIQUE

Lettering is more similar to freehand drawing than it is to writing, so the six fundamental drawing strokes and their directions are basic to lettering. Horizontal strokes are drawn left to right. Vertical, inclined, and curved strokes are drawn downward. If you are left-handed, you can use a system of strokes similar to the sketching strokes that work for you.

Lettering ability has little relationship to writing ability. You can learn to letter neatly even if you have terrible handwriting. There are three necessary aspects of learning to letter:

- Proportions and forms of the letters (to make good letters, you need to have a clear mental image of their correct shape)

- Composition—the spacing of letters and words

- Practice

Practical Tips
Lettering with a pencil

- Use a soft pencil, such as an F, H, or HB. Lettering should be dark and sharp, not gray and blurred.

- Keep your pencil sharp.

- Sharpen wooden pencils to a needle point, then dull the point very slightly.

- Turn the pencil frequently during lettering to wear the lead down uniformly and keep the lettering sharp.

- Don't worry about making the exact letter strokes unless you are having trouble making the letters look right. If you do have trouble, the letter strokes are designed to help you draw uniform and symmetrical letters, and you may want to refer to them.

- Use extremely light, 1/8-inch (3.2-mm) horizontal guidelines to regulate the height of letters. Use a few light vertical or inclined lines randomly placed to help you visually keep the letters uniformly vertical or inclined.

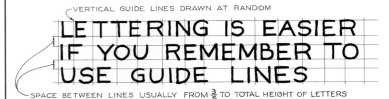

- Draw vertical strokes downward or toward you with a finger movement.

- Draw horizontal strokes from left to right with a wrist movement and without turning the paper.

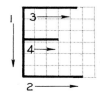

- Draw curved strokes and inclined strokes with a downward motion.

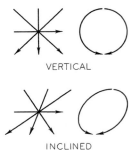

- Since practically all pencil lettering will be reproduced, the letters should be dense black.

Left-handers:

These traditional strokes were designed for right-handed people. Left-handers should experiment with each letter to find out which strokes are best and develop a system of strokes that works best for them.

■ 3.13 VERTICAL LETTERS AND NUMERALS

The proportions of vertical capital letters and numerals are shown in Figure 3.12 in a grid six units high. Numbered arrows indicate the order and direction of strokes. The widths of the letters can be easily remembered: The letter l and the numeral 1 are only a pencil width. The W is eight grid units wide (1-1/3 times its height) and is the widest letter in the alphabet. All the other letters or numerals are either five or six grid units wide, and it is easy to remember the six-unit letters because when assembled they spell TOM Q. VAXY. This means that most letters are as wide as they are tall,

which is probably wider than your usual writing. All numerals except the 1 are five units wide.

Lowercase letters are rarely used in engineering sketches except for lettering large volumes of notes. Vertical lowercase letters are used on map drawings, but very seldom on machine drawings. Lowercase letters are shown in Figure 3.13. The lower part of the letter is usually two-thirds the height of the capital letter.

Find Worksheet 3.5 in the tear-out section and practice lettering vertical capitals on the grid provided. Pay attention to the arrows indicating stroke direction.

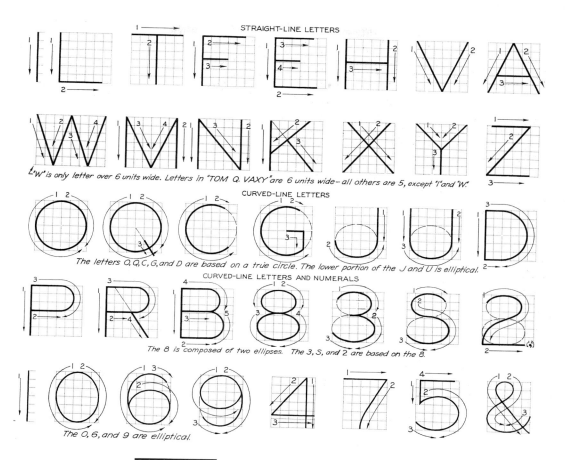

STRAIGHT-LINE LETTERS

"W" is only letter over 6 units wide. Letters in "TOM Q. VAXY" are 6 units wide—all others are 5, except "I" and "W".

CURVED-LINE LETTERS

The letters O, Q, C, G, and D are based on a true circle. The lower portion of the J and U is elliptical.

CURVED-LINE LETTERS AND NUMERALS

The 8 is composed of two ellipses. The 3, S, and 2 are based on the 8.

The O, 6, and 9 are elliptical.

FIGURE 3.12 Vertical Capital Letters and Numerals.

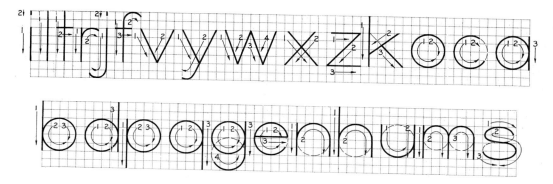

FIGURE 3.13 Vertical Lowercase Letters.

Hands On 3.2
Practice freehand lettering

Use the space provided to repeat each letter or word shown.

A B C D E F G H I J K L M N O P Q R S T U V W X Y Z

A B C D E F G H I J K L M N O P Q R S T U V W X Y Z

PRACTICE LETTERING BY HAND

USING A VARIETY OF SIZES.

ONE-EIGHTH INCH LETTERING IS COMMONLY USED

ON FREEHAND SKETCHES. MAKE LETTERS BLACK

AND CLEAN-CUT — NEVER FUZZY OR INDEFINITE.

$1\frac{1}{2}$ 1.500   45'-6   32°   15.489   1" = 20'   3.75

ONCE YOU HAVE MASTERED VERTICAL LETTERING

ADDING A SLANT FOR INCLINED LETTERS IS EASY.

FIGURE 3.14 Inclined Capital Letters and Numerals.

FIGURE 3.15 Inclined Lowercase Letters.

■ 3.14 INCLINED LETTERS AND NUMERALS

Inclined capital letters and numerals, shown in Figure 3.14, are similar to vertical characters, except for the slope. The slope of the letters is about 68 degrees from the horizontal. While you may practice drawing slanted hand-lettering at approximately this angle, it is important in CAD drawings to always set the amount of incline for the letters at the same value within a drawing so that the lettering is consistent. Inclined lowercase letters, shown in Figure 3.15, are similar to vertical lowercase letters.

■ 3.15 GUIDELINES

Use extremely light horizontal guidelines to keep letter height uniform, as is shown in Figure 3.16. Capital letters are commonly made 1/8 inch (3.2 mm) high, with the space between lines of lettering being from three-fifths to full height of the letters. Lettering size may vary depending on the size of the sheet. (See Chapter 5 for standard sheet sizes and their corresponding lettering heights.) Do not use vertical guidelines to space the letters—this should be done by eye while lettering. Use a vertical guideline at the beginning of a row of text to help you line up the left edges of the following rows, or use randomly spaced vertical guidelines to help you maintain the correct slant.

A simple method of spacing horizontal guidelines is to use a scale and set off a series of 1/8-inch spaces, making both the letters and the spaces between lines of letters 1/8 inch high. Another quick method of creating guidelines is to use a guideline template like the Berol Rapidesign 925 shown in Figure 3.17.

When large and small capitals are used in combination, the small capitals should be three-fifths to two-thirds as high as the large capitals.

Find Worksheet 3.6. Practice lettering typical notes using 1/8-inch guidelines.

■ 3.16 GUIDELINES FOR WHOLE NUMBERS AND FRACTIONS

Beginners should use guidelines for whole numbers and fractions. Draw five equally spaced guidelines for whole numbers and fractions, as shown in Figure 3.18. Fractions are twice the height of the corresponding whole numbers. Make the numerator and the denominator each about three-fourths as high as the whole number to allow enough space between them and the fraction bar. For dimensioning, the most commonly used height for whole numbers is 1/8 inch (3.2 mm), and for fractions 1/4 inch (6.4 mm), as shown in the figure.

Some of the most common errors in lettering fractions are shown in Figure 3.19. To make fractions appear correctly:

- Never let numerals touch the fraction bar.
- Center the denominator under the numerator.
- Never use an inclined fraction bar, except when lettering in a narrow space, as in a parts list.
- Make the fraction bar slightly longer than the widest part of the fraction.

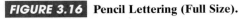

THE IMPORTANCE OF GOOD LETTERING CANNOT BE OVER-EMPHASIZED. THE LETTERING CAN MAKE OR BREAK AN OTHERWISE GOOD DRAWING.

PENCIL LETTERING SHOULD BE DONE WITH A FAIRLY SOFT SHARP PENCIL AND SHOULD BE CLEAN-CUT AND DARK. ACCENT THE ENDS OF THE STROKES.

FIGURE 3.16　Pencil Lettering (Full Size).

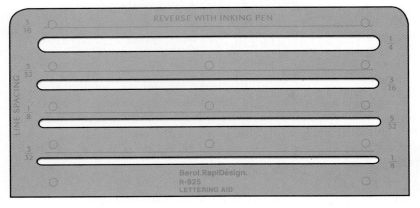

FIGURE 3.17　The Berol Rapidesign 925 template is used to quickly create guidelines for lettering.

Practical Tips
Creating letters that appear stable

- Certain letters and numerals appear top-heavy when they are drawn with equal upper and lower portions like the example below.

$$CGBEKSXZ$$

- To correct this, reduce the size of the upper portion to give a balanced appearance, as in this example.

$$CGBEKSXZ$$

- If you put the central horizontal strokes of the letters B, E, F, and H at midheight, they will appear to be below center.

- To overcome this optical illusion, draw the strokes for B, E, F, and H slightly above the center as you letter, keeping letters uniform, as in the second example above.

- The same practice applies to numerals. The upper example looks top-heavy. Note how the lower example looks more balanced.

$$3852$$
$$3852$$

At left is a good example of uniform lettering.   RELATIVELY

Look at the examples of **what not to do** below. They show a lack of uniformity of:

Style	Relatively
Letter height	RELATIVELY RELATIVELY
Angle	RELATIVELY *RELATIVELY*
Stroke thickness	RELATIVELY RELATIVELY
Letter spacing	RELA TIVELY
Word spacing	NOW IS THE TIME FOR EVERY GOOD PERSON TO COME TO THE AID OF HIS OR HER COUNTRY

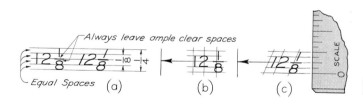

FIGURE 3.18 Guide Lines for Dimension Figures.

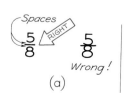

FIGURE 3.19 Common Errors.

GRAPHICS SPOTLIGHT
Sketching and Parametric Modeling

THE DESIGN PROCESS

Using CAD parametric modeling in many ways mirrors the design process. To get the rough ideas down, the designer starts by making hand sketches. Then as the ideas are refined, more accurate drawings are created either with instruments or using CAD. Necessary analysis is performed and in response the design may change. The drawings are revised as needed to meet the new requirements. Eventually the drawings are approved so that the parts may be manufactured.

ROUGH SKETCHES

Using parametric modeling software, initially the designer roughly sketches the basic on the screen. These sketches do not have to have perfectly straight lines or accurate corners. The software interprets the sketch much as you would interpret a rough sketch given to you by a colleague. If the lines are nearly horizontal or vertical, the software assumes that you meant them thus. If the line appears to be perpendicular it is assumed that it is.

CONSTRAINING THE SKETCH

Using a parametric CAD system, you can start by sketching on the computer screen as though you were sketching freehand. Then the two-dimensional sketch is refined by adding geometric constraints, which tell how to interpret the sketch and by adding parametric dimensions, which control the size of sketch geometry.

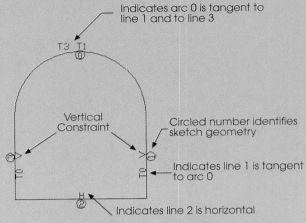

(B) Constrained Sketch

Once the sketch is refined, it can be created as a 3D feature to which other features can be added. As the design changes the dimensions and constraints that control the sketch geometry can be changed, and the parametric model will update to reflect the new design.

When you are creating sketches by hand or for parametric modeling, think about the implications of the geometry you are drawing. Does the sketch imply that lines are perpendicular? Are the arcs you have drawn intended to be tangent or intersecting? When you are creating a parametric model, the software makes assumptions about how you intend to constrain the geometry based on your sketch. You can remove, change, or add new constraints as you wish.

Using AutoCAD Mechanical Desktop software, you can create a rough sketch like the one you see in Fig. A. You select the Profile command to have the software constrain the sketch automatically. The results of profiling the sketch are shown in Fig. B. The symbols show the constraints that were assumed.

The dialog box shown on the next page in Fig. C lists types of geometric constraints you can use to control the sketch geometry in AutoCAD Designer. The dialog box labeled Fig. D on the left shows the constraints that you can use to control the way parts fit together in an assembly.

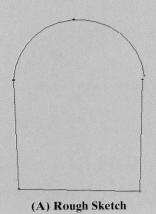

(A) Rough Sketch

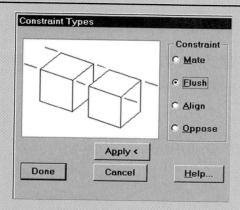

(C) Assembly Constraints

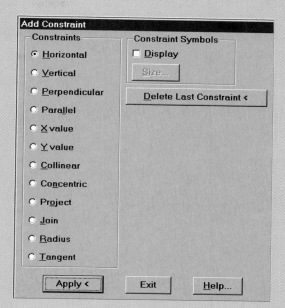

(D) Sketch Constraints

When you have completed the parametric model, you have an "intelligent" part. When design changes are necessary, you can change a dimension or constraint causing the model to automatically update. Orthographic drawings with correctly shown hidden lines and dimensions can be generated automatically. Or the part can be exported for rapid prototyping or manufacture.

■ 3.17 SPACING OF LETTERS AND WORDS

Uniform spacing of letters is a matter of equalizing spaces by eye. The background areas between letters, not the distances between them, should be approximately equal. Equal distances from letter to letter causes the letters to appear unequally spaced. Equal background areas between letters results in an even and pleasing spacing.

Some combinations, such as LT and VA, may even have to be slightly overlapped to secure good spacing. In some cases the width of a letter may be decreased. For example, the lower stroke of the L may be shortened when followed by A. These pairs of letters that need to be spaced extra closely to appear correctly are called *kerned pairs* in typesetting.

Space words well apart, but space letters closely within words. Make each word a compact unit well separated from adjacent words. For either uppercase or lowercase lettering, make the spaces between words approximately equal to a capital O. Be sure to have space between rows of letters, usually equal to the letter height. Rows spaced too closely are hard to read. Rows that are too far apart do not appear related.

 The World Wide Web has resources for lettering and sketching. You can locate vendors for engineering graphics supplies, or look into the history of typography. You can also find more specific information such as type fonts for use with your CAD system. Some engineering reprographic companies offer on-line catalogues and direct ordering.

Check the sites below for engineering graphics supplies and equipment:

- http://www.reprint-draphix.com/
- http://www.eclipse.net/~essco/draft/draft.htm
- http://www.seventen.com/art_eng/index.html

These sites feature typography information:

- http://www.graphic-design.com/type/
- http://www.webcom.com/cadware/letease2.html

To find other sites like these, use keywords like *reprographic supplies* or *engineering type fonts*.

Hands On 3.3
Practice kerning pairs

Using equal spacing from one letter to the next does not actually appear equal, as in this example.

LATHING

literate

Space your lettering so that background areas appear equal, like the example shown below.

LATHING

literate

Letter the following words in the guidelines provided to achieve a visually balanced look. Adjust the space between letters as shown in the examples above.

ELLIPSES

ANCHOR BRACKET

INVOLUTE

UTILITY

EQUILATERAL

VARIGRAPH

SURFACE WAVINESS

DRILL

COUNTERSINK

COUNTERBORE

SPOTFACE

SURFACE FINISH

UNIFIED THREAD

DRAWINGS

ENGINEERING

LOCATIONAL FIT

VISUALIZATION

■ 3.18　TITLES

In most cases, the title and related information are lettered in *title boxes* or title strips, which may be printed directly on the drawing paper or polyester film, as shown in Figure 3.20. The main drawing title is usually centered in a rectangular space, which is easy to do in CAD. When lettering by hand, arrange the title symmetrically about an imaginary centerline, as shown in Figure 3.21. In any kind of title, the most important words are given most prominence by making the lettering larger, heavier, or both. Other data, such as scale and date, can be smaller.

FIGURE 3.20　Centering Title in Title Box.

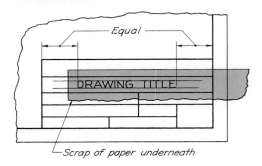

FIGURE 3.21　Balanced Machine-Drawing Title.

TOOL GRINDING MACHINE
TOOL REST SLIDE
SCALE : FULL SIZE
AMERICAN MACHINE COMPANY
NEW YORK CITY

DRAWN BY ____ | CHECKED BY ____

Practical Tips
Balancing titles

Lettering to stop lines

When it is necessary to letter to a stop line, space each letter from right to left, estimating the widths of the letters by eye. Then letter from left or right, and finally erase the spacing marks.

Lettering to center lines

When it is necessary to space letters symmetrically about a centerline—which is frequently the case in titles—number the letters as shown, with the space between words considered as one letter.

Place the middle letter on center, making allowance for narrow letters (e.g., I) or wide letters (e.g., W) on either side. The X in the example is placed slightly to the left of center to compensate for the letter I, which has no width.

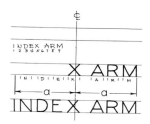

■ KEY WORDS

freehand sketch	centerlines	gothic	spacing
construction lines	shading	roman	guidelines
oblique	sketch	serif	title block
grid paper	proportions	inclined	kerned pairs
hidden lines	lettering	stability	

■ CHAPTER SUMMARY

- Sketching is a quick way of visualizing and solving a drawing problem. It is an effective way of communicating with all members of the design team.
- There are special techniques for sketching lines, circles, and arcs. These techniques should be practiced so they become second nature.
- Moving your thumb up or down the length of a pencil at arms length is an easy method for estimating proportional size.
- Using a grid makes sketching in proportion an easy task. Grid paper comes in a variety of types, including square grid and isometric grid.
- You can sketch circles by constructing a square and locating the four tangent points where the circle touches the square.
- A sketched line does not need to look like a mechanical line. The main distinction between instrumental drawing and freehand sketching is the character or technique of the line work.
- Freehand sketches are made to proportion, but not necessarily to a particular scale.
- Notes and dimensions are added to sketches using uppercase letters drawn by hand.
- The standard shapes of letters used in engineering drawing have been developed to be legible and quick to produce.

■ REVIEW QUESTIONS

1. What are the four standard types of projections?
2. What are the advantages of using grid paper for sketching?
3. What is the correct technique for sketching a circle or an arc?
4. Sketch the alphabet of lines. Which lines are thick? Which are thin? Which are very light and should not reproduce when copied?
5. What is the advantage of sketching an object first before drawing it using CAD?
6. What is the difference between proportion and scale?
7. What font provides the shape of standard engineering lettering?
8. Describe the characteristics of good freehand lettering
9. Why must guidelines always be used for lettering?
10. How are sketches used in the design process?

■ DESIGN PROJECT

Design a new or improved piece of playground, sporting, or recreational equipment. For example, a child's toy could both be educational and recreational. Sketch a layout for a playground, campground, or recreation area that provides for good access, safety, and flow of people in the area. Use good line technique. Using an architectural handbook or the World Wide Web, research standards for how much space to provide around certain pieces of equipment.

■ FREEHAND SKETCHING AND LETTERING PROJECTS

Draw complete horizontal and vertical or inclined guidelines very lightly. Draw the vertical or inclined guidelines through the full height of the lettered area of the sheet. Grid sheets and a lettering guideline sheet are provided for your use in the tear-out section. The grid lines are black so that you can remove a grid sheet and use it under a sheet of white paper to complete sketching or lettering projects. Save the grids for reuse.

Proj. 3.1   Use a sheet of white paper or fade-out grid (using eight squares per inch makes it easy to add lettering). Draw guidelines 1/8 inch high and letter the alphabet three times.

Proj. 3.2   Lay out a sheet like Figure I on the inside front cover. Draw a title block as shown in Figure IX. Neatly letter the information in the title block.

Proj. 3.3   Prepare a sketch to represent the parking meter shown in Figure 3.22 or draw a household object with irregular curved features such as a lamp or your shoe.

Proj. 3.4   Design a neatly sketched logo for an imaginary engineering company. Use good freehand technique and line quality.

Proj. 3.5   Use guidelines for 1/8-inch letters and write the lyrics of your favorite song or an inspirational saying.

Proj. 3.6   Use your CAD system to create a sheet layout like Figure I on the inside front cover. Create a title block like Figure IX or as described by your instructor.

Proj. 3.7   Sketch the free-body diagrams shown in Figure 3.23.

Proj. 3.8   Sketch the objects shown in Figures 3.24–3.29 in proportion. You do not need to make the sketch to any particular scale. Omit the dimensions from your sketch. Use the sheet size from Fig. I (inside front cover) and the title strip from Fig. II.

FIGURE 3.22   Parking Meter.

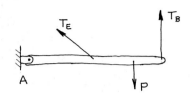

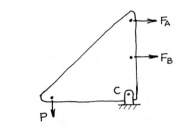

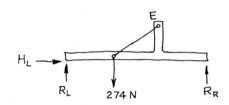

FIGURE 3.23   Free body Diagram Sketches.

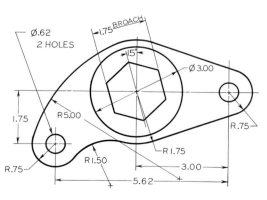

FIGURE 3.24   (Proj. 3.8) Rocker Arm.

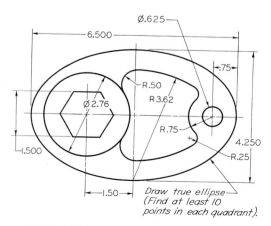

FIGURE 3.25   (Proj. 3.8) Special Cam.

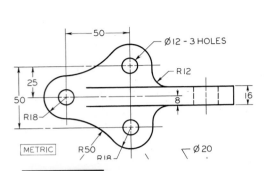

FIGURE 3.26 (Proj. 3.8) Boiler Stay.

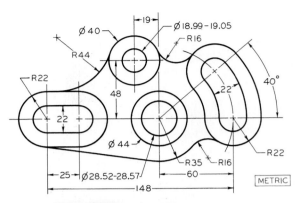

FIGURE 3.28 (Proj. 3.8) Gear Arm.

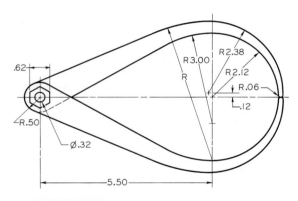

FIGURE 3.27 (Proj. 3.8) Outside Caliper.

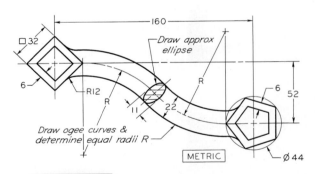

FIGURE 3.29 (Proj. 3.8) Special S-Wrench.

C H A P T E R   4

GEOMETRIC CONSTRUCTIONS AND MODELING BASICS

OBJECTIVES

After studying the material in this chapter, you should be able to:

1. Use terminology for basic drawing constructions.
2. Define tangency and construct tangent lines, circles and arcs.
3. Identify various types of solids.
4. Draw Venn diagrams showing the three Boolean operators.
5. Break down a part into solid features joined by Boolean operators.

OVERVIEW

All traditional drawing and CAD drawing techniques are based on the construction of basic geometric elements. Points, lines, arcs, and circles are the basic elements used to create the most complex 2-D drawings. You must understand basic geometric construction techniques in order to sketch on paper or with CAD, or to apply geometric techniques to solving problems.

Using CAD, you can also create 3-D models of objects that can be used to directly manufacture parts. Solid-modeling CAD techniques require an understanding of geometric solids and how they are created and combined to form more complex parts. This same understanding is useful in interpreting drawings and in visualization. You will probably be using CAD instead of the mechanical drawing techniques illustrated in this section, so using your knowledge of CAD, simply relate these techniques to CAD as necessary. Some geometric constructions are easier using CAD, but for most the approach is the same, except that you use updated tools.

For sketching techniques, keep in mind that your sketch should convey the geometric relationships. For example, lines that appear perpendicular in a sketch will be interpreted as perpendicular unless they are dimensioned otherwise. You can also use the symbol ⊥ between two sketched lines to convey that they are perpendicular. Likewise, objects that appear tangent in a sketch are assumed to be so unless noted otherwise in dimensioning and notes.

Geometric primitives—such as points, lines, circles, and arcs—are basic elements you should be familiar with when sketching or using CAD. A thorough understanding of drawing geometry will allow you to construct clearly understood sketches and accurate CAD geometry.

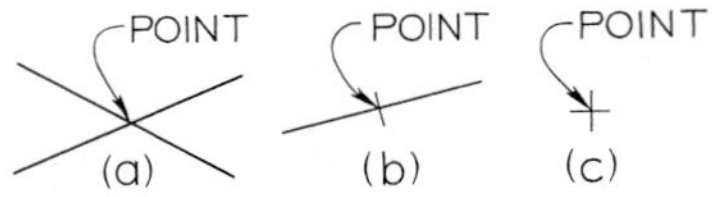

■ 4.1 POINTS AND LINES

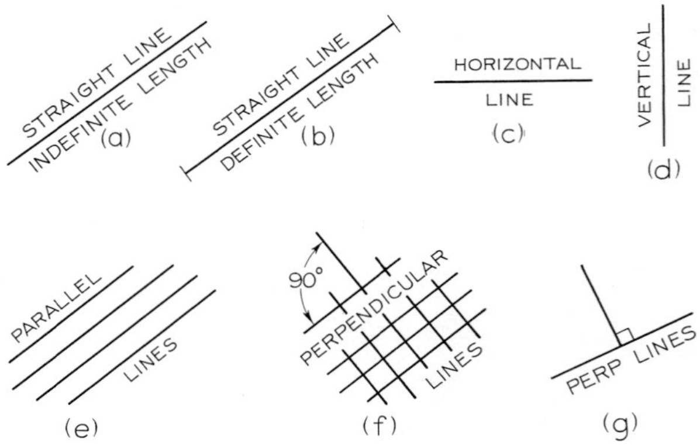

A *point* represents a location in space or on a drawing and has no width, height, or depth. On a sketch a point is represented by the intersection of two lines, by a short crossbar on a line, or by a small cross. Never represent points by a simple dot on the paper, because these are too easily misinterpreted and make the sketch messy and unprofessional. Examples of sketched points are shown in Figure 4.1.

A *line* was defined by Euclid as "that which has length without breadth." A straight line is the shortest

FIGURE 4.1   Points.

FIGURE 4.2   Lines.

distance between two points and is commonly referred to simply as a line. If the line extends indefinitely, you may draw the length to your convenience and leave the endpoints unmarked. If the endpoints of the line are significant, mark them with small crossbars. Common terms used to describe lines are illustrated in Figure 4.2.

Sets of straight lines or curved lines are *parallel* if the shortest distance from each point on one line to the other line is constant. The common symbol for parallel lines is ||, and for perpendicular lines is ⊥. Most CAD systems allow you to specify that two lines will be perpendicular using some kind of snap tool, and they also allow you to easily create parallel lines, either through a point or at a specified distance from another line. When sketching, you may mark two lines with a box at their intersection to indicate that they are perpendicular.

■ 4.2 ANGLES

An *angle* is formed by two intersecting lines. A common symbol for angle is ∠. A common measure is degree of angle. There are 360 degrees (360°) in a full circle, as shown in Figure 4.3. A degree is divided into 60 minutes (60′), and a minute is divided into 60 seconds (60″) The angle designation 37° 26′ 10″ is read 37 degrees, 26 minutes, and 10 seconds. When indicating minutes alone, place 0° in front of the number of minutes, as in 0° 20′. Angles can also be measured in decimal degrees, for example 45.20°. Gradians and radians are also systems used to measure angles.

Different kinds of angles are illustrated in Figure 4.3. Two angles are complementary if they total 90 degrees and are supplementary if they total 180 degrees. You can sketch angles approximately, use a protractor if needed, or use CAD when an accurate drawing is required. Using CAD systems, you can specify the exact angle for a line using a variety of methods, for example decimal degrees; degrees, minutes, and seconds; radians; gradians; and survey bearings.

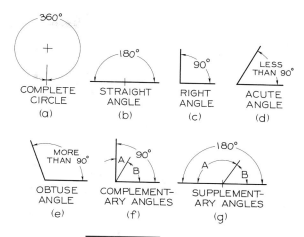

FIGURE 4.3 Angles.

■ 4.3 TRIANGLES

A *triangle* is a plane figure bounded by three straight sides. The sum of the interior angles of a triangle is always 180 degrees. A right triangle has one 90-degree angle, and the square of the hypotenuse is equal to the sum of the squares of the two sides. Any triangle inscribed in a semicircle is a right triangle if the hypotenuse coincides with the diameter. This information can be useful in sketching and constructions. Examples of triangles are shown in Figure 4.4.

■ 4.4 QUADRILATERALS

A *quadrilateral* is a plane figure bounded by four straight sides. If the opposite sides are parallel, the quadrilateral is also a *parallelogram*. Quadrilaterals are shown in Figure 4.5

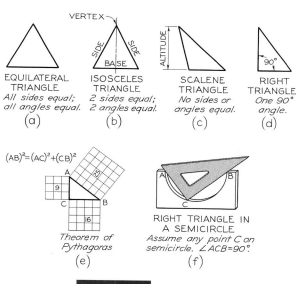

FIGURE 4.4 Triangles.

FIGURE 4.5 Quadrilaterals.

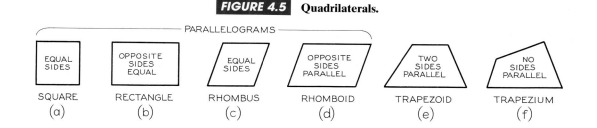

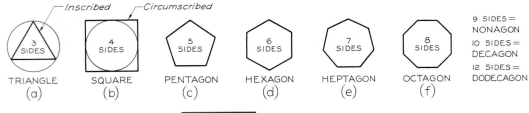

FIGURE 4.6 Regular Polygons.

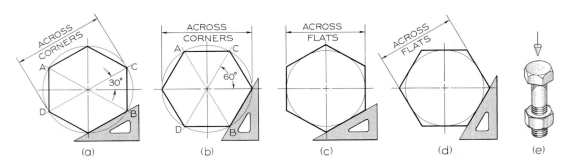

FIGURE 4.7 Inscribed and Circumscribed Hexagons.

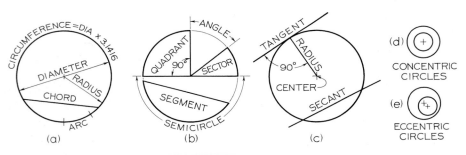

FIGURE 4.8 The Circle.

■ 4.5 POLYGONS

A *polygon* is any plane figure bounded by straight lines. If the polygon has equal angles and equal sides, is called a *regular polygon*. Polygons are shown in Figure 4.6.

INSCRIBED AND CIRCUMSCRIBED SHAPES

Regular polygons are often described and dimensioned by whether they are inscribed in a circle or circumscribed about a circle. Examples of inscribed and circumscribed polygons are shown in Figure 4.7. If a hexagon shape such as a bolt head is inscribed in a circle,

the diameter of the circle will be the dimension across opposite corners of the hexagon. If it is circumscribed about a circle, the diameter of the circle is the distance "across the flats" of the hexagon.

■ 4.6 CIRCLES AND ARCS

A *circle* is a closed curve, all points of which are the same distance from a point called the center. *Circumference* refers to the circle or to the distance around the circle. This distance equals the diameter multiplied by π (called pi, which is approximately 3.1416). Other features of circles are illustrated in Figure 4.8.

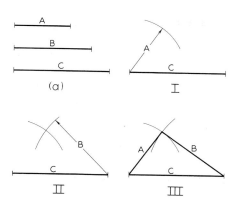

FIGURE 4.9 **Creating a Triangle with the Sides Given.**

■ 4.7 CONSTRUCTIONS AND CAD

Most CAD systems have a set of tools which allow you to quickly and easily complete tasks like finding the midpoint of a line or arc, or drawing a line perpendicular or parallel to another line. These basic operations will not be covered here. Complex constructions still can require a series of steps, the creation of accurate construction geometry, or functions that your CAD systems may not provide. In these cases understanding the basic construction methods is helpful. The following examples feature hand-drafting tools. You can make the connection between these steps and whatever CAD system you may be using. For sketching purposes, you should understand the underlying geometry that is implied in the sketch, but accuracy is not required. You can use symbols or write notes to clarify the sketch if necessary.

■ 4.8 CREATING A TRIANGLE GIVEN LENGTHS OF SIDES

Given the sides A , B, and C, as shown in Figure 4.9:

I.   Draw one side, shown as C, in the desired position. Draw an arc with its radius equal to side A.

II.   Draw a second arc with its radius equal to side B.

III.   Draw sides A and B from the intersection of the arcs, as shown.

A triangle can also be defined by specifying the lengths of two sides and the angle between them, or by specifying the length of one side and the two angles at each end. Because these are easily constructed using CAD or a protractor, they are not shown.

Step by Step 4.1
Bisecting an angle

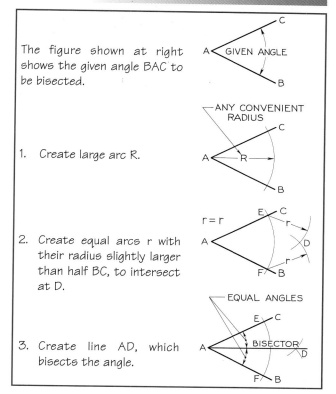

The figure shown at right shows the given angle BAC to be bisected.

1. Create large arc R.

2. Create equal arcs r with their radius slightly larger than half BC, to intersect at D.

3. Create line AD, which bisects the angle.

■ 4.9 CREATING A RIGHT TRIANGLE WITH HYPOTENUSE AND ONE SIDE GIVEN

To create a right triangle with the hypotenuse and one side given as in Figure 4.10, draw a semicircle with a diameter AB equal to the given side S. Using A as the center and the length of R as the radius, draw an arc or circle that intersects the first semicircle to find C. Draw lines AC and CB to complete the right triangle.

FIGURE 4.10 **Creating a Right Triangle.**

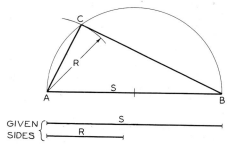

Step by Step 4.2
Dividing a line into equal parts

1. Sketch a vertical construction line at one end of the given line.

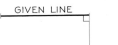
GIVEN LINE

2. Set zero of scale at the other end of the line.

3. Swing scale up until units for the required number of divisions fall on the vertical line (for example, three units to divide into thirds).

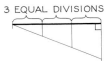

4. Make light marks at each point.

3 EQUAL DIVISIONS

5. Sketch vertical construction lines through each point.

Some practical applications of this method used to sketch (a) screw threads, (b) a framing layout, and (c) stair tread are shown below.

Practical Tip
Proportional parts

Dividing a line into proportional parts

Imagine you need to divide line AB into three parts proportional to 2, 5, and 9.

Sketch a vertical line from point B. Select a scale of convenient size for a total of nine units and set the zero of the scale at A.

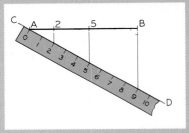

Swing the scale up until the ninth unit falls on the vertical line. Along the scale, mark points for 2, 5, and 9 units, as shown.

Sketch vertical lines through these points as shown.

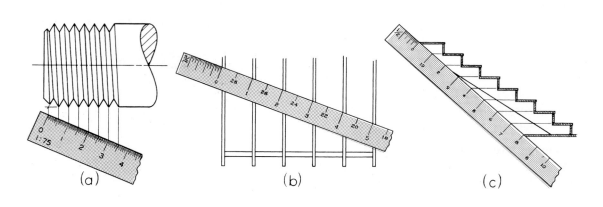

(a)   (b)   (c)

■ 4.10 CONSTRUCTING A PERPENDICULAR BISECTOR

As you might think, a **perpendicular bisector** is a perpendicular line that divides a given line into two equal segments. This is a useful construction because a perpendicular bisector to any chord on a circle passes through the circle's center. Figure 4.11 shows the given a line AB to be bisected with a perpendicular line.

I. From A and A draw equal arcs with radii greater than half AB on both sides of the line.

II. Connect intersections D and E with a straight line. Line DE will intersect line AB at the midpoint, shown as C.

III. Line DE will be perpendicular through the midpoint of AB.

You can quickly create perpendicular bisectors using your CAD system. One way to create a perpendicular bisector is to draw a line perpendicular to the given line from any point. Then move the new perpendicular line to the midpoint of the given line using a tool like snap-to-midpoint.

■ 4.11 CONSTRUCTING CIRCLES

Most CAD programs have easy-to-use tools which will create a circle through three points, find the center of a circle, or create a tangent entity. Being familiar with the following constructions for circles can help you to better interpret drawings, create freehand sketches and produce accurate CAD geometry.

A CIRCLE THROUGH THREE POINTS

I. Let A, B, and C be the three given points not in a straight line, as shown in Figure 4.12. Draw lines AB and BC, which will be chords of the circle.

II. Create perpendicular bisectors EO and DO intersecting at O.

III. With the center at O and the radius defined as OA, OB, or OC, create the required circle through the points.

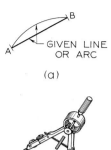

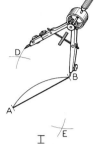

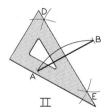

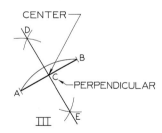

FIGURE 4.11 Bisecting a Line or Circular Arc.

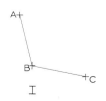

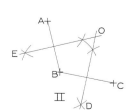

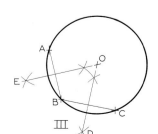

FIGURE 4.12 Creating a Circle Through Three Points.

THE CENTER OF A CIRCLE

This method uses the principle that any right triangle inscribed in a circle cuts off a semicircle. Draw any chord AB, preferably horizontal, as shown Figure 4.13. Create perpendiculars from A and B, cutting the circle at D and E. Make diagonals DB and EA whose intersection C will be the center of the circle.

Another method is to draw any two nonparallel chords and draw perpendicular bisectors. The intersection of the bisectors will be the center of the circle.

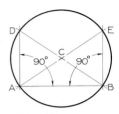

FIGURE 4.13   **The Center of a Circle.**

■ 4.12 TANGENCY

Lines, circles, or arcs lying in the same plane are considered to be *tangent* if they touch each other at only one point, even if they were to be extended. You can check to see if a line is tangent to a circle or arc by imagining it extended. If, when extended, the line would still only touch at one point, it is tangent. Otherwise the line is intersecting. Three-dimensional objects are considered to be tangent if they touch at only one point or only along one line. Figure 4.14 shows examples of tangent two dimensional and three dimensional geometry.

Given a line tangent to a circle or arc, a *radial line* (a line through the center of the circle) is perpendicular to the given line at the point of tangency.

 Tear out Worksheet 4.1. Use it to practice sketching tangent arcs.

FIGURE 4.14   **Tangency.**

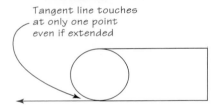
Tangent line touches at only one point even if extended

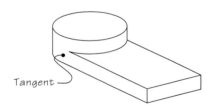
Tangent

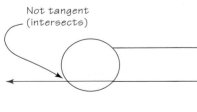
Not tangent (intersects)

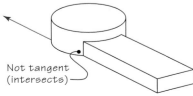
Not tangent (intersects)

DRAWING A CIRCLE TANGENT TO A LINE AT A GIVEN POINT

As shown in Figure 4.15, given a line AB and a point P on the line, a circle of radius R tangent to AB at point P is drawn as follows:

I.   Create a perpendicular to the line through P.
II.   Make an arc having radius R of the required circle along the perpendicular to locate point C.
III.   Make a circle with radius R using C as its center.

DRAWING AN ARC TANGENT TO TWO LINES AT RIGHT ANGLES

I.   Two lines are given at right angles to each other as shown in Figure 4.16.
II.   With given radius R, create an arc intersecting given lines at tangent points T.
III.   With given radius R again, and with points T as centers, create arcs intersecting at C.
IV.   With C as the center and given radius R, create the required tangent arc.

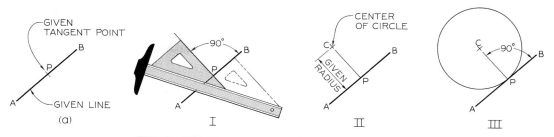

FIGURE 4.15 Creating a Circle Tangent to a Line.

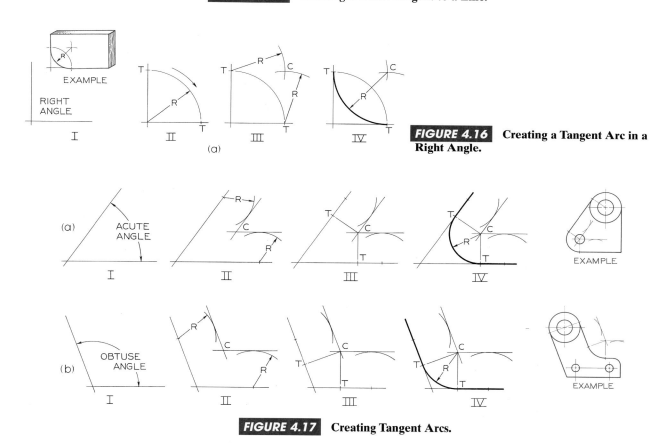

FIGURE 4.16 Creating a Tangent Arc in a Right Angle.

FIGURE 4.17 Creating Tangent Arcs.

Many CAD programs have a fillet command that will quickly create arcs tangent to lines and arcs. Understanding these techniques can still help you develop your sketching ability even if you do not need them to create CAD geometry.

DRAWING AN ARC TANGENT TO TWO LINES AT ACUTE OR OBTUSE ANGLES

I. Given two intersecting non-perpendicular lines as shown in Figure 4.17.

II. Create lines parallel to the given lines, at an equal distance R from them. The intersection of these lines, C, will be the center of the tangent arc.

III. From C, create perpendiculars to the given lines to locate tangent points T.

IV. With C as the center and with given radius R, create the required tangent arc as shown.

Step by Step 4.3
Creating tangent arcs

An arc tangent to a line and through a point:	An arc tangent to a line and through a point:	An arc tangent to an arc and through a point:

An arc tangent to a line and through a point:

Given line AB, point P, and radius R as shown:

1. Create line DE parallel to the given line and distance R from it.

2. From P create an arc with radius R. The point where the arc intersects line DE will be the center of the tangent arc.

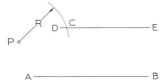

3. Create the tangent arc using C as its center.

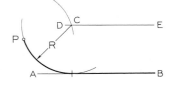

An arc tangent to a line and through a point:

Given line AB, with tangent point Q on the line, and point P as shown:

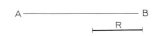

1. Create line PQ, which will be a chord of the required arc.

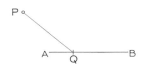

2. Create perpendicular bisector DE, and at Q make a perpendicular to the line to intersect DE at C.

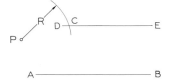

3. Create the tangent arc using C as its center.

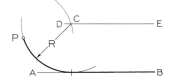

An arc tangent to an arc and through a point:

Given an arc with center Q, point P, and radius R as shown:

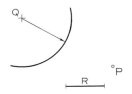

1. From P, create an arc with radius R.

2. From Q, create an arc with its radius equal to that of the given arc plus R.

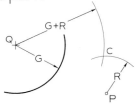

3. The intersection C of the two arcs is the center of the required tangent arc.

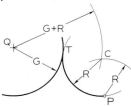

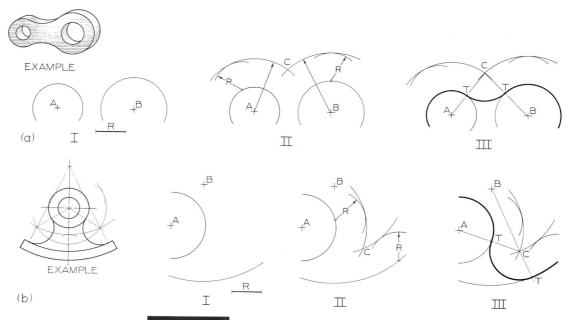

FIGURE 4.18   Creating an Arc Tangent to Two Arcs.

CREATING AN ARC TANGENT TO TWO ARCS

I.   Arcs with centers A and B and required radius R are given, as shown in Figure 4.18.

II. With A and B as centers, make arcs parallel to the given arcs and at a distance R from them; their intersection C is the center of the required tangent arc.

III. Create lines of centers AC and BC to locate points of tangency T, and make the required tangent arc between the points of tangency, as shown.

CREATING A SERIES OF TANGENT ARCS CONFORMING TO A CURVE

First sketch lightly a smooth curve as desired, as shown in Figure 4.19. Find a radius R and a center C, producing an arc AB that closely follows that portion of the curve. The successive centers D, E, and so on will be on lines joining the centers with the points of tangency, as shown in Figure 4.19.

CREATING AN OGEE CURVE CONNECTING TWO PARALLEL LINES

Curves connecting two parallel lines are ogee curves if they each have 90 degrees of arc as shown in Figure 4.20 a and b. Figure 4.20a shows parallel lines NA and BM.

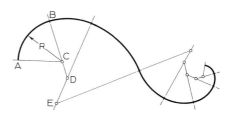

FIGURE 4.19   Creating a Series of Tangent Arcs Conforming to a Curve.

Practical Tip
Various sketching surfaces

As you learn the technique for sketching angles and curves, it is good to practice sketching on notebook and plain typing paper as well as grid paper. You will need to be able to create clear sketches on a variety of different surfaces. If you have access to a white marker board or chalkboard, practice sketching in this vertical orientation also, making large drawings with clear cut lines of chalk or marker. Hold the marker on edge to create thinner lines for hidden and centerlines.

Step by Step 4.4
Creating an arc tangent to an arc and a straight line

Tangent Arcs

The steps for creating an inside arc and an outside arc are essentially the same. The left column of figures illustrates the steps for the inside arc. The right column shows steps for the outside arc.

Inside arc

Outside arc

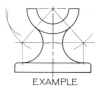
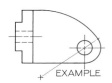

EXAMPLE   EXAMPLE

1. Two examples of an arc with radius G and a straight line AB are given.

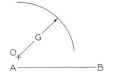

2. Make a straight line and an arc parallel to the given straight line and arc at radius distance R away, to intersect at C, the required center.

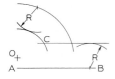

3. From C, make a perpendicular to the given straight line to find one point of tangency T. Join the centers C and O with a straight line to find the other point of tangency T.

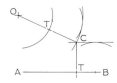
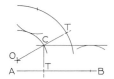

4. With center C and given radius R, create the required tangent arc between the points of tangency.

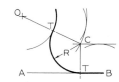
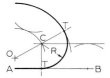

Freehand Constructions

When you are sketching constructions freehand, darken final curves first, then darken connecting lines. It is easier to connect sketched arcs with a line than it is to sketch an accurate curve to connect elements.

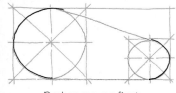
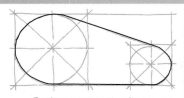

Darken curves first   Darken connecting lines

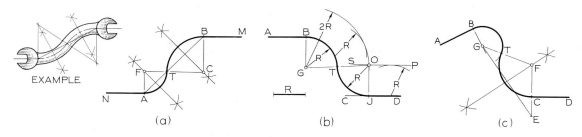

FIGURE 4.20 Curves Connecting Parallel Lines.

To create an ogee curve, create line AB, and assume inflection point T (at the midpoint if two equal arcs are desired). At A and B make perpendiculars AF and BC. Create perpendicular bisectors of AT and BT. The intersections F and C of these bisectors and the perpendiculars are the centers of the required tangent arcs.

In the example shown in Figure 4.20b, AB and CD are the two parallel lines, point B is one end of the curve and R is the given radii. At B, create a perpendicular to AB, make BG equal to R, and create the arc as shown. Make line SP parallel to CD at a distance R from CD. With center G, create the arc of radius 2R, intersecting line SP at O. Create perpendicular OJ to locate tangent point J, and join centers G and O to locate point of tangency T. Using centers G and O and radius R, create the two tangent arcs as shown.

CONNECTING TWO NONPARALLEL LINES

Let AB and CD, shown in Figure 4.20c, be the two nonparallel lines. Make a line perpendicular to AB at B. Select point G on the perpendicular so that BG equals any desired radius, and create the arc as shown. Make a line perpendicular to CD at C and make CE equal to BG. Join G to E and bisect it. The intersection F of the bisector and the perpendicular CE, extended, is the center of the second arc. Join the centers of the two arcs to locate tangent point T, the inflection point of the curve.

■ 4.13 CONSTRUCTING AN ELLIPSE

An ellipse is generated by a point moving so that the sum of its distances from two points (the foci) is constant and equal to the major axis. The long axis of an ellipse is the major axis, and the short axis is the minor axis. Refer to Figure 4.21. The foci E and F are found by locating arcs with radii equal to half the major axis and with a center at the end of the minor axis. Another

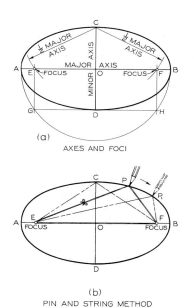

FIGURE 4.21 Ellipse Constructions.

method is to create a semicircle with the major axis as the diameter, and then to make GH parallel to the major axis and GE and HF parallel to the minor axis, as shown. An ellipse may be constructed by placing a looped string around the foci E and F and around C, one end of the minor axis, and moving the pencil point P along its maximum orbit while the string is kept taut.

■ 4.14 FINDING THE AXES OF AN ELLIPSE WITH CONJUGATE DIAMETERS GIVEN

Conjugate diameters AB and CD and the ellipse are given, as shown in Figure 4.22a. With intersection O of the conjugate diameters (center of ellipse) as the center and any convenient radius, create a circle to intersect

Step by Step 4.5
Creating tangent curves

Follow the steps below to learn how to create a curve tangent to the three intersecting lines given.

1. Let AB, BC, and CD be the given lines.

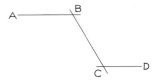

2. Select point of tangency P at any point on line BC.

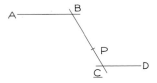

3. Make BT equal to BP, and CS equal to CP, and create perpendiculars at the points P, T, and S.

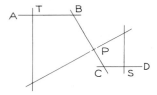

4. Their intersections O and Q are the centers of the required tangent arcs.

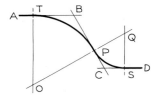

Another example of a curve tangent to three intersecting lines is shown below. The same process was used, but only the final result is shown.

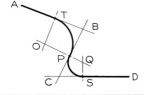

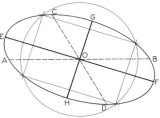

(a) CONJUGATE DIAMETERS AND ELLIPSE ARE GIVEN

FINDING CENTER
(b)

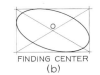

(c) CONJUGATE DIAMETERS GIVEN

FIGURE 4.22 Finding the Axes of an Ellipse.

the ellipse at four points. Join these points with straight lines, as shown. The result will be a rectangle whose sides are parallel to the required major and minor axes. Create the axes EF and GH parallel to the sides of the rectangle.

When only an ellipse is given, as shown in Figure 4.22b, use the following method to find the center of the ellipse: First, make a circumscribing rectangle or parallelogram about the ellipse. Then create diagonals to intersect at center O, as shown. Locate the axes similar to Figure 4.22a.

When the conjugate diameters AB and CD are given, as shown in Figure 4.22c, use the following method: Create a circle with O as the center and CD as the diameter. Through center O and perpendicular to CD, make a line EF. From points E and F (where this perpendicular intersects the circle), create lines FA and EA to form angle FAE. Make the bisector AG of this angle. The major axis JK will be parallel to this bisector, and the minor axis LM will be perpendicular to it. The length AH will be one-half the major axis, and HF one-half the minor axis. The resulting major and minor axes are JK and LM, respectively.

■ 4.15 CREATING A TANGENT TO AN ELLIPSE

CONCENTRIC-CIRCLE CONSTRUCTION

Figure 4.23 shows examples of the construction for drawing a tangent to an ellipse. To create a tangent at any point on an ellipse, such as E, make the ordinate at E intersect a circumscribed circle at V. Create a tangent to the circumscribed circle at V, and extend it to intersect the major axis extended at G. The line GE is the required tangent.

To create a tangent from a point outside the ellipse, such as P, create the ordinate PY and extend it. Make line DP intersect the major axis at X. Create line FX and extend it to intersect the ordinate through P at Q. Then, from the rule of similar triangles, QY:PY = OF:OD. Create a tangent to the circle from Q, find the point of tangency R, and then create the ordinate at R to intersect the ellipse at Z. The line ZP is the required tangent. As a check on the drawing, the tangents RQ and ZP should intersect at a point on the major axis extended. Two tangents to the ellipse can be created from point P.

■ 4.16 CREATING AN APPROXIMATE ELLIPSE

For many purposes, particularly where a small ellipse is required, the approximate circular-arc method is satisfactory. Some CAD systems do not create true ellipses but use this method instead. The approximate ellipse method is shown in Figure 4.24.

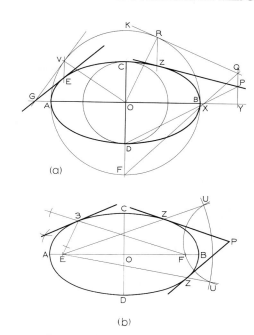

(a)

(b)

FIGURE 4.23 Tangents to an Ellipse.

Given axes AB and CD:

I. Create line AC. With O as the center and OA as the radius, make arc AE. With C as the center and CE as the radius, make arc EF.

II. Create perpendicular bisector GH of the line AF; the points K and J, where it intersects the axes, are centers of the required arcs.

III. Find centers M and L by using OL = OK and OM = OJ. Using centers K, L, M, and J, create circular arcs as shown. The points of tangency T are at the intersections of the arcs on the lines joining the centers.

FIGURE 4.24 Creating an Approximate Ellipse.

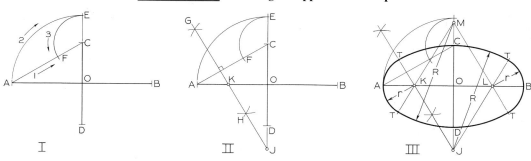

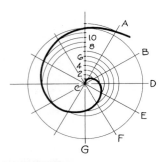

FIGURE 4.25 Spiral of Archimedes.

■ 4.17 DRAWING A SPIRAL OF ARCHIMEDES

To find points on the curve, create lines through the pole C, making equal angles with each other, such as 30-degree angles, as shown in Figure 4.25. Beginning with any one line, measure any distance, such as 2 mm; measure twice that distance on the next line, three times on the third, and so on. Draw a smooth curve through the points you have determined.

■ 4.18 DRAWING A HELIX

A *helix* is generated by a point moving around and along the surface of a cylinder or cone with a uniform angular velocity about the axis, with a uniform linear velocity about the axis, and with a uniform velocity in the direction of the axis. A cylindrical helix is generally known simply as a helix. The distance measured parallel to the axis traversed by the point in one revolution is called the lead. The helix finds many applications in industry, as in screw threads, worm gears, conveyors, spiral stairways, and so on. The stripes of a barber pole are helical in form. Helical shapes can be created using NC machines, moving the cutter in the X, Y, and Z directions at the same time.

If the cylindrical surface on which a helix is generated is rolled out onto a plane, the helix becomes a straight line, as shown in Figure 4.26a. The portion below the helix becomes a right triangle, the altitude of which is equal to the lead of the helix; the length of the base is equal to the circumference of the cylinder. A helix of this type can be defined as the shortest line on the surface of a cylinder connecting two points not on the same element.

To create the helix, first create two views of the cylinder on which the helix is generated, as shown in Figure 4.26b. Divide the circle of the base into any number of equal parts. On the rectangular view of the cylinder, set off the lead and divide it into the same number of equal parts as the base. Number the divisions as shown (in this case 16). When the generating point has moved one sixteenth of the distance around the cylinder, it will have risen one sixteenth of the lead; when it has moved halfway around the cylinder, it will have risen half the lead; and so on. Points on the helix are found by projecting up from point 1 in the circular view to line 1 in the rectangular view, from point 2 in the circular view to line 2 in the rectangular view, and so on.

Figure 4.26b is a right-hand helix. In a left-hand helix, shown in Figure 4.26c, the visible portions of the curve are inclined in the opposite direction-that is, downward to the right. A right-hand helix can be converted into a left-hand helix by interchanging the visible and hidden lines.

The construction for a right-hand conical helix is shown in Figure 4.26d.

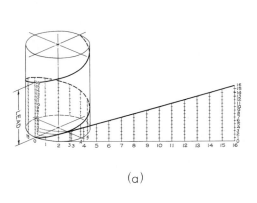

(a)

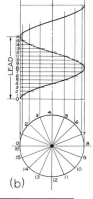

(b)

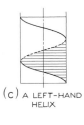

(c) A LEFT-HAND HELIX

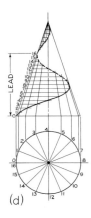

(d)

FIGURE 4.26 Helix.

■ 4.19 DRAWING AN INVOLUTE

An *involute* is the path of a point on a string as the string unwinds from a line, polygon, or circle. The involute of a circle is used in the construction of involute gear teeth. In this system, the involute forms the face and a part of the flank of the teeth of gear wheels; the outlines of the teeth of racks are straight lines.

TO DRAW AN INVOLUTE OF A LINE

Let AB be the given line. With AB as the radius and B as the center, make semicircle AC, as shown in Figure 4.27a. With AC as the radius and A as the center, create semicircle CD. With BD as the radius and B as the center, make semicircle DE. Continue similarly, alternating centers between A and B, until you have completed the required size for the figure.

TO DRAW AN INVOLUTE OF A TRIANGLE

Let ABC be the given triangle. With CA as the radius and C as the center, make arc AD, shown in Figure 4.27b. With BD as the radius and B as the center, create the arc DE. With AE as the radius and A as the center, make arc EF. Continue similarly until the figure is of the required size.

TO DRAW AN INVOLUTE OF A SQUARE

Let ABCD be the given square. With DA as the radius and D as the center, create the 90-degree arc AE, shown in Figure 4.27c. Proceed as for the involute of a triangle until you have completed the required size for the figure.

TO DRAW AN INVOLUTE OF A CIRCLE

You can think of a circle as a polygon with an infinite number of sides (Figure 4.27d). The involute is constructed by dividing the circumference into a number of equal parts and creating a tangent at each division point. Along each tangent, mark the length of the corresponding circular arc. Create the required curve through the points you have located on several tangents.

An involute can be generated by a point on a straight line that is rolled on a fixed circle (Figure 4.27e).

Practical Tip
Using ellipse templates

Thin plastic ellipse templates, like the one shown, may be used to save time in drawing ellipses. The templates are identified by the ellipse angle (the angle at which a circle is tipped to appear as an ellipse).

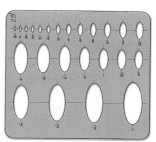

ELLIPSE TEMPLATE

Ellipse templates are generally available in ellipse angles at 5-degree intervals, such as 15 degrees, 20 degrees, and 25 degrees. In this example, the angle between the line of sight and the edge view of the plane of the circle is about 49 degrees. For this you would use a 50-degree ellipse template, since it is the closest size. The 50-degree template provides a variety of sizes of 50-degree ellipses; select the one that fits.

If the ellipse angle is not easily determined, you can always look for the ellipse that is approximately as long and as fat as the ellipse to be drawn.

Points on the required curve may be determined by measuring equal distances 0-1, 1-2, 2-3, and so on, along the circumference, and drawing a tangent at each division point. Then proceed as explained in Figure 4.27d.

FIGURE 4.27 Involutes.

(a)

(b)

(c)

(d)

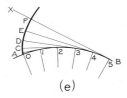

(e)

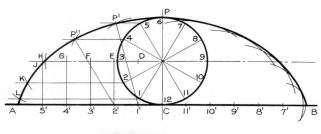

FIGURE 4.28 Cycloid.

■ 4.20 DRAWING A CYCLOID

A *cycloid* can be generated by tracing a point P in the circumference of a circle that rolls along a straight line.

Given the generating circle and the straight line AB tangent to it, make the distances CA and CB each equal to the semicircumference of the circle. Divide these distances and the semicircumference into the same number of equal parts (six, for instance) and number them consecutively, as shown in Figure 4.28. Suppose the circle rolls to the left; when point 1 of the circle reaches point 1′ of the line, the center of the circle will be at D, point 7 will be the highest point of the circle, and the generating point 6 will be at the same distance from the line AB as point 5 is when the circle is in its central position. To find the point P′, create a line through point 5 parallel to AB and intersect it with an arc from the center D having a radius equal to that of the circle. To find point P″, create a line through point 4 parallel to AB and intersect it with an arc from the center E, with a radius equal to that of the circle. Points J, K, and L are found in a similar manner.

Another method you can use is shown in the right half of Figure 4.28. With center at 11′ and the chord 11-6 as the radius, create an arc. With 10′ as the center and the chord 10-6 as the radius, create an arc. Continue similarly with centers 9′, 8′, and 7′. Make the required cycloid tangent to these arcs.

You can use either method; however, the second is shorter. The line joining the generating point and the point of contact for the generating circle is a normal of the cycloid (think about the method used to create the tangent arcs in the method previously described). The lines 1′-P″ and 2′-P′, for instance, are normals. This property makes the cycloid suitable for the outlines of gear teeth.

■ 4.21 EPICYCLOIDS AND HYPOCYCLOIDS

If the generating point P is on the circumference of a circle that rolls along the convex side of a larger circle, the curve generated is an *epicycloid*, as shown in Figure 4.29a. If the circle rolls along the concave side of a larger circle, the curve generated is a *hypocycloid*, as shown in Figure 4.29b. These curves are drawn in a manner similar to the cycloid. These curves are used to form the outlines of certain gear teeth and are, therefore, of practical importance in machine design.

■ 4.22 SOLIDS

Solids bounded by plane surfaces are called *polyhedra*. Examples are shown in Figure 4.30. The surfaces are called *faces*, and if the faces are equal regular polygons, the solids are called *regular polyhedra*.

A *prism* has two bases, which are parallel equal polygons, and three or more lateral faces, which are parallelograms. A triangular prism has a triangular base, a rectangular prism has rectangular bases, and so on. If the bases are parallelograms, the prism is a *parallelepiped*. A right prism has faces and lateral edges perpendicular to the bases; an oblique prism has faces and laterals edge oblique to the bases. If one end is cut off to form an end that is not parallel to the bases, the prism is said to be *truncated*.

A *pyramid* has a polygon for a base and triangular lateral faces intersecting at a common point called the *vertex*. The centerline from the center of the base

FIGURE 4.29 Epicycloid and Hypocycloid.

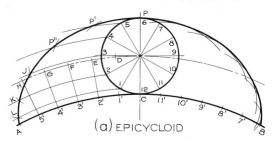

(a) EPICYCLOID

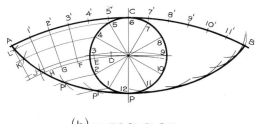

(b) HYPOCYCLOID

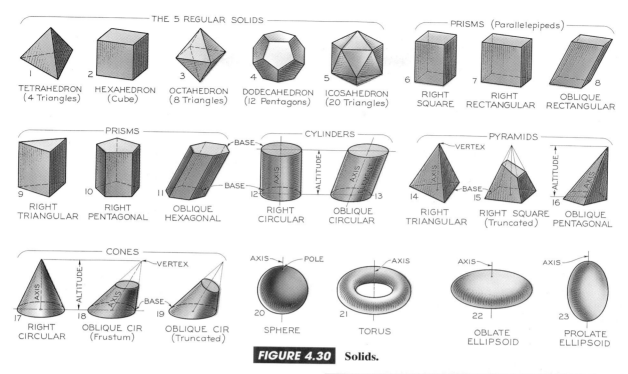

FIGURE 4.30 Solids.

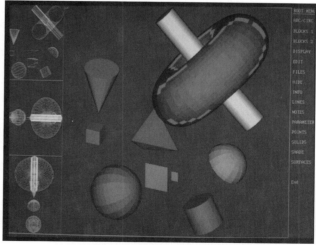

FIGURE 4.31 CAD-generated Solids.

to the vertex is the *axis*. If the axis is perpendicular to the base, the pyramid is a right pyramid; otherwise it is an oblique pyramid. A triangular pyramid has a triangular base, a square pyramid has a square base, and so on. If a portion near the vertex has been cut off, the pyramid is truncated, or it is referred to as a frustum.

A *cylinder* is generated by a straight line, called the generatrix, moving in contact with a curved line and always remaining parallel to its previous position or to the axis. Each position of the generatrix is called an element of the cylinder.

A *cone* is generated by a straight line moving in contact with a curved line and passing through a fixed point, the vertex of the cone. Each position of the generatrix is an element of the cone.

A *sphere* is generated by a circle revolving about one of its diameters. This diameter becomes the axis of the sphere, and the ends of the axis are *poles* of the sphere.

A *torus,* which is shaped like a doughnut, is generated by a circle (or other curve) revolving about an axis that is eccentric to the curve.

An *oblate* or *prolate ellipsoid* is generated by revolving an ellipse about its minor or major axis, respectively.

■ 4.23 SOLID MODELING

CAD systems that offer *solid modeling* usually have options or commands for creating complex solids using *primitive* shapes, including boxes, prisms, cylinders, spheres, cones, tori, and sometimes wedges and pyramids. Figure 4.31 shows an example of some solids generated using CAD. Many CAD systems also have a primitive command which creates any regular solid.

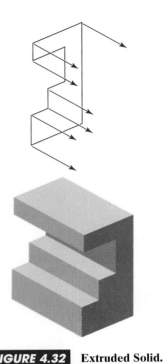

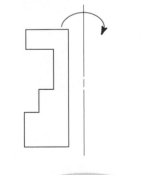

FIGURE 4.32   Extruded Solid.

FIGURE 4.33   Revolved Solid.

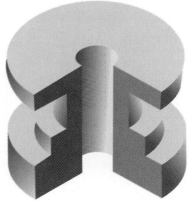

If there is not a specific command to create the solid you want, you usually have the option of creating new solid objects through processes called extrusion and revolution (both generally referred to as sweeping solids)

Extrusion is named for the manufacturing process which forms material by forcing it through a shaped opening. You can also think of extrusion as taking the cross-sectional shape of the object and sweeping it along a path to enclose a solid volume. Features which have a continuous cross section along a straight axis can be created using extrusion. Some CAD software can only extrude shapes along a straight-line path, while others can do straight or curved paths. Most have the option to taper the extruded feature as it is created. An example of an extruded solid model is shown in Figure 4.32.

Revolution is the process of forming a solid by revolving the cross-sectional shape of the object along a circular path to enclose a solid volume. Many objects that cannot be created by extrusion can be created by revolution. A solid object created by revolving a shape by 270 degrees is shown in Figure 4.33.

■ 4.24 BOOLEAN OPERATORS

Complex solid models can be formed by joining primitives and solids formed by extrusion and revolution using Boolean operators. *Boolean operators* are named for 18th century mathematician and logician, Charles Boole. Most CAD programs that allow solid modeling support three Boolean operators: *union* (sometimes called addition), *difference* (sometimes called subtraction), and *intersection*. Venn diagrams are often used to show how sets are joined together using Boolean operators. Figure 4.34 shows Venn diagrams for union, difference, and intersection. Boolean operators can also be used to join solids you create by extruding or revolving other solids or solid primitives.

The union of solid A and solid B forms a single new solid that is their combined volume without any duplication where they have overlapped. Solid A difference solid B is similar to subtracting B from A. The order of the operation does make a difference in the result (unlike union). For A difference B, any volume from solid B that overlaps solid A is eliminated and the result forms a new solid. A intersect B results in a new solid where only the volume common both to solid A and solid B is retained. You can use these primitives, extrusion, revolution, and Boolean operators to create solid models for a variety of objects. Objects that cannot be created in this way are warped surfaces, such as those on the exteriors of automobiles and airplanes.

Venn diagrams depicting Boolean operators

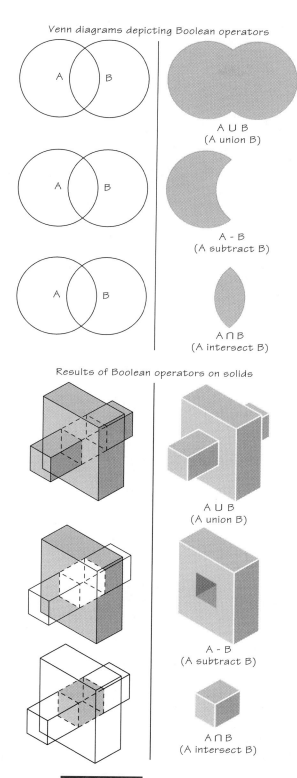

A ∪ B
(A union B)

A − B
(A subtract B)

A ∩ B
(A intersect B)

Results of Boolean operators on solids

A ∪ B
(A union B)

A − B
(A subtract B)

A ∩ B
(A intersect B)

FIGURE 4.34 Boolean Operators.

Practical Tips
Solid modeling

Begin with simplified shapes

Keep your solid models general at first. Don't model small fillets and rounds until the final stage of the design. This prevents the need to make repetitive changes and also keeps the file size smaller, so that you can work faster.

Screw thread is not usually modeled because it adds greatly to the complexity of the CAD file. It is usually represented in a simplified form. Screw thread specifications are usually given in a note.

Since it is always good practice to back up your files in progress, keeping your working files a manageable size saves back-up time as well.

Use Boolean intersection to save time modeling parts

Beginning CAD users sometimes overlook the usefulness of the Boolean intersection operation. You can save time creating many parts by extruding standard engineering views and intersecting them.

For example, first draw the basic shape seen from the top and extrude it to create a solid.

Then draw the shape seen from the front and extrude it to create another solid. The Boolean intersection of these two solids will often create the part you need.

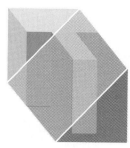

Two solids to be intersected

You can also extrude three views, for example, top, front, and side, and intersect all three to create more complex models.

Keep in mind that there are many shapes that you might be able to model with this process.

Boolean intersection

GRAPHICS SPOTLIGHT

From Art to Part

SINGLE DATABASE

You can use a single CAD database to design, document, analyze, create prototypes, and directly manufacture finished parts for your design. The term *art to part* is sometimes used to describe a CAD database being utilized for many or all of these purposes.

SKETCHING FREEDOM

Initial ideas for the design are frequently sketched freehand, as shown in Fig. A. While generating ideas for the design, it is important to be able to quickly generate creative ideas without the confines of using the computer. Sketching is still generally the best tool to help in this process.

INTELLIGENT MODELS

After generating the initial ideas, the best alternatives for the design are developed further. At this point, the engineer may create rough 3D drawing geometry like you see in Fig. B, perhaps using a parametric modeling software. *Parametric modeling* uses variables to constrain the shape of the geometry. Using parametric modeling the designer roughly sketches initial shapes and applies drawing dimensions and constraints to create models that have "intelligence." Later the

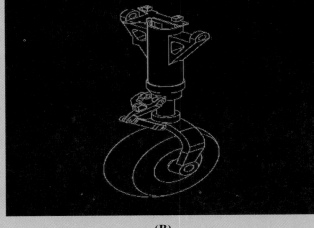

(B)

designer can change the dimensions and constraints as the design is refined so that new models do not have to be created for each design change. Realistic renderings of the model help you visualize the design.

OPTIMIZING THE DESIGN

You can export the refined model directly into a Finite Element Analysis (FEA) program to perform structural, thermal, and modal analysis as shown in Fig. C. The parametric model can easily be changed if the

(A)

(C)

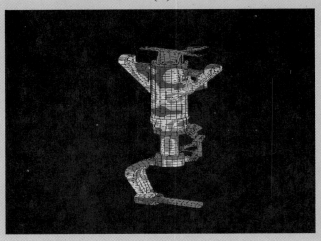

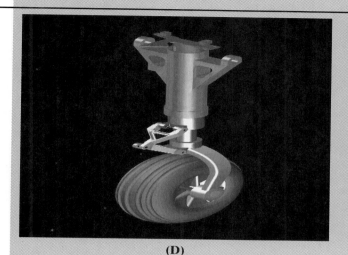

(D)

(E)

analysis shows that the initial design will not meet requirements. Simulation programs may even animate the performance and function of the system before a prototype is ever constructed. The tolerances and fits between mating parts can be checked within the parametric modeling and design software. Fig. D shows a shaded 3D model that closely resembles the final part.

RAPID PROTOTYPING

While refining the design ideas, engineers often work concurrently with manufacturing to determine the best ways to make and assemble the necessary parts. After several cycles of refining, analyzing, and synthesizing the best ideas, the final design is ready to go into production. Rapid prototyping systems allow parts to quickly be generated directly from the 3D models for mockup and testing. Fig. E shows the prototyped part.

When the design is approved the finished parts can be created using numerical controlled machines which get their tool paths directly from the 3D models.

GET NEW PRODUCTS TO MARKET QUICKLY

The necessary documentation for the design, manuals, brochures and other literature can be created directly from the same geometry used for design and manufacturing. Shortened design cycle time, improved communication, better opportunity to analyze and make design changes, are all advantages for companies using integrated CAD software for the design, documentation, and manufacture of their products.

Photographic material reprinted with the permission from and under the copyright of Autodesk, Inc.

■ KEY WORDS

line	triangle	prism	poles
point	perpendicular bisector	parallelepiped	torus
tangent	radial line	truncated	oblate ellipsoid
parallel	helix	pyramid	solid modeling
angle	involute	sphere	primitive
polygon	cycloid	vertex	extrusion
regular polygon	epicycloid	axis	revolution
circle	hypocycloid	cylinder	Boolean operator
circumference	polyhedra	cone	union
quadrilateral	faces	vertex	difference
parallelogram	regular polyhedra	sphere	intersection

Hands On 4.1
Using Boolean operators

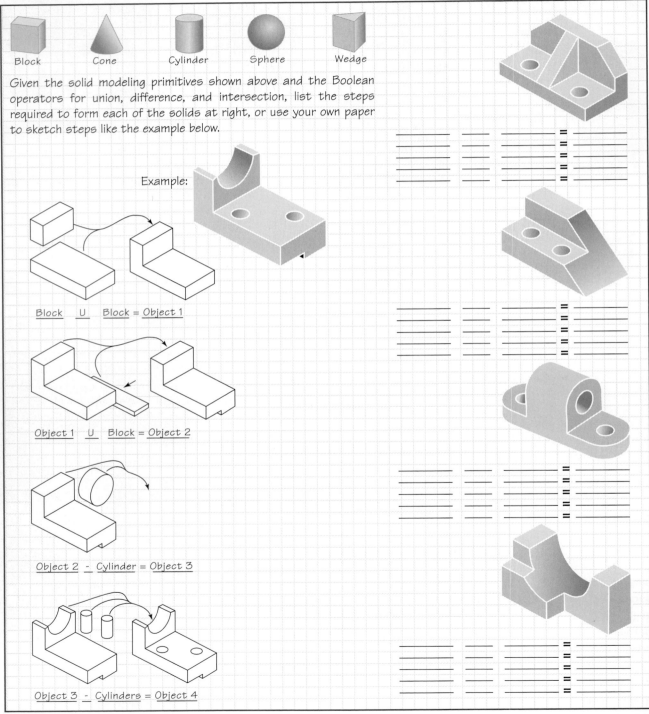

Block   Cone   Cylinder   Sphere   Wedge

Given the solid modeling primitives shown above and the Boolean operators for union, difference, and intersection, list the steps required to form each of the solids at right, or use your own paper to sketch steps like the example below.

Example:

Block   U   Block = Object 1

Object 1   U   Block = Object 2

Object 2   -   Cylinder = Object 3

Object 3   -   Cylinders = Object 4

■ CHAPTER SUMMARY

- Understanding basic geometric construction techniques and terminology is fundamental to the success of both traditional drawing and CAD drawing.
- All drawings are made up of points, lines, arcs, and circles drawn at various sizes and constructed with specific orientations to each other.
- The advantage that CAD provides for geometric construction is in drawing precision. Only you know where a point, line, arc, or circle needs to be drawn.

It is just as easy to create poor drawing geometry using CAD as it is by hand.

- It is important to show tangencies correctly in your hand sketches and CAD drawings.
- Solid modeling techniques using CAD require an understanding of geometric solids and Boolean operators. Using good strategies, you can create more-effective solid models using CAD.

■ REVIEW QUESTIONS

1. How can you divide a line into equal parts using sketching? Using your CAD system?
2. How many ways can an arc be tangent to one line? To two lines? To a line and an arc? To two arcs? Draw an example of each.
3. Sketch an equilateral triangle, a right triangle, and an isosceles triangle.
4. Given a line, sketch another line (1) parallel to the first, (2) perpendicular to the first. Then draw a horizontal line through the intersection of the two lines. Finally, draw a vertical line through the intersection.
5. Sketch a Venn diagram showing the three Boolean operators.
6. What solid modeling primitives does your CAD system provide, if any?
7. What is the difference between extrusion and revolution?

■ DESIGN PROJECT

A need exists for home lighting that is attractive, inexpensive, and easy to install and use. Design a unique wall-mounted or free-standing lamp that makes use of elegant lines and smooth curves. Consider different mechanisms for operating the lamp and creative use of materials. Strive for a finished design with attention to function and style. Represent your design accurately, using the geometric construction techniques you learned in Chapter 4.

■ GEOMETRIC CONSTRUCTION PROJECTS

Geometric constructions should be done very accurately, using CAD if available. In solving the following problems, use Layout A-2 or Layout A4-2 (adjusted) inside the front cover. Set up each problem so as to make the best use of the space available, to present the problem to best advantage, and to produce a pleasing appearance. Letter the principal points of all constructions in a manner similar to the various illustrations in this chapter. You may be able to show four problems per sheet. Many problems are dimensioned in the metric system. Your instructor may ask you to convert the remaining problems to metric measure. Since many of the problems in this chapter are of a general nature, they can also be solved on most computer graphics systems.

Proj. 4.1   Draw an inclined line AB 65 mm long and bisect it.

Proj. 4.2   Draw any angle with the vertex at C and bisect it.

Proj. 4.3   Draw any inclined line EF. Given distance GH is 5 inches or 42 mm. Draw a line parallel to EF and at the distance GH from it.

Proj. 4.4   Draw the line JK 95 mm long and divide it into five equal parts. Draw a line LM 58 mm long and divide it into three equal parts.

Proj. 4.5   Draw a line OP 92 mm long and divide it into three proportional parts in the ratio of 3 : 5 : 9.

Proj. 4.6   Draw a triangle with sides of 76 mm, 85 mm, and 65 mm. Bisect the three interior angles. The bisectors should meet at a point. Draw the inscribed circle with the point as the center.

Proj. 4.7   Draw a right triangle having a hypotenuse 65 mm and one leg 40 mm, and draw a circle through the three vertexes.

Proj. 4.8   Draw an inclined line QR 84 mm long. Select a point P on the line 32 mm from Q and create a perpendicular through point P. Create point S 45.5 mm from the line, and draw a perpendicular from S to the line.

Proj. 4.9   Draw an equilateral triangle having 63.5-mm sides. Bisect the interior angles. Draw the inscribed circle using the intersection of the bisectors as the center.

Proj. 4.10   Draw a 54-mm-diameter circle. Inscribe a square in the circle, and circumscribe a square on the circle.

Proj. 4.11   Draw a 65-mm-diameter circle. Inscribe a regular pentagon, and join the vertexes to form a five-pointed star.

Proj. 4.12   Draw a 65-mm-diameter circle. Inscribe a hexagon, and circumscribe a hexagon.

Proj. 4.13   Draw a square with 63.5-mm sides, and inscribe an octagon.

Proj. 4.14   Draw a 58-mm-diameter circle. Assume a point S on the left side of the circle and draw a tangent at that point. Assume a point T to the right of the circle 50 mm from its center. Draw two tangents to the circle through that point.

Proj. 4.15   Draw a horizontal centerline. Then draw two circles with 50-mm diameters and 38-mm diameters, respectively, with centers 54 mm apart. Locate the circles so that the construction will be centered in the space. Draw open-belt tangents to the circles.

Proj. 4.16   Do the same as for Problem 4.15 except draw crossed-belt tangents to the circle.

Proj. 4.17   Draw a vertical line VW near the middle of your drawing area. Locate a point P 44 mm farther to the right and 25 mm down from the top of the space. Draw a 56-mm-diameter circle through P, tangent to VW.

Proj. 4.18   Draw a vertical line XY near the middle of your drawing area. Locate point P 44 mm farther to the right and 25 mm down from the top of the space. Locate point Q on line XY and 50 mm from P. Draw a circle through P and tangent to XY at Q.

Proj. 4.19   Draw a 64-mm-diameter circle with center C 16 mm directly to left of the center of your space. Assume point P at the lower right and 60 mm from C. Draw an arc with a 25-mm-radius through P and tangent to the circle.

Proj. 4.20   Draw a vertical line and a horizontal line, each 65 mm long. Draw an arc with a 38-mm radius tangent to the lines.

Proj. 4.21   Draw two intersecting lines making an angle of 60 degrees with each other. Assume a point P on one line at a distance of 45 mm from the intersection. Draw an arc tangent to both lines with one point of tangency at P.

Proj. 4.22   Draw a vertical line AB 32 mm from the left side of your space. Draw an arc of 42-mm radius with its center 75 mm to the right of the line and in the lower right portion of your space. Draw an arc of 25-mm radius tangent to AB and to the first arc.

Proj. 4.23   With centers 20 mm up from the bottom of your space and 86 mm apart, draw arcs of radii 44 mm and 24 mm, respectively. Draw an arc of 32-mm radius tangent to the two arcs.

Proj. 4.24   Draw two parallel inclined lines 45 mm apart. Choose a point on each line and connect them with an ogee curve tangent to the two parallel lines.

Proj. 4.25   Using the center of the space as the pole, draw a spiral of Archimedes with the generating point moving in a counterclockwise direction and away from the pole at the rate of 25 mm in each convolution.

Proj. 4.26   Through the center of your space, draw a horizontal centerline and on it construct a right-hand helix 50 mm in diameter, 64 mm long, and with a lead of 25 mm. Draw only a half-circular end view.

Proj. 4.27   Draw the involute of an equilateral triangle with 15-mm sides.

Proj. 4.28   Draw the involute of a 20-mm-diameter circle.

Proj. 4.29   Draw a cycloid generated by a 30-mm-diameter circle rolling along a horizontal straight line.

Proj. 4.30   Draw and epicycloid generated by a 38-mm-diameter circle rolling along a circular arc with a radius of 64 mm.

Proj. 4.31   Draw a hypocycloid generated by a 38-mm-diameter circle rolling along a circular arc with a radius of 64 mm.

Proj. 4.32   Using Layout A-2 or A4-2 (adjusted), draw the spanner shown. Omit dimensions and notes unless assigned (Figure 4.35).

Proj. 4.33   Using Layout A-2 or A4-2 (adjusted), draw the shaft hanger casting in Figure 4.36. Omit dimensions and notes unless assigned.

Proj. 4.34   Using Layout A-2 or A4-2 (adjusted), draw the shift lever in Figure 4.37. Omit dimensions and notes unless assigned.

Proj. 4.35   Using Layout A-2 or A4-2 (adjusted), draw the form roll lever in Figure 4.38. Omit dimensions and notes unless assigned.

Proj. 4.36   Using Layout A-2 or A4-2 (adjusted), draw the press base in Figure 4.39. Omit dimensions and notes unless assigned.

Proj. 4.37   Using Layout A-2 or A4-2 (adjusted), draw the photo floodlight reflector in Figure 4.40. Omit dimensions and notes unless assigned.

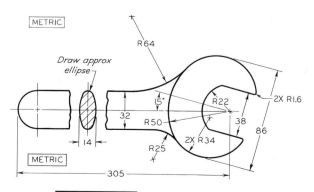

FIGURE 4.35   (Proj. 4.32) Spanner.

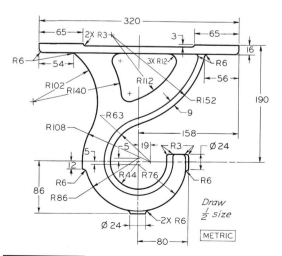

FIGURE 4.36   (Proj. 4.33) Shaft Hanger Casting.

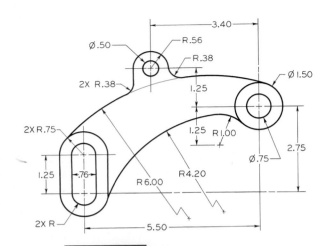

FIGURE 4.37   (Proj. 4.34) Shift Lever.

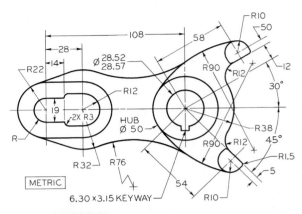

FIGURE 4.38   (Proj. 4.35) Form Roll Lever.

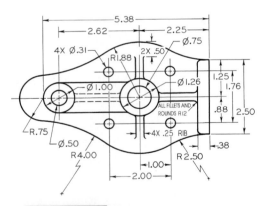

FIGURE 4.39   (Proj. 4.36) Press Base.

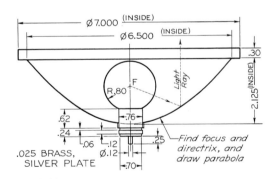

FIGURE 4.40   (Proj. 4.37) Photo Floodlight Reflector.

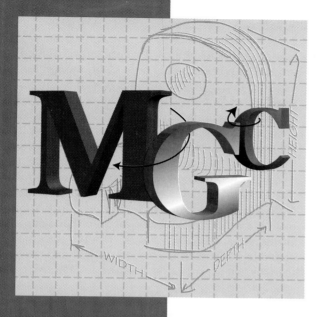

C H A P T E R 5

MULTIVIEW SKETCHING AND PROJECTION

OBJECTIVES

After studying the material in this chapter, you should be able to:

1. Sketch and arrange six standard views of an object.

2. Sketch any three views using proper conventions, placement, and alignment.

3. Read and measure with an architects', an engineers', or a metric scale.

4. Transfer dimensions between views.

5. Center a three-view sketch on the drawing medium.

6. Describe first- and third-angle projection.

7. Identify and project normal, inclined, and oblique surfaces in all views.

8. Sketch positive and negative cylinders in all views.

9. Plot conic sections and irregular curves in all views.

10. Understand drawing conventions hole treatments and machine processes.

11. Apply conventions to revolve of ribs, spokes, and webs.

OVERVIEW

A view of an object is called a projection. By projecting multiple views from different directions in a systematic way, you can completely describe the shape of 3-D objects. There are many conventions that you must learn in order to create sketches and drawings that can be interpreted by others. The standard published in ANSI/ASME Y14.3M-1994 is common in the United States, where third-angle projection is used. Europe, Asia, and many other places use the first-angle projection system.

In order to create and interpret drawings, you need to know how to create projections and to understand the standard arrangement of views. You must also understand the geometry of solid objects and how to visualize an object given a sketch or drawing. Understanding whether surfaces are normal, inclined, or oblique in orientation can help you to visualize objects. Common features such as vertices, edges, contours, holes, fillets, and rounds are shown in a standard way. These details should be shown clearly by choosing an appropriate scale.

■ 5.1 VIEWS OF OBJECTS

A photograph shows an object as it appears to the observer, but not necessarily as it is. It cannot describe the object accurately, no matter what distance or which direction it is taken from, because it does not show the exact shapes and sizes of the parts. It would be impossible to create an accurate three dimensional model of an object using only a photograph for reference because it shows only one view. It is a two dimensional representation of a three dimensional object.

In engineering and other fields, a complete and clear description of the shape and size of an object is necessary to be sure that it is manufactured exactly as the designer intended. To provide this information, a number of views, systematically arranged, are used. This system of views is called *multiview projection.* Each view provides certain definite information. For example, a front view shows the true shape and size of surfaces that are parallel to the front of the object. An example showing the direction of sight and the resulting front view projection is shown in Figure 5.1.

■ 5.2 THE SIX STANDARD VIEWS

Any object can be viewed from six mutually perpendicular directions, as shown in Figure 5.2. These are called the six principal views. Three of the views are aligned with the other three and show essentially the same information about the object, except that they are viewed from the exact opposite direction. For instance, the top view is opposite the bottom view, the left-side view is opposite the right-side view, and the front view is opposite the rear view.

You can think of the six views as what an observer would see by moving around the object. As shown in Figure 5.3, the observer can walk around a house and view its front, sides, and rear. You can imagine the top view as seen by an observer from an airplane and the bottom or "worm's-eye view" as seen from underneath.

FIGURE 5.1 Front View of an Object.

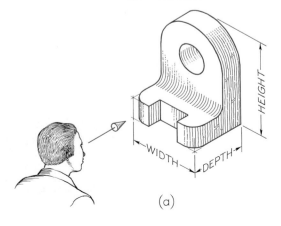

(a)

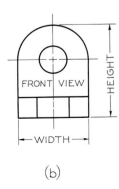

(b)

The term plan may also be used for the top view. The term elevation is used for all views showing the height of the building. These terms are regularly used in architectural drawing and occasionally in other fields.

You can also produce different views by revolving the object, as shown in Figure 5.4. First, hold the object in the front view position (shown as a). To get the top view (shown as b), tilt the object toward you to bring the top of the object into your view. To get the right-side view (shown as c), you would begin by holding the object's front view facing you and revolve it to bring the right side toward you. To obtain views of the rear, bottom, or right-side, you would simply turn the object to bring those sides toward you.

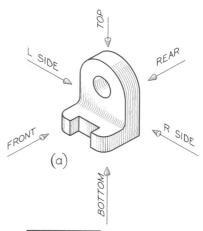

FIGURE 5.2 The Six Views.

FIGURE 5.3 Six Views of a House.

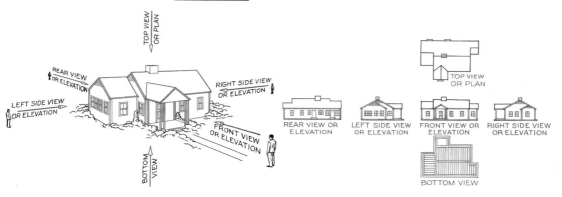

FIGURE 5.4 Revolving the Object to Produce Views.

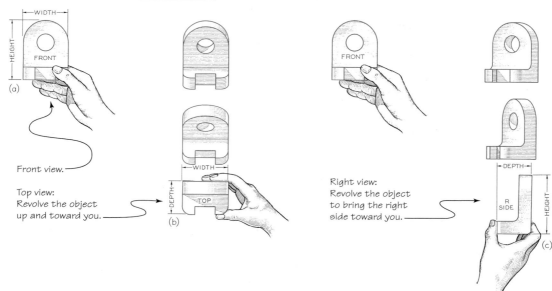

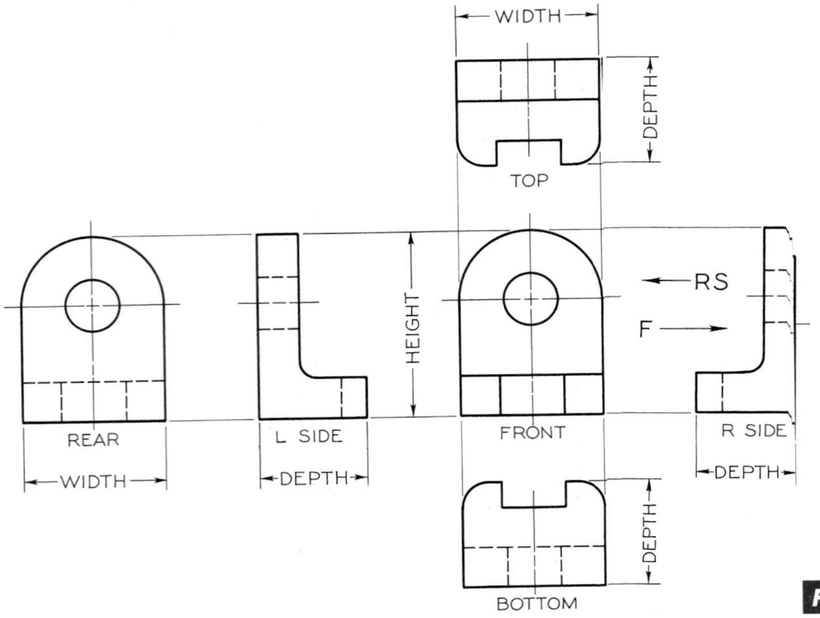

FIGURE 5.5 Standard Arrangement of Views.

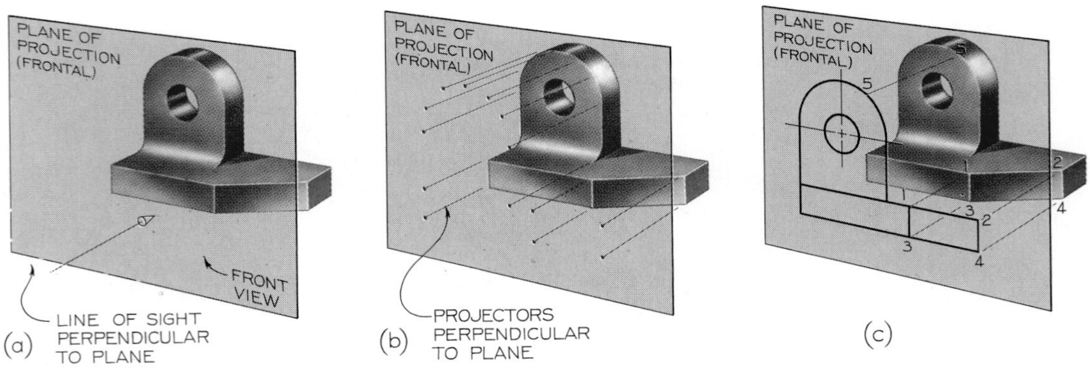

FIGURE 5.6 Projection of an Object.

To make drawings easier to read, the views are arranged on the paper in a standard way. The views in Figure 5.5 show the American National Standard arrangement. The top, front, and bottom views align vertically. The rear, left-side, front, and right-side views align horizontally. To draw a view out of place is a serious error and is generally regarded as one of the worst possible mistakes in drawing.

■ 5.3 PRINCIPAL DIMENSIONS

The three principal dimensions of an object are *width*, *height*, and *depth*. In technical drawing, these fixed terms are used for dimensions shown in certain views, regardless of the shape of the object. The terms length

and thickness are not used because they cannot be applied in all cases.

The front view shows only the height and width of the object and not the depth. In fact, any principal view of a 3-D object shows only two of the three principal dimensions; the third is found in an adjacent view. Height is shown in the rear, left-side, front, and right-side views. Width is shown in the rear, top, front, and bottom views. Depth is shown in the left-side, top, right-side, and bottom views.

■ 5.4 PROJECTION METHOD

Figure 5.6 shows how to create a front view of an object using a multiview projection. Imagine a sheet of

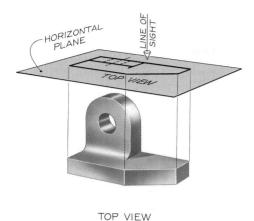

TOP VIEW
(a)

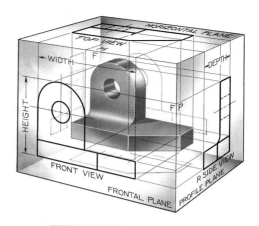

The Glass Box.

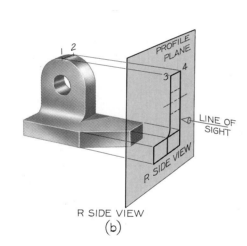

R SIDE VIEW
(b)

FIGURE 5.7 **Top and Right-Side Views.**

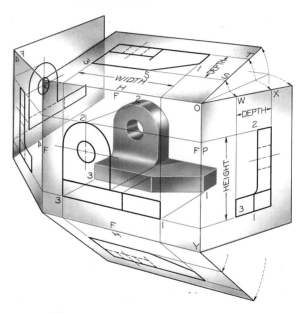

FIGURE 5.9 **Unfolding the Glass Box.**

glass parallel to the front surfaces of the object. This represents the ***plane of projection***. The outline on the plane of projection shows how the object appears to the observer. In ***orthographic*** projection, rays (or ***projectors***) from all points on the edges or contours of the object extend parallel to each other and perpendicular to the plane of projection.

Similar examples of the top and side views are shown in Figure 5.7. The plane of projection on which the front view is projected is called the frontal plane, the plane on which the top view is projected is the horizontal plane, and the plane on which the side view is projected is the profile plane.

■ 5.5 THE GLASS BOX

One way to understand the standard arrangement of views on the sheet of paper is the ***glass box***. If planes of projection were placed parallel to each principal face of the object, they would form a box, as shown in Figure 5.8. The outside observer would see six standard views of the object through the sides of this imaginary glass box.

To organize the views of a 3-D object on a flat sheet of paper, imagine the six planes of the glass box unfolded to lie flat, as shown in Figure 5.9. Think of all planes except the rear plane as hinged to the frontal plane. The rear plane is usually hinged to the left-side

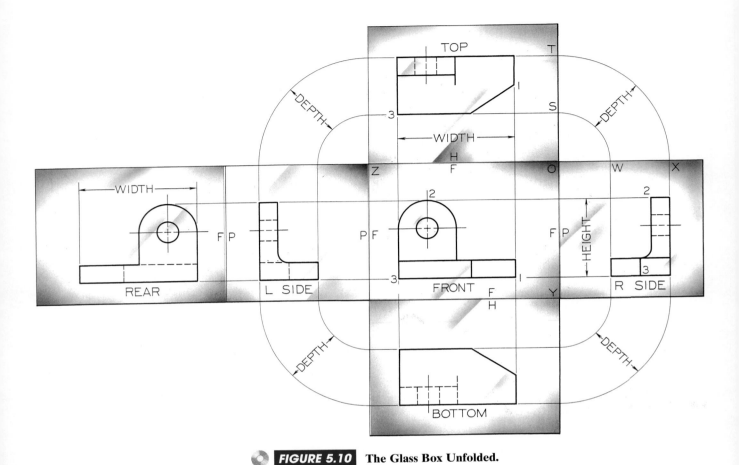

 FIGURE 5.10 The Glass Box Unfolded.

plane. Each plane folds out away from the frontal plane. The representation of the hinge lines of the glass box in a drawing are known as folding lines. The positions of these six planes after they have been unfolded are shown in Figure 5.10.

Carefully identify each of these planes and corresponding views with the planes' original position in the glass box.

Find Worksheet 5.1, the glass box, in the tear-out section. Follow the directions on the sheet to visualize the glass box and the concept of folding lines.

In Figure 5.9, lines extend around the glass box from one view to another on the planes of projection. These are the projectors from a point in one view to the same point in another view. The size and position of the object in the glass box does not change. This explains why the top view is the same width as the front view and why it is placed directly above the front

view. The same relation exists between the front and bottom views. Therefore, the front, top, and bottom views all line up vertically and are the same width. The rear, left-side, front, and right-side views all line up horizontally and are the same height.

Objects do not change position in the box, so the top view must be the same distance from the folding line OZ as the right side view is from the folding line OY. The bottom and left-side views are the same distance from their respective folding lines as are the right-side and the top views. The top, right-side, bottom, and left-side views are all the same distance from the respective folding lines and show the same depth.

The front, top, and right-side views of the object shown in the previous figures are shown in Figure 5.11a with folding lines between the views. These folding lines correspond to the hinge lines of the glass box. The H/F folding line, between the top and front views, is the intersection of the horizontal and frontal

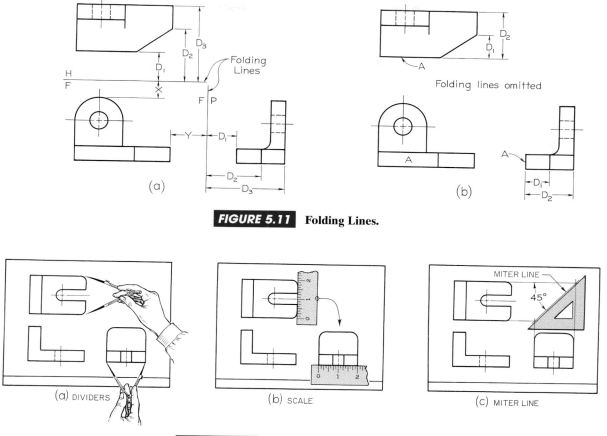

FIGURE 5.11 Folding Lines.

FIGURE 5.12 Transferring Depth Dimensions.

planes. The F/P folding line, between the front and side views, is the intersection of the frontal and profile planes. While you should understand folding lines, particularly because they are useful in solving problems in descriptive geometry, they are usually left off the drawing, as shown in Figure 5.11b. Instead of using the folding lines as reference lines for marking depth measurements in the top and side views, you may use the front surface A of the object as a reference line. This way, D1, D2, and all other depth measurements correspond in the two views as if folding lines were used.

■ 5.6 SPACING BETWEEN VIEWS
Spacing between views is a mainly a matter of appearance. Views should be spaced well apart, but close enough to appear related to each other. You may need to leave space between the views to add dimensions.

■ 5.7 TRANSFERRING DEPTH DIMENSIONS
The depth dimensions in the top and side views must correspond point-for-point. When using CAD or instruments, transfer these distances accurately.

You can transfer dimensions between the top and side views either with dividers or with a scale, as shown in Figures 5.12a and 5.12b. Marking the distances on a scrap of paper and using it like a scale to transfer the distance to the other view works well when sketching.

You may find it convenient to use a 45-degree miter line to project dimensions between top and side views, as shown in Figure 5.12c. Because the miter line is drawn at 45 degrees, or Y = X, depths shown vertically in the top view Y can be transferred to be shown as horizontal depths in the side view X and vice versa.

Tear out Worksheet 5.2. Use it to practice transferring depth dimensions.

Step by Step 5.1
Using a miter line

Given two completed views you can use a miter line to transfer the depths and draw the side view of the object shown at right.

1. Locate the miter line a convenient distance away from the object to produce the desired spacing between views.

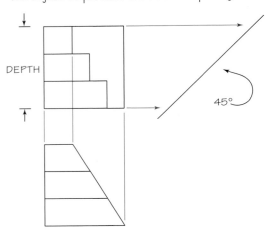

2. Sketch light lines projecting depth locations for points to miter line and then down into side view as shown

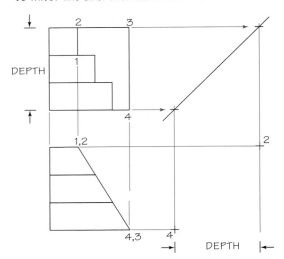

3. Project the remaining points.

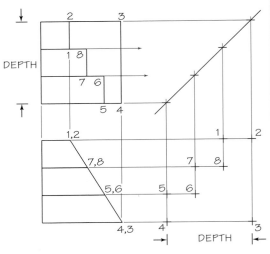

4. Draw view locating each vertex of surface on projection line and miter line.

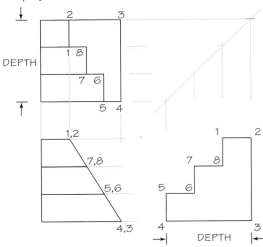

To move the right-side view to the right or left, or move the top view upward or downward by moving the miter line closer of further from the view. You don't need to draw continuous lines between the top and side views via the miter line. Instead, make short dashes across the miter line and project from these. The 45° miter-line method is also convenient for transferring a large number of points, as when plotting a curve.

■ 5.8 NECESSARY VIEWS

The right- and left-side views are essentially mirror images of each other, only with different lines appearing hidden. Hidden lines use a dashed-line pattern to represent portions of the object that are not directly visible from that direction of sight. Both the right and left views do not need to be shown, so usually the right-side view is drawn. This is also true of the top and bottom views, and of the front and rear views. The top, front, and right-side views, arranged together, are shown in Figure 5.13. These are called the three regular views because they are the views most frequently used.

A sketch or drawing should only contain the views needed to clearly and completely describe the object. These minimally required views are referred to as the necessary views. Choose the views that have the fewest hidden lines and show essential contours or shapes most clearly. Complicated objects may require more than three views or special views such as partial views.

Many objects may need only two views to clearly describe their shape. If an object requires only two views and the left-side and right-side views show the object equally well, use the right-side view. If an object requires only two views and the top and bottom views show the object equally well, choose the top view. If only two views are necessary and the top view and

right-side view show the object equally well, choose the combination that fits best on your paper. Some examples are shown in Figure 5.14.

Often, a single view supplemented by a note or by lettered symbols is enough. Objects that can be shown using a single view usually have a uniform thickness, like the shim shown in Figure 5.15a. One view of the shim plus a note indicating the thickness as 0.25 mm is sufficient. In Figure 5.15b, the left end is 65 mm square, the next portion is 49.22 mm in diameter, the next is 31.75 mm in diameter, and the portion with the thread is 20 mm in diameter—as indicated in the note. Nearly all shafts, bolts, screws, and similar parts are represented by single views and notes.

FIGURE 5.13   **The Three Regular Views.**

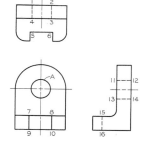

FIGURE 5.14   Choice of Views to Fit Paper.

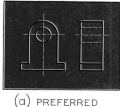 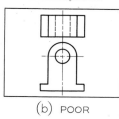 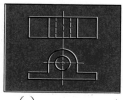 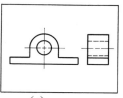

(a) PREFERRED   (b) POOR   (c) PREFERRED   (d) POOR

FIGURE 5.15   One-View Drawings of a Shim and a Connecting Rod.

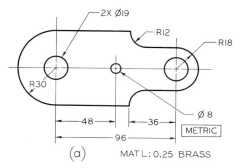

(a)   MAT L: 0.25 BRASS

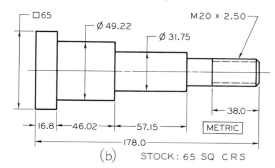

(b)   STOCK: 65 SQ CRS

FIGURE 5.16   Six Views of a Compact Automobile.

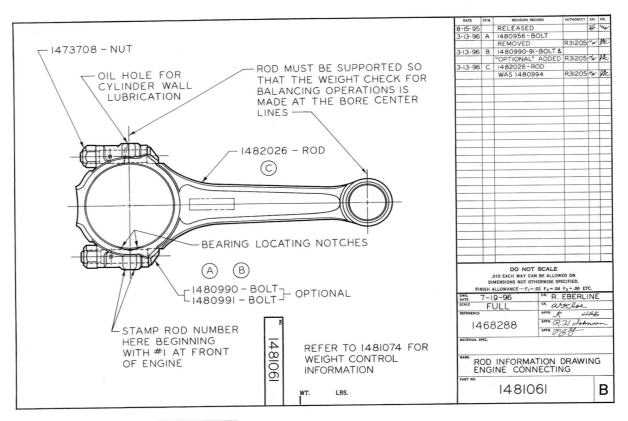

FIGURE 5.17   An Assembly Drawing of a Connecting Rod.

■ 5.9   ORIENTATION OF THE FRONT VIEW

Six views of a compact automobile are shown in Figure 5.16. The view chosen for the front view in this case is the side, not the front, of the automobile. In general, the front view should show the object in its operating position, particularly for familiar objects such as a house and an automobile. When possible, a machine part is drawn in the position it occupies in the assembly. Usually screws, bolts, shafts, tubes, and other elongated parts are drawn in a horizontal position. For example, the automobile connecting rod shown in Figure 5.17 is drawn horizontally on the sheet.

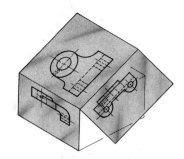

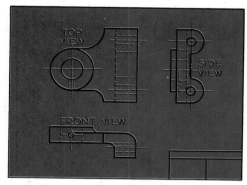

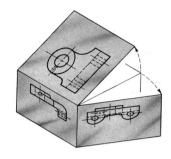

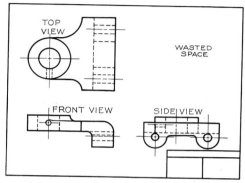

(a) POOR ARRANGEMENT OF VIEWS

(b) APPROVED ARRANGEMENT OF VIEWS

FIGURE 5.18   **Position of Side View.**

■ 5.10 ALTERNATE POSITIONS OF VIEWS

If three views of a wide, flat object are drawn using the conventional arrangement of views, a large part of the paper can't be used, as shown in Figure 5.18a. Better use of the available space sometimes makes the use of a reduced scale unnecessary. In this case you can think of the profile (side view) hinged to the horizontal plane (top view) instead of to the frontal plane (front view) so that the side view is beside the top view when unfolded, as shown in Figure 5.18b. Notice that the side view is rotated 90 degrees from its orientation in the side view when it is in this placement.

You can use the box you cut out earlier to help visualize this arrangement. Cut the views apart between the front and side planes and tape the box together so that the side plane is hinged to the top plane. Fold the box and visualize how an object placed inside the box would appear. Now unfold the box again. Notice that now you can directly project the depth dimension from the top view into the side view.

If necessary, it is also acceptable to place the side view horizontally across from the bottom view. In this case the profile plane is considered hinged to the bottom plane of the projection. Similarly, the rear view may be placed directly above the top view or under the bottom view, if necessary. The rear plane is considered

hinged to the horizontal or bottom plane and then rotated to coincide with the frontal plane. Try these arrangements using your paper box.

 Tear out Worksheet 5.3. Use it to practice blocking in a multiview drawing.

■ 5.11 VISUALIZATION

Along with the basic understanding of the system for projecting views, you must be able interpret multiple views and picture the object shown. In addition to being an indispensable skill to help you capture and communicate your ideas, technical sketching is also a way for others to present their ideas to you.

Even experienced engineers and designers can't always look at a multiview sketch and instantly visualize the object represented. You must learn to study the sketch and interpret the lines in a logical way in order to piece together a clear idea of the whole. This process is sometimes called visualization.

■ 5.12 SURFACES, EDGES, AND CORNERS

To effectively create and interpret multiview projections, you have to consider the elements that make up most solids. *Surfaces* form the boundaries of solid objects. A *plane* (flat) surface may be bounded by straight lines, curves, or a combination of them.

EV = Edge View
TS = True Size
FS = Foreshortened

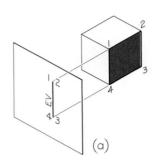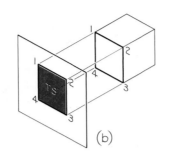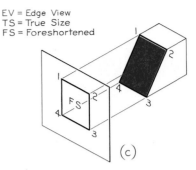

(a) (b) (c)

FIGURE 5.19 Projections of Surfaces.

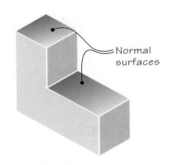

FIGURE 5.20 Normal Surfaces.

FIGURE 5.21 Inclined Surface.

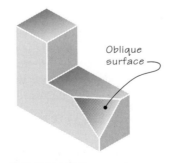

FIGURE 5.22 Oblique Surface.

■ 5.13 VIEWS OF SURFACES

A plane surface that is perpendicular to a plane of projection appears on edge as a straight line (Figure 5.19a). If it is parallel to the plane of projection, it appears true size (Figure 5.19b). If it is angled to the plane of projection, it appears foreshortened or smaller than its actual size (Figure 5.19c). A plane surface always projects either on edge (appearing as a single line) or as a surface (showing its characteristic shape) in any view. It can appear foreshortened, but it can never appear larger than its true size in any view.

There are special terms used for describing a surface's orientation to the plane of projection. The three orientations that a plane surface can have to the plane of projection are normal, inclined, and oblique. Understanding these terms can help you picture and describe objects.

■ 5.14 NORMAL SURFACES

A *normal surface* is parallel to a plane of projection. It appears true size and true shape on the plane to which it is parallel, and as a vertical or a horizontal line on adjacent planes of projection. Figure 5.20 shows an illustration of normal surfaces.

■ 5.15 INCLINED SURFACES

An *inclined surface* is perpendicular to one plane of projection, but inclined or tipped to adjacent planes. An inclined surface projects on edge on the plane to which it is perpendicular and appears foreshortened on planes to which it is inclined. An inclined surface is shown in Figure 5.21. The degree of foreshortening is proportional to the inclination. While the surface may not appear true size in any view, it will have the same characteristic shape and the same number of edges in the views in which you see its shape.

 Tear out Worksheet 5.4. Use it to practice drawing projections in an inclined plane.

Practice identifying normal surfaces on Cad drawings. You can download orthographic views of subjects that show many normal surfaces at the following web sites: http://www.constructionsite.come/harlen/8001-81.htm. http://user.mc.net/hawk/cad.htm.

■ 5.16 OBLIQUE SURFACES

An oblique surface is tipped to all principal planes of projection. Since it is not perpendicular to any projection plane, it cannot appear on edge in any standard view. Since it is not parallel to any projection plane, it cannot appear true size in any standard view. An oblique surface always appears as a foreshortened surface in all three standard views. Figure 5.22 shows an oblique surface.

■ 5.17 EDGES

The intersection of two plane surfaces on the object produces an *edge,* which is represented by a straight line in the drawing. The edge is common to both surfaces and forms a boundary for each surface. If an edge is perpendicular to a plane of projection, it appears as a point; otherwise it appears as a line. If an edge is parallel to the plane of projection, it shows true length; if not parallel, it appears foreshortened. A straight line always projects as a straight line or as a point. The terms *normal, inclined,* and *oblique* are also used to describe the relationship of edges to the planes of projection.

■ 5.18 NORMAL EDGES

A *normal edge* is a line that is perpendicular to a plane of projection. It appears as a point on that plane of projection and as a true-length line on adjacent planes of projection. Figure 5.23 shows normal edges and their projections.

Hands On 5.1
Normal and oblique surfaces

Oblique surface C appears in the top view and front view with its vertices labeled 1–2–3–4.

• Locate the same vertices and number them in the side view.

• Shade oblique surface C in the side view. (Note that any surface appearing as a line in any view cannot be an oblique surface.)

• How many inclined surfaces are there on the part shown?_____

• How many normal surfaces?_____

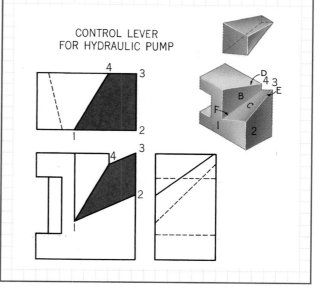

CONTROL LEVER FOR HYDRAULIC PUMP

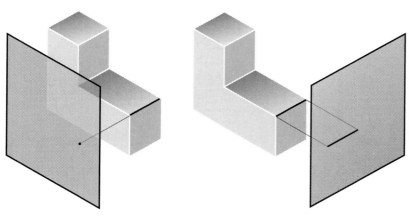

FIGURE 5.23 Projections of a Normal Edge.

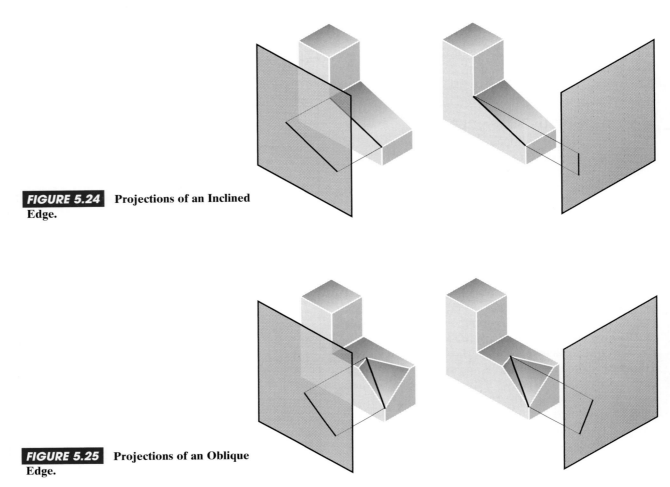

FIGURE 5.24 Projections of an Inclined Edge.

FIGURE 5.25 Projections of an Oblique Edge.

■ 5.19 INCLINED EDGES

An *inclined edge* is parallel to one plane of projection but inclined to adjacent planes. It appears as a true-length line on the plane to which it is parallel and as a foreshortened line on adjacent planes. The true-length view of an inclined line always appears as an angled line, but the foreshortened views appear as either vertical or horizontal lines. Figure 5.24 shows an inclined edge projected.

■ 5.20 OBLIQUE EDGES

An *oblique edge* is tipped to all planes of projection. Since it is not perpendicular to any projection plane, it cannot appear as a point in any standard view. Since it is not parallel to any projection plane, it cannot appear true-length in any standard view. An oblique edge appears foreshortened and as an angled line in every view. Figure 5.25 shows an oblique edge projected.

■ 5.21 ANGLES

If an angle is in a normal plane—in other words, is parallel to a plane of projection—the angle will be shown true size on the plane of projection to which it is parallel, as shown in Figure 5.26. If the angle is in an inclined plane, it may be projected either larger or smaller than the true angle, depending on its position. In Figure 5.26b the 45-degree angle is shown oversize in the front view, while in Figure 5.26c the 60-degree angle is shown undersize in both views.

A 90-degree angle will be projected true size, even if it is in an inclined plane, provided that one leg of the angle is a normal line. In Figure 5.26d the 60-degree angle is projected oversize and the 30-degree angle is projected undersize. Try this on your own using your own 30-degree or 60-degree triangle as a model, or even using the 90-degree corner of a sheet of paper. Tilt the triangle or paper to obtain an oblique view.

Practical Tip
Parallel planes

Parallel lines in space will be projected as parallel lines in any view unless they lie in the same plane and coincide to appear as a single line.

The figures at right show three views of an object after a plane has been passed through the points A, B, and C.

Only points that lie in the same plane are joined.

In the front view, you would join points A and C, which are in the same plane, extending the line to P on the extended vertical front edge of the block.

In the side view, you would join P to B, and in the top view join B to A.

The remaining lines are drawn parallel to lines AP, PB, and BA.

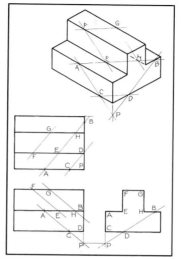
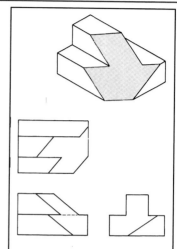

Another example that illustrates parallel lines intersected by a plane is shown at right. The projected views show that when two lines are parallel in space, their projections are parallel.

In figure b, the top plane of the object intersects the front and rear planes, to create parallel edges 1–2 and 3–4.

This is an example of the special case in which the two lines appear as points in one view and coincide as a single line in another. While this is a special case, it should not be considered an exception to the rule.

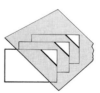
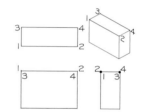

Parallel planes intersected by another plane
(a)

Lines 1,2 & 3,4 parallel, & parallel to horizontal plane
(b)

FIGURE 5.26 Angles.

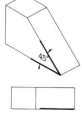

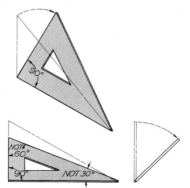

(a) ANGLE IN NORMAL PLANE

(b) ANGLE IN INCLINED PLANE

(c) ANGLE IN INCLINED PLANE

(d) PROJECTIONS OF THE ANGLES OF THE 30°x60° TRIANGLE.

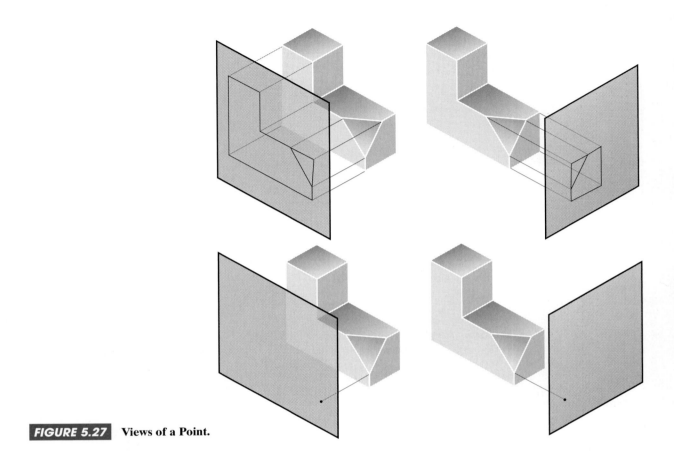

FIGURE 5.27   Views of a Point.

■ 5.22 VERTICES

A corner, or point, is the common intersection of three or more surfaces. A point appears as a point in every view. An example of a point on an object is shown in Figure 5.27.

■ 5.23 MEANING OF POINTS

A point located in a sketch can represent two things on the object:

- A vertex
- The point view of an edge (two vertices lined up one directly behind the other)

■ 5.24 MEANING OF LINES

A straight visible or hidden line in a sketch has three possible meanings, which are illustrated in Figure 5.28:

- An edge (intersection) between two surfaces
- The edge view of a surface
- The limiting element of a curved surface

Since no shading is used on a working drawing, you have to examine all the views to determine the meaning of the lines. For example, in Figure 5.28 you might believe the line AB at the top of the front view is the edge view of a flat surface if you were to look at only the front and top views. When you look at the right-side view, you can see that there is actually a curved surface on top of the object. Similarly, you might believe the vertical line CD in the front view is the edge view of a plane surface if you look at only the front and side views. However, an examination of the top view reveals that the line actually represents the intersection of an inclined surface.

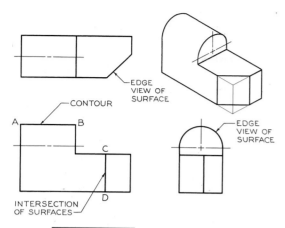

FIGURE 5.28   Meaning of Lines.

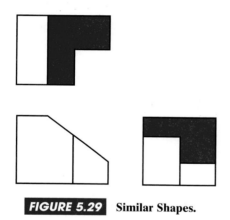

FIGURE 5.29   Similar Shapes.

■ 5.25 SIMILAR SHAPES OF SURFACES

If a flat surface is viewed from several different positions, each view will show the same number of sides and a similar shape. This repetition of shapes is useful in analyzing views. For example, the L-shaped surface shown in Figure 5.29 appears L-shaped in every view in which it does not appear as a line. A T-shaped, U-shaped, or hexagonal surface will in each case have the same number of sides and vertices and the same characteristic shape whenever it appears as a surface.

■ 5.26 INTERPRETING VIEWS

One method of interpreting sketches is to reverse the mental process used in projecting them. The views of an angle bracket are shown in Figure 5.30a.

Hands On 5.2
Adjacent views

In the top view shown here, lines divide the view into three adjacent areas. No two adjacent areas lie in the same plane because each line represents an edge (or intersection) between surfaces. While each area represents a surface at a different level, you can't tell whether A, B, or C is the highest surface or what shape the surfaces may be until you see the other necessary views of the object.

The same reasoning applies to the adjacent areas in any given view. Since an area or surface in a view can be interpreted in different ways, other views are necessary to determine which interpretation is correct.

Below are various shapes that the top view above might represent. Match the descriptions below to the shapes they describe at right. On your own paper, make a rough sketch of the front view for each description. Sketch two more possible interpretations for this top view and write their descriptions.

- Surface B is highest, and C and A are both lower

- One or more surfaces is cylindrical

- One or more surfaces is inclined

- Surface A is highest, and surfaces B and C are lower

- Surface A is highest and B is lower than C

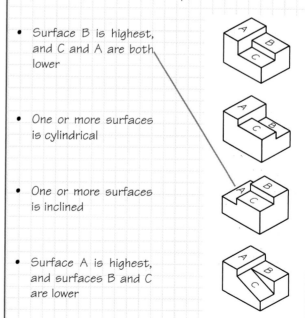

Hands On 5.3
Identifying surfaces

Each set of three views shows a shaded surface that is on edge or appears as a straight line in one view and shows its characteristic shape in the other two views. Shade in the surface in the other view where you see its shape and highlight the same surface where it appears on edge. Note how recognizing the similar surfaces helps you create a better mental picture of the three dimensional object.

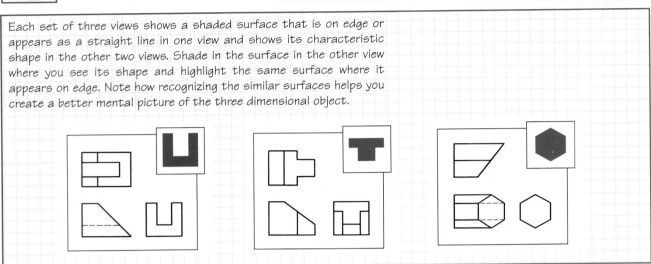

The front view, shown in Figure 5.30I, shows that the object's L-shape, its height and width, and the thickness of its members. The meaning of the hidden lines and centerlines is not yet clear, and you don't know the object's depth.

The top view, shown in Figure 5.30II, shows that the horizontal member is rounded at the right end and has a round hole. Some kind of slot is indicated at the left end. The depth and width of the object are shown.

The right-side view, shown in Figure 5.30III, shows that the left end of the object has rounded corners at the top and has an open-end slot in a vertical position. The height and depth of the object are shown.

In Figure 5.30 each view provides certain definite information regarding the shape of the object, and all are necessary to visualize the object completely.

■ 5.27 MODELS
One of the best aids to visualization is an actual model of the object. Models don't necessarily need to be made accurately or to scale. They may be made of any convenient material, such as modeling clay, soap, wood, wire, or Styrofoam, or any material that can easily be shaped, carved, or cut. Some examples of soap models are shown in Figure 5.31.

FIGURE 5.30 Visualizing from Given Views.

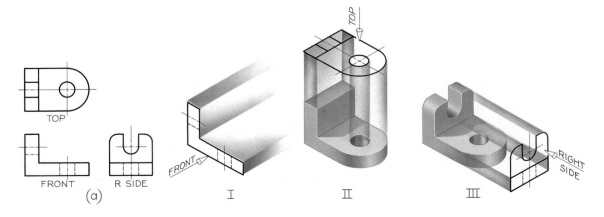

Step by Step 5.2
Reading a drawing

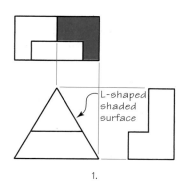 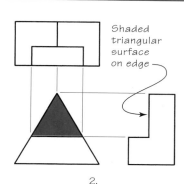 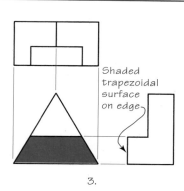

1.   2.   3.

1. Visualize the object shown by the three views at left. Since no lines are curved, we know that the object is made up of plane surfaces.

   The shaded surface in the top view is a six-sided L-shape. Since you do not see its shape in the front view—and every surface either appears as its shape or as a line—it must be showing on edge as a line in the front view. The indicated line in the front view also projects to line up with the vertices of the L-shaped surface.

   Because we see its shape in the top view and because it is an angled line in the front view, it must be an inclined surface on the object. This means it will show its foreshortened shape in the side view as well, appearing L-shaped and six-sided. The L-shaped surface in the right-side view must be the same surface that was shaded in the top view.

2. In the front view we see the top portion as a triangular-shaped surface, but no triangular shapes appear in either the top or the side view. The triangular surface must appear as a line in the top view and in the side view. Sketch projection lines from the vertices of the surface where you see its shape. The same surface in the other views must line up along the projection lines. In the side view, it must be the line indicated. That can help you to identify it as the middle horizontal line in the top view.

3. The trapezoidal-shaped surface shaded in the front view is easy to identify, but there are no trapezoids in the top and side views. Again the surface must be on edge in the adjacent views.

4. On your own, identify the remaining surfaces using the same reasoning. Which surfaces are inclined, which are normal. Are there any oblique surfaces?

FIGURE 5.31 Soap Models.

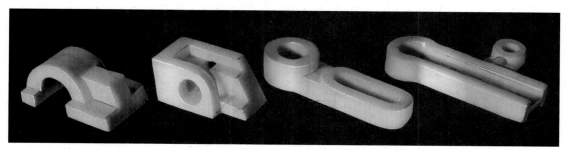

Step by Step 5.3
Making a model

Follow these steps for making a soap or clay model from projected views:

1. First, look at the three views of the object given. Make your block of clay to the same principal dimensions (height, width, and depth) as shown in the views.

2. Score lines on the frontal surface of your clay block to correspond with those shown on the front view in the drawing. Then do the same for the top and right-side views.

3. Slice straight along each line scored on the clay block to get a 3-D model that represents the projected views.

FIGURE 5.32 Precedence of Lines.

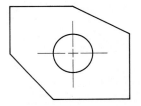

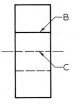

■ 5.28 PROJECTING A THIRD VIEW

Ordinarily when you are designing a product or system, you have a good mental picture of what the object you are sketching will look like from different directions. However, skill in projecting a third view can be useful for two reasons: One is that views must be shown in alignment in the drawing and projected correctly. The second is that practice in projecting a third view from two given views is an excellent way to develop your visual abilities.

Numbering the vertices on the object makes projecting a third view easy. Points that you number on the drawing represent points on the object where three surfaces come together to form a vertex (and sometimes a point on a contour or the center of a curve).

Once you have located a point in two drawing views, its location in the third view is known. In other words, if a point is located in the front and top view, its location in the side view is a matter of projecting the height of the point in the glass box from the front view and the depth of the point in the glass box from the top view.

In order to number the points or vertices on the object and show those numbers in different views, you need to be able to identify surfaces on the object. Then project (or find) the points in each new view surface-by-surface. You can use what you know about edges and surfaces to identify surfaces on the object when you draw views. This will help you to interpret drawings created by others as well as know how to project your own drawings correctly.

■ 5.29 SHOWING OTHER FEATURES

So far this chapter has presented information on how to project and interpret plane surfaces. Now you are ready to begin representing surfaces that are hidden from the direction of sight and surfaces that are curved.

PRECEDENCE OF LINES Visible lines, hidden lines, and centerlines (showing the axis of symmetry for contoured shapes, like holes) often coincide on a drawing and you have to decide which line to show. A visible line always takes precedence over and covers up a centerline or a hidden line when they fall over the top of each other in a view, as shown at A and B in Figure 5.32. A hidden line takes precedence over a centerline, as shown at C. At A and C the ends of the centerline are shown separated from the view by short gaps, but the centerline can be left off entirely.

G R A P H I C S   S P O T L I G H T
Multiview Projection from a 3D Model

3D VISUALIZATION

Viewing a three-dimensional object is a complex process, most of which you never need to think about in daily life because your brain is so effective for organizing and interpreting three-dimensional information.

When you create a three-dimensional model using a CAD system you control the factors that determine how the model appears on the screen. Several factors determine how the model appears; its distance from you (the zoom), the direction or angle from which you are looking at it, even the lighting. It can be challenging sometimes to produce the view you want. We are not used to giving specific directions about how to view everyday objects; we just move to change our view point.

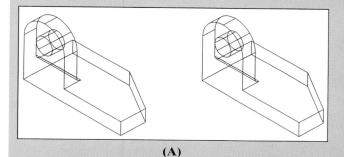

(A)

Some CAD systems automatically show the object in a standard arrangement of views for an engineering drawing. In others, you must control every factor, the distance from the object (or zoom), the viewing angle, the lighting, and where each view is placed on the monitor. Other CAD systems allow you to choose between standard arrangements and controlling views yourself.

WIREFRAMES VIEWING

Most 3D CAD systems draw the object on the screen as a wireframe drawing. They are called wireframe because the intersections or surfaces on the object are represented with lines, circles, and arcs on the screen, resulting in the appearance of a sculpture made of wires. It can be very difficult to tell whether you are

(B)

looking at a wireframe model from above or below. Figure A shows examples of two different wireframe drawings. They appear the same, but they are actually two different models viewed from two different viewing directions. Can you tell the difference between them?

When you are creating a 3D model on a computer, the screen is flat and only flat 2D views of the object can be seen on the screen. You must interpret multiple views of the object to understand the 3D model, the same way you do with paper drawings. Fortunately most 3D CAD systems can make this easier by allowing you to shade or render the views of the object on your screen to make visualization easier.

The same two models are shown in Fig. B rendered so that you easily interpret the viewing direction. As you can tell, the model on the right was viewed from below. Yet it appears the same as the mirror image of the model as viewed from above.

WITH 3D CAD ANY VIEW CAN BE CREATED

Once you have learned how to create 3D objects using your CAD system, you can produce any view you would like of the object. This can save you a lot of time in creating drawings. You can generate any view you want by viewing the object from the proper direction. You may want to view the object from an angle to show an angled or oblique surface true size. While this may be a time consuming process to draw a 2D projection, you can usually do it in one or two steps using 3D CAD. In addition to saving time, 3D modeling can be a useful visualization tool. Shaded models are often easier for people unfamiliar with engineering drawings to interpret.

Step by Step 5.4
Projecting a third view

Follow the steps to project a third view.

The figure at right is a pictorial drawing of an object that has three necessary views. It has numbers on it identifying each corner (vertex) of the object and letters identifying some of the major surfaces. You are given the top and front view. You will use point numbers to project the side view.

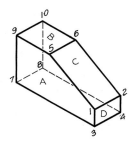

1. To number points effectively you first need to identify surfaces and interpret the views that are given. First label visible surfaces that have an easy to identify shape in one view. Then locate that same surface in the adjacent view. (The surfaces on the pictorial object have been labeled to make it easy.)

2. Surface A in the front view is a normal surface. It will appear as a horizontal line in the top view. The two rectangular surfaces, B and C in the top view are a normal surface and an inclined surface. They will show as a horizontal line and an inclined line in the front view, respectively.

3. Once you have identified the surfaces, label the vertices of a surface that has an easy to recognize shape, in this case surface A.

   Label its vertices with numbers at each corner as shown. If a point is directly visible in the view, place the number outside the corner

   If the point not directly visible in that view, place the numeral inside the corner. Using the same numbers to identify the same points in different views will help you to project known points in two views to unknown positions in a third view.

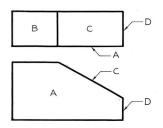

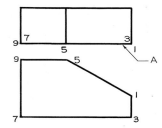

4. Continue on surface by surface until you have numbered all of the vertices in the given views as shown below. Do not use two different numbers for the same vertex as it can be confusing.

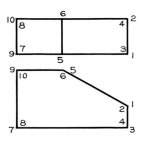

5. Try to visualize the right-side view you will create. Then construct the right-side view point by point, using very light lines. Locate point 1 in the side view by drawing a light horizontal projection line from point 1 in the front view. Use the edge view of surface A in the top view as a reference plane to transfer the depth location for point 1 to the side view as shown in the figure below.

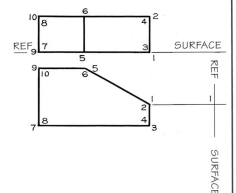

Follow steps 6 through 9, shown on the next page ⟶

Step by Step 5.4
Projecting a third view, cont.

6. Project points 2, 3, and 4 in a similar manner to complete the vertical end surface of the object.

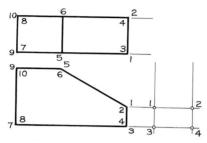

7. Project the remaining points using the same method, proceeding surface by surface.

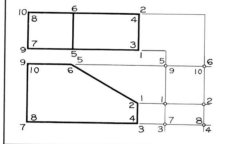

8. Use the points that you have projected into the side view to draw the surfaces of the object as in the example at right.

   If surface A extended between points 1-3-7-9-5 in the front view where you can see its shape clearly, it will extend between those same points in every other view.

   When you connect these points in the side view, they form a vertical line.

   This makes sense, because A is a normal surface. As is the rule with normal surfaces, you will see its shape in one standard view (the front in this case) and it will appear as a horizontal or vertical line in the other views.

   Continue connecting vertices to define the surfaces on the object, to complete the third view.

9. Inspect your drawing to see if all of the surfaces are shown and darken the final lines.

   You should also consider the visibility of surfaces. Surfaces that are hidden behind other surfaces should be shown with hidden lines.

   You will learn more about using hidden lines next.

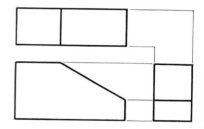

■ 5.30 HIDDEN LINES

One advantage of views over photographs is that each view can show the entire object from that viewing direction. While a photograph shows only the visible surface of an object, a view shows the object all the way through, as if it were transparent. Thick, dark lines are used to represent features of the object that are directly visible. Dashed hidden lines are used to represent features that would be hidden behind other surfaces.

Remember to choose views that show features with visible lines when you can. Use hidden lines wherever they are needed to make the drawing clear.

You can save time and reduce clutter by leaving out hidden lines that aren't necessary as long as you are certain that the remaining hidden lines describe the object clearly and completely. In a note, let the reader know that the lines were left out intentionally and that it is not a drawing error.

Sketch hidden lines by eye using thin dark dashes about 5 mm long and spaced about 1 mm apart. Hidden lines should be as dark as other lines in the drawing, but drawn with a thin line.

When hidden lines intersect each other in the drawing, their dashes should meet. In general, hidden lines should intersect neatly with visible lines at the edge of an object.

Some of the practices you will learn for representing intersections of hidden lines with other lines may be difficult to follow when using CAD. When using CAD, you should adjust the line patterns so that the hidden lines in your drawing have the best appearance possible. Step by Step 5.5 provides additional details about showing hidden line intersections correctly.

Tear out Worksheet 5.5. Use it to practice drawing with hidden lines.

Step by Step 5.5
Correct and incorrect practices for hidden lines

- Make a hidden line join a visible line, except when it causes the visible line to extend too far, as shown here.

- Leave a gap whenever a hidden line is a continuation of a visible line.

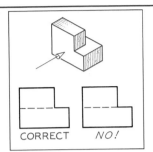

- Make hidden lines intersect to form L and T corners.

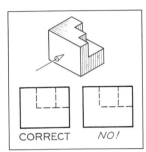

- Make a hidden line "jump" a visible line when possible.

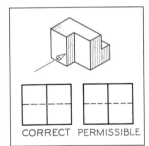

- Draw parallel hidden lines so that the dashes are staggered, as in bricklaying.

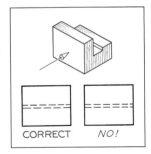

- When two or three hidden lines meet at a point, join the dashes, as shown for the bottom of this drilled hole.

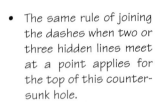

- The same rule of joining the dashes when two or three hidden lines meet at a point applies for the top of this countersunk hole.

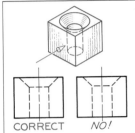

- Hidden lines should not join visible lines when this makes the visible line extend too far.

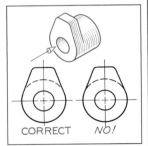

- Draw hidden arcs like the upper example. Not like the example below.

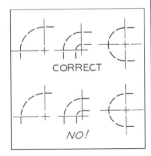

Accent Dashes

Accent the beginning and end of each dash by pressing down on the pencil. Make hidden lines as tidy as you can so they are easy to interpret. Be sure to make hidden line dashes longer than gaps so they clearly represent lines.

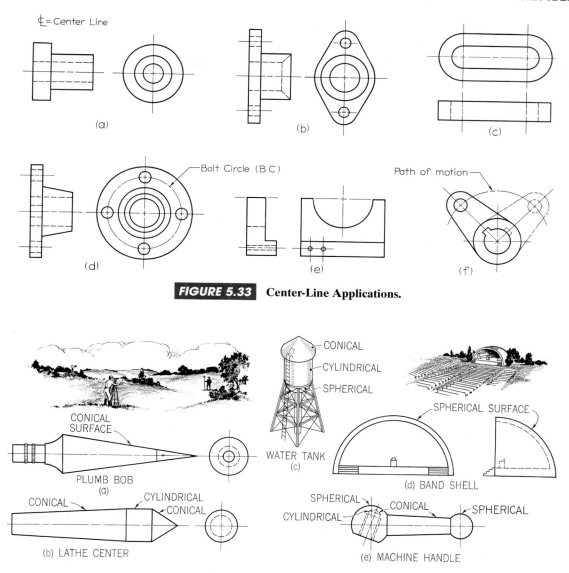

FIGURE 5.33 Center-Line Applications.

FIGURE 5.34 Curved Surfaces.

■ 5.31 CENTERLINES

Centerlines (symbol: ₵) are used to indicate symmetrical axes of objects or features, bolt circles, and paths of motion. Typical applications are shown in Figure 5.33. Centerlines are useful in dimensioning. They are not needed on unimportant rounded or filleted corners or on other shapes that are self-locating.

■ 5.32 CURVED SURFACES

Some examples of the most common rounded surfaces found in engineering—the cylinder, cone, and sphere—are shown in Figure 5.34. The cylinder is the most common rounded shape. Cylindrical shapes are easily produced with common manufacturing equipment such as drills and lathes.

 Tear out Worksheet 5.6. Use it to practice with centerlines.

Step by Step 5.6
Sketching centerlines

- As in the examples shown, sketch a single centerline in the longitudinal view, and crossed centerlines in the circular view.

- Make sure the small dashes cross at the intersections of the centerlines.

- Make centerlines extend uniformly about 8 mm outside the feature for which they are drawn.

- The long dashes of centerlines may vary from 20 to 40 mm or more in length, depending on the size of the drawing.

- Make the short dashes about 5 mm long, with spaces of about 2 mm.

- Always start and end centerlines with long dashes.

- Always leave a gap when a centerline forms a continuation of a visible or hidden line.

- Make centerlines thin enough to contrast well with the visible and hidden lines, but dark enough that they can be seen clearly and reproduce well.

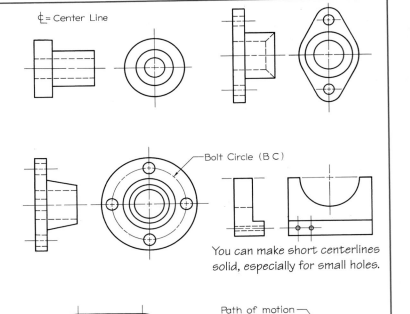

₵ = Center Line

Bolt Circle (B C)

You can make short centerlines solid, especially for small holes.

Path of motion

Hands On 5.4
Practice sketching centerlines

Practice sketching centerlines
on the features shown.

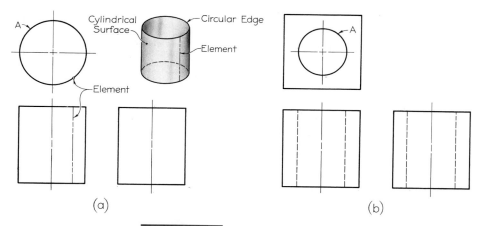

FIGURE 5.35   Cylindrical Surfaces.

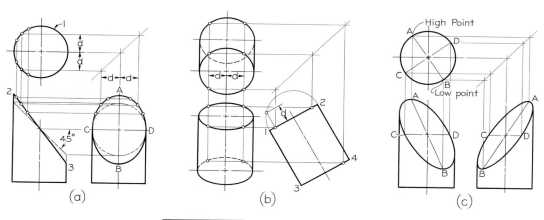

FIGURE 5.36   Deformities of Cylinders.

■ 5.33 CYLINDRICAL SURFACES

Three views of a right-circular cylinder are shown in Figure 5.35a. The single cylindrical surface has flat top and bottom surfaces, which are bounded by circular edges, the only actual edges on the cylinder.

Figure 5.35b shows three views of a cylindrical hole in a right square prism. The cylinder is represented on a drawing by its circular edges and the limiting contour elements. An element is a straight line on the cylindrical surface, parallel to the axis. An element of the cylinder is labeled in the pictorial view of the cylinder in Figure 5.35a. The circular edges of the cylinder appear in the top views as circles A, and in the front and side views as horizontal lines.

Elements of the cylinder appear as points in the top view. The contour elements in the side views appear as points in the top views.

■ 5.34 CYLINDERS AND ELLIPSES

If a cylinder is cut by an inclined plane, as shown in Figure 5.36a, the inclined surface is bounded by an ellipse. This ellipse will appear as a circle in the top view, as a straight line in the front view, and as an ellipse in the side view. Note that circle 1 would remain a circle in the top view regardless of the angle of the cut. If the cut is 45 degrees from the horizontal, it will also appear as a circle in the side view.

When a circular shape is shown inclined in another view, project it into the adjacent view as shown in Figure 5.36 and use the techniques you learned in Chapter 4 to sketch the ellipse. The standard views do not show the true size and shape of inclined surfaces such as the ellipse in Figure 5.36. You will learn in Chapter 8 how to create auxiliary views to show the true size and shape of inclined surfaces like these.

Step by Step 5.7
Space curves

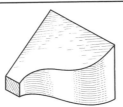

Follow the steps below in plotting a space curve (an irregular curve).

1. Establish the views of a space curve by identifying arbitrary points along the curve, where its shape shows clearly. In this case it is the top view.

2. Project those same points to the adjacent view.

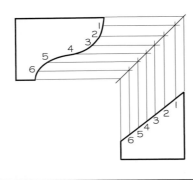

3. Project each point from the two established views into the remaining view. The point will be located where the projection lines intersect.

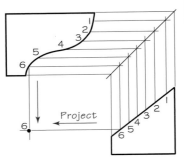

4. Sketch the resulting curve through the points. Identify and project the remaining surfaces.

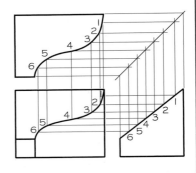

■ 5.35 INTERSECTIONS AND TANGENCIES

No line is drawn where a curved surface is tangent to a plane surface, as in Figure 5.37a, but when a curved surface intersects a plane surface, as in Figure 5.37b, a definite edge is formed. If curves join each other and plane surfaces smoothly (i.e., they are tangent), no line is drawn to show where they come together, as is shown in Figure 5.37c. If a combination of curves creates a vertical surface, as in Figure 5.37d, the vertical surface is shown as a line.

When plane surfaces join a contoured surface, they don't show a line if they are tangent, but do show a line if they intersect. Examples of planes joining contoured surfaces are given in Figures 5.37e–5.37f.

Figure 5.38a shows an example of a small cylinder intersecting a large cylinder. When the intersection is small, its curved shape is not plotted accurately because it adds little to the sketch or drawing for the time it takes. Instead it is shown as a straight line. When the intersection is larger, it can be approximated by drawing an arc with the radius the same as that of the large cylinder, as shown in Figure 5.38b. Large intersections can be plotted accurately by selecting points along the curve to project, as shown in Figure 5.38c. When the cylinders are the same diameter, their

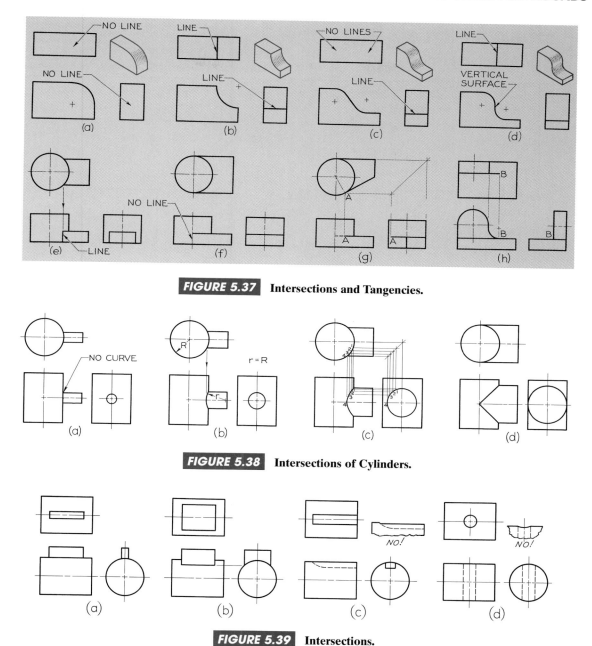

FIGURE 5.37   Intersections and Tangencies.

FIGURE 5.38   Intersections of Cylinders.

FIGURE 5.39   Intersections.

intersection appears as straight lines in the adjoining view, as shown in Figure 5.38d.

Figure 5.39a and 5.39b show similar examples of a narrow prism intersecting a cylinder. Figure 5.39c and 5.39d show the intersections of a keyseat and cylinder and a small hole and cylinder.

■ 5.36 FILLETS AND ROUNDS

A rounded interior corner is called a *fillet*. A rounded exterior corner is called a *round* (Figure 5.40a). Sharp corners are usually avoided in designing parts to be cast or forged because they are difficult to produce and can weaken the part.

Step by Step 5.8
Representing holes

Here are the methods for representing most common types of machined holes. In general, you do not need to give instructions to the machinist describing how to form the holes. Unless you require a specific process, it is usually best to let the experienced manufacturing technician determine which will be the most efficient and cost-effective way to produce the part. If you do need a specific process, use notes to tell what to do and in which order to do it.

- Always specify hole sizes by diameter—never by radius.

- If a hole is to be drilled with the upper part enlarged conically to a specified angle and diameter—called countersinking—the angle is commonly 82 degrees, but it is drawn at 90 degrees for simplicity.

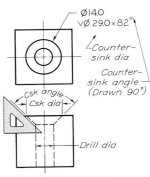

DRILLED & COUNTER-SUNK HOLE

- A hole that goes through a member is called a through hole.

- If a hole is to be drilled with the upper part enlarged cylindrically to a specified diameter, usually to produce a smooth bearing surface (called spotfacing), the depth usually is not specified but is left to the shop to determine For average cases, draw the depth as 1.5 mm (1/16")

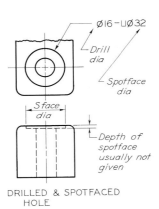

DRILLED & SPOTFACED HOLE

- A hole with a specified depth is called a blind hole. In a blind hole, the depth includes the cylindrical portion of the hole only. The point of the drill leaves a conical bottom in the hole, drawn at approximately a 30° angle.

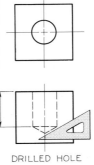

DRILLED HOLE

- If a hole is to be drilled and the upper part is to be enlarged cylindrically to a specified diameter and depth, called counterboring, it should be specified like this:

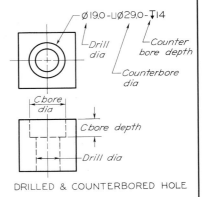

DRILLED & COUNTERBORED HOLE

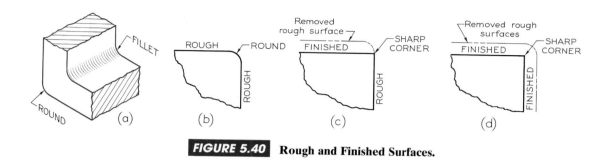

FIGURE 5.40 Rough and Finished Surfaces.

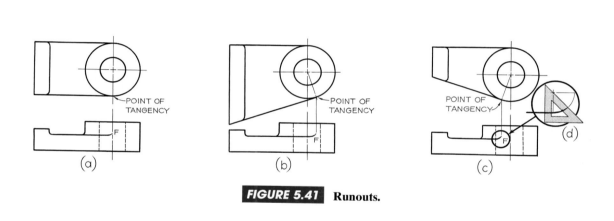

FIGURE 5.41 Runouts.

Two intersecting rough surfaces produce a rounded corner (Figure 5.40b). If one of these surfaces is machined, as shown in Figure 5.40c, or if both surfaces are machined, as shown in Figure 5.40d, the corner becomes sharp. In drawings, a rounded corner means that both intersecting surfaces are rough. A sharp corner means that one or both surfaces are machined. Do not shade fillets and rounds on multiview drawings. The presence of the curved surfaces is indicated only where they appear as arcs, unless it is done to call attention to them, as shown in Figure 5.44.

■ 5.37 RUNOUTS

Small curves called *runouts* are used to represent fillets that connect with plane surfaces tangent to cylinders, as shown in Figure 5.41a–5.41d. The runouts, labelled F, should have a radius equal to that of the fillet and a curvature of about one-eighth of a circle, as shown in Figure 5.41d.

The runouts from different filleted intersections will appear differently due to the shapes of the horizontal intersecting members. Figure 5.42 shows more examples of conventional representations for fillets, rounds, and runouts. In Figures 5.42e and 5.42f the runouts differ because the top surface of the web is flat in Figure 7.42e, while the top surface of the web in Figure 5.42f is considerably rounded.

When two different sizes of fillets intersect, the direction of the runout is dictated by the larger fillet, as shown in Figures 5.42g and 5.42j.

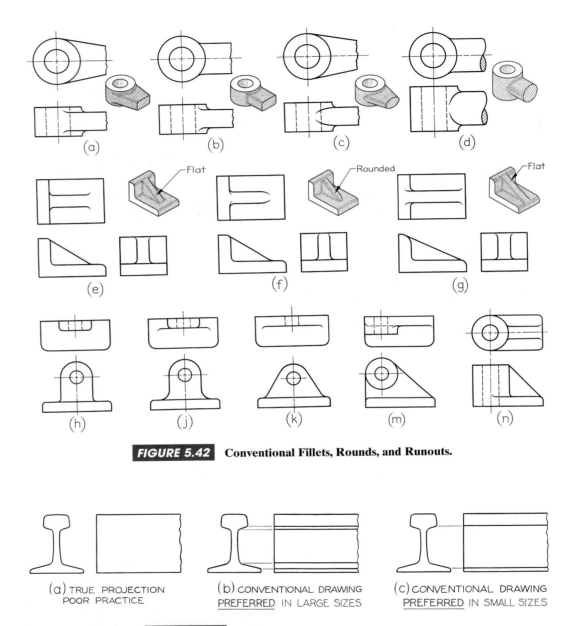

FIGURE 5.42   Conventional Fillets, Rounds, and Runouts.

FIGURE 5.43   Conventional Representation of a Roll.

■ 5.38 CONVENTIONAL EDGES

Rounded and filleted intersections eliminate sharp edges and can make it difficult to present the shape clearly. In some cases, as shown in Figure 5.43a, the true projection may be misleading. Added lines depicting rounded and filleted edges, as shown in Figures 5.43b and 5.43c, give a clearer representation, even though it is not the true projection. Project the added lines from the intersections of the surfaces as if the fillets and rounds were not present.

Figure 5.44 shows top views for each given front view. The first set of top views have very few lines,

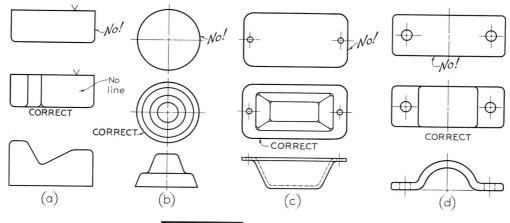

FIGURE 5.44   Conventional Edges.

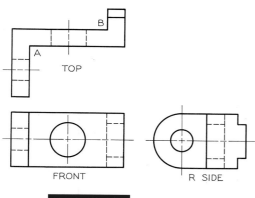

FIGURE 5.45   Three Views.

even though they are the true projection. The second set of top views, where lines are added to represent the rounded and filleted edges, are quite clear. Note the use of small Y's where rounded or filleted edges meet a rough surface. If an edge intersects a finished surface, no Y is shown.

■ 5.39   NECESSARY VIEWS

What are the absolute minimum views required to completely define an object? For example, in Figure 5.45, the top view might be omitted, leaving only the front and right-side views. However, it is more difficult to read the two views or visualize the object because the characteristic Z shape of the top view is omitted. In addition, you must assume that corners A and B in the top view are square and not rounded. In this example, all three views are necessary.

■ 5.40   PARTIAL VIEWS

A view may not need to be complete, but only need to show what is necessary to clearly describe the object. This is called a partial view and is used to save sketching time. You can use a break line to limit the partial view, as shown in Figure 5.46a, or limit a view by the contour of the part shown, as shown in Figure 5.46b. If the view is symmetrical, you can draw a half-view on one side of the centerline, as shown in Figure 5.46c, or break out a partial view, as shown in Figure 5.46d. The half-views should be the near side, as shown.

When drawing a partial view, do not place a break line where it will coincide with a visible or hidden line, as this may cause the drawing to be misinterpreted.

Occasionally the distinctive features of an object are on opposite sides. In either complete side view there will be a considerable overlapping of shapes. In

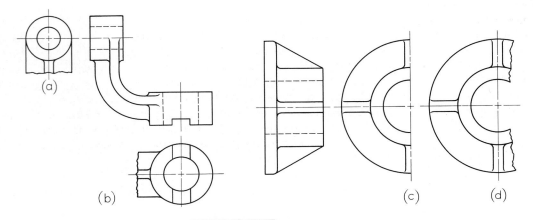

FIGURE 5.46   Partial Views.

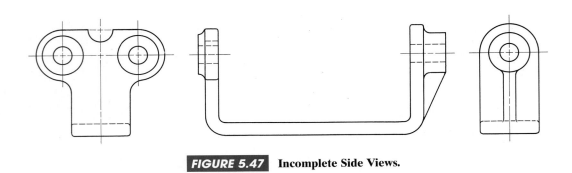

FIGURE 5.47   Incomplete Side Views.

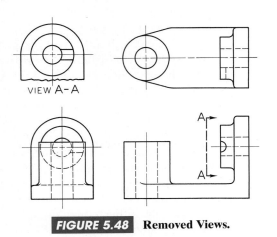

FIGURE 5.48   Removed Views.

cases like this, two side views are often the best solution, as shown Figure 5.47. The views are partial views, and certain visible and hidden lines have been omitted for clarity.

■ 5.41 REMOVED VIEWS

A removed view is a complete or partial view removed to another place on the sheet so that it no longer is in direct projection with any other view, as shown in Figure 5.48. A removed view may be used to show a feature of the object more clearly, possibly to a larger scale, or to save drawing a complete regular view. A viewing-plane line is used to indicate the part being viewed. The arrows at the corners show the direction of sight. The removed views should be labeled view A–A or view B–B and so on; the letters refer to those placed at the corners of the viewing-plane line.

Hands On 5.5
Sketching three views

A pictorial sketch of a lever bracket that requires three views is shown. Follow the steps to sketch the three views:

1. Block in the enclosing rectangles for the three views. You can either use overall proportions by eye or if you know the dimensions you can use your scale to sketch accurately sized views. Spacing your views equally from the edge of the rectangle and from each other, sketch horizontal lines to establish the height of the front view and the depth of the top view. Sketch vertical lines to establish the width of the top and front views and the depth of the side view. Make sure that this is in correct proportion to the height, and remember to maintain a uniform space between views. Remember that the space between the front and right-side view is not necessarily equal. Transfer the depth dimension from the top view to the side view, use the edge of a strip of paper or a pencil as a measuring stick. The depth in the top and side views should always be equal.

2. Block in all details lightly.

3. Sketch all arcs and circles lightly.

4. Darken all final lines.

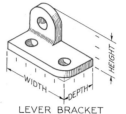

LEVER BRACKET

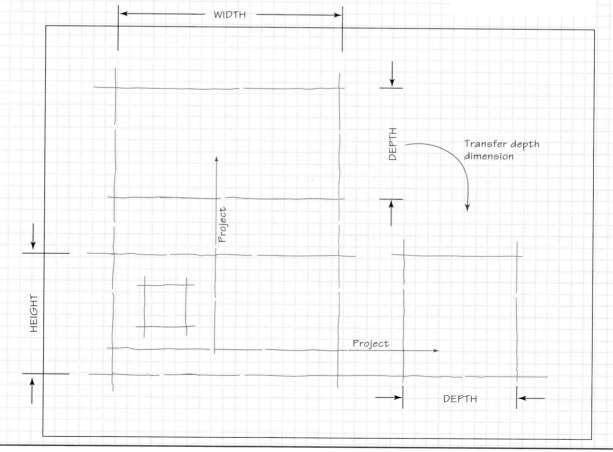

Step by Step 5.9
Eliminating unnecessary views

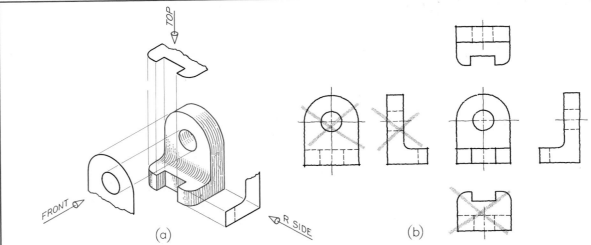

(a)

(b)

Three distinctive features of this object need to be shown on the drawing:

1. The rounded top and hole, seen from the front

2. The rectangular notch and rounded corners, seen from the top

3. The right angle with filleted corner, seen from the side

Both the front and rear views show the true shapes of the hole and the rounded top, but the front view is preferred because it has no hidden lines. Cross out the rear view.

The top and bottom views show the rectangular notch and rounded corners, but the top view is preferred because it has fewer hidden lines. Cross out the bottom view.

Both the right-side and left-side views show the right angle with the filleted corner. In fact, in this case the side views are identical, except reversed, so the right-side view is chosen. Cross out the left-side view.

In this example the three remaining views should be the top, front, and right-side—the three regular views.

■ 5.42 ALIGNMENT OF VIEWS

Always draw views in the standard arrangement shown in Figure 5.2 to be sure that your drawings are not misinterpreted. Figure 5.49a shows an offset guide that requires three views. Their correct arrangement is shown in Figure 5.49b. The top view should be directly above the front view, and the right-side view directly to the right of the front view—not out of alignment, as in Figure 5.49c. Never draw the views in reversed positions, with the bottom over the front view or the right-side to the left of the front view, as shown in Figure 5.49d. Even though the views do line up with the front view, this arrangement could be misread.

After design sketches are completed, you will usually follow them with detailed CAD drawings. In finished CAD drawings you should apply the same rules

for arranging views, clearly depicting the subject of the drawing, using the proper line patterns and line weights, and following all of the necessary standards as used in manually created drawings (Figure 5.50). Many programs allow you to select a standard arrangement of views produced directly from your 3-D CAD model. Because CAD makes it easy to move whole views, it is tempting to place views where they fit on the screen or plotted sheet and not in the standard arrangement. This is not an acceptable practice.

■ 5.43 FIRST- AND THIRD-ANGLE PROJECTION

As you saw earlier in this chapter, you can think of the system of projecting the views as unfolding a glass box made from the viewing planes. There are two main

Hands On 5.6
Cross out the unnecessary views

Using Step by Step 5.9 as an example, look carefully at the projections in the space at right and decide which views are necessary to accurately describe the object.

Cross out any views that are unnecessary.

FIGURE 5.49   Position of Views.

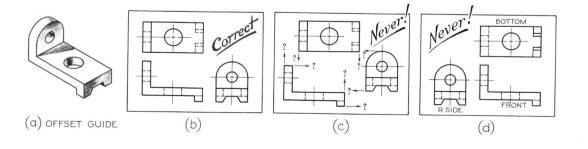

(a) OFFSET GUIDE   (b)   (c)   (d)

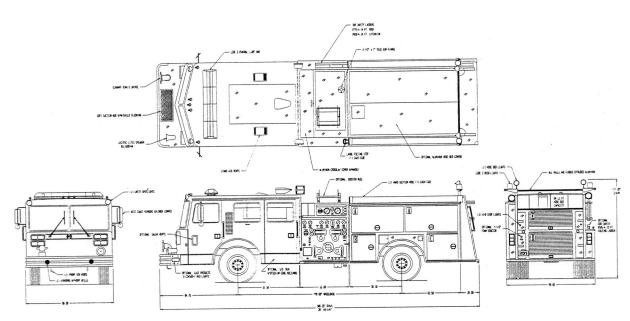

FIGURE 5.50   **Multiview CAD Assembly Drawing of a MAXIM Fire Truck.** *Courtesy of CADKEY*

systems used for projection and unfolding the views: *third-angle projection*, which is used in the United States, Canada, and some other countries, and *first-angle projection*, which is primarily used in Europe and Asia. Because of the global nature of engineering careers, you should thoroughly understand both methods. Figure 5.51 shows a comparison between first-angle orthographic projection and third-angle orthographic projection. Difficult interpretation of the drawing and even manufacturing errors may result when a first-angle drawing is confused with a third-angle drawing.

To avoid misunderstanding, international projection symbols have been developed to distinguish between first-angle and third-angle projections on drawings. The symbol in Figure 5.51 shows two views of a truncated cone. You can examine the arrangement of the views in the symbol to determine whether first- or third-angle projection was used. On international drawings you should be sure to include this symbol.

To understand the two systems, think of the vertical and horizontal planes of projection, shown in Figure 5.52a, as indefinite in extent and intersecting at 90 degrees with each other; the four angles produced are called the first, second, third, and fourth angles (similar to naming quadrants on a graph.) The profile plane is placed so that it intersects these two planes at 90 degrees. If the object to be drawn is placed below the horizontal plane and behind the vertical plane, as in the glass box you saw earlier, the object is said to be in the third angle. In third-angle projection, the views are produced as if the observer is outside, looking in.

If the object is placed above the horizontal plane and in front of the vertical plane, the object is in the

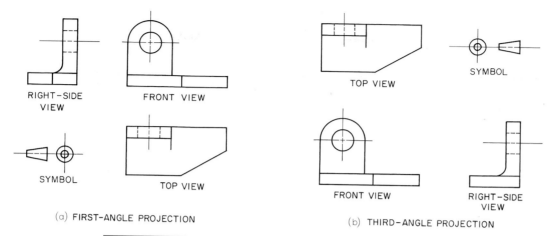

RIGHT−SIDE VIEW

FRONT VIEW

TOP VIEW

SYMBOL

TOP VIEW

SYMBOL

FRONT VIEW

RIGHT−SIDE VIEW

(a) FIRST−ANGLE PROJECTION

(b) THIRD−ANGLE PROJECTION

FIGURE 5.51   First-Angle Projection Compared to Third-Angle Projection.

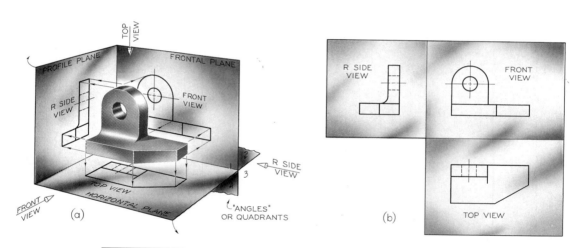

FIGURE 5.52   First-Angle Projection. An object that is above the horizontal plane and in front of the vertical plane is in the first angle. An observer looks through the object to the planes of projection.

first angle. In first-angle projection the observer looks through the object to the planes of projection. The right-side view is still obtained by looking toward the right side of the object, the front by looking toward the front, and the top by looking down toward the top; but the views are projected from the object onto a plane in each case. The biggest difference between third-angle projection and first-angle projection is in how the planes of the glass box are

unfolded, as shown in Figure 5.52b. In first-angle projection, the right-side view is to the left of the front view, and the top view is below the front view, as shown.

You should understand the difference between the two systems and know the symbol that is placed on drawings to indicate which has been used. Keep in mind that you should use third-angle projection throughout this book.

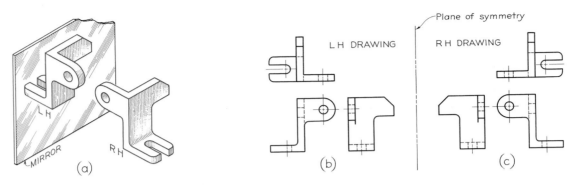

FIGURE 5.53 Right-Hand and Left-Hand Parts.

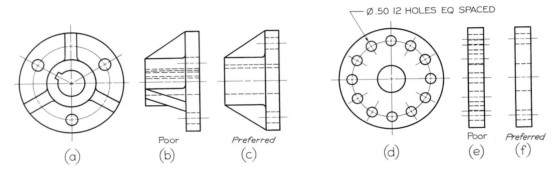

FIGURE 5.54 Revolution Conventions.

■ 5.44 RIGHT-HAND AND LEFT-HAND PARTS

Often individual parts function in pairs where opposite parts are similar. But opposite parts can rarely be exactly alike. For example, the right-front fender of an automobile cannot be the same shape as the left-front fender. A left-hand part is not simply a right-hand part turned around; the two parts are mirror images and are not interchangeable.

On sketches and drawings a left-hand part is noted as LH, and a right-hand part as RH. In Figure 5.53a, the part in front of the mirror is a right-hand part, and the image shows the left-hand part. No matter how the object is turned, the image will show the left-hand part. Figures 5.53b and 5.53c show left-hand and right-hand drawings of the same object.

Ordinarily you draw only one of two opposite parts and label the one that is drawn with a note, such as LH PART SHOWN, RH OPPOSITE. If the opposite-hand shape is not clear, you should make a separate sketch or drawing to show it clearly and completely.

■ 5.45 REVOLUTION CONVENTIONS

Regular multiview projections are sometimes awkward, confusing, or actually misleading. For example, Figure 5.54a shows an object that has three triangular ribs, three holes equally spaced in the base, and a keyway. The right-side view is a regular projection and is not recommended—the lower ribs appear in a foreshortened position, the holes do not appear in their true relation to the rim of the base, and the keyway is projected as a confusion of hidden lines.

The conventional method shown in Figure 5.54c is preferred because it is simpler to read and requires less time to sketch. Each of the features mentioned has been revolved in the front view to lie along the vertical centerline, from where it is projected to the correct side view.

Figures 5.54d and 5.54e show regular views of a flange with many small holes. The hidden holes are confusing and take unnecessary time to show. The preferred representation in Figure 5.54f shows the holes revolved for clarity.

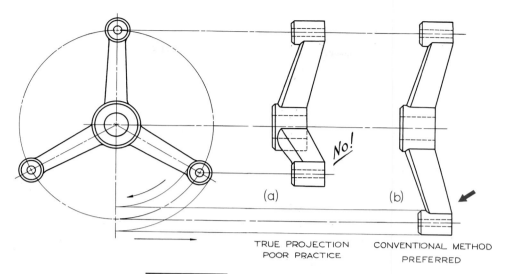

FIGURE 5.55   Revolution Conventions.

Figure 5.55 shows a regular projection with a confusing foreshortening of the inclined arm. To make the object's symmetry clear, the lower arm is revolved to line up vertically in the front view so that it projects the true length in the side view, shown in Figure 5.55b.

Revolutions of the type discussed here are frequently used in connection with sectioning. Such sectional views are called aligned sections.

■ 5.46 FITTING VIEWS ON PAPER

With the many details that must be shown to represent complex objects clearly, you need to use paper *sheet sizes* and drawing scales that allow your information to come across clearly to the reader. For example, if the part you are drawing is very small, you may need to show it oversized. If you are drawing a large system, you may need to represent the details at one scale and the overall plan at a smaller scale.

■ 5.47 DRAWING MEDIUMS

You should find a drawing medium that works for you. There are many different types of paper, vellum, and other tracing papers available. Vellum and blue fade-out grid paper are popular and come in a variety of sizes. Some companies provide vellum, paper, and engineering notebooks in standard sizes printed with the company's border and title block.

Polyester film, or *Mylar*, is a superior drafting material because it erases cleanly, is transparent, is durable, and has high-dimensional stability, not stretching or shrinking. Many companies plot their

Nearest International Size a (millimeter)	Standard U.S. Size a (inch)
A4 210 × 297	A 8.5 × 11.0
A3 297 × 420	B 11.0 × 17.0
A2 420 × 594	C 17.0 × 22.0
A1 594 × 841	D 22.0 × 34.0
A0 841 × 1189	E 34.0 × 44.0

[a] ANSI Y14.1m-1992.

TABLE 5.1   *Sheet Sizes.*

final CAD drawings on polyester film and store them in a central location where blueprints or copies can be made quickly for distribution.

■ 5.48 STANDARD SHEETS

Paper and Mylar comes in rolls and standard sheet sizes. Two systems of sheet sizes, together with length, width, and letter designations, are listed by ANSI, as shown in Table 5.1.

The use of the basic sheet size, 8-1/2 x 11 inches or 210 x 297 mm, and multiples thereof permits filing of small tracings and of folded prints in standard files with or without correspondence. These sizes can be cut without waste from the standard rolls of paper, cloth, or film.

For layout designations, title blocks, revision blocks, and a list of materials blocks see the sheet layouts at the front of this book.

(a) Metric Scale

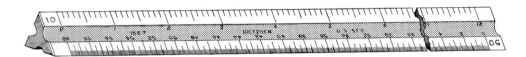

(b) Engineers' Scale

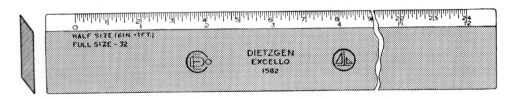

(c) Decimal Scale

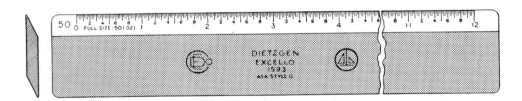

(d) Mechanical Engineers' Scale

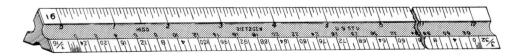

(e) Architects' Scale

FIGURE 5.57   Types of Scales.

FIGURE 5.56 **Printed Circuit Board.** *United Nations/Guthrie.*

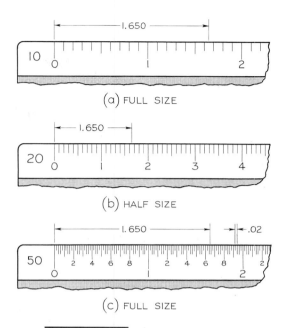

FIGURE 5.58 Decimal Dimensions.

■ 5.49 SCALE

A drawing of an object may be the same size as the object (full size), or it may be larger or smaller than the object. The scale you choose depends on the size of the object and the size of the sheet of paper you will use. For example, a machine part may be drawn half size, a building may be drawn 1/48th size, a map may be drawn 1/1,200th size, or a printed circuit board may be drawn four times its size (Figure 5.56).

Measuring scales are used in making technical drawings full size, enlarged, or reduced. Figure 5.57 shows (a) the *metric scale*, (b) the *engineers' scale,* (c) the *decimal scale,* (d) the *mechanical engineers' scale*, and (e) the *architects' scale*. On full-divided scales the basic units are subdivided throughout the length of the scale. On open divided scales, such as architects' scales, only the end unit is subdivided.

■ 5.50 DECIMAL SCALES

On the full-size decimal scale, each inch is divided into fiftieths of an inch, or .02 inch (Figure 5.58c), and on the half- and quarter-size scales, the inches are compressed to half size or quarter size and then are divided into 10 parts, so that each subdivision stands for .1 inch. You will learn more about the decimal inch system of dimensioning in Chapter 9. Engineers' scales are a type of decimal scale. Their units of 1 inch are broken down into 10, 20, 30, 40, 50, and 60 parts, making it useful for decimal dimensions. To measure 1.65 inch

full size on the 10-scale use one main division plus six subdivisions (Figure 5.58a). To show the same dimension half scale, use the 20-scale, which is half of the 10-scale (Figure 5.59b). To show it quarter size, use the 40-scale.

The engineers' scale is also used in drawing maps to scales of 1 in. = 50 ft., 1 in. = 500 ft., and 1 in. = 5 miles and in drawing stress diagrams or other graphical constructions to such scales as 1 in. = 20 lb. and 1 in. = 4,000 lb.

■ 5.51 METRIC SCALES

Metric equivalents are listed below:

$$1 \text{ mm} = 1 \text{ millimeter } (\tfrac{1}{1000} \text{of a meter})$$

$$1 \text{ cm} = 1 \text{ centimeter } (\tfrac{1}{100} \text{ of a meter})$$

$$= 10 \text{ mm}$$

$$1 \text{ dm} = 1 \text{ decimeter } (\tfrac{1}{10} \text{ of a meter})$$

$$= 10 \text{ cm} = 100 \text{ mm}$$

$$1 \text{ m} = 1 \text{ meter}$$

$$= 100 \text{ cm} = 1000 \text{ mm}$$

$$1 \text{ km} = 1 \text{ kilometer} = 1000 \text{ m}$$

$$= 100,000 \text{ cm} = 1,000,000 \text{ mm}$$

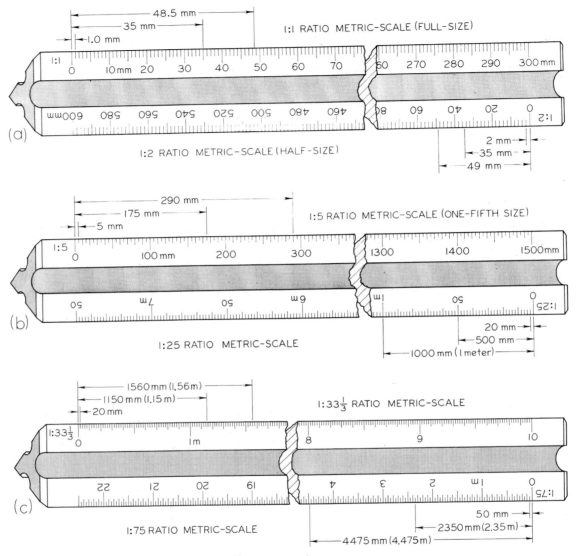

FIGURE 5.59 Metric Scales.

Metric scales are available in flat and triangular styles with a variety of scale graduations. The triangular scale illustrated in Figure 5.59 has one full-size scale and five reduced-size scales, all fully divided. Using these scales, you can make a drawing full size, enlarged size, or reduced size.

FULL SIZE The 1:1 scale, shown in Figure 5.59a, is full size, and each division is 1 mm in width, with the divisions on the scale at 10-mm intervals. The same scale is also convenient for ratios of 1:10, 1:100, 1:1,000, and so on.

HALF SIZE The 1:2 scale, shown in Figure 5.59a, is one-half size, and each division equals 2 mm, with

divisions on the scale at 20-unit intervals. This scale is also convenient for ratios of 1:20, 1:200, 1:2,000, and so on.

The remaining four scales on this triangular metric scale include the typical scale ratios of 1:5, 1:25, 1:33-1/3, and 1:75, shown in Figures 5.59a and 5.59b. These ratios may also be enlarged or reduced as desired by multiplying or dividing by a factor of 10. Metric scales are also available with other scale ratios for specific drawing purposes.

The metric scale is used in map drawing, in drawing force diagrams and machine parts, or for other things that use units such as 1 mm = 1 kg and

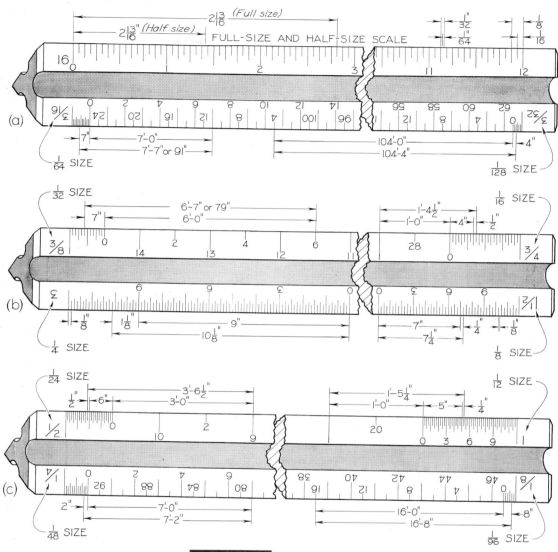

FIGURE 5.60 Architects' Scales.

1 mm = 500 kg. Many of the dimensions in the illustrations and problems in this text are given in metric units.

Dimensions that are given in inches and feet may be changed to metric values using the conversion 1 inch = 25.4 mm. Decimal equivalent tables can be found inside the back cover of this book.

■ 5.52 INCH-FOOT SCALES

Several scales based on the inch-foot system of measurement are still in use today, although the metric system is more common and internationally accepted.

■ 5.53 ARCHITECTS' SCALES

Architects' scales are used for drawing buildings, piping systems, and other large structures. The full-size scale shown in Figure 5.60a is also useful in drawing relatively small objects.

The architects' scale has one full-size scale and 10 overlapping reduced-size scales. You can use these scales to make a drawing various sizes from full size to 1/128 size. *In all the reduced scales the major divisions represent feet and their subdivisions represent inches and fractions thereof.* For example, the scale marked "1/2" means 1/2 in. = 1 ft., *not* 1/2 in. = 1 in.

Hands On 5.7
Sketching to scale

Measure the dimensions of the objects below in metric units. Sketch each object at the scale listed below. Specify the scale correctly on your drawings. Before you sketch, choose which of the boxes at right will best fit each item when it is sketched to the specified scale.

- Sketch the paper clip at 2 : 1 scale.

- Sketch the push pin at 5: 1 scale.

- Sketch the drink cup at 1 : 2 scale.

■ 5.54 MECHANICAL ENGINEERS' SCALES

The objects represented in technical drawings vary in size from small parts, an inch or smaller in size, to very large equipment or parts. By drawing these objects full size, half size, quarter size, or eighth size, you can fit them on a standard sheet of paper. Mechanical engineers' scales are divided into units representing inches to full size, half size, quarter size, or eighth size, as shown in Figure 5.57b. To scale a drawing to one-half size, use the mechanical engineers' scale marked "half size," which is graduated so that every half-inch represents one inch.

Triangular combination scales include the full- and half-size mechanical engineers' scales, several architects' scales, and an engineers' scale.

■ 5.55 SPECIFYING THE SCALE ON A DRAWING

For technical drawings, the scale indicates the ratio of the size of the drawn object to its actual size, no matter what unit of measurement is used. Letter scales as follows: FULL SIZE or 1 : 1, HALF SIZE or 1 : 2, and so on. Specify enlargement as 2 : 1 or 2×, 3 : 1 or 3×, 5 : 1 or 5×, 10 : 1 or 10×, and so on. The preferred metric scale ratios are 1 : 1, 1 : 2, 1 : 5, 1 : 10, 1 : 20, 1 : 50, 1 : 100, and 1 : 200.

Map scales are indicated in terms of fractions, such as Scale 1/62500, or graphically, such as:

400   0   400   800 Ft

Practical Tips
Accurate measurements

- The methods you use to take measurements can affect the accuracy of your drawing, because even a small error will get bigger when it is sized up.

- Place the scale on the drawing with the edge parallel to the line on which the measurement is to be made. Use a sharp pencil to make a short dash at right angles to the scale.

- You can avoid cumulative errors in distances that are all end-to-end by adding each successive measurement to the preceding one instead of moving the scale each time. This way several minor errors won't add up to one major one.

■ KEY WORDS

multiview projection	alignment of views	oblique edge	third-angle projection
depth	surface	counterbore	first-angle projection
height	plane	countersink	sheet size
width	normal surface	spotface	mylar
orthographic	inclined surface	blind hole	engineers' scale
plane of projection	oblique surface	through hole	mechanical engineers' scale
projectors	edge	fillet	decimal scale
fold lines	normal edge	round	architects' scale
glass box	inclined edge	runout	metric scale

■ CHAPTER SUMMARY

- Orthographic drawings are the result of projecting the image of a 3-D object onto one of six standard planes of projection. The six standard views are often thought of as an unfolded "glass box." The arrangement of the views in relation to one another is important. Views must project to line up with adjacent views, so that any point in one view projects to line up with that same point in the adjacent view. The standard arrangement of views shows the top, front, and right side of the object.

- Visualization is an important skill for engineers. You can build your visual abilities through practice and through understanding terminology describing objects. For example, surfaces can be normal, inclined, or oblique. Normal surfaces appear true size in one principal view and as an edge in the other two principal views. Inclined surfaces appear as an edge view in one of the three principal views. Oblique surfaces do not appear as an edge view in any of the principal views.

- Conventions define usual practices for the representation of features such as holes, bosses, ribs, webs, spokes, fillets, and rounds. Choice of scale is important for representing objects clearly on the drawing medium.

- Creating CAD drawings involves applying the same concepts as paper drawing. The main difference is that drawing geometry is stored more accurately using a computer than any hand drawing. CAD drawing geometry can be reused in many ways and plotted to any scale as necessary.

■ REVIEW QUESTIONS

1. Sketch the symbol for third-angle projection.
2. List the six principal views of projection.
3. Sketch the top, front, and right-side views of an object of your design having normal, inclined, and oblique surfaces.
4. In a drawing that shows the top, front, and right-side view, which two views show depth? Which view shows depth vertically on the sheet? Which view shows depth horizontally on the drawing sheet?
5. What is the definition of a normal surface? An inclined surface? An oblique surface?
6. What are three similarities between using a CAD program to create 2-D drawing geometry and sketching on a sheet of paper? What are three differences?
7. What dimensions are the same between the top and front view: width, height, or depth? Between the front and right side view? Between the top and right-side view?
8. List two ways of transferring depth between the top and right-side views.
9. If surface A contained corners 1, 2, 3, 4, and surface B contained corners 3, 4, 5, 6, what is the name of the line where surfaces A and B intersect?
10. If the top view of an object shows a drilled through hole, how many hidden lines would be necessary in the front view to describe the hole?

■ MULTIVIEW PROJECTION PROJECTS

The following problems are intended to be sketched freehand on graph paper or plain paper. Sheet layouts such as A-1, found in the back of this book, are suggested, but your instructor may prefer a different sheet size or arrangement. Use metric or decimal-inch, as assigned. The marks shown indicate rough units of either 1/2 inch and 1/4 inch, or 10 mm and 5 mm. All holes are through holes.

For the following problems, use a layout similar to Figure 5.61 or 5.62

■ DESIGN PROJECT

Portable devices such as CD players, telephones and laptop computers are increasingly popular. Design another portable item that you think would be useful or improve on an existing product. In your design, strive to incorporate convenience, function, and durability with an appealing appearance. Consider lightweight materials and mechanisms for collapsing and carrying the item. Represent your idea using orthographic projection.

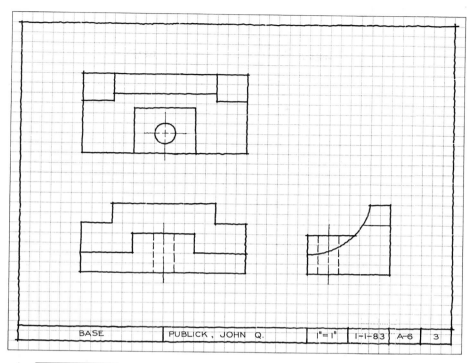

| BASE | PUBLICK , JOHN Q. | 1"=1" | 1-1-83 | A-6 | 3 |

FIGURE 5.61 Suggested Layout for Freehand Sketch (Layout A–2 or A4–2 adjusted).

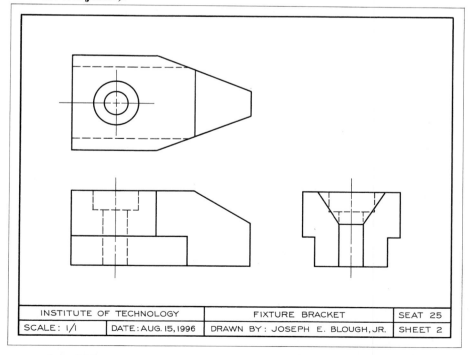

| INSTITUTE OF TECHNOLOGY | | FIXTURE BRACKET | SEAT 25 |
| SCALE: 1/1 | DATE: AUG. 15, 1996 | DRAWN BY: JOSEPH E. BLOUGH, JR. | SHEET 2 |

FIGURE 5.62 Suggested Layout for Mechanical Drawing (Layout A–3 or A4–3 adjusted).

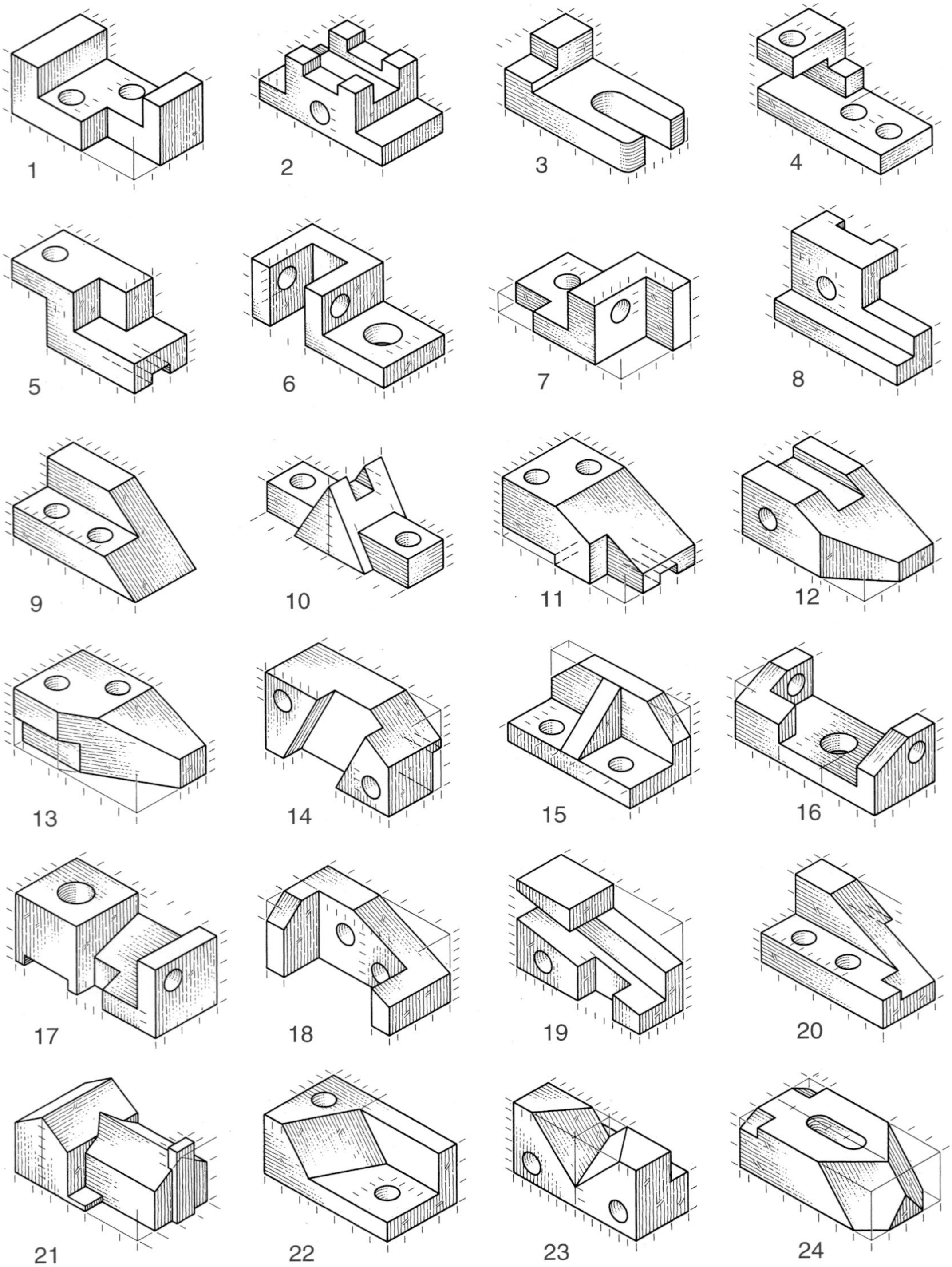

FIGURE 5.63   Multiview Sketching Problems. Sketch necessary views on graph paper or plain paper, two problems per sheet. The units shown may be either .500 and .250 or 10 mm and 5 mm. See instructions on page 160. All holes are through holes.

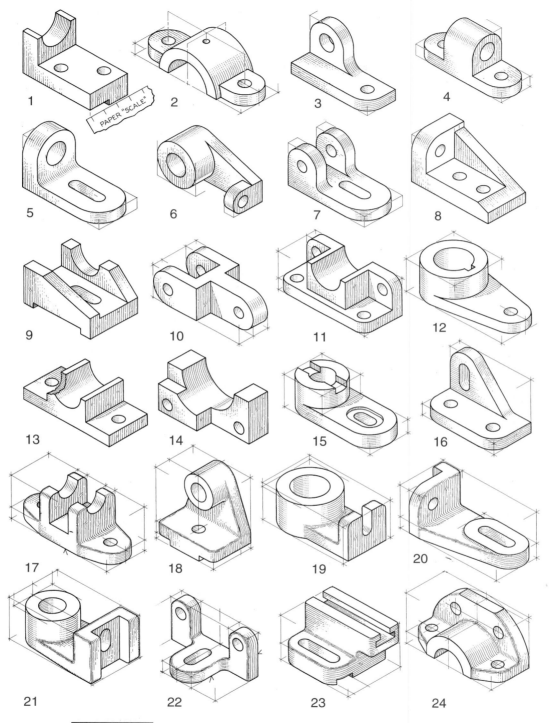

FIGURE 5.64 Multiview Sketching Problems. Sketch necessary views, on graph paper or plain paper, two problems per sheet. Prepare paper scale with divisions equal to those in Prob. 1, and apply to problems to obtain approximate sizes. Let each division equal either .500 or 10 mm on your sketch. See instructions on page 160.

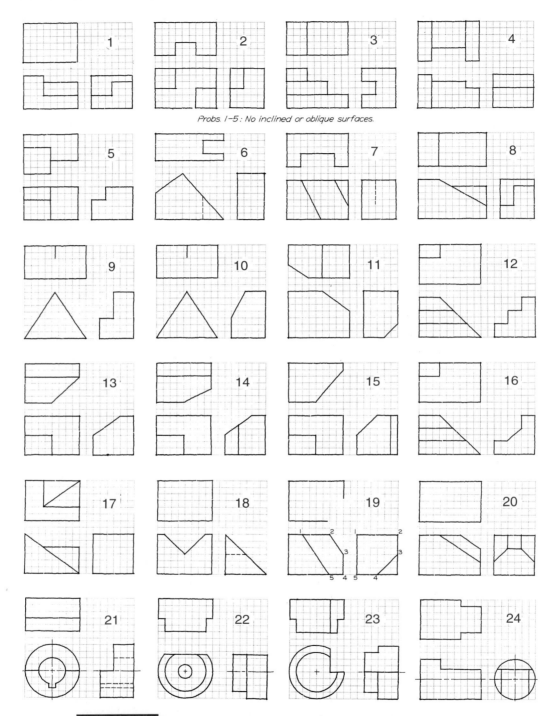

Probs. 1–5: No inclined or oblique surfaces.

FIGURE 5.65 Missing-Line Sketching Problems. (1) Sketch given views, on graph paper or plain paper, two problems per sheet. Add missing lines. The squares may be either .250 or 5 mm. See instructions on page 160. (2) Sketch in isometric on isometric paper or in oblique on cross-section paper.

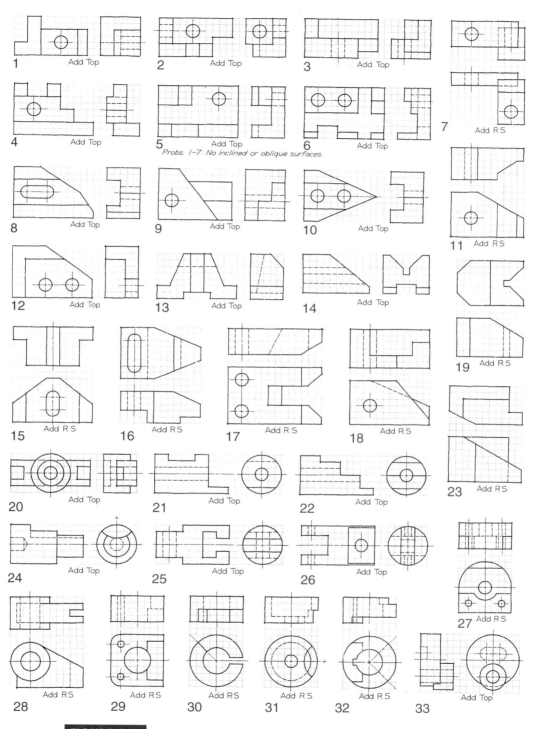

FIGURE 5.66 Third-View Sketching Problems. Sketch the two given views and add the missing views, as indicated. The squares may be either .250 or 5 mm. See instructions on page 160. The given views are either front and right-side views or front and top views. Hidden holes with center lines are drilled holes.

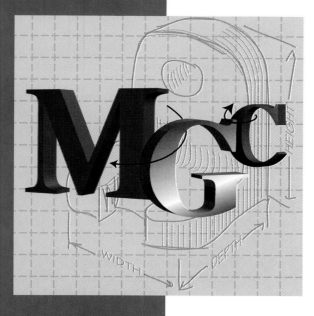

C H A P T E R   6

PICTORIAL SKETCHING

OBJECTIVES

After studying the material in this chapter, you should be able to:

1. Describe the differences between multiview projection, axonometric projection, oblique projection, and perspective.

2. List the advantages of multiview projection, axonometric projection, oblique projection, and perspective.

3. Create an isometric sketch given a multiview drawing.

4. Measure along each isometric axis.

5. Sketch inclined and oblique surfaces in isometric.

6. Sketch angles, ellipses, and irregular curves in isometric.

7. Describe how an oblique projection is created.

8. Sketch cavalier and cabinet oblique drawings.

9. Know how to place circles when creating an oblique drawing.

10. Describe why CAD software does not automatically create oblique drawings.

11. Sketch one- and two-point perspectives.

OVERVIEW

Multiview drawings makes it possible to accurately represent complex objects through a series of views, where each view only shows two out of three principal dimensions, not showing length, width, and height together. **Pictorial drawings** (ANSI/ASME Y14.4M–1989 (R1994)), which look more like a picture than a multiview drawing, are used to quickly communicate your ideas, to develop your own thoughts during the design process, and sometimes to clarify otherwise difficult-to-read drawings. Pictorial drawings can be easily understood without technical training.

Various types of pictorial drawing are used extensively in catalogs, sales literature, and technical work. For example, pictorial drawings are used to obtain patents, to show process piping design, and to show machine, structural, architectural, and furniture design. Pictorial drawings created using computer 3-D modeling techniques can be used very effectively to present design ideas, to help market the product, and to help visually inspect fits, assembly, and other aspects of the design. Pictorial drawings are not usually dimensioned because they do not show the object accurately. Pictorial sketches are an invaluable aid in the design process.

■ 6.1　METHODS OF PROJECTION

Chapter 1 described the principal types of projection systems—parallel and perspective—which can be further broken down into subtypes, as shown in Figure 6.1. Pictorial sketches can be created either using perspective projection or parallel projection. In this chapter you will learn to create isometric pictorial sketches, oblique pictorial sketches, and perspective sketches. The four principal types of projection are illustrated in Figure 6.2, and all except the regular multiview projection are pictorial types, since they show several sides of the object in a single view. In all cases, the views are formed by projection, which was described in previous chapters. Only in perspective projection do the projectors converge to a point; all of the other types shown are parallel projection.

In both **multiview projection** and **axonometric projection**, the visual rays are parallel to each other and perpendicular to the plane of projection. Therefore, both are classified as **orthographic projections**.

In **oblique projection**, the visual rays are parallel to each other but oblique (not at 90 degrees) to the

plane of projection. This produces a pictorial view where the front surface is usually shown true size and other surfaces foreshortened. Oblique pictorials are easy to draw and have the advantage that circular shapes in the true-size view are not distorted into ellipses. But oblique views are not very realistic because that particular view is never seen in real life.

In **perspective sketches** the visual rays extend to the observer's eye from all points of the object to form a cone of rays. This type of pictorial drawing is the most realistic, but the most difficult to draw. CAD systems often can generate isometric and **perspective views** automatically from 3-D models.

■ 6.2　AXONOMETRIC PROJECTION

The distinguishing feature of axonometric projection, as compared to multiview projection, is that the object is inclined to the plane of projection, as in Figure 6.2b. Since the object is inclined to the projection plane, the lengths of lines, the sizes of angles, and the general proportions depend on the exact orientation the object has with respect to the plane of projection. Three axonometric projections of a cube are shown in Figure 6.3.

In these cases the edges of the cube are inclined to the plane of projection and are therefore foreshortened. The degree of foreshortening of any line depends on its angle with the plane of projection; the greater the angle, the greater the foreshortening. The three edges of the cube that meet at the corner nearest the observer are considered the **axonometric axes**. After the degree of foreshortening for each axis has been determined, scales can be constructed for measuring along these edges or along any other edges parallel to them. As shown in Figure 6.3, axonometric projections are classified as **isometric projection** (all axes equally foreshortened), **dimetric projection** (two axes equally foreshortened), and **trimetric projection** (all three axes foreshortened differently, requiring different scales for each axis).

The term **isometric** means equal measure. For isometric pictorial drawings, the object is tilted to the viewing plane so that all of the object's principal dimensions are equally foreshortened in each axis direction. Isometric pictorials are relatively easy to sketch, and special sketching paper with an isometric grid is available

■ 6.3　ISOMETRIC SKETCHING

When a surface on the object is tipped with respect to the viewing plane, it will appear foreshortened in stan-

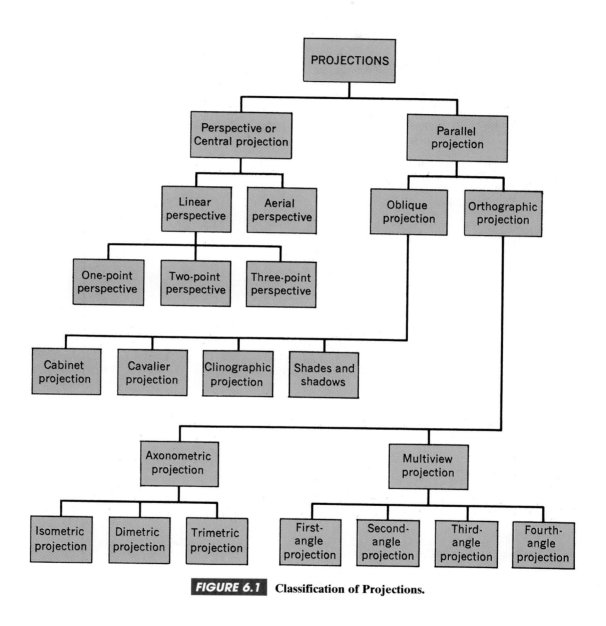

FIGURE 6.1 Classification of Projections.

dard views. When you are sketching isometric views, you don't need to bother calculating how much the surface should be foreshortened. Just sketch it the actual size. Since all normal surfaces on the object are foreshortened equally, the sketch will still appear in proportion. This is called an ***isometric sketch*** or drawing. When an isometric view is prepared using a foreshortened isometric scale, or when the object is actually projected on a plane of projection, it is called an isometric projection.

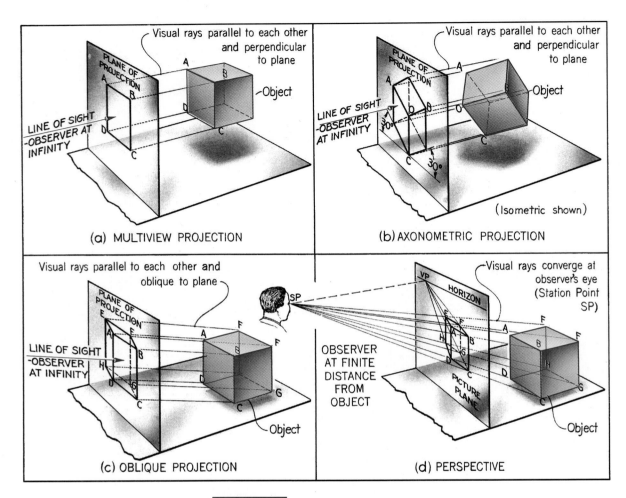

FIGURE 6.2　Four Types of Projection.

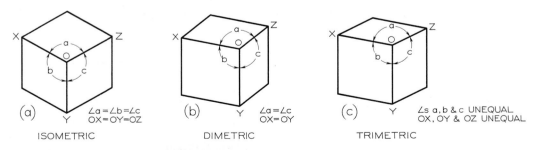

FIGURE 6.3　Axonometric Projections.

Figure 6.4 shows the contrast between an isometric sketch and an isometric projection. (You will learn how to project auxiliary views in Chapter 9.) An isometric projection is a second auxiliary view. The isometric sketch is about 25 percent larger than the isometric projection, but the pictorial value is obviously the same in both. Isometric sketches are much easier to make and just as useful as an isometric projection.

When you are creating isometric sketches, you do not always have to make accurate measurements

Step by Step 6.1
Isometric sketching from an object

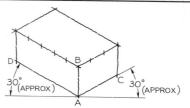

Positioning the object

To make an isometric sketch from an actual object, first hold the object in your hand and tilt it toward you, as shown in the illustration at right. In this position the front corner will appear vertical. The two receding bottom edges and those edges parallel to them should appear to be at about 30 degrees with horizontal. The steps for sketching the object are as follows:

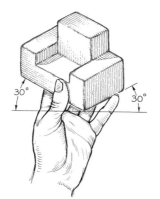

1. Sketch the enclosing box lightly, making AB vertical and AC and AD approximately 30 degrees with horizontal. These three lines are the isometric axes. Make AB, AC, and AD approximately proportional in length to the actual corresponding edges on the object. Sketch the remaining lines parallel to these three lines.

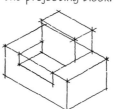

2. Block in the recess and the projecting block.

3. Darken all final lines.

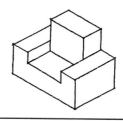

locating each point in the sketch exactly. Instead, keep your sketch in proportion. If the height of the object appears to be twice the depth, make it appear so in your sketch. Use the techniques you learned in Chapter 3 to help you estimate distances and sketch in proportion. Many of the examples that follow discuss

using accurate measurements, but when you are sketching, the most important thing is that the overall proportions of the object and the relationships between the details are shown clearly and correctly. Isometric pictorials are great for showing piping layouts and structural designs.

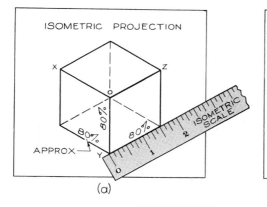

(a)

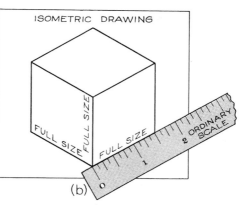

(b)

FIGURE 6.4 Isometric and Ordinary Scales.

Practical Tip
Isometric views in CAD

When you plot an isometric view of a 3-D model using CAD, it will be foreshortened because it is an exact projection.

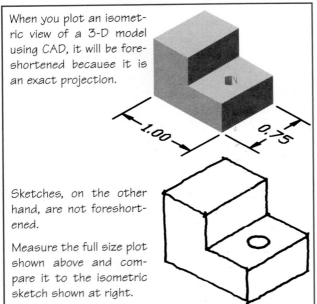

Sketches, on the other hand, are not foreshortened.

Measure the full size plot shown above and compare it to the isometric sketch shown at right.

■ 6.4 NORMAL AND INCLINED SURFACES IN ISOMETRIC VIEW

Making an isometric sketch of an object having only normal surfaces is shown in Figure 6.5. Notice that all measurements are made parallel to the main edges of the enclosing box—that is, parallel to the isometric axes. No measurement along a diagonal (nonisometric line) on any surface or through the object can be measured directly. The object may be drawn in the same position by beginning at the corner Y, or any other corner, instead of at the corner X.

Making an isometric sketch of an object which has inclined surfaces (and oblique edges) is shown in Figure 6.6. Notice that inclined surfaces are located by offset, or coordinate measurements along the isometric lines. For example, distances E and F are used to locate the inclined surface M, and distances A and B are used to locate surface N.

■ 6.5 OBLIQUE SURFACES IN ISOMETRIC VIEW

Oblique surfaces in isometric view may be drawn by finding the intersections of the oblique surface with

FIGURE 6.5 **Isometric Drawing of Normal Surfaces.**

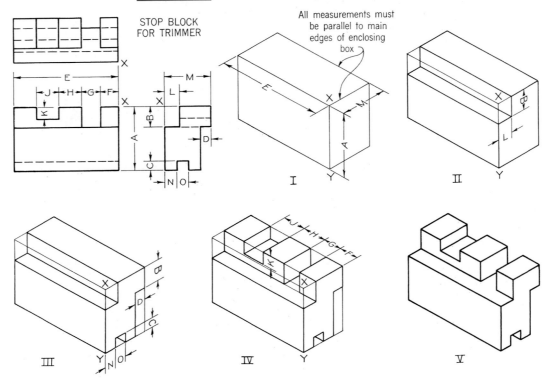

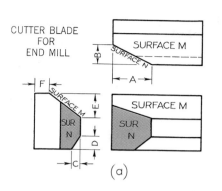

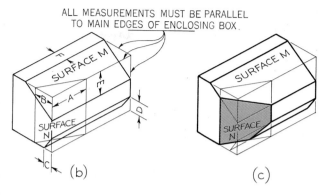

FIGURE 6.6   Inclined Surfaces in Isometric.

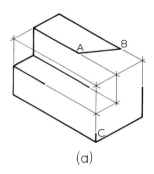

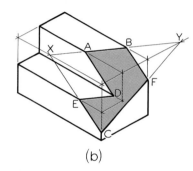

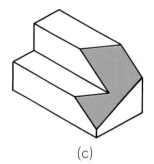

FIGURE 6.7   Oblique Surfaces in Isometric.

isometric planes. For example, in Figure 6.7a the oblique plane contains points A, B, and C. Locate the plane by extending line AB to X and Y, points that share isometric planes with point C, as shown in Figure 6.7b. Sketch lines XC and YC to locate points E and F. Then draw lines AD and ED, using the rule that parallel lines on the object remain parallel in any view. The completed drawing is shown in Figure 6.7c.

■ 6.6 OTHER POSITIONS OF THE ISOMETRIC AXES

Orient your drawing on the isometric axes in the way that shows the part most clearly, without the need for hidden lines. Figure 6.8 shows an isometric view of a birdhouse in two different orientations. The view from below shows more of the details of its construction and therefore may be a better view, depending on your purpose. If the object is particularly long, orient it horizontally for best effect, as shown in Figure 6.9.

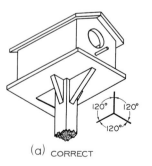
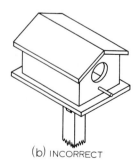

FIGURE 6.8   **An Object Naturally Viewed from Below.**

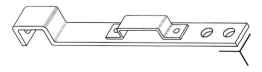

FIGURE 6.9   **Long Axis Horizontal.**

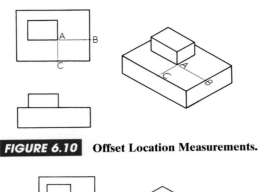

FIGURE 6.10 Offset Location Measurements.

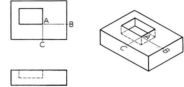

FIGURE 6.11 Offset Location Measurements.

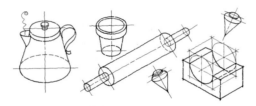

FIGURE 6.12 Isometric Ellipses.

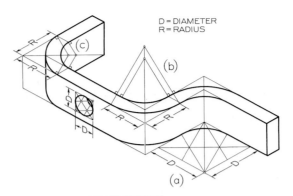

FIGURE 6.13 Arcs in Isometric.

FIGURE 6.14 Instrumaster Isometric Template.

■ 6.7 OFFSET MEASUREMENTS

You can locate a new point from an existing corner, as illustrated in Figures 6.10 and 6.11. Box in the main block first, then sketch the offset lines CA and BA to locate corner A of the small block or the rectangular recess. These are called *offset measurements*, and since they are parallel to normal edges on the object, they will be parallel to the same edges in the isometric.

 Worksheet 6.1 provides beginning practice in isometric sketching and offset measurements.

■ 6.8 ISOMETRIC ELLIPSES

When objects with cylindrical or conical shapes are placed in isometric or other oblique positions, the circles are seen at an angle and appear as ellipses, as shown in Figure 6.12.

When sketching isometric ellipses, keep in mind that the major axis of the ellipse is always at right angles to the centerline of the cylinder, and the minor axis is at a right angle to the major axis and coincides with the centerline.

■ 6.9 ARCS IN ISOMETRIC VIEW

Figure 6.13 shows an isometric view of an object that has rounded corners. To sketch arcs in an isometric view use the radius and block in the ellipse along the appropriate isometric axis lines. In this case the radius R is measured from the construction corner to locate the center of the elliptical arc. Notice that the radius R does not remain a constant value when showing the arc in an isometric view.

Worksheet 6.2 provides practice in sketching ellipses and elliptical arcs for isometric pictorials.

■ 6.10 ISOMETRIC ELLIPSE TEMPLATES

The template shown in Figure 6.14 combines the angles, foreshortened isometric scales, and ellipses on the same instrument. The ellipses are provided with markings to coincide with the isometric centerlines of the holes—a convenient feature in isometric drawing.

Step by Step 6.2
Sketching isometric ellipses

Two views of a block with a large cylindrical hole are shown. The steps in sketching the object are as follows:

1. Sketch the block and the enclosing parallelogram for the ellipse, making the sides of the parallelogram parallel to the edges of the block and equal in length to the diameter of the hole. Draw diagonals to locate the center of the hole and then draw centerlines AB and CD. Points A, B, C, and D will be midpoints of the sides of the parallelogram, and the ellipse will be tangent to the sides at those points. The major axis will be on the diagonal EF, which is at right angles to the centerline of the hole, and the minor axis will fall along the short diagonal. Sketch long, flat elliptical sides CA and BD, as shown.

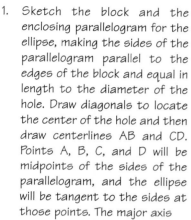

2. Sketch short, small-radius arcs CB and AD to complete the ellipse. Avoid making the ends of the ellipse squared off or pointed like a football.

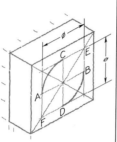

3. Sketch lightly the parallelogram for the ellipse that lies in the back plane of the object, and sketch the ellipse in the same manner as the front ellipse.

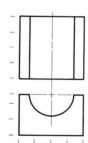

4. Draw lines GH and JK tangent to the two ellipses. Darken all final lines.

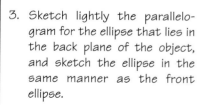

Another method for determining the back ellipse is shown at right:

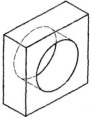

1. Select points at random on the front ellipse and sketch lines equal in length to the depth of the block.

2. Sketch the ellipse through the ends of the lines.

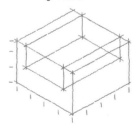

Two views of a bearing with a semicylindrical opening are shown below. The steps in sketching are as follows:

1. Block in the object, including the rectangular space for the semicylinder.

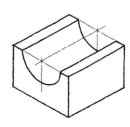

2. Block in the box enclosing the complete cylinder. Sketch the entire cylinder lightly.

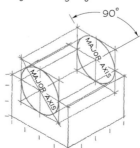

3. Darken all final lines, showing only the lower half of the cylinder.

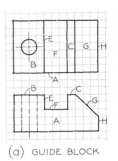
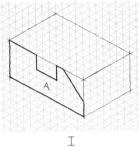
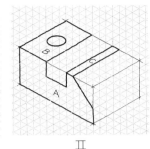
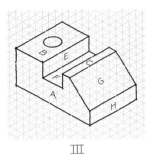

(a) GUIDE BLOCK   I   II   III

FIGURE 6.15 Sketching on Isometric Paper.

■ 6.11 SKETCHING ON ISOMETRIC PAPER

Two views of a guide block are shown in Figure 6.15a. The steps in sketching illustrate the use of isometric paper. Start with individual planes or faces to build a pictorial from given views:

1. Sketch the isometric of the enclosing box, counting off the isometric grid spaces to equal the corresponding squares on the given views, as shown in Figure 6.15I. Sketch surface A, as shown.

2. Sketch additional surfaces B and C and the small ellipse, as shown in Figure 6.15II.

3. Sketch additional surfaces E, F, G, and H to complete the sketch, as shown in Figure 6.15III.

■ 6.12 HIDDEN LINES

Hidden lines are omitted in pictorial sketches unless they are needed to make the drawing clear. Figure 6.16 shows a case in which hidden lines are needed. In this case, the hidden lines show a projecting part that cannot be clearly represented without them.

■ 6.13 CENTERLINES

Centerlines are drawn in a pictorial sketch if they are needed to indicate symmetry or if they are needed for dimensioning. Use centerlines sparingly and omit them in cases of doubt. Using too many centerlines makes the drawing appear confusing. Figure 6.16 shows an example in which centerlines are needed for dimensioning purposes.

■ 6.14 NONISOMETRIC LINES

Since the only lines of an object that are drawn true length in an isometric drawing are the isometric axes or lines parallel to them, nonisometric lines cannot be measured directly. For example, in Figure 6.17 inclined lines BA and CA are shown in true length (54 mm) in

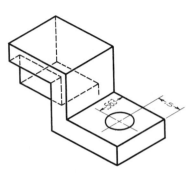

FIGURE 6.16 Use of Hidden Lines.

the top view, but since they are not parallel to the isometric axes they are not true length in the isometric sketch. Use box construction and offset measurements to draw nonisometric lines like these. The distances 44 mm, 18 mm, and 22 mm can be measured directly along isometric lines, as shown in Figure 6.17I. The nonisometric distance (54 mm) cannot be measured directly, but you can determine dimension X from the top view. This dimension is parallel to an isometric axis and can be measured in the isometric sketch, as shown in Figure 6.17II. The remaining distances (24 mm and 9 mm) are parallel to isometric lines which can be measured directly, as shown in Figure 6.17III.

 People in various fields use pictorial drawings to visualize objects before they exist in reality. The World Wide Web is one of the places you can find great examples of pictorial drawing. You can find samples of pictorial drawings produced by the NASA Lewis Graphics team at this URL: http://www.lerc.nasa.gov./www/STI/graphics/samillus.htm.

Hands On 6.1
Box Construction

Rectangular objects are easily drawn using box construction, where you imagine the object to be enclosed in a rectangular box whose sides coincide with the main faces of the object.

Sketch the object shown in two views by imagining it is enclosed in a construction box.

- A construction box and distances a, b, c, d, e, and f have been blocked in for you.

- Refer to the given views and finish constructing the features of the object.

- Darken your final lines.

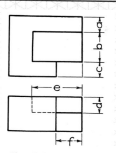

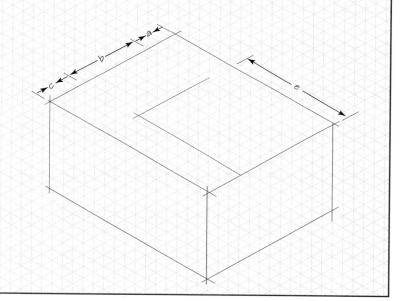

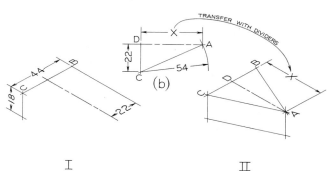

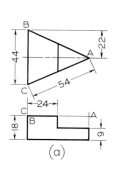

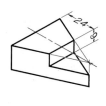

FIGURE 6.17 Nonisometric Lines (metric dimensions).

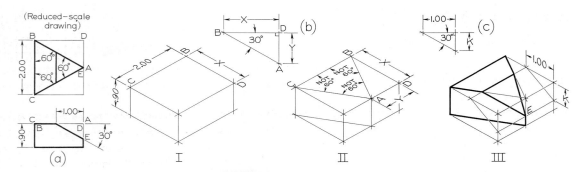

FIGURE 6.18   Angles in Isometric.

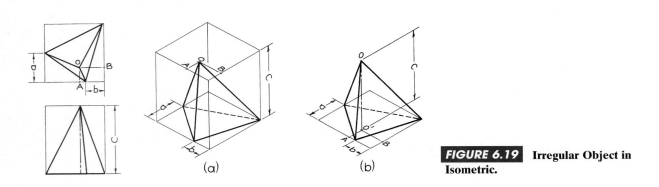

FIGURE 6.19   Irregular Object in Isometric.

■ 6.15 ANGLES IN ISOMETRIC VIEW

A regular protractor cannot be used to measure angles in isometric.* Angular measurements must be converted to linear measurements along isometric lines. Angles project true size only when the plane of the angle is parallel to the plane of projection. An angle may project larger or smaller than true size, depending on its position. Since the various surfaces of the object are usually inclined to the plane of projection in an isometric sketch, angles generally will not project true size. For example, in the multiview drawing in Figure 6.18a none of the three 60-degree angles will be 60 degrees in the isometric drawing. Use a scrap of paper and measure each angle in the isometric in Figure 6.18II and compare these measurements to the true 60 degrees. No two angles are the same; two are smaller and one is larger than 60 degrees.

To show the angle, first sketch the enclosing box from the given dimensions, as shown in Figure 6.18I, except for X, which is not given. To find X, draw full size triangle BDA as seen in the top view shown in Figure 6.18b. Use distance X to complete the enclosing box in the isometric sketch.

Find distance Y using the same reasoning and locate it in the isometric as shown in Figure 6.18II. The completed isometric is shown in Figure 6.18III, where point E is located using distance K.

■ 6.16 IRREGULAR OBJECTS

If the general shape of an object is not somewhat rectangular, it may be drawn using box construction. As shown in Figure 6.19, various points of the triangular base are located using offsets a and b along the edges of the bottom of the construction box. The vertex is located by means of offsets OA and OB on the top of the construction box.

It is not always necessary to sketch the complete construction box. You can sketch the triangular base using only the bottom of the box, as shown in Figure 6.19b. The vertex O' on the base can then be located by offsets O'A and O'B, as shown, and measurement C can be used to draw vertical line O'O.

---

* Isometric protractors for setting off angles on isometric surfaces are available from drafting supplies dealers.

Hands On 6.2
Angles in isometric

Two views of an object to be drawn in isometric are given.

Point A can easily be located in the isometric by measuring .88 inches down from point 0, as shown.

However point B is dimensioned by the 30-degree angle. To locate B in the isometric you must find dimension X.

You can solve this problem graphically by using CAD to draw a triangle with angles of 30 degrees and 90 degrees and a side of .88 inches and then listing the length for dimension X.

Measure distance X in the true-size graphical solution shown at right, or use trigonometry to find the distance. Once you have found dimension X, use it to locate point B in the isometric sketch as shown and finish the sketch the bottom of the page.

You can also sketch the isometric so that it appears proportionate and, if necessary, dimension the angle as 30 degrees to call attention to it.

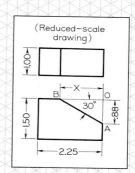

(Reduced−scale drawing)

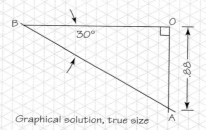

Graphical solution, true size

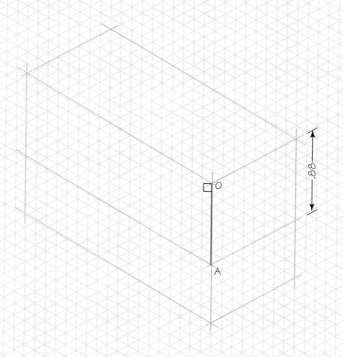

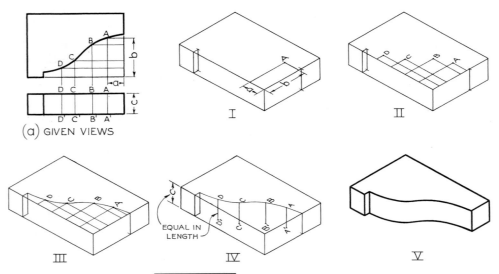

FIGURE 6.20　Curves in Isometric.

■ 6.17　CURVES IN ISOMETRIC VIEW

Draw curves in isometric sketches by using a series of offset measurements. Select any number of points, such as A, B, and C, randomly along the curve, as shown in the top view in Figure 6.20a. The more points used, the greater the accuracy.

Use measurements a and b in the isometric to locate point A on the curve, as shown in Figure 6.20I. Locate points B, C, and D in a similar manner, as shown in Figure 6.20II. Sketch a light freehand curve smoothly through the points, as shown in Figure 6.20III. The lower curve is located directly under A, B, C, and D by drawing vertical lines equal to the height of the block (c), as shown in Figure 6.20IV. Darken the final sketch, as in Figure 6.20V.

■ 6.18　SCREW THREADS IN ISOMETRIC

Parallel partial ellipses are used to represent the crests only of a screw thread in isometric as shown in Figure 6.21. The ellipses may be sketched freehand or with an ellipse template.

FIGURE 6.21　Screw Threads in Isometric.

■ 6.19　THE SPHERE IN ISOMETRIC

The isometric projection of a sphere is a circle whose diameter is the major axis of the isometric ellipse. Think about an isometric drawing of any curved surface as the envelope of all lines that can be drawn on that surface. For a sphere, the great circles (those cut by any plane through the center) are lines on the surface. Since all great circles—except those that are perpendicular to or parallel to the plane of projection—are shown as ellipses having equal major axes, their envelope is a circle whose diameter is the major axis of the ellipses.

Figure 6.22a shows two views of a sphere enclosed in a construction cube. In Figure 6.22I the cube is drawn in, together with the isometric of a great circle that is parallel to one face of the cube. In Figure 6.22II the result is an isometric sketch, and its diameter is the square root of 3/2 times the actual diameter of the sphere. The isometric projection of the sphere, shown in Figure 6.22III, is a circle whose diameter is equal to the true diameter of the sphere.

■ 6.20　ISOMETRIC DIMENSIONING

Isometric drawings are not usually dimensioned because they do not show object features at true size. (You will learn more about dimensioning techniques in Chapter 9.) ANSI has approved two dimensioning methods—namely, the pictorial plane, or aligned, system and the unidirectional system—both shown in Figure 6.23. Note that vertical lettering is used in both

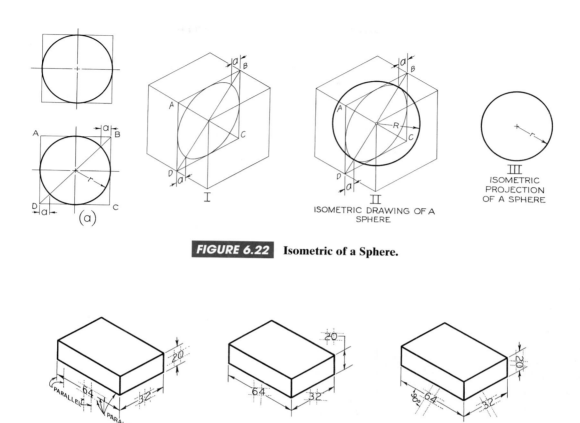

FIGURE 6.22 Isometric of a Sphere.

(a) ALIGNED   (b) UNIDIRECTIONAL   (c) INCORRECT

FIGURE 6.23 Numerals and Arrowheads in Isometric (metric dimensions).

systems; inclined lettering is not recommended for pictorial dimensioning. In the aligned system, shown in Figure 6.23a, the extension lines, dimension lines, and lettering are all drawn in the isometric plane of one face of the object. The "horizontal" guidelines for the lettering are drawn parallel to the dimension line, and the "vertical" guidelines are drawn parallel to the extension lines. The barbs of the arrowheads should line up parallel to the extension lines.

In the unidirectional system, shown in Figure 6.23b, the extension lines and dimension lines are all drawn in the isometric plane of one face of the object,

and the barbs of the arrowheads should line up parallel to the extension lines. However, the lettering for the dimensions is vertical and reads from the bottom of the drawing. This simpler system of dimensioning is often used on pictorials for production purposes.

Figure 6.23c shows the poor appearance resulting from vertical guidelines for the letters perpendicular to the dimension lines, causing dimensions to be neither in an isometric plane nor read vertically from the bottom of the drawing. Notice the awkward appearance of the 20-mm dimension. Figure 6.24 shows correct and incorrect isometric dimensioning.

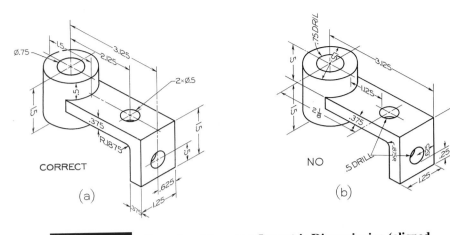

FIGURE 6.24   Correct and Incorrect Isometric Dimensioning (aligned system).

■ 6.21 EXPLODED ASSEMBLIES

Exploded assemblies are often used in design presentations, in catalogs, in sales literature, and in the shop to show all the parts of an assembly and how they fit together. They may be drawn by any of the pictorial

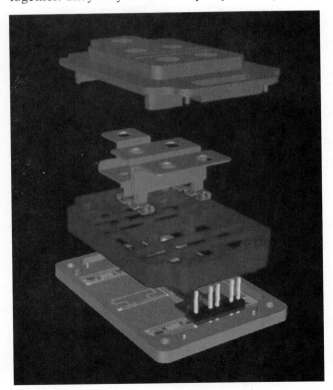

FIGURE 6.25   Exploded Assembly of Power Module for Zero Emission Automobiles. *Courtesy of SDRC, Milford, OH.*

methods, including the isometric (Figure 6.25). (Chapter 12 presents more information on assembly drawings.)

■ 6.22 USING CAD

Pictorial drawings of all sorts can be created using 3-D CAD. To create pictorials using 2-D CAD you would use projection techniques similar to those presented in this chapter. The advantage of 3-D CAD is that once you make a 3-D model of a part or assembly you can change the viewing direction at any time for orthographic, isometric, or perspective views. You can also apply different materials to the drawing objects and shade them to produce a high degree of realism in the pictorial view, as shown in Figure 6.25.

■ 6.23 OBLIQUE PICTORIALS

Oblique projection provides an easy method for drawing circular features that are parallel to the plane of projection. With oblique projection, the front view is the same as the front view in a multiview drawing. Circles and angles parallel to the projection plane are true size and shape and are therefore easy to construct. Oblique views are not as realistic as isometric views because the depth appears distorted, and CAD is not typically used to create oblique views since better-appearing isometric drawings can be created easily from 3-D models. While circular shapes are easy to sketch in the front oblique plane, they appear elliptical in the top or side views. Oblique views are primarily a sketching technique used when the majority of circular shapes appear in the front view or when the object can be rotated in order to position circles in the front view.

Step by Step 6.3
Oblique sketching on grid paper

Ordinary grid paper is convenient for oblique sketching. Two views of a bearing bracket are shown below. The dimensions can be determined by counting the squares, and receding lines can easily be drawn at 45° by sketching diagonally through the grid squares.

1. Lightly sketch the enclosing box construction for the front view. To establish the depth at t reduced scale, sketch receding lines diagonally through half as many grid squares as the number shown in the side view.

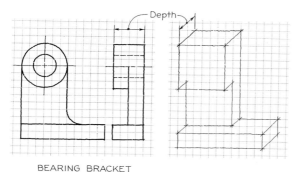

BEARING BRACKET

2. Sketch all arcs and circles.

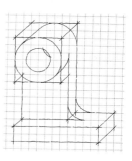

3. Darken final lines.

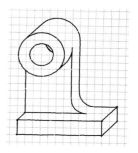

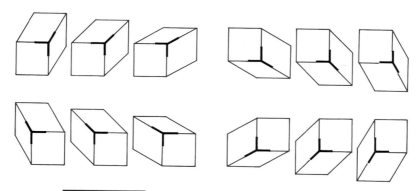

FIGURE 6.26 Variation in Direction of Receding Axis.

■ 6.24 CHOOSING THE ANGLE FOR RECEDING LINES

The receding lines may be drawn at any convenient angle. Some examples are shown in Figure 6.26. The angle that should be used depends on the shape of the object and the location of its features. For example, in Figure 6.27a a large angle was used because it gives a better view of the rectangular recess on the top, while in Figure 6.27b a small angle was chosen to show a similar feature on the side.

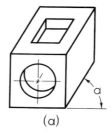

(a)   (b)

FIGURE 6.27 Angle of Receding Axis.

■ 6.25 LENGTH OF RECEDING LINES

Oblique sketches present an unnatural appearance, with more or less serious distortion, depending on the object shown. The appearance of distortion may be reduced by decreasing the length of the receding lines. In Figure 6.28 a cube with a hole through the front is shown in five oblique drawings with varying degrees of foreshortening of the receding lines.

When the receding lines are true length—that is, when the projectors make an angle of 45 degrees with the plane of projection—the oblique drawing is called a *cavalier projection*. Cavalier projections originated in the drawing of medieval fortifications and were made on horizontal planes of projection. On these fortifications the central portion was higher than the rest, and it was called cavalier because of its dominant position.

When the receding lines are drawn to half size, the drawing is commonly known as a *cabinet projection*. The term is attributed to the early use of this type of oblique drawing in the furniture industries. Figure 6.29 shows a comparison of cavalier projection and cabinet projection.

A striking example of the unnatural appearance of an oblique drawing when compared with the natural appearance of a perspective is shown in Figure 6.30. This example demonstrates that long objects should not be shown in oblique sketches with the long dimension receding from your viewpoint.

Other practices are similar to isometric sketching. If the object does not lend itself to easily being shown in an oblique sketch, you should consider using isometric sketching to show it instead.

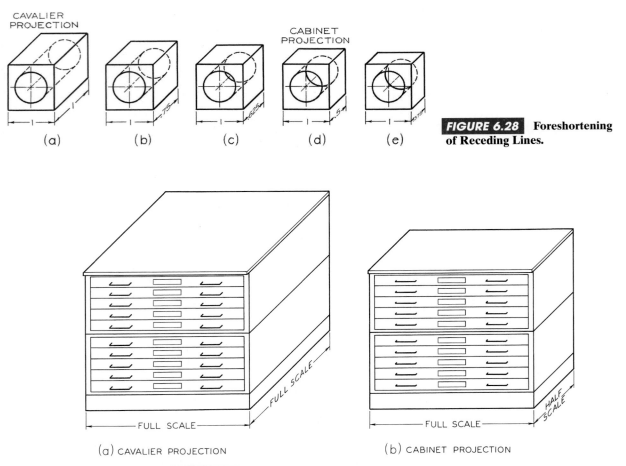

FIGURE 6.28 Foreshortening of Receding Lines.

(a) CAVALIER PROJECTION   (b) CABINET PROJECTION

FIGURE 6.29 Comparison of Cavalier and Cabinet Projections.

Hands On 6.3
Create an oblique sketch of an object

Directions:

Hold the object in your hand.

Orient the object so that most or all of the circular shapes are toward you. This way they will appear as true circles and arcs in the oblique sketch.

1. Block in the front face of the bearing as if you were sketching a front view.

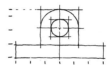

2. Sketch receding lines parallel to each other and at any convenient angle (say 30 or 45 degrees). Decide on the depth you will show. The depth lines may be shown full length, but three-quarters or one-half size produces a more natural appearance. Block in the back shape of the object.

Estimate depth

Any angle

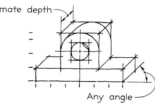

3. Darken the final lines.

Use this space to create an oblique sketch of a household object such as your clock, doorknob, or microwave, or sketch another object as provided by your instructor.

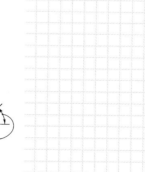

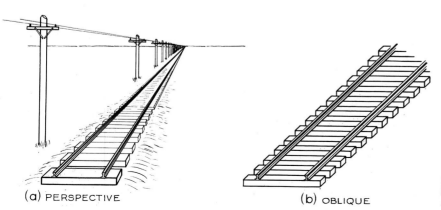

(a) PERSPECTIVE

(b) OBLIQUE

FIGURE 6.30 Unnatural Appearance of Oblique Drawing.

FIGURE 6.31   A CAD-produced Perspective of an Airport. *The material has been reprinted with permission from and under the copyright of Autodesk, Inc.*

■ 6.26 PERSPECTIVE SKETCHING

Perspective pictorials most closely approximate the view produced by the human eye. Figure 6.31 shows a perspective view of an airport produced using CAD. Photographs too show perspective. Perspective pictorials are important in architecture, industrial design, and illustration. Engineers also often need to show pictorial representation of objects and should understand the basic principles of perspective (see ANSI/ASME Y14.4M–1989 (R1994)). While perspective views are time consuming to sketch, they are easy to create from 3-D CAD models.

Unlike axonometric projection, perspective causes parallel edges to converge at vanishing points. The three types of perspective are one-point, two-point, and three-point perspective, depending on the number of vanishing points.

■ 6.27 GENERAL PRINCIPLES

A perspective involves four main elements: (1) the observer's eye, (2) the object being viewed, (3) the plane of projection, and (4) the projectors from all points on the object to the observer's eye.

In Figure 6.32 the observer is shown looking along a boulevard and through an imaginary plane of projection, the *picture plane*. The position of the observer's eye is called the *station point*, and the lines from the station point to the points in the scene are the *projectors* or *visual rays*. Collectively the points where the visual rays pierce the picture plane are the perspective view of the object as seen by the observer, which is shown in Figure 6.33.

Notice how each succeeding lamp post, as it is farther from the observer, appears smaller than the preceding one. A lamp post at an infinite distance from the observer would appear as a point on the picture plane. A lamp post in front of the picture plane would be projected taller than its actual size, and a lamp post in the picture plane would be projected in true length.

The line representing the *horizon* is the edge view of the *horizon plane*, which is parallel to the *ground plane* and passes through the station point. The horizon is the line of intersection of this plane with the picture plane and represents the eye level of the observer. The ground plane is the edge view of the ground that the objects are sitting on. The *ground line* is the intersection of the ground plane with the picture plane.

Notice that lines that are parallel to each other but not parallel to the picture plane—such as curb lines, sidewalk lines, and lines along the tops and bottoms of the lamp posts—all converge toward a single point on the horizon—the *vanishing point* of the lines. The first

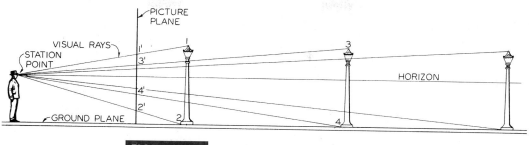

FIGURE 6.32 Looking Through the Picture Plane.

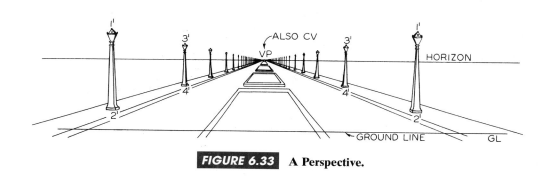

FIGURE 6.33 A Perspective.

rule of perspective is that all parallel lines that are not parallel to the picture plane vanish at a single vanishing point, and if these lines are parallel to the ground, the vanishing point will be on the horizon. Parallel lines that are also parallel to the picture plane, such as the lamp posts, remain parallel and do not converge toward a vanishing point.

■ 6.28 THE THREE TYPES OF PERSPECTIVES

Perspective drawings are classified according to the number of vanishing points required, which in turn depends on the position of the object with respect to the picture plane.

If the object sits with one face parallel to the plane of projection, only one vanishing point is required. The result is a *one-point perspective*, or *parallel perspective*.

If the object sits at an angle with the picture plane but with vertical edges parallel to the picture plane, two vanishing points are required, and the result is a *two-point perspective*, or an *angular perspective*. This is the most common type of perspective drawing.

If the object sits so that no system of parallel edges is parallel to the picture plane, three vanishing points are necessary, and the result is a three-point perspective.

■ 6.29 ONE-POINT PERSPECTIVE

To sketch a one-point perspective view, orient the object so that a principal face is parallel to the picture plane. If desired, this face can be placed in the picture plane. The other principal face is perpendicular to the picture plane, and its lines will converge toward a single vanishing point.

■ 6.30 TWO-POINT PERSPECTIVE

Two-point perspective is more true to life than one-point perspective. To sketch a two-point perspective, orient the object so that principal edges are vertical and therefore have no vanishing point; edges in the other two directions have vanishing points. Two-point perspectives are especially good for representing buildings and large civil structures, such as dams or bridges.

Step by Step 6.4
One-point perspective

To sketch the bearing in one-point perspective—that is, with one vanishing point—follow the steps illustrated below.

1. Sketch the true front face of the object, just as in oblique sketching. Select the vanishing point for the receding lines. In most cases it is desirable to place the vanishing point above and to the right of the picture, as shown, although it can be placed anywhere in the sketch. However, if the vanishing point is placed too close to the center, the lines will converge too sharply and the picture will be distorted.

VP
+

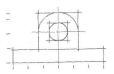

2. Sketch the receding lines toward the vanishing point.

VP

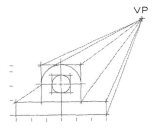

3. Estimate the depth to look good and sketch in the back portion of the object. Note that the back circle and arc will be slightly smaller than the front circle and arc.

VP
Estimate depth

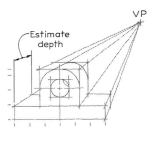

4. Darken all final lines. Note the similarity between the perspective sketch and the oblique sketch earlier in the chapter.

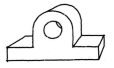

■ 6.31 THREE-POINT PERSPECTIVE

In *three-point perspective*, the object is placed so that none of its principal edges is parallel to the picture plane. Each of the three sets of parallel edges has a separate vanishing point. In this case, use a picture plane that is approximately perpendicular to the centerline of the cone of visual rays.

Figure 6.34 shows the construction of a three-point perspective. Think of the paper as the picture plane, with the object behind the paper and placed so that all its edges make an angle with the picture plane. Locate the vanishing points P, Q, and R by drawing lines from a station point in space parallel to the principal axes of the object and finding their piercing points in the picture plane.

The imaginary corner O is in the picture plane and may coincide with CV, but as a rule the front corner is placed at one side near the center of vision for the drawing (CV), determining how close the observer is to being directly in front of this corner.

In this method the perspective is drawn directly from measurements and not projected from views. The dimensions of the object are given by the three views, and these will be used on the measuring lines GO, EO, and OF. The measuring lines EO and OF are drawn parallel to the vanishing trace PQ, and the measuring line GO is drawn parallel to RQ. These measuring lines are actually the lines of intersection of principal surfaces of the object, extended, with the picture plane. Since these lines are in the picture plane, true measurements of the object can be set off along them.

Three measuring points—M_1, M_2, and M_3—are used in conjunction with the measuring lines. To find M_1, revolve triangle CV–R–Q about RQ as an axis. Since it is a right triangle, it can be constructed true size using a semicircle, as shown. With R as the center and R–SP_1 as the radius, make arc SP_1–M_1, as shown. M_1 is the measuring point for the measuring line GO. Measuring points M_2 and M_3 are found in a similar manner.

Height dimensions are measured full size or to any desired scale along measuring line GO at points 3, 2, and 1. From these points, lines are drawn to M_1, and heights on the perspective are the intersections of these lines with the perspective front corner OT of the object. Similarly, the true depth of the object is set off on measuring line EO from 0 to 5, and the true width is set off on measuring line OF from 0 to 8. Intermediate points can be constructed in a similar manner.

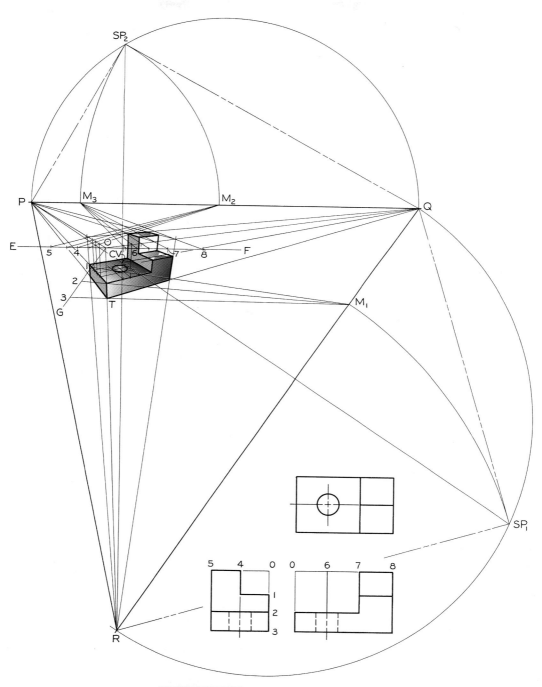

FIGURE 6.34 Three-Point Perspective.

Step by Step 6.5
Two-point perspective

To sketch a desk using two vanishing points, follow these steps:

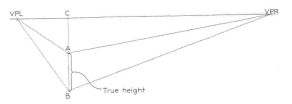

1. As shown above, sketch the front corner of the desk at true height. Locate two vanishing points (VPL and VPR) on a horizon line (at eye level). Distance CA may vary—the greater it is, the higher the eye level will be and the more we will be looking down on top of the object. A rule of thumb is to make C–VPL one-third to one-fourth of C–VPR.

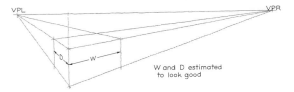

2. Estimate depth and width, and sketch the enclosing box.

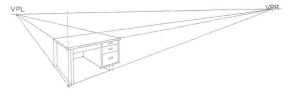

3. Block in all details. Note that all parallel lines converge toward the same vanishing point.

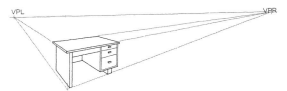

4. Darken all final lines. Make the outlines thicker and the inside lines thinner, especially where they are close together.

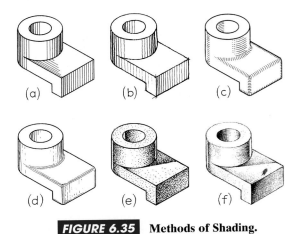

FIGURE 6.35 Methods of Shading.

■ 6.32 SHADING

Shading can make it easier to visualize pictorial drawings—such as display drawings, patent drawings, and catalog drawings. Ordinary multiview and assembly drawings are not shaded. The shading should be simple, reproduce well, and produce a clear picture. Some of the common types of shading are shown in Figure 6.35. Two methods of shading fillets and rounds are shown in Figures 6.35c and 6.35d. Shading produced with dots is shown in Figure 6.35e, and pencil tone shading is shown in Figure 6.35f. Pencil tone shading used in pictorial drawings on tracing paper reproduces well only when making blueprints, not when using a copier.

Examples of line shading on pictorial drawings used in industrial sales literature are shown in Figures 6.36 and 6.37.

■ 6.33 COMPUTER GRAPHICS

Perspective drawings, which provide pictorials most resembling photographs, are also the most time-consuming to draw. CAD programs are available that produce either wireframe representations (Figure 6.38) or solid perspective representations, with the user selecting viewing distance, focal point, z-axis convergence, and arc resolution scale. Historically, perspectives have seen far greater application in architectural than in engineering drawing. Now the availability of these computer graphics routines makes perspective drawing a viable alternative for the drafter wishing to employ a pictorial representation of an object.

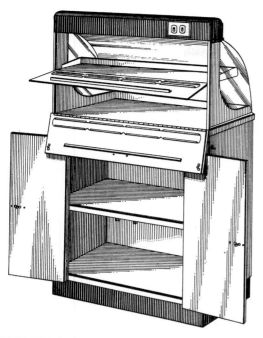

FIGURE 6.36 Surface Shading Applied to Pictorial Drawing of Display Case.

FIGURE 6.37 A Line-Shaded Drawing of an Adjustable Support for Grinding. *Courtesy of A.M. Byers Co.*

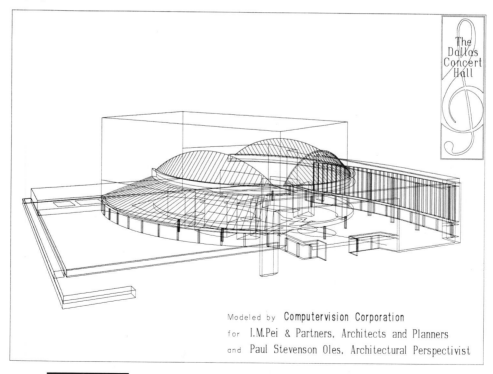

FIGURE 6.38 Perspective Drawing Produced by Using the Computervision Designer System for Building and Management (BDM).

Courtesy of Computervision Corporation, a subsidiary of Prime Computer, Inc.

GRAPHICS SPOTLIGHT

3D Pictorials Aid Designers of Future Electric Cars

They rode on rockets to the moon and proved they can generate enough electricity to power an automobile. But fuel cells are still years away from widespread use in the automobile industry.

Standing in their path are high costs and the problems of a complex technology. Together, these forces ensure large numbers of fuel-cell-powered vehicles will not hit showroom floors for a decade or more.

"There's a lot of serious work going into it, and there's a lot of potential there," says Bernard Robertson, DaimlerChrysler's vice president of engineering technologies. "It's just that the challenges are pretty formidable."

FUEL CELLS POSE THORNY PROBLEMS FOR DESIGN TEAM

A fuel cell uses sophisticated membranes to strip electrons from hydrogen atoms creating a charge imbalance and electrical current. The cell recombines the hydrogen with oxygen to form water vapor. DaimlerChrysler is trying to lead the industry in fuel cells by forming joint ventures with Ford Motor Co. and Ballard Powers Systems, Inc. of Vancouver, British Columbia, a supplier of fuels cell stacks. The goal is to be the first to manufacture complete fuel-cell powertrains for sale in the world market.

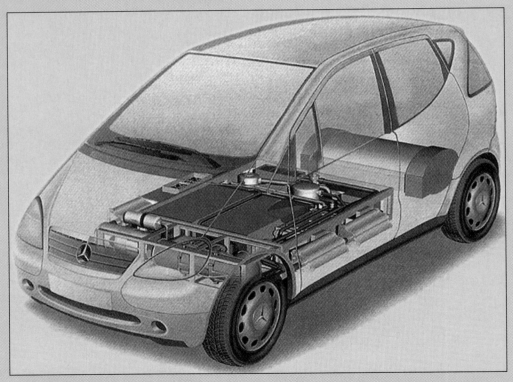

3D pictorial of Necar 4 fuel-cell-powered car. Such 3D images aid designers in overcoming current design problems.

The task is daunting. At DaimlerChrysler's fuel-cell development center near Stuttgart, Germany, 900 technical people are devoted exclusively to fuel-cell research. Necar 4, the fourth generation of the center's New Electric Car series of fuel-cell concepts, was unveiled to the public in March 1999. Necar is technically impressive because it crams the entire fuel-cell system in a 6-inch-deep space under the floor, but also illustrated the shortcomings of current fuel-cell technology.

DaimlerChrysler officials admit the vehicle is overweight by more than 600 pounds and is astronomically expensive. A mass-produced fuel cell system would cost $30,000 using today's technology, although Necar 4's hand-built engine is estimated to have cost $350,000. Gasoline engines typically cost $3,000. Specialized hardware is what drives the fuel cell's costs. The largest expense is the row of electricity-conducting bipolar plates in the fuel cell stack. These plates, made from ultra hard carbon-graphite, have dozens of intricate channels that must be individually cut by computer-controlled machine tools. For maximum efficiency, the channels must be machined to the high tolerance usually reserved for jet engine turbines.

CAD HELPS DESIGNER SEE POSSIBILITIES

With all of this precision machining involved, the engineers at DaimlerChrysler have been assigned two separate tasks: (1) develop a fuel cell that is small and light enough to meet size and weight restrictions; and (2) develop manufacturing equipment capable of mass producing such a product at a reasonable cost.

Since making even one model of such a cell or vehicle is cost restrictive, the design teams have turned to computer-design and computer-manufacturing software programs for help. 3D pictorials let them see not only how they can better design the fuel cell for each vehicle, but what is needed in the way of retooling for mass production of such cells. By keeping costs down, engineers are hopeful that they will be able to design and develop an efficient electric car by the year 2004. Retooling of assembly lines could take a few years longer, but DaimlerChrysler is confident that they will have an efficient, reliable, fuel-cell powered vehicle on the world market within the next decade.

Adapted from "Fuel Cells Still Pose Thorny Problems," by Aaron Robinson, "Automotive News," March 29, 1999.

■ KEY WORDS

pictorial drawing	isometric projection	perspective pictorial	ground line
multiview projection	dimetric projection	picture plane	vanishing point
axonometric projection	trimetric projection	station point	one-point perspective
orthographic projection	isometric	projectors	parallel perspective
oblique projection	isometric sketch	visual rays	two-point perspective
perspective sketch	offset measurements	horizon	angular perspective
perspective view	cavalier projection	horizon plane	three-point perspective
axonometric axes	cabinet projection	ground plane	

■ CHAPTER SUMMARY

- Axonometric projection is a method of creating a pictorial representation of an object. It shows all three dimensions of length, width, and height in one view.

- Isometric projection is the easiest of the axonometric projections to draw and is therefore the most common pictorial drawing technique.

- The spaces between the axes of an isometric drawing each are 120 degrees. Isometric axes are drawn at 30 degrees to the horizontal and vertical.

- The only lines on an isometric drawing that are equally foreshortened are lines parallel to the three isometric axes.

- Inclined surfaces and oblique surfaces must be determined by plotting the endpoints of each edge of the surface along isometric axis lines.

- A common method of drawing an object in isometric is by creating as isometric box and drawing the features of the object within the box.

- Unlike perspective drawing, in which parallel lines converge at a vanishing point, parallel lines remain parallel in axonometric drawings.

- Oblique projection makes it easy to sketch objects that show circular shapes and other details parallel to the front view.

- Cavalier projection shows the depth of the object full size and receding lines at 45 degrees.

- Cabinet projection shows receding lines half size, often using 30 degrees for their angle.

- Perspective views are most like the view seen by the human eye.

- There are three types of perspective projection: one-point, two-point, and three-point perspective.

- In perspective projection, parallel edges converge at one or more vanishing points, replicating the image of objects seen by the human eye.

- The location and relationship between the vanishing points, the picture plane, and the object determine the appearance of the perspective view.

- In one-point perspective, the object is placed so that a principal surface of the object is parallel to the picture plane.

- In two-point perspective, the object is placed so that only principal edges of the object are oriented vertically, but major surfaces are not parallel to the picture plane.

- In three-point perspective, the object is placed so that none of the three primary axes of the object are parallel to the picture plane.

■ REVIEW QUESTIONS

1. Why is isometric drawing more common than perspective drawing in engineering?

2. What are the differences between axonometric projection and perspective?

3. At what angles are the isometric axes drawn?

4. What are the three views that are typically shown in an isometric drawing?

5. What is the primary advantage of an oblique projection?

6. Which is the most realistic: isometric, perspective, or oblique?

7. Why are oblique drawings seldom created using CAD?

8. What is the primary advantage of a perspective projection?

9. Why is perspective projection rarely used in engineering?

10. What is the purpose of the picture plane?

11. What is the station point?

12. How does the distance between the station point and the ground line affect the final perspective drawing?

13. What is the relationship between the station point and the horizon?

■ AXONOMETRIC PROJECTS

Figures 6.39–6.45 consist of problems to be drawn axonometrically. Use either isometric or oblique sketching as assigned by your instructor.

Choose an appropriate sheet layout and scale. Use isometric or plain drawing paper.

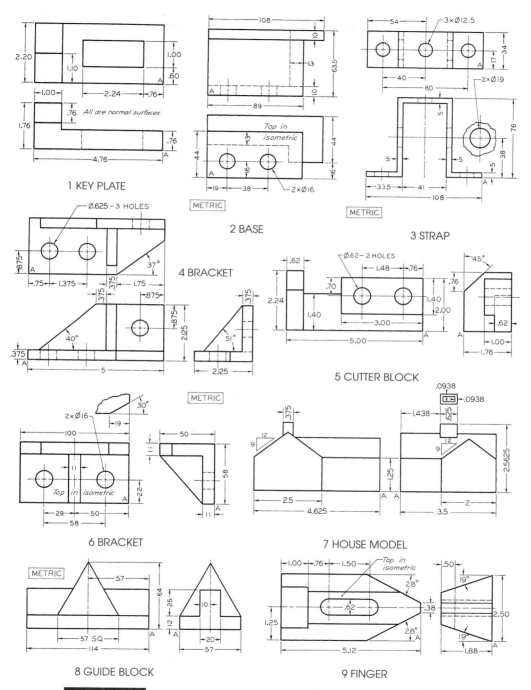

1 KEY PLATE

All are normal surfaces

METRIC

2 BASE

METRIC

3 STRAP

4 BRACKET

Ø.625 – 3 HOLES

5 CUTTER BLOCK

Ø.62 – 2 HOLES

6 BRACKET

2×Ø16

Top in isometric

METRIC

7 HOUSE MODEL

8 GUIDE BLOCK

METRIC

57 SQ

9 FINGER

Top in isometric

FIGURE 6.39 (1) Make freehand isometric sketches, with axes chosen to show the objects to best advantage.

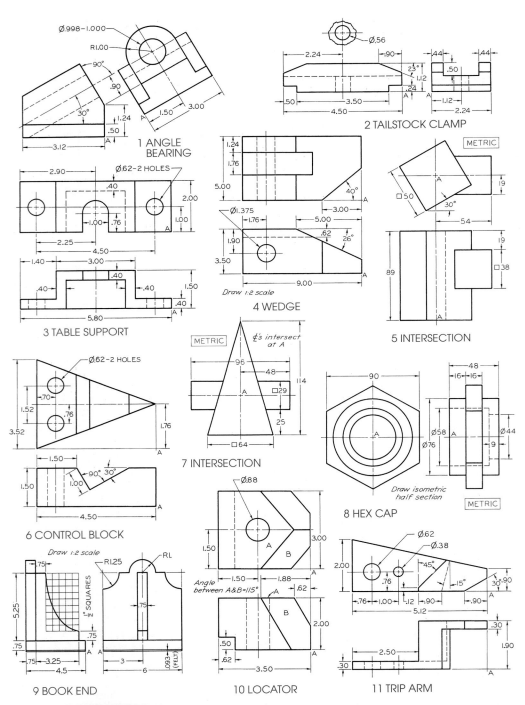

FIGURE 6.40   Make freehand isometric or oblique sketches, as assigned. If dimensions are required, refer to §6.20.

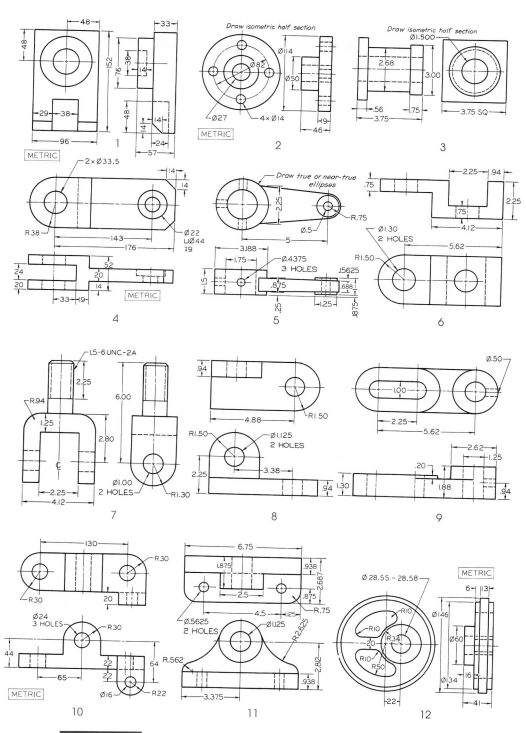

FIGURE 6.41 Make freehand isometric or oblique sketches, as assigned. If dimensions are required, refer to §6.20.

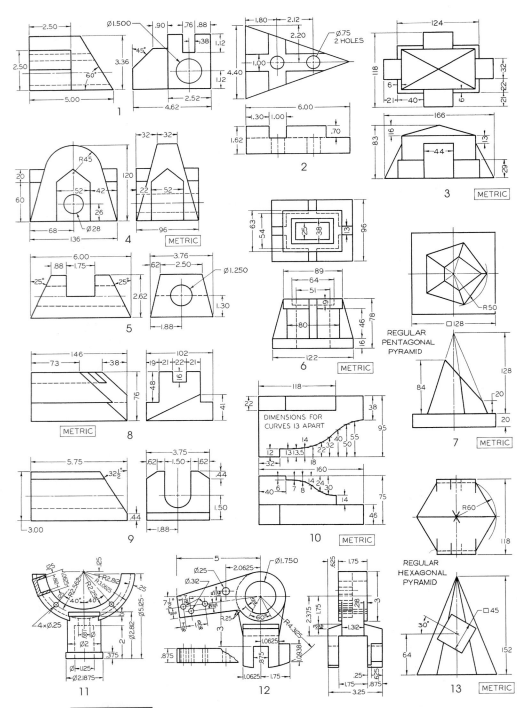

FIGURE 6.42 Make freehand isometric or oblique sketches, as assigned. If dimensions are required, refer to §6.20.

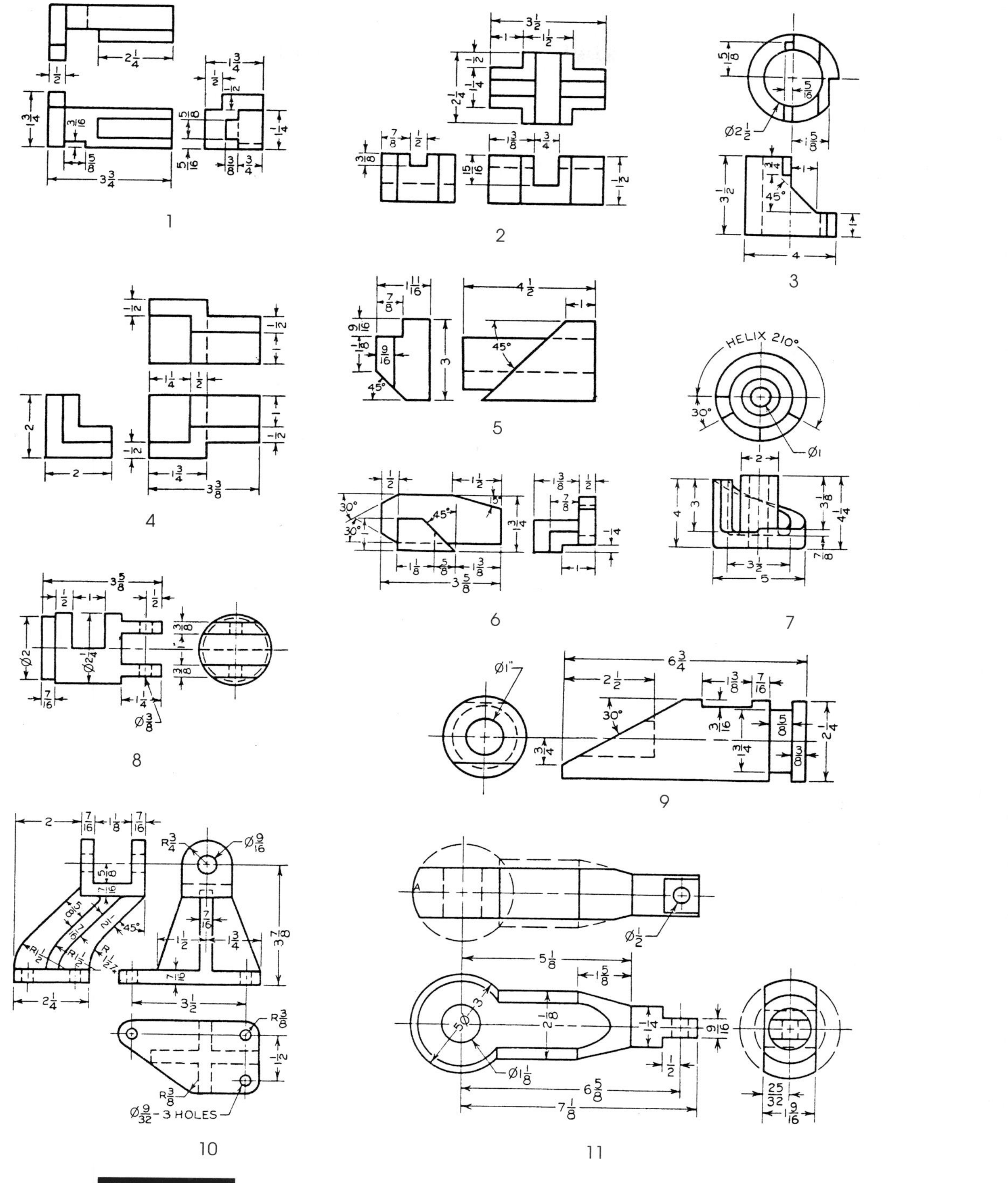

FIGURE 6.43   Make freehand isometric or oblique sketches, as assigned. If dimensions are required, refer to §6.20.

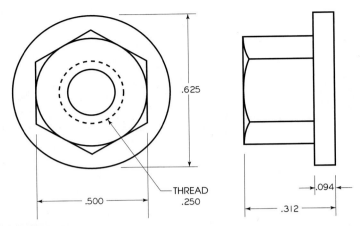

FIGURE 6.44   *Nylon Collar Nut.* **(1) Make isometric freehand sketch. (2) Make isometric drawing with instruments, using Size A or A4 sheet or Size B or A3 sheet, as assigned.**

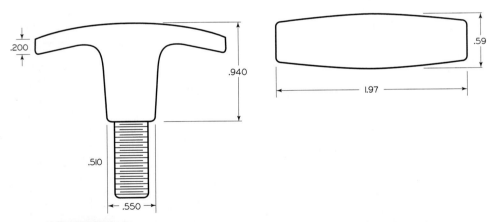

FIGURE 6.45   *Plastic T-Handle Plated Steel Stud.* **(1) Make dimetric drawing with instruments, using Size A or A4 sheet as assigned. (2) Make trimetric drawing, using instruments. Use Size A or A4 sheet as assigned.**

■ PERSPECTIVE PROJECTS

Draw two orthographic views and a perspective of each figure shown in 6.46 and 6.47. Select an appropriate sheet size and scale.

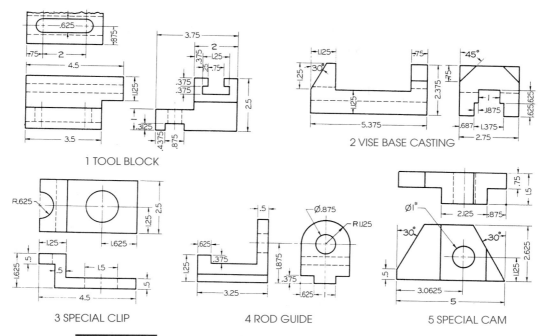

1 TOOL BLOCK

2 VISE BASE CASTING

3 SPECIAL CLIP

4 ROD GUIDE

5 SPECIAL CAM

FIGURE 6.46   Draw two views and a perspective of assigned problem. Omit dimensions. Select sheet size and scale.

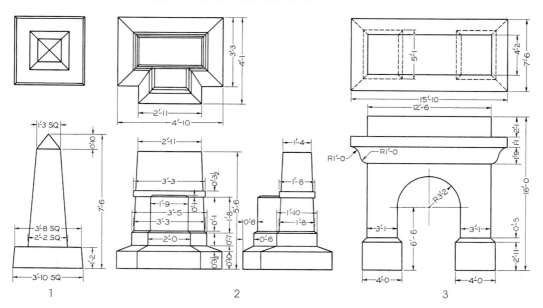

FIGURE 6.47   Draw two views and a perspective of assigned problem. Omit dimensions. Select sheet size and scale.

■ DESIGN PROJECT

Use pictorial drawing to show your design of a better mousetrap. Decide whether you want your mousetrap to be low-priced and efficient or feature-rich and imaginative. Consider a traditional size, large-scale, or miniature approach. Use an appropriate scale to show all features clearly.

C H A P T E R   7

SECTIONAL VIEWS

OBJECTIVES

After studying the material in this chapter, you should be able to:

1. Understand the meaning of sections and cutting-plane lines.
2. Identify seven types of sections.
3. Draw a sectional view given a two-view drawing.
4. Demonstrate the proper techniques for sectioning ribs, webs, and spokes.
5. Demonstrate the proper technique for aligned sections.
6. Recognize the correct hatching patterns, or section lining, for 10 different materials.
7. Draw correct conventional break symbols for elongated objects.

OVERVIEW

So far you have learned the basic methods for representing objects through views or projections. By drawing carefully selected views, you can describe complicated designs. However hidden lines showing interior features are often difficult to interpret. Sectional views—often called cross sections or simply sections—show such interiors by imagining the object sliced through, much as you cut through an apple or melon. You can use sections when you need to clearly show the internal structure of complex objects which otherwise would require many hidden lines. There are many types of sectional views. You should become familiar with the different types and know when to use them. Sections often replace one of the primary views in the drawing.

The cutting-plane line shows where the object is hypothetically cut. Cross hatching, sometimes called section lining, accentuates the solid parts of the object cut by the cutting plane. Lines that were previously hidden may become exposed by sectioning the object. Lines that remain hidden lines are not usually drawn in sectional views. Special conventions—like not hatching ribs, webs, and other similar features—are used to make sections easy to interpret. Creating a sectional view can be a complicated operation when you are using CAD programs. You should thoroughly understand the concepts described in this chapter to be able to create clear, easy-to-interpret sectional views using CAD. See ANSI/ASME Y14.2M–1992 and Y14.3M–1994 for complete standards for multiview and sectional-view drawings.

■ 7.1 SECTIONING

To produce a *sectional view*, a *cutting plane* is imagined cutting through the part, as shown in Figure 7.1a. Picture the two halves of the object pulled apart, exposing the interior construction, as shown in Figure 7.1b. In this case, you will look toward the left half of the object in the section view. In other words, the direction of sight for the sectional view is toward the left half, with the right half mentally discarded. In this case, the sectional view replaces the right-side view.

■ 7.2 FULL SECTIONS

The section produced by cutting through the entire object is called a *full section*, as shown in Figure 7.2c. Compare this section with the left-side view in Figure 7.2a and notice how much better the sectional view shows the interior detail. (You would not normally

show the left-side view in this case because it duplicates information better shown by the section. It is shown here just as an example.) Notice that the right half of the object is not removed anywhere except in the sectional view itself. It is only imagined to be removed to produce the section. The cross-hatched areas, created with thin parallel lines equally spaced by eye, represent the solid portions that have been cut. Parts of the object now visible in back of the cutting plane are shown but are not cross hatched.

■ 7.3 THE CUTTING PLANE

The *cutting-plane line*, seen in the front view in Figure 7.2b, is shown as a special line pattern. The arrows at the ends of the cutting-plane line indicate the direction of sight for the sectional view. In most cases, the location of the cutting plane is obvious from the section itself, so the cutting-plane line may be omitted, but it should be shown wherever necessary for clarity. It is shown in Figure 7.2 for illustration only. The cutting plane is indicated in a view adjacent to the sectional view. You can think of the cutting-plane line as showing the edge view of the cutting plane. When a cutting-plane line coincides with a centerline, the cutting-plane line takes precedence.

■ 7.4 CUTTING PLANE LINE PATTERNS

Figure 7.3 shows two line patterns used to show the cutting-plane line. One style uses equal dashes each about 6 mm (1/4 inch) or more long plus the arrowheads. The other uses alternating long dashes and pairs of short dashes plus the arrowheads. This pattern has been in general use for a long time. Draw both line types as thick lines, similar to the thickness of visible lines in your drawing. Arrowheads indicate the direction in which the cut object is viewed. One variation of the cutting-plane line shows just the ends of the cutting plane with the arrows. When using this style, leave a slight gap between the cutting plane and the object, as shown in Figure 7.4c. This style is useful for complicated drawings where showing the entire cutting plane would obscure part of the drawing. Capital letters are used at the ends of the cutting-plane line when needed to identify the indicated section, such as in drawings of multiple sections or removed sections, which are described later in this chapter.

■ 7.5 INTERPRETING CUTTING PLANES AND SECTIONS

Sectional views often replace standard views. In Figure 7.4a the cutting plane is a frontal plane—that is, parallel to the front view—and appears as a line in the top

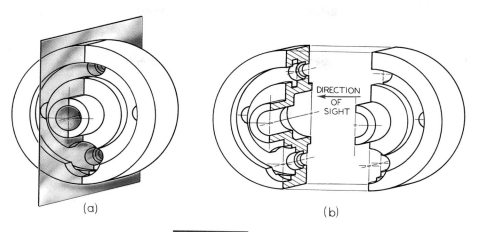

(a)   (b)

FIGURE 7.1   A Section.

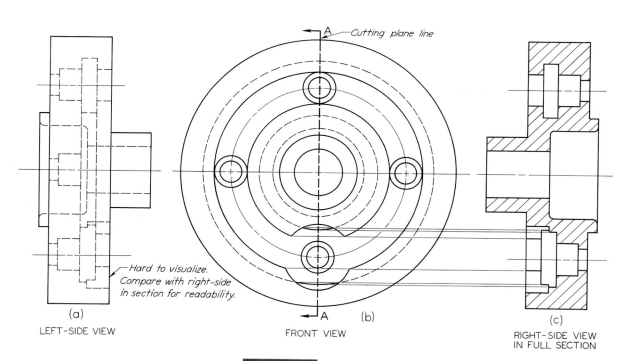

Cutting plane line

Hard to visualize.
Compare with right-side
in section for readability.

(a)   (b)   (c)
LEFT-SIDE VIEW   FRONT VIEW   RIGHT-SIDE VIEW
                                                   IN FULL SECTION

FIGURE 7.2   Full Section.

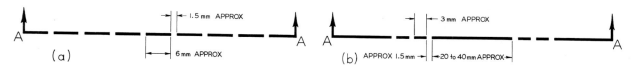

1.5 mm APPROX   3 mm APPROX

(a)   6 mm APPROX   (b)   APPROX 1.5 mm   20 to 40 mm APPROX

FIGURE 7.3   Cutting Plane Lines (Full Size).

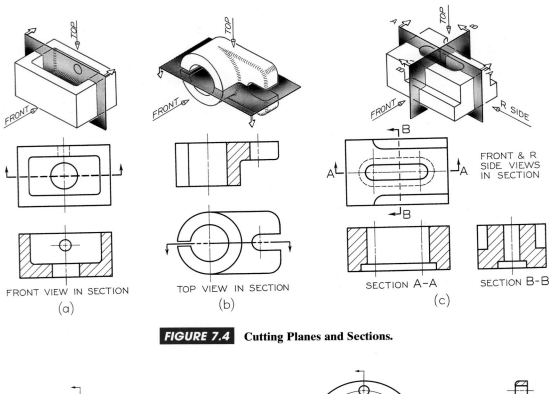

FIGURE 7.4 Cutting Planes and Sections.

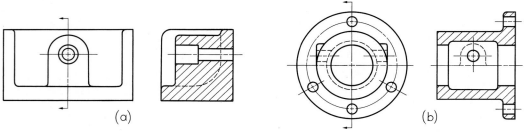

FIGURE 7.5 Hidden Lines in Sections.

view. The front half of the object is imagined removed. The arrows at the ends of the cutting-plane line point in the direction of sight for a front section. Note that the arrows do not point in the direction of withdrawal of the removed portion, but instead point toward the remaining portion of the object and indicate the direction you will look to draw the sectional view. The result is called a front section, or front view in section, since it replaces the front view in the drawing.

In Figure 7.4b the cutting plane is a horizontal plane, appearing as a line in the front view. The upper half of the object is imagined removed. The arrows point toward the lower half in the same direction of sight as for a top view. The result is a top view in section.

In Figure 7.4c two cutting planes are shown—one a frontal plane, the other a profile plane, or parallel to a side view—and both appear on edge in the top view. Each section is completely independent of the other and drawn as if the other were not present. For section A–A, the front half of the object is imagined removed, the back half is viewed in the direction of the arrows for a front view, and the result is a front section. For section B–B, the right half of the object is imagined removed and the left half is viewed in the direction of the arrows, producing a right-side section. Preferably, the cutting-plane lines are drawn through an exterior view (in this case the top view) instead of through a sectional view.

The cutting-plane lines in Figure 7.4 are shown for purposes of illustration only. They are generally omitted in cases such as these, in which the location of the cutting plane is obvious.

Keep the following things in mind when drawing sectional views:

- *Visible edges and contours behind the cutting plane should be shown;* otherwise a section will appear to be made up of disconnected and unrelated parts. Occasionally, however, visible lines behind the cutting plane are not necessary for clarity and can be omitted.

- *Hidden lines should be omitted in sectional views.* Sections are used primarily to replace hidden-line representations, which take extra time to draw and can be confusing. Sometimes hidden lines are necessary for clarity—for example, when a feature on the object would not be clearly defined in another view. Showing some hidden lines in the section may make it possible to omit a view, in which case they should be shown. An example is given in Figure 7.5.

- *A section-lined area is always completely bounded by a visible outline—never by a hidden line.* In every case the cut surfaces and their boundary lines will be visible because they are now the closest portion of the object. Also, a visible line can never cut across a section-lined area because all of the cutting plane area is in a single plane. (You will learn to use offset and aligned cutting planes later in this chapter. Even in those cases, the cut surface is imagined to lie in the same plane.)

- *The cross hatching in all hatched areas must be parallel in a sectional view of a single object,* alone or in assembly. Using hatching in different directions indicates different parts, as when two or more parts are adjacent in an assembly drawing.

Hands On 7.1
Find the errors in section views

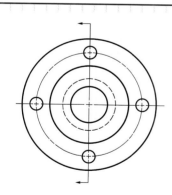

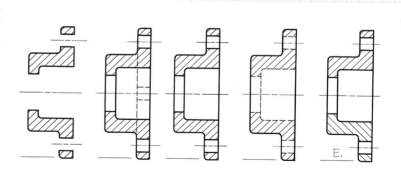

Directions:

Each of the drawings shown at right represents a section view of the front view above. One is drawn correctly and the others are incorrect for various reasons. Match the description of what is wrong by writing its letter in the space provided. One has been done for you.

A. Missing lines of object that are visible behind cutting plane

B. Correct

C. Hatched areas are visible, never bounded by hidden lines

D. Hidden lines are not usually shown

E. Hatching always runs a single direction on a single part

Hands On 7.2

Cutting plane lines and sectional views

Directions

Of the cutting plane lines and corresponding sectional views shown below, two are incorrect and one is correct.

1. Use the pictorial views at right to help you determine which drawing shows the correct relationship between the cutting plane line and the corresponding sectional view. Write "correct" on the line next to it.

2. Next to the drawings that show an incorrect relationship, write "incorrect." Sketch them correctly in the space provided.

Refer to Section 7.5 for help interpreting cutting planes and sections.

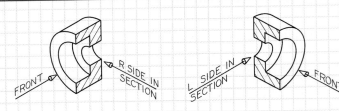

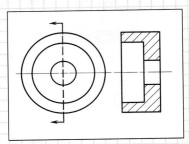

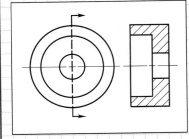

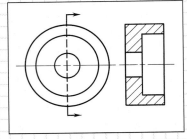

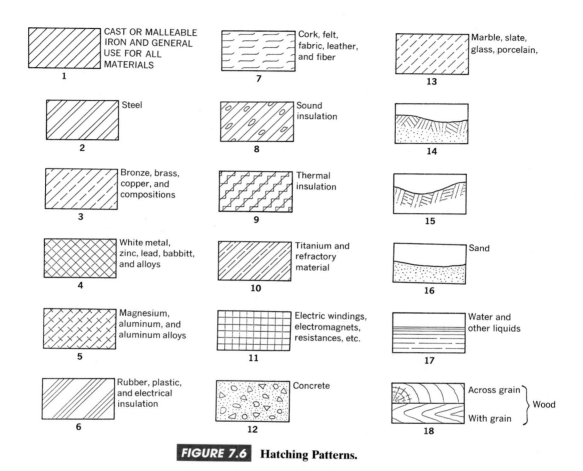

FIGURE 7.6 Hatching Patterns.

■ 7.6 CROSS-HATCHING

Cross-hatching patterns, shown in Figure 7.6, are used to represent general material types, such as cast iron, brass, and steel. Because there are today so many different types of materials, each with so many subtypes, a general name or symbol is not descriptive enough. For example, there are hundreds of different kinds of steel. Therefore, detailed specifications of material are usually provided in a note or in the title strip, and the general-purpose crosshatch for cast iron is used for all materials in detail drawings.

Different cross-hatching symbols may be used in assembly drawings when it is desirable to distinguish between materials; otherwise, the general-purpose symbol is used. CAD programs usually include a library of cross hatching patterns, making it easy to indicate various types of material.

By eye, space the cross-hatching as evenly as possibly, from approximately 1.5 mm (1/16 inch) to 3 mm (1/8 inch) or more apart depending on the size of the drawing or of the sectioned area. For most drawings, the spacing should be about 2.5 mm (3/32 inch) or

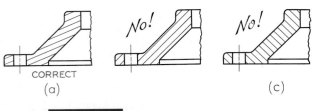

FIGURE 7.7 Direction of Hatching.

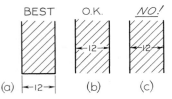

FIGURE 7.8 Dimensions and Hatching. Metric.

slightly more apart. As a rule, space the lines as generously as possible and yet close enough to distinguish clearly the sectioned areas.

Make hatching lines at 45 degrees with horizontal, unless they look better at a different angle. For example, in Figure 7.7 cross-hatching at 45 degrees with horizontal would be nearly parallel or nearly perpendicular to a prominent feature. In this case, the hatching appears better drawn at 30 degrees, 60 degrees, or some other angle.

Dimensions should be kept off hatched areas, but when this is unavoidable the cross-hatching should be omitted where the dimension figure is placed. An example of omitting hatching where dimensions are necessary is shown in Figure 7.8.

For large areas of hatching, use outline hatching. In outline hatching, portions of the hatching are left out and only the hatching near boundaries is drawn. Make sure that the drawing can still be read clearly. A good example is shown in Figure 7.9.

FIGURE 7.9 Design Layout.

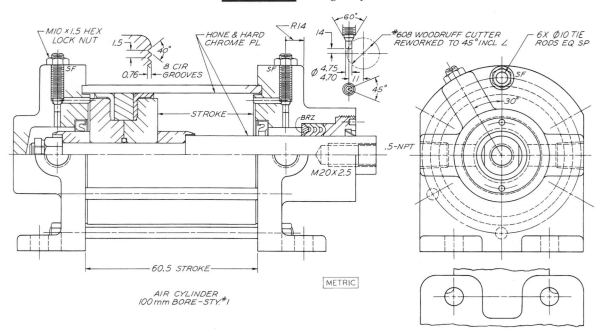

Hands On 7.3
Hatching

Directions

What is wrong in each hatching sample below? Look for hints in the Practical Tips at right. Write answers in the spaces provided. The first one is done for you.

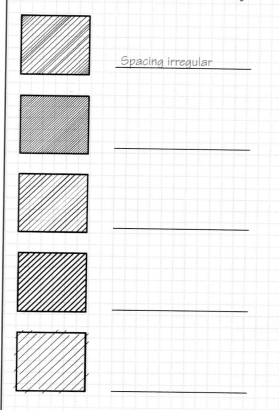

Spacing irregular

_____

_____

_____

_____

Practical Tips
Hatching methods

The correct method of drawing cross-hatching is shown at right. Draw the cross-hatching with a sharp, medium-grade pencil such as H or 2H. Here are some helpful hints:

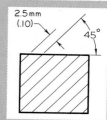

2.5mm (.10)   45°

- After the first few lines have been drawn, look back repeatedly at the original spacing to avoid gradually increasing or decreasing the intervals.

- Beginners almost invariably draw hatch lines too close together. This not only takes longer, but makes the least inaccuracy in spacing obvious.

- Hatching should be uniformly thin, not varying in thickness. There should be a noticeable contrast in thickness of the visible outlines and the hatching.

- Hatching should not be too thick.

- Avoid running cross-hatching beyond the visible outlines or stopping the lines too short.

Practice

Practice your hatching technique in the boxes below

Then add hatching to the drawing at right.

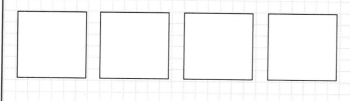

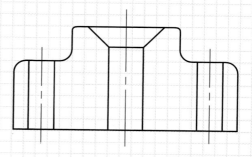

■ 7.7 VISUALIZING A SECTION

Since the purpose of a sectional view is to eliminate hidden lines and clearly show interior features of an object, the cutting plane should be placed so that it reveals the details that you want to show.

Before you can draw a full section, it is necessary to visualize how the object will look when the cutting plane has sliced though it. Step by Step 7.1 details the process used to visualize and draw a section of a drilled and counterbored collar.

Identifying the surfaces of the object in each view is an important step that makes the process of projecting the section view less confusing. Determine the location for the cutting plane, and imagine that the portion of the object between you and the cutting plane has been removed.

To locate the points you will use to create a projection of the section, identify where the cutting plane passes through solid parts of the object, starting with the outer surfaces of the object, which you know to be solid. Project the points that bound those solid areas through which the cutting plane passes. Use hatching to represent the cut surfaces in the section view. Remember to also show any parts of the object that are visible behind the cutting plane. Do not show hidden lines.

 On the World Wide Web you can find examples of cutaway views of such diverse subjects as complex heating systems, architectural products, the Mir space station, and even a human body.

* http://www.dryair.com/dahp6b.jpg
* http://www.hebel.com/cutaway.htm
* http://shuttle.nasa.gov/sts-71/pob/sts71/slmir/cutaway.html
* http://ucarwww.ucar.edu/staffnotes/12.94/vizmanvid.mpg

You can also find hatching patterns, tips and shareware for use with your CAD system, at sites like the following:

* http://www.cadsyst.com

 Use Worksheet 7.1 to practice creating a full section view.

■ 7.8 HALF SECTIONS

Symmetrical objects can be sectioned using a cutting plane passing halfway through an object, resulting in a *half section*. A half section exposes the interior of one

Step by Step 7.1
Visualizing a full section

Step 1: Choose a cutting plane

The illustration below shows two views of a collar to be sectioned. It has a drilled and counterbored hole. To produce a clear section showing both the counterbored hole and the smaller hole near the top of the object, choose the cutting plane to pass through the vertical center line in the front view and imagine the right half of the object removed.

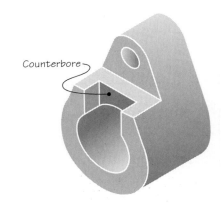

Step 2: Identify the surfaces

Below is a pictorial drawing of the remaining half. The first step in projecting the section view is making sure that you interpret the object correctly. Identifying the surfaces on the object can help. Surfaces R, S, T,U and V have been labeled on the given views and the pictorial view. Which surface is R in the front view? Which surface is U in the top view? Are they normal, inclined, or oblique surfaces. Can you identify the counterbore in each view?

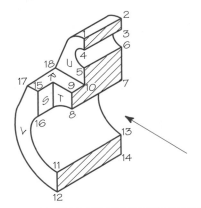

Step by Step 7.1
Visualizing a full section, cont.

Step 3: Drawing the section view

To draw the section view, omit the portion of the object in front of the cutting plane. You will only be drawing the portion that remains.

Determine which are solid parts of the object the cutting plane will pass through. Hint: The outside of an object can never be a hole; it must be solid.

The points which will be projected to create the section view have been identified for you in the example shown.

The three surfaces produced by the cutting plane are bounded by points 1-2-3-4 and 5-6-7-8-9-10 and 13-14-12-11. These are shown hatched.

Each section-lined area is completely enclosed by a boundary of visible lines. In addition to the cut surfaces, the sectional view shows all visible parts behind the cutting plane.

No hidden lines are shown. However, the corresponding section shown in this step is incomplete because visible lines are missing.

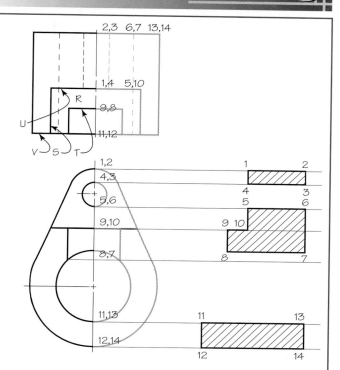

Step 4: Projecting the visible lines

From the direction the section is viewed, the top surface (V) of the object appears in the section as a visible line (12-11-16-15-17).

The bottom surface of the object appears similarly as 14-13-7-6-3-2. The bottom surface of the counterbore appears in the section as line 19-20.

Also, the back half of the counterbore and the drilled hole will appear as rectangles in the section at 19-20-15-16 and 3-4-5-6. These points must also be projected. The finished view is shown at right.

Notice that since all cut surfaces are part of the same object, the hatching must all run in the same direction.

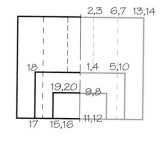
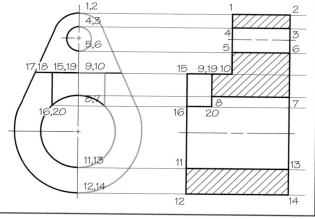

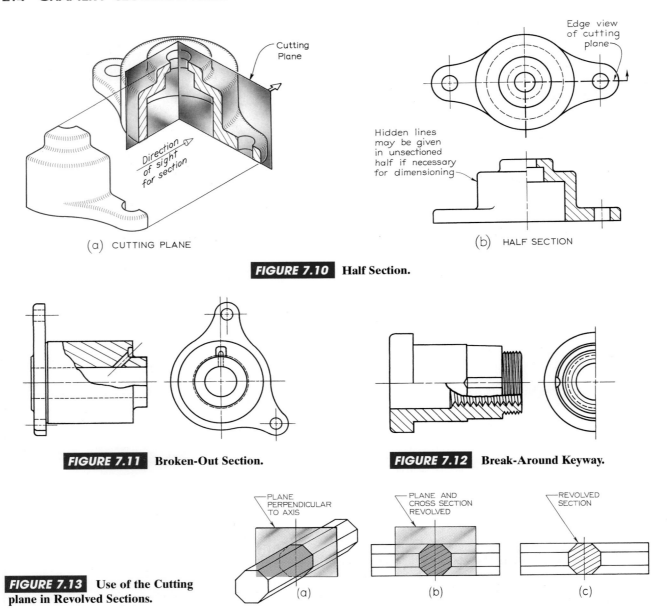

(a) CUTTING PLANE

(b) HALF SECTION

FIGURE 7.10 Half Section.

FIGURE 7.11 Broken-Out Section.

FIGURE 7.12 Break-Around Keyway.

FIGURE 7.13 Use of the Cutting plane in Revolved Sections.

half of the object and still shows the exterior of the other half. An example is shown in Figure 7.10. Half sections are not widely used in detail drawings—those showing dimensioned views of single parts—because it can be hard to fully dimension a part when internal features are only shown in the sectioned half. Half sections, however, are very useful in undimensioned assembly drawings (which you will learn about in Chapter 12) because they show both internal and external construction on the same view.

In general, hidden lines are omitted from both halves of a half section. However, they may be used in

the unsectioned half if necessary for dimensioning. As shown in Figure 7.10b, the American National Standards Institute recommends a centerline for the division line between the sectioned and unsectioned halves of a half section. A broken-out section representation may also be used.

■ 7.9 BROKEN-OUT SECTIONS
Often, only a partial section of a view is needed to expose interior shapes. Such a section, limited by a break line, is called a *broken-out section*. In Figure 7.11, a full or half section is not necessary. A broken-out

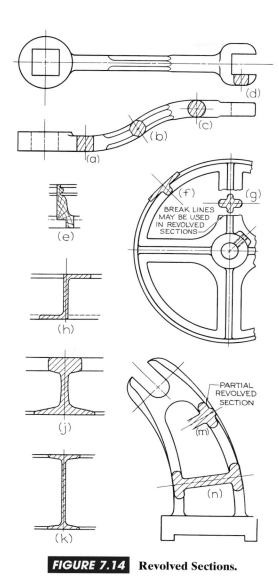

FIGURE 7.14   Revolved Sections.

- The visible lines adjacent to a revolved section may be broken out if desired, as shown below.

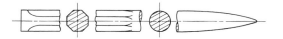

- Remove all original lines covered by the revolved section. Correct and incorrect examples are shown below.

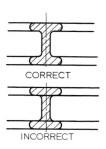

- The true shape of a revolved section should be retained after the revolution of the cutting plane, regardless of the direction of the lines in the view. Correct and incorrect examples are shown below.

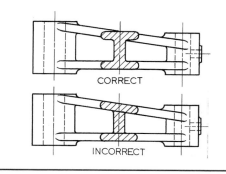

section is sufficient to explain the construction. In Figure 7.12, a half section would have caused the removal of half the keyway. The keyway is preserved by breaking out around it. In this case, the section is limited partly by a break line and partly by a centerline.

■ 7.10 REVOLVED SECTIONS

The shape of the cross-section of a bar, arm, spoke, or other elongated object may be shown by means of a **revolved section**. Revolved sections are made by assuming a plane perpendicular to the centerline or axis of the bar or other object, and then revolving the plane through 90 degrees about a centerline at right angles to the axis. Figure 7.13 depicts the process of

creating a revolved section. A number of examples of revolved sections are shown in Figure 7.14.

■ 7.11 REMOVED SECTIONS

A **removed section** is a section that is not in direct projection from the view containing the cutting plane—it is located somewhere else on the drawing. If you must locate sections in a removed position, their orientation should remain the same as if located adjacent to the view with the cutting-plane line. Removed sections should not be rotated (which is sometimes tempting to

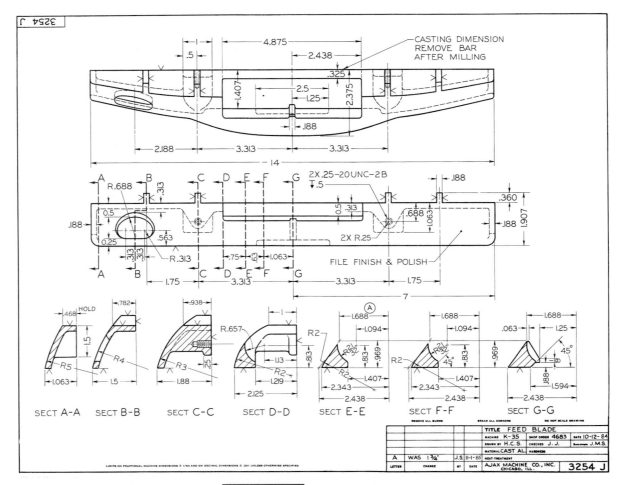

FIGURE 7.15 Removed Sections.

make them fit on the drawing sheet better) as this makes the section difficult to interpret. A removed section should be placed so that it no longer lines up in projection with any other view. It should be separated clearly from the standard arrangement of views. Whenever possible, removed sections should be on the same sheet as the regular views. Figure 7.15 shows correctly drawn removed sections.

Label removed sections, such as SECTION A–A and SECTION B–B, corresponding to the letters at the ends of the cutting-plane line. Arrange removed sections in alphabetical order from left to right on the sheet. Section letters should be used in alphabetical order, but the letters I, O, and Q should not be used because they are easily confused with the numerals 1

and 0. If you must place a section on a different sheet, cross-reference the related sheet. A note should be given below the section title, such as SECTION B–B ON SHEET 4, ZONE A3. Place a similar note on the sheet where the cutting-plane line is shown, with a leader pointing to the cutting-plane line and referring to the sheet on which the section will be found.

A removed section is often a partial section and is frequently drawn to an enlarged scale, as shown in Figure 7.16. This shows small detail clearly and provides sufficient space for dimensioning. The enlarged scale should be indicated below the section title. Sometimes it is convenient to place removed sections on centerlines extended from the section cuts, as shown in Figure 7.17.

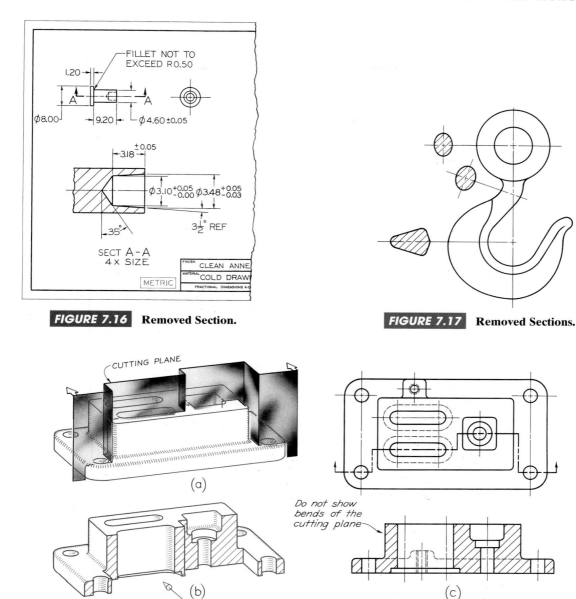

FIGURE 7.16   Removed Section.

FIGURE 7.17   Removed Sections.

FIGURE 7.18   Offset Section.

■ 7.12 OFFSET SECTIONS

In sectioning irregular objects, you may show features that do not lie in a straight line by offsetting, or bending the cutting plane. Such a section is called an ***offset section***. In Figure 7.18a the cutting plane is offset in several places to include the hole at the left end, one of the parallel slots, the rectangular recess, and one of the holes at the right end. The front portion of the object is then imagined to be removed, as shown in Figure 7.18b. The path of the cutting plane is shown by the cutting-plane line in the top view of

Figure 7.18c, and the resulting offset section is shown in the front view. The offsets or bends in the cutting plane are all 90 degrees and are never shown in the sectional view.

Figure 7.18 also illustrates how hidden lines in a section eliminate the need for an additional view. If hidden lines were not shown, an extra view would be needed to show the small boss on the back.

Figure 7.19 shows an example of multiple offset sections. Notice that the visible background shapes, without hidden lines, appear in each sectional view.

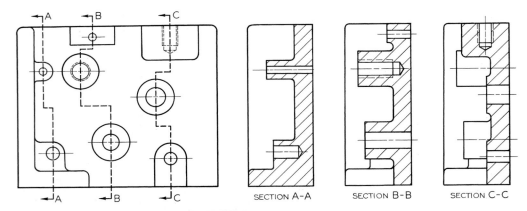

FIGURE 7.19 **Three Offset Sections.**

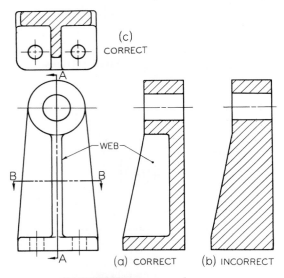

FIGURE 7.20 **Webs in Section.**

■ 7.13 RIBS IN SECTIONS

To avoid a false impression of thickness and solidity, ribs, webs, gear teeth, and other similar flat features are not sectioned, even though the cutting plane passes along the center plane of the feature. For example, in Figure 7.20 the cutting plane A–A passes through the long dimension of the vertical web, or rib, but as Figure 7.20a shows, the web is not section lined. Such thin features should not be section lined, even though the cutting plane passes through them. The incorrect section, shown in Figure 7.20b, gives the false impression of thickness or solidity.

If the cutting plane passes crosswise through a rib or any thin member, as does the plane B–B in Figure 7.20, the member should be section lined in the usual manner, as shown in Figure 7.20c.

If a rib is not section lined, it can be difficult to tell that the rib is present, as shown in Figure 7.21b. It is difficult to distinguish between open spaces (labeled B) and ribs (labeled A). In such cases, alternately spaced hatching of the ribs should be used, as shown in Figure 7.21c.

 Worksheet 7.2. provides practice in sectioning a part with ribs.

■ 7.14 ALIGNED SECTIONS

To include in a section certain angled features, the cutting plane may be bent to pass through those features. The plane and features are then imagined to be revolved into the original plane. For example, in Figure 7.22 the cutting plane bends to pass through the angled arm and then is aligned to a vertical position, where it is projected to the sectional view.

In Figure 7.23a the cutting plane is bent so that both a drilled and a counterbored hole will be included in the sectional view. The correct sectional view, shown in Figure 7.23b, is clearer and more complete than a full section, shown in Figure 7.23c. The angle of revolution should always be less than 90 degrees.

In Figure 7.24a the projecting lugs are not sectioned for the same reason that ribs are not sectioned. In Figure 7.24b the projecting lugs are located so that the cutting plane passes through them crosswise, therefore, they are sectioned.

Use Worksheet 7.3. to practice creating an aligned section.

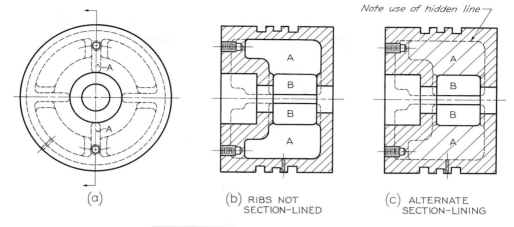

FIGURE 7.21   Alternate Hatching.

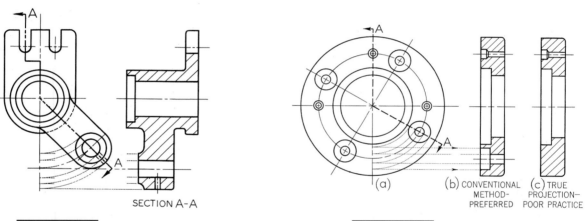

FIGURE 7.22   Aligned Section.

FIGURE 7.23   Aligned Section.

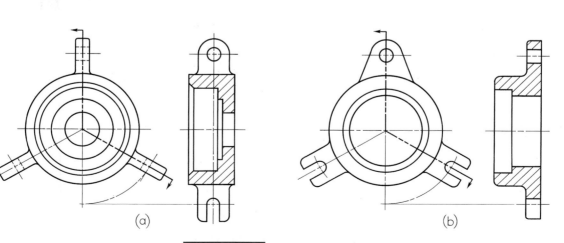

FIGURE 7.24   Aligned Sections.

GRAPHICS SPOTLIGHT

Modeling Irregular Surfaces

(A)

CAD-DESIGNED SKI GRIP

Designers at Life-link International in Jackson Hole, Wyoming, used CAD surface modeling to design and adjustable ski pole grip. The ski pole grip they wanted had to be ergonomic—it had to fit the human hand comfortably. It also needed to have the correct fits with other parts in the assembly for an adjustable-length ski pole used by advanced skiers in changing terrain and snow conditions. Inside the grip, a cam and collet fit together, letting the user turn the grip and extend the length of the pole by 2 inches.

NO MANUAL DRAWINGS CREATED

According to Rick Liu, product development manager at Life-link International, the complex shape of the ski grip, shown in Figure A, made it a good candidate for design using CAD surface modeling. To design and manufacture the part using traditional hand drawings, many cross sections of the shape would have been necessary. Each cross section would then have to be carefully interpreted to create the mold. Instead, CAD surface modeling and direct numerically controlled (NC) machining were used to eliminate the step of interpreting the manual drawings.

REFINING THE SURFACE MODEL

Liu began by drawing cross sections of the grip using AutoCAD. These were used to create a rough AutoCAD surface model. A company in California refined the model for Lifelink. In the last minutes of transferring the refined model to Lifelink via modem, an earthquake flattened the computer where the file was stored. Luckily the model had finished transferring.

The refined model shown in Figure B was sent to Jungst Scientific in Bozeman, Montana, for direct machining. A prototype was created by exporting the surface model to a computer running Gibbs system software. Using the Gibbs software, Jungst generated the tool path for the NC milling machine to make the prototype. NC machines must receive instructions that they can interpret, usually called g-codes. These instructions are interpreted into machine code specific to the machine. The first prototype was accepted as designed, with no modifications necessary.

DIRECT MACHINING

An injection mold was needed to manufacture the finished ski grip. To create a mold, Jungst machined carbon blanks of the grip directly from the model. The carbon blanks were used as electrodes for electro-discharge machining (EDM) to burn the cavities in the shape of the grip in the aluminum mold base. Using EDM, several carbon blanks are needed for each cavity created. EDM can be used to machine very hard materials and produce accurate, intricate shapes.

Next, Jungst designed the ejectors and cooling for the mold. Fits inside the grip for the height adjustment cam were critical. Because plastics shrink as they cool, special cooling was designed to cool the cavity inside the grip where the cam fits before the outer portion of the grip cooled. This kept the hole where the cam fits from shrinking unevenly.

MASS PRODUCTION

With the mold complete, the skip grips could be manufactured using an injection mold press owned by a company in Colorado. Finally, they are assembled in Bozeman, Montana, and sold to skiers internationally.

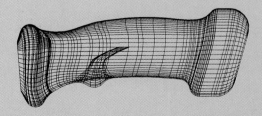

(B)

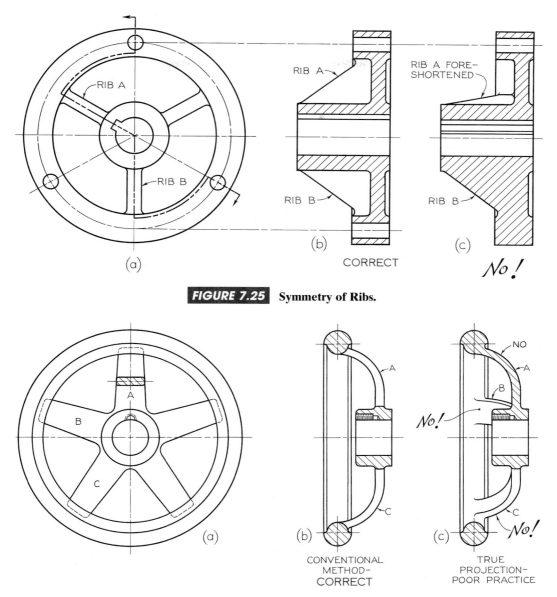

FIGURE 7.25   Symmetry of Ribs.

FIGURE 7.26   Spokes in Section.

Another example involving rib sectioning and aligned sectioning is shown in Figure 7.25. In the circular view, the cutting plane is offset in circular-arc bends to include the upper hole and upper rib, the keyway and center hole, the lower rib, and one of the lower holes. These features are imagined to be revolved until they line up vertically and are then projected from that position to obtain the section shown in Figure 7.25b. Note that the ribs are not sectioned. If a regular full section of the object were drawn without using the conventions discussed here, the resulting section, shown in Figure 7.25c, would be incomplete and confusing and would take more time to draw.

In sectioning a pulley or any spoked wheel, as shown in Figure 7.26a, it is standard practice to revolve the spokes if necessary, such as when there is an odd number, and not to section-line the spokes, as shown in Figure 7.26b. If the spoke is sectioned, the section gives a false impression of a continuous metal shell, as shown in Figure 7.26c. If the lower spoke is not revolved, it will be foreshortened in the sectional view, in which it presents an amputated and misleading appearance.

Figure 7.26 also illustrates correct practice in omitting visible lines in a sectional view. Notice that spoke B is omitted in Figure 7.26b. If it is included, as shown in Figure 7.26c, the spoke is foreshortened, making it difficult and time-consuming to draw and confusing to the reader of the drawing.

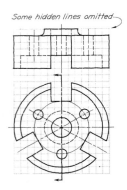

FIGURE 7.27 **Symmetrical Part in Which Features Should Not Be Shown Revolved When Sectioned.**

Remember, do not revolve features unless clarity is improved. In some cases revolving the features results in a loss of clarity. Figure 7.27 shows an example in which revolution should not be used.

■ 7.15 PARTIAL VIEWS

If space is limited on the paper or to save drafting time, *partial views* may be used in connection with sectioning. Only half views are shown in the top view in Figure 7.28a and 7.28b. In each case the back half of the object in the circular view is shown to expose the back portion of the object for viewing in section. When drawing a partial view where the adjacent view will not be sectioned, the front part of the object is shown to provide the reason for the hidden lines in the adjacent view.

Another method of drawing a partial view is to break out much of the circular view, retaining only those features that are needed for minimum representation, as shown in Figure 7.28c.

■ 7.16 INTERSECTIONS IN SECTIONING

Where an intersection is small or unimportant in a section, it is standard practice to disregard the true projection of the figure of intersection, as shown in Figures 7.29a and 7.29c. Larger figures of intersection may be

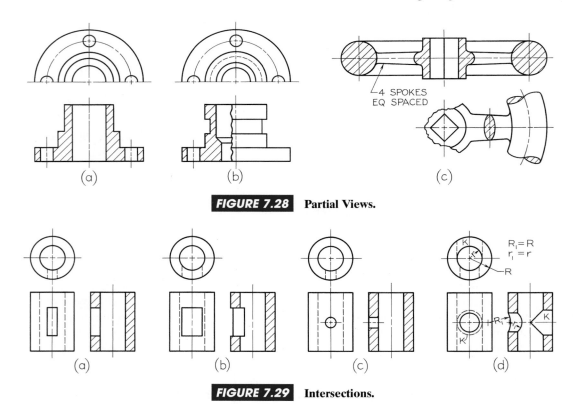

FIGURE 7.28 Partial Views.

FIGURE 7.29 Intersections.

projected, as shown in Figure 7.29b, or approximated by circular arcs, as shown for the smaller hole in Figure 7.29d. Note that the larger hole K is the same diameter as the vertical hole. In such cases the curves of intersection, or ellipses, appear as straight lines, as shown.

■ 7.17 CONVENTIONAL BREAKS

To shorten a view of an elongated object, whether in section or not, *conventional breaks* are recommended, as shown in Figure 7.30. For example, the two views of a garden rake in Figure 7.31a are drawn to a small scale to fit them on the paper. In Figure 7.31b, the handle is "broken," a long central portion is removed, and the rake is then drawn to a larger scale, producing a much clearer delineation. Parts to be broken must have the same section throughout, or if tapered, they must have a uniform taper. Note in Figure 7.31b that the full-length dimension is given, just as if the entire rake were shown.

The breaks used on cylindrical shafts or tubes are often referred to as S-breaks and are usually sketched freehand. Excellent S-breaks are also obtained with an S-break template. Breaks for rectangular metal and wood sections are always drawn freehand, as shown in Figure 7.30.

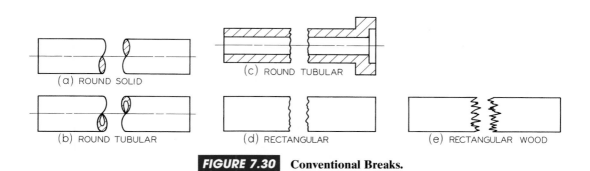

(a) ROUND SOLID   (c) ROUND TUBULAR

(b) ROUND TUBULAR   (d) RECTANGULAR   (e) RECTANGULAR WOOD

FIGURE 7.30 Conventional Breaks.

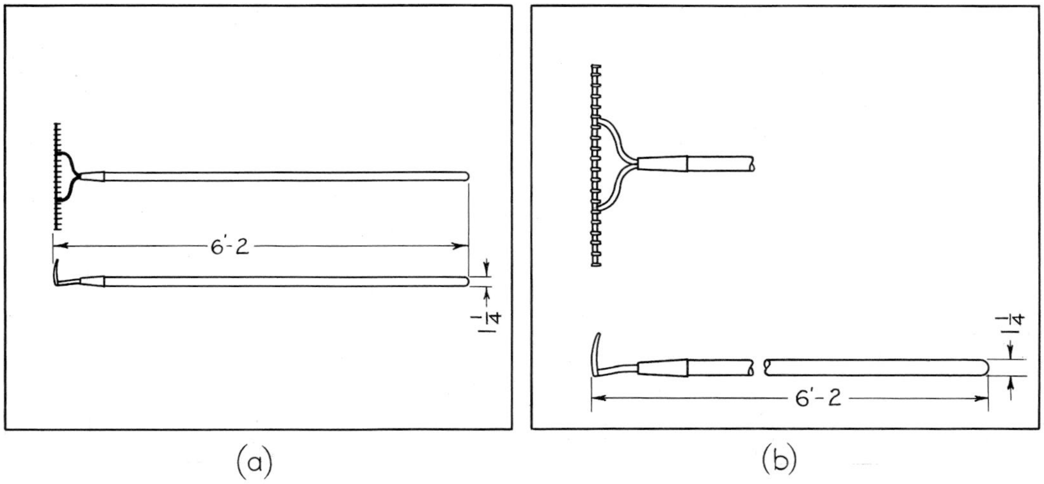

FIGURE 7.31 Use of Conventional Breaks.

(a)   (b)

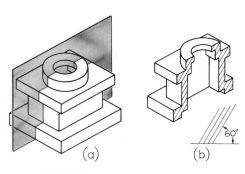

FIGURE 7.32 Isometric Full Section.

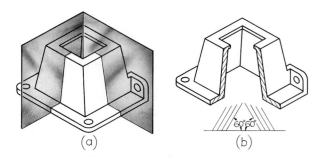

FIGURE 7.33 Isometric Half Section.

■ 7.18 ISOMETRIC SECTIONING

You can create pictorial sectional views by showing the cut object in an isometric or oblique view and hatching the cut surfaces. An *isometric full section* is shown in Figure 7.32. It is usually best to draw the cut surface first and then to draw the portion of the object that lies behind the cutting plane.

An *isometric half section* is shown in Figure 7.33. To sketch isometric half-sections, it is usually easiest to first sketch the entire object and then the cut surfaces. Since only a quarter of the object is removed in a half section, the resulting pictorial sketch is more useful than a full section to show both the exterior and interior shapes. Isometric broken-out sections are also sometimes used.

Hatching in an isometric sketch is similar to that in a multiview sketch. Showing it at an angle of 60 degrees with horizontal is recommended, but the direction should be changed if this would be parallel to major visible lines.

■ 7.19 OBLIQUE SECTIONS

You can also show pictorial sections in oblique view, especially to show interior shapes. An *oblique half section* is shown in Figure 7.34. *Oblique full sections* are seldom used because they do not show enough of the exterior shapes. In general, oblique sections are similar to isometric sections.

FIGURE 7.34 Oblique Half Section.

■ 7.20 COMPUTER GRAPHICS

You can create 2-D and 3-D sectional views using CAD. Most CAD systems have a hatch command to generate the hatch patterns to fill an area automatically. A wide variety of hatch patterns are generally available to show materials such as steel, bronze, sand, and concrete.

Creating a full-sectional view from a 3-D model is generally very easy. You only need to define the cutting plane. Often the hatching for the cut surfaces is generated automatically. Sectioned views other than full sections can be more difficult to create. To create good sectional drawings using CAD, as shown in Figure 7.35, you should have a clear understanding of the standards for showing sectional views.

■ KEY WORDS

sectional view	half section	aligned sections	oblique half section
cutting plane	broken-out section	partial views	oblique full section
full section	revolved section	conventional breaks	
cutting-plane line	removed section	isometric full section	
cross-hatching	offset section	isometric half section	

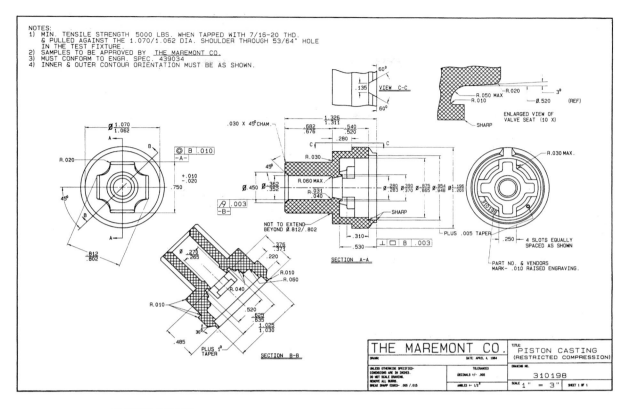

FIGURE 7.35 Detail Drawing Produced by Using the VersaCAD Advanced System. *Courtesy of Versa CAD*

■ CHAPTER SUMMARY

- Sectioning is a technique where the object is hypothetically cut to expose interior details that would otherwise be shown using hidden lines.

- Sections show internal details without the need for hidden lines.

- Objects are imagined to be cut along the cutting-plane line.

- The solid part of the object cut by the cutting plane is shown hatched—with thin lines drawn at 45 degrees—in the sectional view.

- In a sectional view, many hidden lines are replaced by object lines since internal surfaces are exposed.

- Section-lining symbols can be used to indicate the general material of the object.

- Ribs, webs, and spokes are not shown with hatching when the cutting plane passes through them lengthwise.

- Symmetrical features like spokes and webs are revolved so the sectional view shows the true relationships.

- Conventional breaks are used to represent various objects in shortened form.

■ REVIEW QUESTIONS

1. What does the cutting-plane line represent?
2. Sketch the section-line patterns for 10 different materials.
3. List seven different types of sections and an example of when each is drawn.
4. Which sectional views are used to replace an existing primary view? Which sectional views are used in addition to the primary views?
5. How much of an object is imagined to be cut away in a half section?
6. What type of line is used to show the boundary of a broken-out section?
7. Why are hidden lines generally omitted in a sectional view?
8. Why are some symmetrical features, like spokes and webs, revolved in the sectional view?
9. Why is a rib outlined with object lines and not hatched?

■ SECTIONING PROJECTS

Any of the following projects, shown in Figures 7.36-7.66 may be drawn freehand or by using CAD. However, the projects in Figure 7.36 are especially suitable for sketching on 8-1/2 x 11-inch graph paper with appropriate grid squares. Two problems can be drawn on one sheet, using layout A-1, with borders drawn freehand. The projects may be sketched on plain drawing paper. If metric or decimal dimensions are required, you should first study the chapter on dimensioning.

■ DESIGN PROJECT

Design an input device. It can be an ergonomically improved all-purpose mouse or a highly specialized controller for use with your favorite game or application. Keep in mind features such as durability, comfort and precision. Use section views to show the interior clearly.

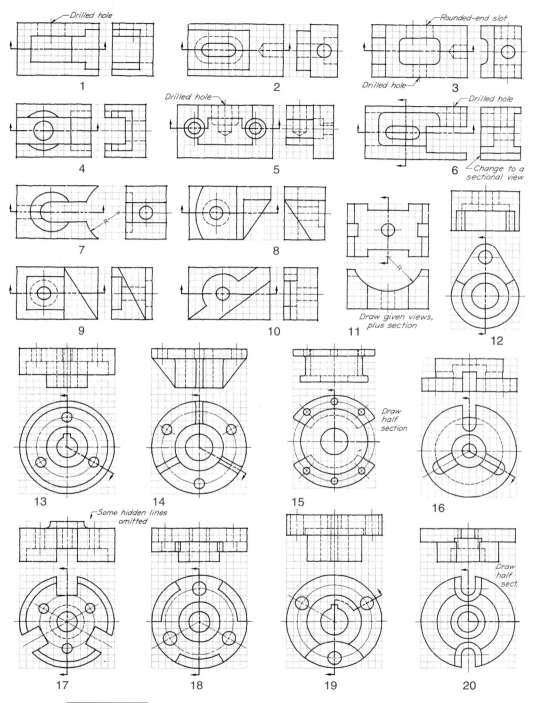

FIGURE 7.36 Freehand Sectioning Problems. Using Layout A–1 or A4–1 adjusted (freehand) on graph paper or plain paper, two problems per sheet, sketch views with sections as indicated. Each grid square = 6 mm (1/4″). In Probs. 1–10, top and right-side views are given. Sketch front sectional views and then move right-side views to line up horizontally with front sectional views. Omit cutting planes except in Probs. 5 and 6.

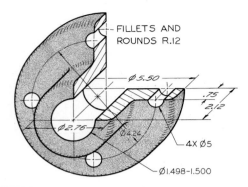

FIGURE 7.37 Bearing. Draw necessary views, with full section (Layout A–3).

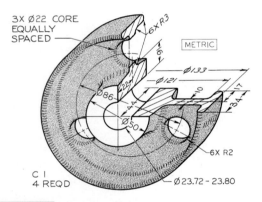

FIGURE 7.38 Truck Wheel. Draw necessary views, with half section (Layout A–3).

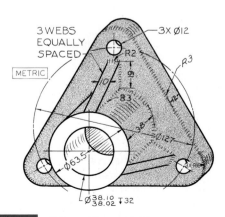

FIGURE 7.39 Column Support. Draw necessary views, with full section (Layout A–3).

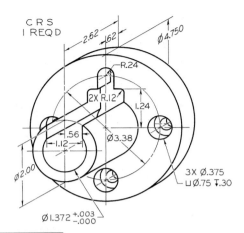

FIGURE 7.40 Centering Bushing. Draw necessary views, with full section (Layout A–3).

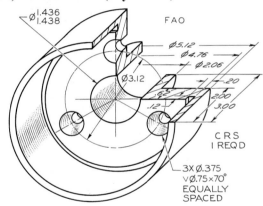

FIGURE 7.41 Special Bearing. Draw necessary views, with full section (Layout A–3).

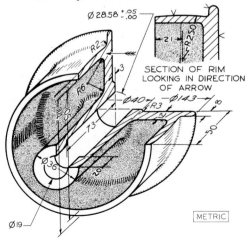

FIGURE 7.42 Idler Pulley. Draw necessary views, with full section (Layout A–3).

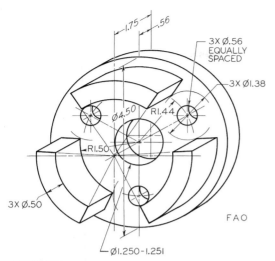

FIGURE 7.43 Cup Washer. Draw necessary views, with full section (Layout A–3 or A4–3 adjusted).

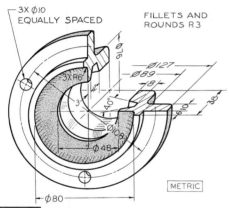

FIGURE 7.44 Fixed Bearing Cup. Draw necessary views, with full section (Layout A–3 or A4–3 adjusted).

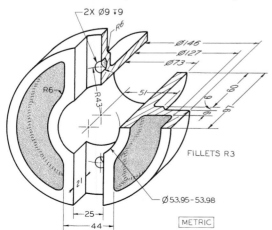

FIGURE 7.45 Stock Guide. Draw necessary views, with half section (Layout B–4 or A3–4 adjusted).

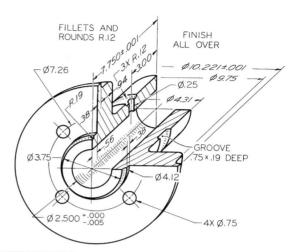

FIGURE 7.46 Bearing. Draw necessary views, with half section. Scale: half size (Layout B–4 or A3–4 adjusted).

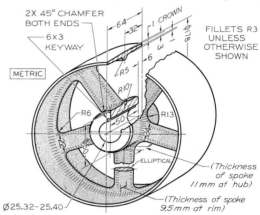

FIGURE 7.47 Pulley. Draw necessary views, with full section, and revolved section of spoke (Layout B–4 or A3–4 adjusted).

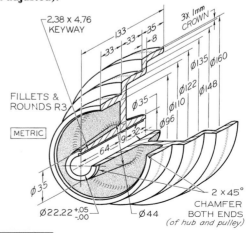

FIGURE 7.48 Step-Cone Pulley. Draw necessary views, with full section (Layout B–4 or A3–4 adjusted).

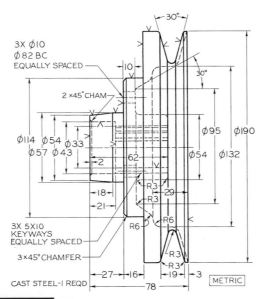

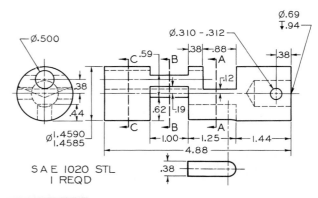

FIGURE 7.49 Sheave. Draw two views, including half section (Layout B–4).

FIGURE 7.50 Operating Valve. Given: Front, left-side, and partial bottom views. Required: Front, right-side, and full bottom views, plus indicated removed sections (Layout B–4).

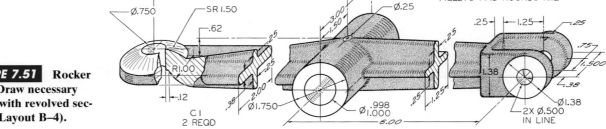

FIGURE 7.51 Rocker Arm. Draw necessary views, with revolved sections (Layout B–4).

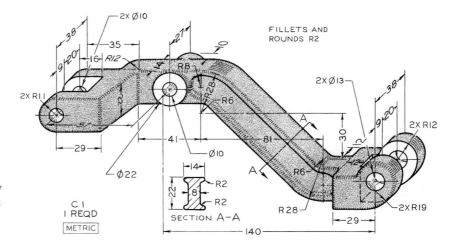

FIGURE 7.52 Dash Pot Lifter. Draw necessary views, using revolved section instead of removed section (Layout B–4).

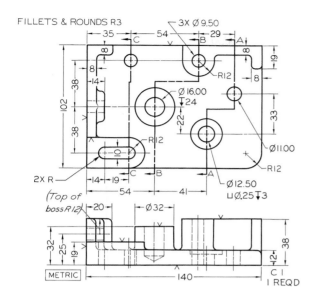

FIGURE 7.53 Adjuster Base. Given: Front and top views. Required: Front and top views and sections A–A, B–B, and C–C. Show all visible lines (Layout B–4).

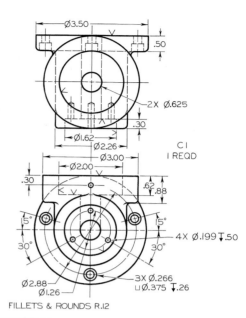

FIGURE 7.55 Hydraulic Fitting. Given: Front and top views. Required: Front and top views and right-side view in full section (Layout B–4).

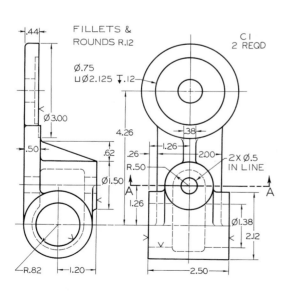

FIGURE 7.54 Mobile Housing. Given: Front and left-side views. Required: Front view, right-side view in full section, and removed section A–A (Layout B–4).

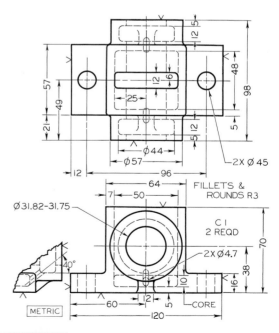

FIGURE 7.56 Auxiliary Shaft Bearing. Given: Front and top views. Required: Front and top views and right-side view in full section (Layout B–4).

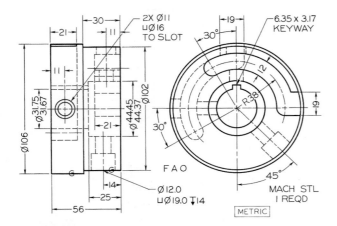

FIGURE 7.57 Traverse Spider. Given: Front and left-side views. Required: Front and right-side views and top view in full section (Layout B–4 or A3–4 adjusted).

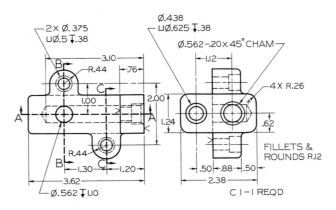

FIGURE 7.59 Bracket. Given: Front and right-side views. Required: Take front as new top; then add right-side view, front view in full section A–A, and sections B–B and C–C (Layout B–4 or A3–4 adjusted).

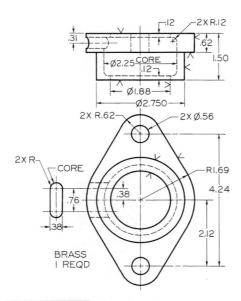

FIGURE 7.58 Gland. Given: Front, top, and partial left-side views. Required: Front view and right-side view in full section (Layout A–3 or A4–3 adjusted).

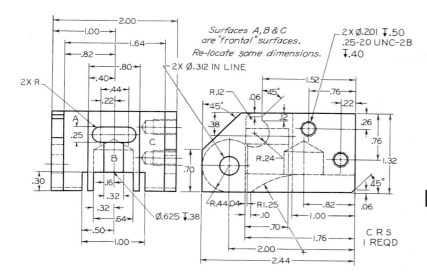

FIGURE 7.60 Cocking Block. Given: Front and right-side views. Required: take front as new top view; then add new front view, and right-side view in full section. Draw double size on Layout C–4 or A2–4 adjusted.

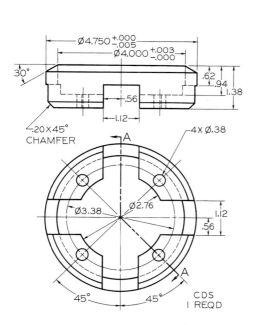

FIGURE 7.61 Packing Ring. Given: Front and top views. Required: Front view and section A–A (Layout A–3 or A4–3 adjusted).

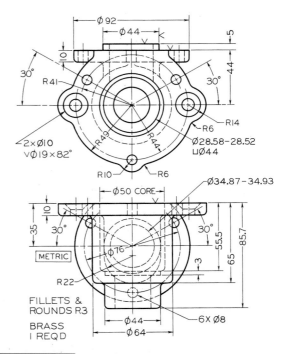

FIGURE 7.62 Strainer Body. Given: Front and bottom views. Required: Front and top views and right-side view in full section (Layout C–4 or A2–4).

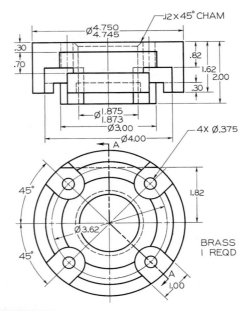

FIGURE 7.63 Oil Retainer. Given: Front and top views. Required: Front view and section A–A (Layout B–4 or A3–4 adjusted).

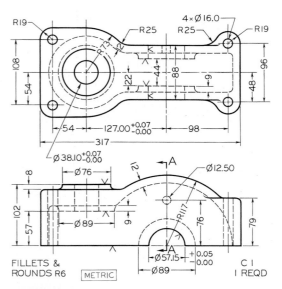

FIGURE 7.64 Gear Box. Given: Front and top views. Required: Front in full section, bottom view, and right-side section A–A. Draw half size on Layout B–4 or A3–4 (adjusted).

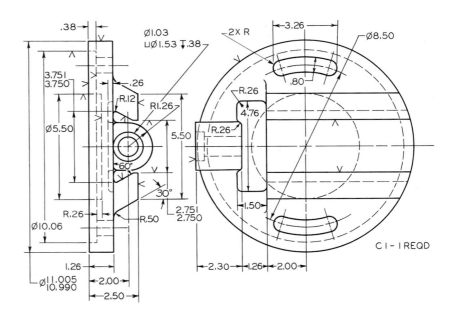

FIGURE 7.65 Slotted Disk for Threading Machine. Given: Front and left-side views. Required: Front and right-side views and top full-section view. Draw half size on Layout B–4 or A3–4 (adjusted).

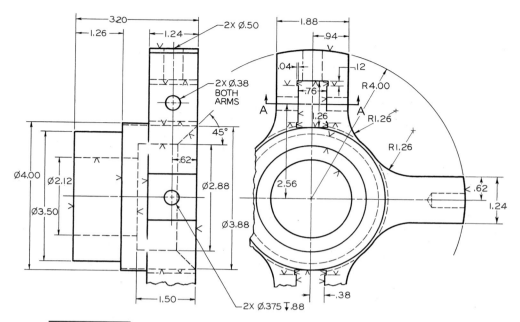

FIGURE 7.66 Web for Lathe Clutch. Given: Partial front and left-side views. Required: Full front view, right-side view in full section, and removed section A–A (Layout C–4 or A2–4).

C H A P T E R 8

AUXILIARY VIEWS, DEVELOPMENTS, AND INTERSECTIONS

OBJECTIVES

After studying the material in this chapter, you should be able to:

1. Create an auxiliary view from any orthographic projection using sketching or CAD.
2. Draw folding lines or reference-plane lines between any two adjacent views.
3. Construct depth, height, or width auxiliary views.
4. Plot curves in auxiliary views.
5. Construct partial auxiliary views.
6. Create auxiliary sectional views.
7. Produce views to show the true length of a line, point view of a line, edge view of a surface, and true size view of a surface.
8. Construct the development of prisms, pyramids, cylinders, and cones.
9. Use triangulation to transfer surface shapes to a development.
10. Create the development of transition pieces.
11. Graphically solve for the intersection of solids.

OVERVIEW

Inclined planes and oblique lines do not appear true size or true length in any of the principal planes of projection. To show the true length of an oblique line or the true size of an inclined plane, an auxiliary view must be created. The principles for creating auxiliary views are the same whether you are using traditional drawing, sketching, or CAD: A line of sight and reference plane are defined. With traditional drawing, the view is manually created along line-of-sight projectors. With CAD drawing, the computer generates the view automatically if a 3-D model of the object was originally created.

■ INTRODUCTION

Many objects are shaped such that their principal faces are not parallel to the regular planes of projection. For example, in Figure 8.1a the base of the design for the bearing is shown in its *true size* and shape, but the rounded upper portion is situated at an angle and does not appear true size and shape in any of the three regular views. To show the true circular shapes, use a direction of sight perpendicular to the planes of those curves, as shown in Figure 8.1b. The result is known as an *auxiliary view*. This view, together with the top view, completely describes the object. The front and right-side views are not necessary.

■ 8.1 DEFINITIONS

Any view obtained by a projection on a plane other than the horizontal, frontal, and profile projection planes is an auxiliary view. A *primary auxiliary view* is projected onto a plane that is perpendicular to one of the principal planes of projection and is inclined to

the other two. A *secondary auxiliary view* is projected from a primary auxiliary view onto a plane that is inclined to all three principal projection planes.

■ 8.2 THE AUXILIARY PLANE

The object shown in Figure 8.2a has an inclined surface (P) that does not appear in its true size and shape in any regular view. To show the inclined surface true size the direction of sight must be perpendicular to the inclined plane. Or using the glass box model, the auxiliary plane is aligned parallel to the inclined surface P to give a true size view of it. The auxiliary plane in this case is perpendicular to the frontal plane of projection and hinged to it.

 Find Worksheet 8.1 (the auxiliary viewing plane glass box) in the tear-out section. Cut out the model of the glass box with an auxiliary viewing plane and follow the directions on the sheet. Use this model as a visual aid to help you understand the relationship between the basic views and auxiliary views. Answer the questions on the sheet as you read through the chapter.

The horizontal and auxiliary planes are unfolded into the plane of the front view, as shown in Figure 8.2b. Drawings do not show the planes of the glass box, but you can think of *folding lines* (H/F and F/1) representing the hinges joining the planes. The folding lines themselves are usually omitted in the actual drawing. Inclined surface P is shown in its true size and shape in the auxiliary view. Note that both the top and auxiliary views show the depth of the object. One dimension of the surface is projected directly from the front view, and the depth is transferred from the top view.

As you learned in Chapter 5, the locations of the folding lines depends on the size of the glass box and

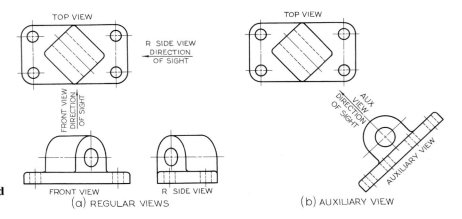

FIGURE 8.1 Regular Views and Auxiliary Views.

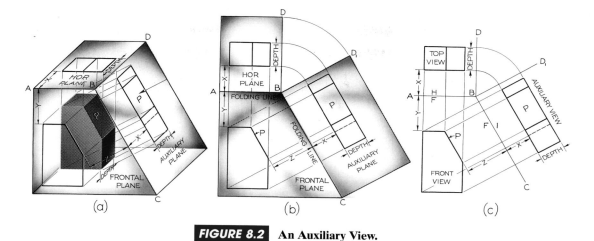

FIGURE 8.2 An Auxiliary View.

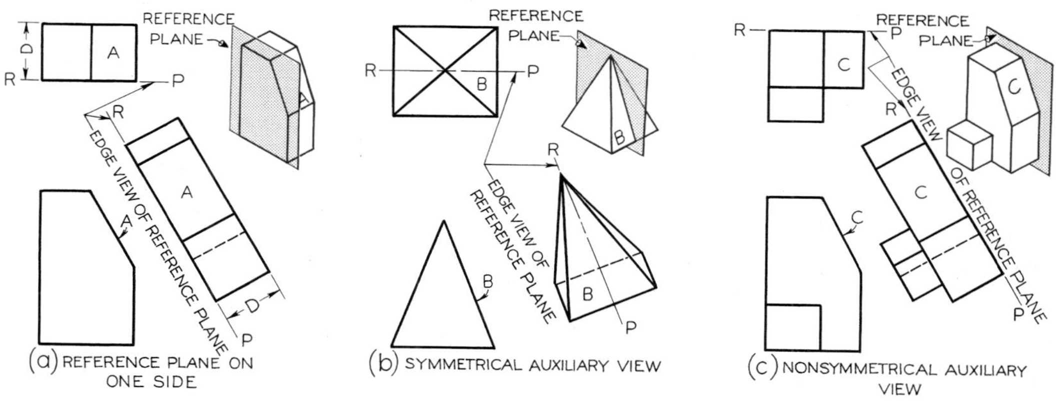

(a)   (b)   (c)

REFERENCE PLANE

REFERENCE PLANE

REFERENCE PLANE

(a) REFERENCE PLANE ON ONE SIDE

(b) SYMMETRICAL AUXILIARY VIEW

(c) NONSYMMETRICAL AUXILIARY VIEW

FIGURE 8.3 Position of the Reference Plane.

the location of the object within it. If the object is further down in the box, distance Y is increased. If the object is moved back in the box, distances X increase but are still equal. If the object is moved to the left inside the glass box, distance Z is increased.

■ 8.3 REFERENCE PLANES

In the auxiliary view shown in Figure 8.2c, the folding line represents the edge view of the frontal plane of projection. In this case, the frontal plane is used for transferring distances—that is, depth measurements—from the top view to the auxiliary view.

Instead of using one of the planes of projection, you can use a *reference plane* parallel to the plane of projection and touching or cutting through the object. For example, in Figure 8.3a a reference plane is aligned

with the front surface of the object. This plane appears on edge, or as a line, in the top and auxiliary views. The two reference lines are used in the same manner as folding lines. Dimensions D in the top and auxiliary views are equal. The advantage of the reference-plane method is that fewer measurements are required because some points of the object lie in the reference plane. Make the reference plane using light lines similar to construction lines.

You can use a reference plane that coincides with the front surface of the object, as shown in Figure 8.3a. When an object is symmetrical, it is useful to select the reference plane to cut through the object, as shown in Figure 8.3b. This way you only have to make half as many measurements to transfer dimensions because they are the same on each side of the reference plane.

Step by Step 8.1
Sketching an auxiliary view

The following steps are used to draw an auxiliary view using folding lines. In this example you are given the top and front views shown.

1. To draw an auxiliary view showing the true size and shape of inclined surface P, the direction of sight must be perpendicular to the edge view of the inclined surface. To produce this view, draw the folding line H/F between the views at right angles

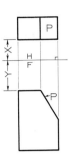

to the projection lines. The distance you select to place the folding line away from the view does not really matter. It just represents the relationship between the location of the object and the planes of the glass box.

Note: In the following steps, use triangles (see Practical Tips, facing page) to draw lines parallel to or perpendicular to the inclined face, or use freehand sketching techniques.

2. Establish a direction of sight perpendicular to surface P. Draw light projection lines from the front view (parallel to the arrow) perpendicular to surface P.

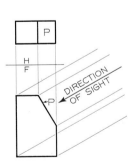

3. Draw folding line F/1 for the auxiliary view at right angles to the projection lines and at any convenient distance from the front view.

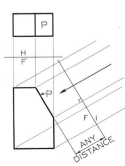

4. Identify the surfaces on the object mentally or by labeling them. Number the vertices of the object to make it easy to project the vertices to the auxiliary view. In this case, the auxiliary view shows the height and

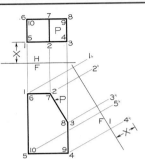

depth of the object. Locate each point on its projection line to transfer its height to the adjacent auxiliary view. Then transfer the depth locations for the points by measuring the distance from the fold line to the point in the top view and transferring it on the correct projection line to the auxiliary view.

5. Locate all points the same distances from folding line F/1 as they are from folding line H/F in the top view. For example, points 1 to 5 are distance X from the folding lines in both the top and auxiliary views, and points 6 to 10 are

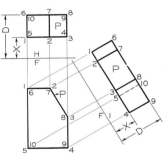

distance D from the corresponding folding lines. Since the object is viewed in the direction of the arrow, edge 5–10 will be hidden in the auxiliary view.

Note: If a projection line crosses a portion of the object, the line in the projected view is often hidden behind another surface. If a projection line does not cross the view of the object, the resultant surface will be visible.

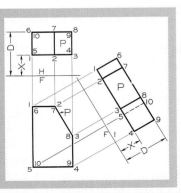

Practical Tips
Drawing auxiliary views

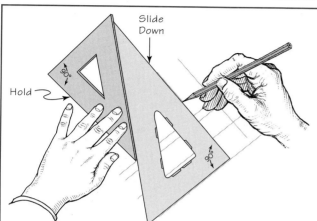

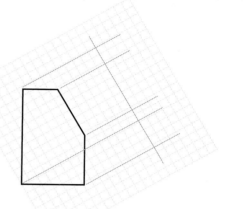

Using triangles

You can use two triangles to quickly draw parallel and perpendicular lines for accurate sketches.

- Place two triangles together so that the 90-degree corners are on the outside.

- Slide them around on your drawing until the outer edge of one is along the line in the drawing to which you want to sketch parallel.

- Holding down the back triangle, slide the other triangle along it.

- Draw parallel lines along one edge of the triangle. Draw perpendicular lines along the other edge.

- This technique works well as an addition to freehand sketching when you want to show an auxiliary view.

Using grid paper

You can use grid paper to help sketch auxiliary views by orienting the lines of the grid paper underneath your vellum or other semitransparent drawing sheet so that the grid is parallel to the inclined edge in the drawing. Use the grid to help sketch lines parallel and perpendicular to the edge in question.

Using CAD

Most CAD systems allow you to rotate the grid or to create a new coordinate system (often called the user coordinate system) so that it aligns with the inclined surface. If you are using 3-D CAD, you can create auxiliary views by viewing the object perpendicular to the surface you want to show true size.

If you are using CAD, you can draw half of the view and then mirror the object. You can also use the back surface of the object, as shown in Figure 8.3c, or any intermediate point that would be advantageous.

Position the reference plane so it is convenient for transferring distances. Remember the following:

1. Reference lines, like folding lines, are always at right angles to the projection lines between the views.

2. A reference plane appears as a line in two alternate views, never in adjacent views.

3. Measurements are always made at right angles to the reference lines or parallel to the projection lines.

4. In the auxiliary view, all points are at the same distances from the reference line as the corresponding points are from the reference line in the alternate view, or the **second previous view**.

Hands On 8.1
Projecting auxiliary views using a reference plane

The object has been numbered for you in the pictorial view at right. To create the auxiliary view:

1. Draw two views of the object and determine the direction of sight needed to produce a view which will show the true size of surface A. This step has been completed for you.

2. Next sketch projection lines parallel to the direction of sight. Some of them have been drawn in for you. Complete all of the projection lines.

3. Establish a reference plane parallel to the back surface of the object. The reference lines in the top and auxiliary views are at right angles to the projection lines and have been drawn for you. These are the edge views of the reference plane.

4. Draw auxiliary view of surface A. It will be true size and shape because the direction of sight was taken perpendicular to that surface. Transfer depth measurements from the top view to the auxiliary view with dividers or a scale. Each point in the auxiliary view will be on its projection line from the front view and will be the same distance from the reference line as it is in the top view to the corresponding reference line. Points 1, 2, and 7 have been projected for you. Finish projecting points 5 and 8. Draw surface A true size in the auxiliary view by connecting the vertices in the same order as they are shown connecting in the top view (1-7-8-5-2-1).

5. Complete the auxiliary view by adding other visible edges and surfaces of the object. Each numbered point in the auxiliary view lies on its projection line from the front view and is the same distance from the reference line as it is in the top view. Note that two surfaces of the object appear as lines in the auxiliary view.

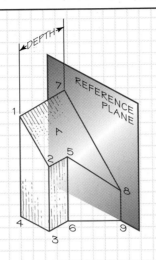

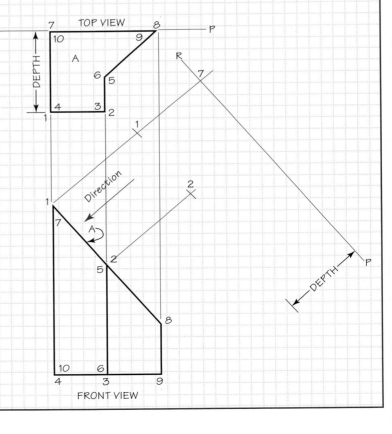

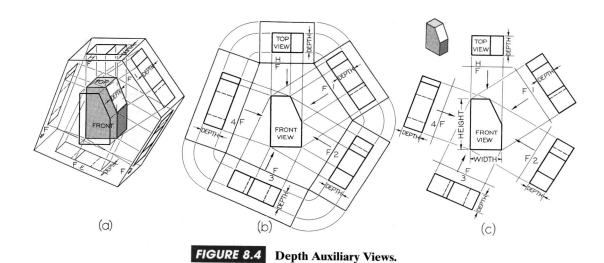

FIGURE 8.4   Depth Auxiliary Views.

■ 8.4 CLASSIFICATION OF AUXILIARY VIEWS

Auxiliary views are named for the principal dimension shown in the auxiliary view. For example, the auxiliary views in Figure 8.4 are depth auxiliary views because they show the object's depth. Any auxiliary view projected from the front view, also known as a front adjacent view, is a depth auxiliary view.

Similarly, any auxiliary view projected from the top view, also known as a top adjacent view, is a height auxiliary view; and any auxiliary view projected from a side view, also known as a side adjacent view, is a width auxiliary view.

■ 8.5 DEPTH AUXILIARY VIEWS

An infinite number of auxiliary planes can be hinged perpendicular to the frontal plane (F) of projection. Five such planes are shown in Figure 8.4a. The horizontal plane is included to show that it is similar to the others. All these views show the object's depth and therefore are all depth auxiliary views.

The unfolded auxiliary planes, shown in Figure 8.4b, show how depth dimensions are projected from the top view to all auxiliary views. The arrows indicate the directions of sight.

The complete drawing, with the outlines of the planes of projection omitted, is shown in Figure 8.4c. Note that the front view shows the height and the width of the object, but not the depth. *The principal dimension shown in an auxiliary view is the one not shown in the adjacent view from which the auxiliary view was projected.*

■ 8.6 HEIGHT AUXILIARY VIEWS

An infinite number of auxiliary planes can be hinged perpendicular to the horizontal plane (H) of projection. Several are shown in Figure 8.5a. The front view and all these auxiliary views show the height of the

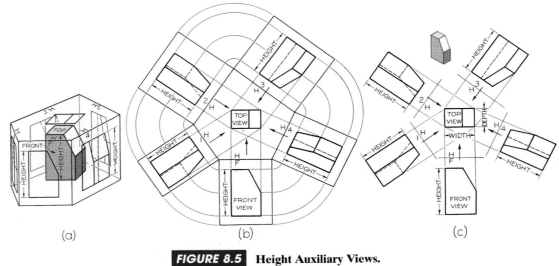

FIGURE 8.5 Height Auxiliary Views.

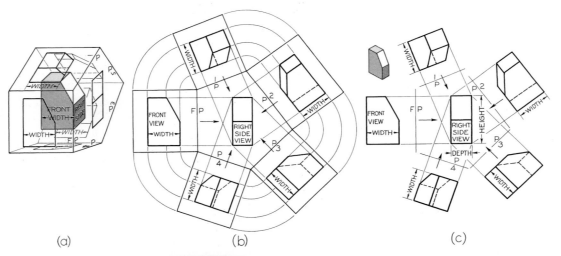

FIGURE 8.6 Width Auxiliary Views.

object. Therefore, all these auxiliary views are height auxiliary views.

The unfolded projection planes are shown in Figure 8.5b, and the complete drawing is shown in Figure 8.5c. Note that in the top view, the only dimension *not shown* is height.

■ 8.7 WIDTH AUXILIARY VIEWS

An infinite number of auxiliary planes can also be hinged perpendicular to the profile plane (P) of projection. Some are shown in Figure 8.6a. The front view and all these auxiliary views are width auxiliary views.

The unfolded planes are shown in Figure 8.6b, and the complete drawing is shown in Figure 8.6c.

In the right-side view, from which the auxiliary views are projected, the only dimension *not shown* is width.

■ 8.8 PLOTTED CURVES AND ELLIPSES

When a cylinder is cut by an inclined plane, the inclined surface is elliptical in shape. To show the true size, create an auxiliary view where the direction of sight is perpendicular to the edge view of the inclined surface. The result is an ellipse shown true size and shape in the auxiliary view. The major axis of the ellipse is true length in the front view, although the ellipse itself is on edge. The minor axis is equal to the diameter of the cylinder.

Step by Step 8.2
Showing an inclined elliptical surface true size

Given the front and side views shown, project an auxiliary view showing true size of the elliptical surface:

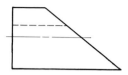

Practical Tip
Creating ellipses in CAD

Since the major and minor axes are known, you can quickly create similar ellipses using CAD by locating the major and minor axes or the center and axes. For hand-sketching you may want to use an ellipse template.

1. Since this is a symmetrical object, use a reference plane through the center of the object, as shown.

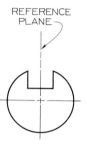

2. Select points on the circle in the side view.

3. Locate the same points on the inclined surface and the left-end surface.

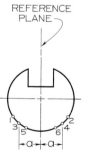
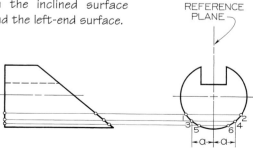

4. Project each point to the auxiliary view along its projection line.

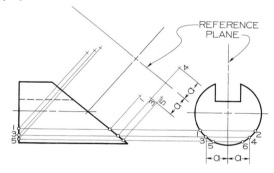

5. Transfer distances from the side view to the auxiliary view. Because the object is symmetrical, two points can be located with each measurement, as shown for points 1–2, 3–4, and 5–6. Project enough points to sketch the curves accurately.

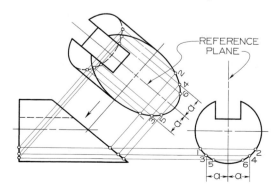

Hands On 8.2
Plotting curves

The auxiliary view shows the true size and shape of the inclined cut through a piece of molding. The method of plotting points is similar to that explained for the ellipse in Step-by-Step 8.2.

1. Identify some points along the curve shown in the side view. This step has been completed for you.

2. Locate those same points in the front view. The curved shape is the inclined surface. Some of these have been located for you.

3. Project the points into the auxiliary view. The reference plane has been located for you and a few points projected.

4. Finish projecting all of the points on the inclined surface and draw its true shape in the auxiliary view.

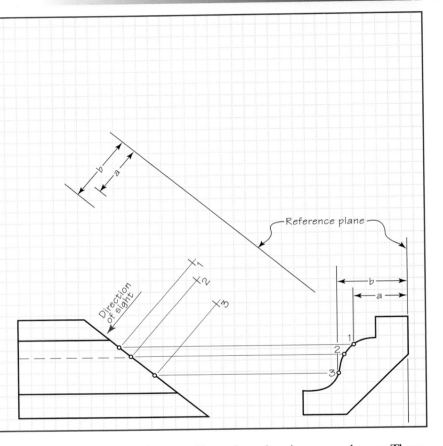

■ 8.9 REVERSE CONSTRUCTION

To complete the regular views, it is often necessary to first construct an auxiliary view where critical dimensions will be shown true size. For example, in Figure 8.7a, the upper portion of the right-side view cannot be constructed until the auxiliary view is drawn. First, points are established on the curves and then projected back to the front view, as shown.

In Figure 8.7b, the 60-degree angle and the location of line 1–2 in the front view are given. To locate line 3–4 in the front view and lines 2–4, 3–4, and 4–5 in the side view, it is necessary to first construct the 60-degree angle in the auxiliary view and project back to the front and side views, as shown.

■ 8.10 PARTIAL AUXILIARY VIEWS

The use of an auxiliary view often makes it possible to omit one or more regular views. In Figure 8.8, three

complete auxiliary-view drawings are shown. These drawings take a lot of time to create and may even be confusing because of the clutter of lines. However, no view can be completely eliminated.

Partial views are often sufficient and easier to read. Partial regular views and *partial auxiliary views* are shown in Figure 8.9. Usually a break line is used to indicate the imaginary break in the views. *Do not draw a break line coinciding with a visible line or hidden line.*

So that partial auxiliary views, which are often small, do not appear "lost" and not related to any view, connect them to the views from which they project, either with a center line or with one or two thin projection lines.

■ 8.11 HALF AUXILIARY VIEWS

If an auxiliary view is symmetrical, and if it is necessary to save space on the drawing or to save time, only

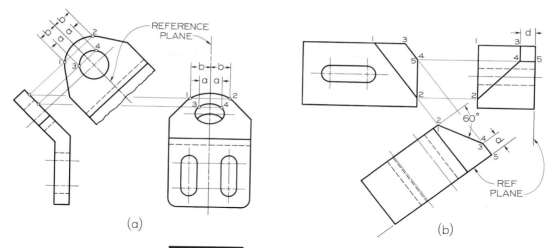

FIGURE 8.7 Reverse Construction.

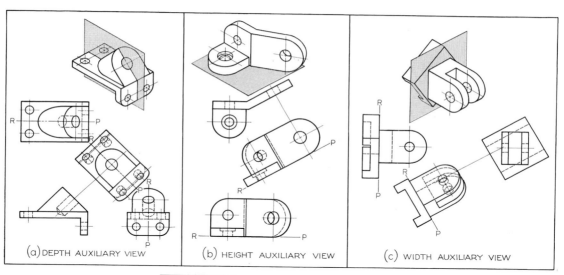

(a) DEPTH AUXILIARY VIEW   (b) HEIGHT AUXILIARY VIEW   (c) WIDTH AUXILIARY VIEW

FIGURE 8.8 Primary Auxiliary Views.

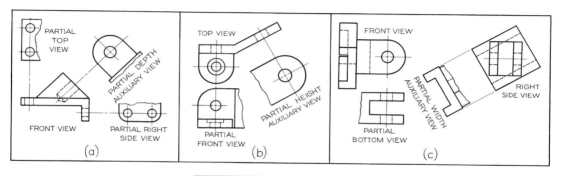

FIGURE 8.9 Partial Views.

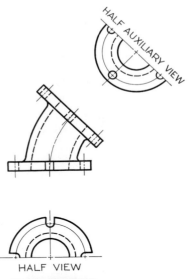

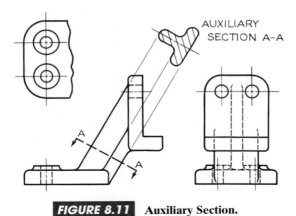

HALF VIEW

FIGURE 8.10 Half Views.

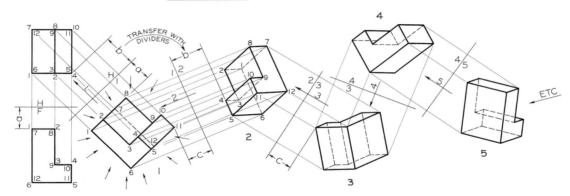

AUXILIARY
SECTION A-A

FIGURE 8.11 Auxiliary Section.

half of the auxiliary view may be drawn, as shown in Figure 8.10. In this case, half of a regular view is also shown since the bottom flange is also symmetrical. Note that in each case the near half is shown.

■ 8.12 HIDDEN LINES IN AUXILIARY VIEWS

Generally, hidden lines should be omitted in auxiliary views, unless they are needed to clearly communicate the drawings intent. For practice, show all hidden lines, especially if the auxiliary view of the entire object is shown. Later, when you are familiar with drawing auxiliary views, omit hidden lines when they do not add needed information to the drawing.

■ 8.13 AUXILIARY SECTIONS

An *auxiliary section* is simply an auxiliary view in section. A typical auxiliary section is shown in Figure 8.11. In this example, there is not sufficient space for a revolved section, although a removed section could have been used instead of an auxiliary section. Note the cutting-plane line and the terminating arrows that indicate the direction of sight for the auxiliary section. In an auxiliary section drawing, the entire portion of the object behind the cutting plane may be shown, or the cut surface alone may be shown.

■ 8.14 SUCCESSIVE AUXILIARY VIEWS

Up to this point you have been learning to project primary auxiliary views—that is, auxiliary views projected from one of the principal views. In Figure 8.12, auxiliary view 1 is a primary auxiliary view projected from the top view.

From primary auxiliary view 1, a secondary auxiliary view 2 can be drawn; then from it a *third auxiliary view* 3, and so on. An infinite number of such successive auxiliary views may be drawn. However,

FIGURE 8.12 Successive Auxiliary Views.

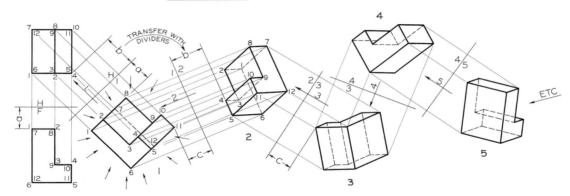

secondary auxiliary view 2 is not the only one that can be projected from primary auxiliary view 1. As shown by the arrows around view 1, an infinite number of secondary auxiliary views, with different lines of sight, may be projected. Any auxiliary view projected from a primary auxiliary view is a secondary auxiliary view. Furthermore, any succeeding auxiliary view may be used to project an infinite series of views from it.

In this example, folding lines are more convenient than reference-plane lines. In auxiliary view 1, all numbered points of the object are the same distance from folding line H/1 as they are in the front view from folding line H/F. These distances, such as distance a, are transferred from the front view to the auxiliary view.

To draw the secondary auxiliary view 2, ignore the front view and focus on the sequence of three views: the top view, view 1, and view 2. Draw light projection lines parallel to the direction of sight desired for view 2. Draw folding line 1/2 perpendicular to the projection lines and at any convenient distance from view 1. Transfer the distances measured from folding line H/1 to locate all points in view 2. For example, transfer distance b to locate points 4 and 5 from folding line 1/2. Connect points to draw the object and determine visibility. The closest corner (11) in view 2 will be visible, and the one farthest away (1) will be hidden, as shown.

To draw views 3, 4, and so on, use a similar process. Remember to use the correct sequence of three views.

■ 8.15 USES OF AUXILIARY VIEWS

Generally, auxiliary views are used to show the true shape or true angle of features that appear distorted in the regular views. Auxiliary views are often used to produce views which show the following:

1. True length of line
2. Point view of line
3. Edge view of plane
4. True size of plane

You can use the ability to generate views that show the specific things listed above to solve a variety of engineering problems. Descriptive geometry is the term for using accurate drawings to solve engineering problems. An accurate CAD drawing database can be used to solve many engineering problems when you understand the four basic views from descriptive geometry. Using 3-D CAD, you can often model objects accurately and query the database for lengths and angles. Even so, you will often need the techniques

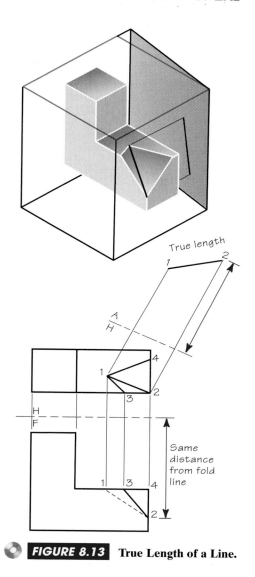

FIGURE 8.13   **True Length of a Line.**

described below to produce views which will help you visualize, create, or display 3-D drawing geometry.

■ 8.16 TRUE LENGTH OF LINE

As shown in Figure 8.13, a line will show true length in a plane of projection which is parallel to the line. In other words, a line will show true length in an auxiliary view where the direction of sight is perpendicular to the line. To show a line true length, make the fold line parallel to the line you want to show true length in the auxiliary view. Whenever a line is parallel to the fold line between two views, it will be true length in the adjacent view.

Step by Step 8.3
Showing the true length of a hip rafter

The top and front views of the hip rafter (line 1–2) are shown. Use an auxiliary view to show the line true length.

1. Choose the direction of sight to be perpendicular to line 1–2 (front view).

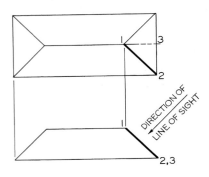

2. Draw the H/F folding line between the top and front view, as shown.

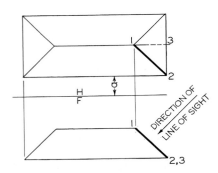

3. Draw the F/1 folding line parallel to line 1-2 and any convenient distance from line 1–2 (front view).

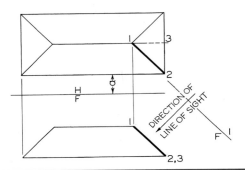

4. Draw projection lines from points 1, 2, and 3 to begin creating the auxiliary view.

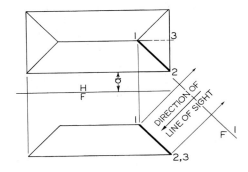

5. Transfer points 1 and 2 to the auxiliary view at the same distance from the folding line as they are in the top view, and along their respective projection lines. The hip rafter (line 1–2) is shown true length in the auxiliary view. Also, triangle 1–2–3 in the auxiliary view shows the true size and shape as that portion of the roof because the direction of sight for the auxiliary view is perpendicular to triangle 1–2–3.

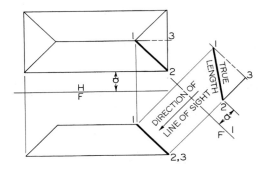

Hands On 8.3
Viewing a line as a point

Draw a line in a plane—for example, a straight line on a sheet of paper. Then tilt the paper to view the line as a point. You will see that when the line appears as a point, the plane containing it appears as a line. (Since your paper will end up being viewed on edge, it may be a little hard to see the line when it is oriented correctly.)

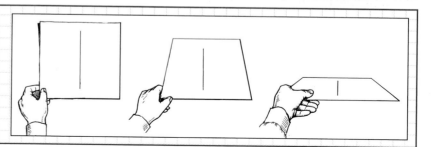

■ 8.17 POINT VIEW OF A LINE

As shown in Figure 8.14, a line will show as a point view when projected to a plane perpendicular it. To show the point view of a line, choose the direction of sight parallel to the line where it is true length.

Refer to Figure 8.15 for the following steps:

1. Choose the direction of sight to be parallel to line 1–2.

2. Draw folding line H/F between the top and front view, as shown.

3. Draw folding line F/1 perpendicular to line 1–2, where it is true length, and any convenient distance from line 1–2, (front view).

4. Draw projection lines from points 1 and 2 to begin creating the auxiliary view.

5. Transfer points 1 and 2 to the auxiliary view at the same distance from the folding line as they are in the top view and along their respective projection lines. They will line up exactly with each other to form a point view of the line.

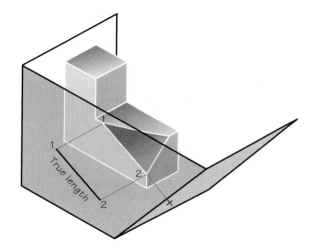

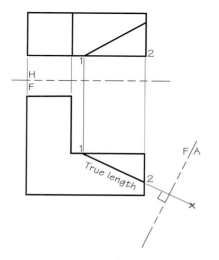

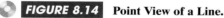

FIGURE 8.14 Point View of a Line.

FIGURE 8.15 Point View of a Line.

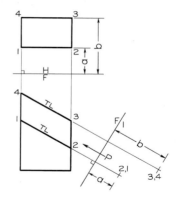

■ 8.18 EDGE VIEW OF A PLANE

As shown in Figure 8.16, a plane will show on edge in a plane of projection which shows a point view of any line that lies entirely within the plane. To get the point view of a line, the direction of sight must be parallel to the line where it is true length. To show the edge view of a plane, choose the direction of sight parallel to a true-length line lying in the plane.

Refer to Figure 8.17 for the following steps

1. Choose the direction of sight to be parallel to line 1–2 in the front view where it is already shown true length.
2. Draw folding line H/F between the top and front view, as shown.
3. Draw folding line F/1 perpendicular to true-length line 1–2 and any convenient distance.
4. Draw projection lines from points 1, 2, 3, and 4 to begin creating the auxiliary view.
5. Transfer points 1, 2, 3, and 4 to the auxiliary view at the same distance from the folding line as they are in the top view and along their respective projection lines. Plane 1–2–3–4 will appear on edge in the finished auxiliary view.

■ 8.19 TRUE SIZE OF AN OBLIQUE SURFACE

As shown in Figure 8.18, a plane will show true size when the plane of projection is parallel to it. *To show the true size view of a plane, choose the direction of sight perpendicular to the edge view of the plane.* You have already practiced showing inclined surfaces true size using this method, where the edge view is already given. But to show an oblique surface true size, you need to construct a second auxiliary view.

FIGURE 8.16 **Edge View of a Plane.**

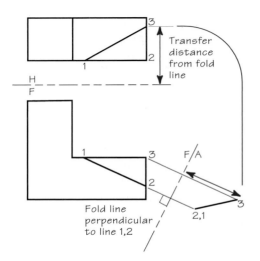

FIGURE 8.17 **Edge View of a Surface.**

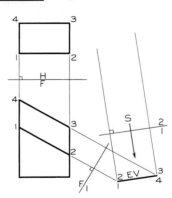

To show the true size and shape of an oblique surface, such as surface 1–2–3–4 in Figure 8.18, create a second auxiliary view. In this example folding lines are used, but you can achieve the same results for all of the preceding examples using reference lines.

1. Draw the auxiliary view showing surface 1–2–3–4 on edge, as explained previously.

2. Create a second auxiliary view with the line of sight perpendicular to the edge view of plane 1–2–3–4 in the primary auxiliary view. Project lines parallel to the arrow. Draw folding line 1/2 perpendicular to these projection lines at a convenient distance from the primary auxiliary view.

3. Draw the secondary auxiliary view. Transfer the distance to each point from folding line F/1 to the second auxiliary view—for example, dimensions c and d. The true size (TS) of the surface 1–2–3–4 is shown in the secondary auxiliary view since the direction of sight is perpendicular to it.

FIGURE 8.18 True Size of an Oblique Surface.

■ 8.20 DIHEDRAL ANGLES

The angle between two planes is called a ***dihedral angle***. Auxiliary views often need to be drawn to show dihedral angles true size, mainly for dimensioning purposes. In Figure 8.19a, a block with a V-groove is shown where the dihedral angle between inclined surfaces A and B is shown true size in the front view.

In Figure 8.19b, the V-groove on the block is at an angle to the front surface so that the true dihedral angle is not shown. Assume that the actual angle is the same as in Figure 8.19a. Does the angle show larger or smaller than in Figure 8.19a? To show the true dihedral angle, the line of intersection (in this case, 1–2) must appear as

a point. Since the line of intersection for the dihedral angle is in both planes, showing it as a point will produce a view which shows both planes on edge. This will give you the true-size view of the dihedral angle.

In Figure 8.19a, line 1–2 is the line of intersection of planes A and B. Now, line 1–2 lies in both planes at the same time; therefore, a point view of this line will show both planes as lines, and the angle between them is the dihedral angle between the planes. *To get the true angle between two planes, find the point view of the line intersection of the planes.*

In Figure 8.19c, the direction of sight is parallel to line 1–2 so that line 1–2 appears as a point, planes A and B appear as lines, and the true dihedral angle is shown in the auxiliary view.

FIGURE 8.19 Dihedral Angles.

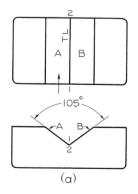

(a)

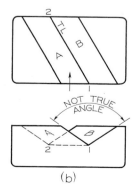

(b)

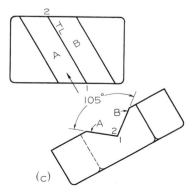

(c)

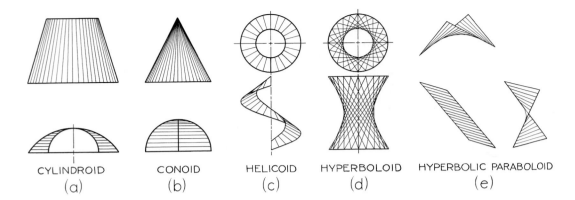

| CYLINDROID | CONOID | HELICOID | HYPERBOLOID | HYPERBOLIC PARABOLOID |
| (a) | (b) | (c) | (d) | (e) |

FIGURE 8.20 **Warped Surfaces.**

■ 8.21 DEVELOPMENTS AND INTERSECTIONS

A *development* is a flat representation or pattern that when folded together creates a 3-D object. An *intersection* is the result of two objects that intersect each other. Sheet-metal construction is the most common application for developments and intersections. The development of surfaces, such as those found in sheet-metal fabrication, is a flat pattern that represents the unfolded or unrolled surface of the form. The resulting flat pattern gives the true size of each connected area of the form so that the part or structure can be fabricated. Auxiliary views are a primary tool used in creating developments. Many specialized software packages are available to automate creating developments and intersections. You can also apply what you have learned about auxiliary views to create developments and intersections using your CAD system.

■ 8.22 TERMINOLOGY

The following terminology describes objects and concepts used in developments and intersections:

A *ruled surface* is one that may be generated by sweeping a straight line, called the *generatrix*, along a path, which may be straight or curved. Any position of the generatrix is an *element* of the surface. A ruled surface may be a plane, a single-curved surface, or a warped surface.

FIGURE 8.21   **Development of Surfaces.**

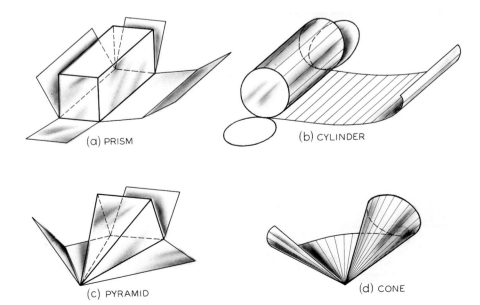

(a) PRISM

(b) CYLINDER

(c) PYRAMID

(d) CONE

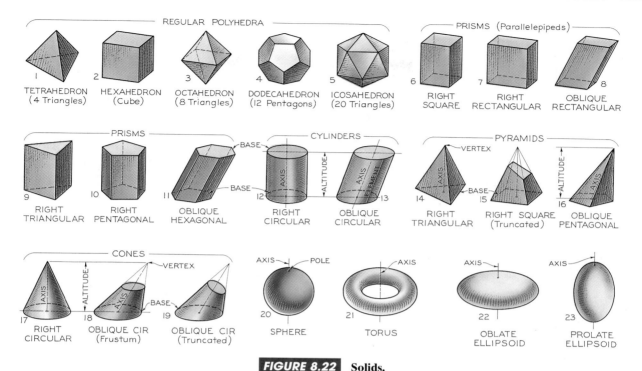

FIGURE 8.22   Solids.

A *plane* is a ruled surface that is generated by a line, one point of which moves along a straight path while the generatrix remains parallel to its original position. Many geometric solids are bounded by plane surfaces.

A *single-curved surface* is a developable ruled surface; that is, it can be unrolled to coincide with a plane. Any two adjacent positions of the generatrix lie in the same plane. Examples are the cylinder and the cone.

A *warped surface* is a ruled surface that is not developable. Some examples are shown in Figure 8.20. No two adjacent positions of the generatrix lie in a flat plane. Warped surfaces cannot be unrolled or unfolded to lie flat. Many exterior surfaces on an airplane or automobile are warped surfaces.

A *double-curved surface* is generated by a curved line and has no straight-line elements. A surface generated by revolving a curved line about a straight line in the plane of the curve is called a *double-curved surface of revolution*. Common examples are the sphere, *torus*, *ellipsoid*, and *hyperboloid.*

A *developable surface* may be unfolded or unrolled to lie flat. Surfaces composed of single-curved surfaces, of planes, or of combinations of these types are developable. Warped surfaces and double-curved surfaces are not directly developable. They may be developed by approximating their shape using developable surfaces. If the material used in the actual manufacturing is sufficiently pliable, the flat sheets may be stretched, pressed, stamped, spun, or otherwise forced to assume the desired shape. Nondevelopable surfaces are often produced by a combination of developable surfaces, which are then formed slightly to produce the required shape. Figure 8.21 shows examples of developable surfaces.

■ 8.23 SOLIDS

Polyhedra are solids which are bounded entirely by plane surfaces—for example cubes, pyramids, and prisms. Convex solids are ones which don't fold back on themselves; in other words they have no dents. Convex solids with all equal faces are called *regular polyhedra*. Examples of regular solids are shown in Figure 8.22. The plane surfaces that bound polyhedra are called the faces of the solid. Lines of intersection of faces are edges of the solids.

A solid generated by revolving a plane figure about an axis in the plane of the figure is a *solid of revolution*. Solids bounded by warped surfaces have no group name. The most common example of such solids is the screw thread.

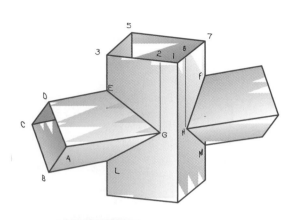

FIGURE 8.23   **Intersecting Prisms.**

■ 8.24 PRINCIPLES OF INTERSECTIONS

A typical need for an accurate drawing showing the intersections of planes and solids is in the cutting of openings in roof surfaces for flues and stacks; in wall surfaces for pipes, chutes, and so on; and in the building of sheet-metal structures such as tanks and boilers. In such cases, you generally need to determine the true size and shape of the intersection of a plane and one of the more common geometric solids. Figure 8.23 shows an example where you would need to determine the intersection of a solid and a plane to create the correctly shaped opening in the vertical prism—the main flue—where the horizontal prism joins it.

For solids bounded by plane surfaces, you need only find the points of intersection of the edges of the solid with the plane and to join these points, in consecutive order, with straight lines.

For solids bounded by curved surfaces, it is necessary to find the points of intersection of several elements of the solid with the plane and to trace a smooth curve through these points. The intersection of a plane and a circular cone is called a *conic section*. Some typical conic sections are shown in Figure 8.24.

 ## ■ 8.25 DEVELOPMENTS

The development of a surface is that surface laid out on a plane. Practical applications of developments occur in sheet-metal work, stone cutting, pattern making, packaging, and package design.

Single-curved surfaces and the surfaces of polyhedra can be developed. Developments for warped surfaces and double-curved surfaces can only be approximated.

In sheet-metal layout, extra material must be provided for laps and seams. If the material is heavy, the thickness may be a factor, and the crowding of metal in bends must be considered. You must also take stock sizes into account and make layouts to economize the use of material and labor. In preparing developments, it is best to put the seam at the shortest edge and to attach the bases at edges where they match; this will economize in soldering, welding, and riveting.

It is common practice to draw development layouts with the *inside surfaces up*. In this way, all fold lines and other markings are related directly to inside measurements, which are the important dimensions in all ducts, pipes, tanks, and other vessels. In this position they are also convenient for use in the fabricating shop.

■ 8.26 HEMS AND JOINTS FOR SHEET METAL AND OTHER MATERIALS

Figure 8.25 shows a wide variety of hems and joints used in fabricating sheet-metal parts and other items. Hems are used to eliminate the raw edge as well as to stiffen the material. Joints and seams may be made for sheet metal by bending, welding, riveting, and soldering and for package materials by gluing and stapling.

You must add material for hems and joints to the layout or development. The amount you add depends on the thickness of the material and the production equipment. A good place to find more information is to talk to manufacturers. They can be extremely helpful in identifying specifications related to the exact process you will use in designing a part.

 A good way to locate manufacturers and products is through the on-line Thomas Register: http://www.thomasregister.com/index.html

■ 8.27 FINDING THE INTERSECTION OF A PLANE AND A PRISM AND DEVELOPING THE PRISM

The true size and shape of the intersection of a plane and a prism is shown in the auxiliary view in Figure 8.26. The length AB is the same as AB in the front view, and the width AD is the same as AD in the top view.

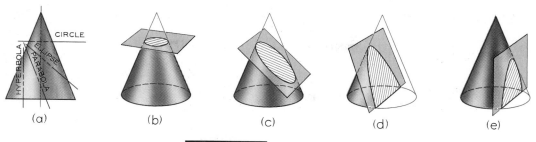

FIGURE 8.24   Conic Sections.

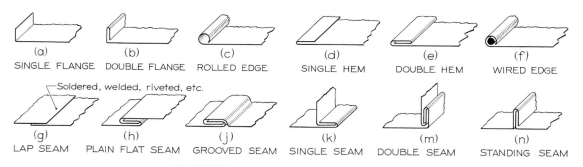

FIGURE 8.25   Sheet-Metal Hems and Joints.

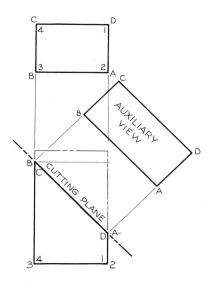

FIGURE 8.26   Auxiliary View Showing True Size and Shape of the Intersection of a Plane and a Prism.

Step by Step 8.4
Developing a prism

These are the steps to create the development for the prism shown.

1. Draw the stretchout line, which represents the axis along which the part is unfolded or unrolled. On the stretchout line, transfer the true sizes of the faces 1–2 and 2–3, which are shown true length in the top view. Remember that a line appears true length when the view is perpendicular to the line. In other words, when a line is parallel to the fold line between views, the line is true length in the adjacent view.

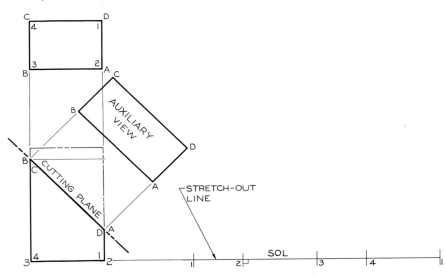

2. Where two surfaces join, draw perpendiculars to the stretchout line and transfer the true height of each respective edge. The front view shows the true heights in this case. Project the heights from the front view, as shown. Complete the development of these surfaces using straight lines to join the points you have plotted. Identify other surfaces that are connected to these and attach their true sizes to the development of the lower base and the upper base. Use an auxiliary view to find the true size of the surface and then draw it in place.

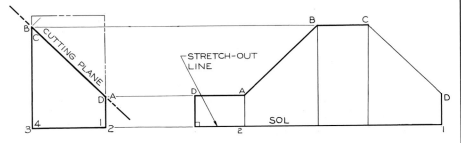

3. When you finish, you will have drawn the development of the entire prism, as shown. If needed, add tabs so that there is material to connect the surfaces when folded up.

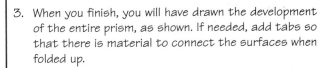
Find Worksheet 8.2 in the tear-out section and cut out the development of the prism. Fold it up according to the directions and use it to help you visualize this development.

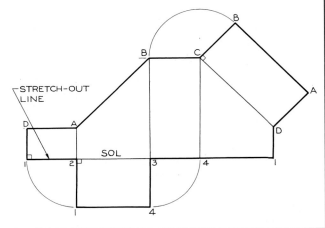

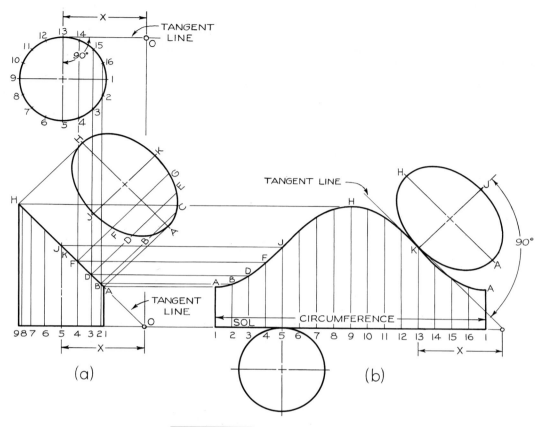

FIGURE 8.27 Plane and Cylinder.

■ 8.28 FINDING THE INTERSECTION OF A PLANE AND A CYLINDER AND DEVELOPING THE CYLINDER

The intersection of a plane and a cylinder is an ellipse whose true size is shown in the auxiliary view of Figure 8.27. The steps for developing a cylinder are as follows:

1. Draw elements of the cylinder. It is usually best to divide the base of the cylinder into equal parts, shown in the top view and then projected into the front view.

2. In the auxiliary view, the widths BC, DE, and so on are transferred from the top view at 2–16, 3–15, respectively, and the ellipse is drawn through these points, as you practiced earlier in this chapter. The major axis AH shows true length in the front view, and the minor axis JK shows true length in the top view. You can use this information to quickly draw the ellipse using CAD.

3. Draw the **stretchout line** for the cylinder. It will be equal to the circumference of the base, whose length is determined by the formula $C = p\,d$ (where d = diameter).

4. Divide the stretchout line into the same number of equal parts as the circumference of the base and draw an element through each division perpendicular to the line.

5. Transfer the true height by projecting it from the front view, as shown.

6. Draw a smooth curve through the points A, B, D, and so on.

7. Draw the tangent lines and attach the bases as shown.

 Find Worksheet 8.3 in the tear-out section. Fold it up according to the directions and use it to help you visualize the development of a cylinder.

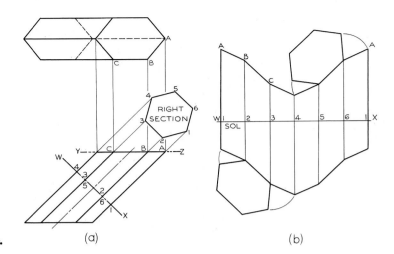

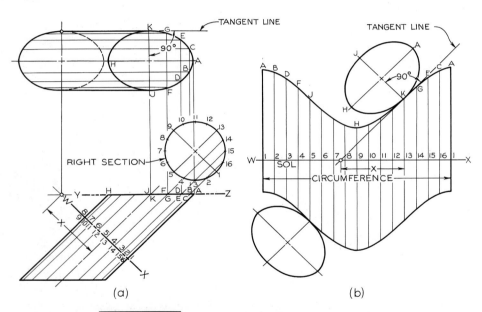

FIGURE 8.28 Plane and Oblique Prism.   (a)   (b)

FIGURE 8.29 Plane and Oblique Circular Cylinder.

■ 8.29 MORE EXAMPLES OF DEVELOPMENTS AND INTERSECTIONS

TAKING A PLANE AND AN OBLIQUE PRISM AND DEVELOPING THE PRISM

The intersection of a plane and an oblique prism is shown in Figure 8.28a. Where the plane is normal to the prism formed by plane WX (called a right section) it appears as a regular hexagon as shown in the auxiliary view labeled RIGHT SECTION. The oblique section cut by horizontal plane YZ is shown true size in the top view.

The development for this oblique prism is shown in Figure 8.28b. Use the right section to create stretchout line WX. On the stretchout line, set off the true widths of the faces 1–2, 2–3, and so on, which are shown true size in the auxiliary view. Draw perpendiculars through each division. Transfer the true heights of the respective edges, which are shown true size in the front view. Join the points A, B, C, and so on with straight lines. Finally attach the bases, which are shown in their true sizes in the top view, along an edge.

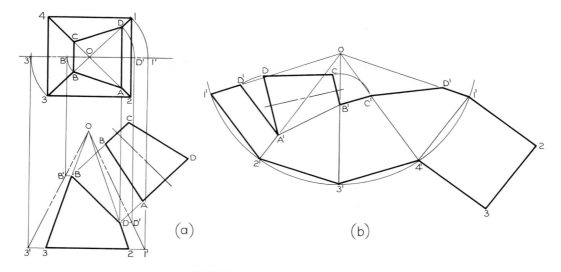

FIGURE 8.30 Plane and Pyramid.

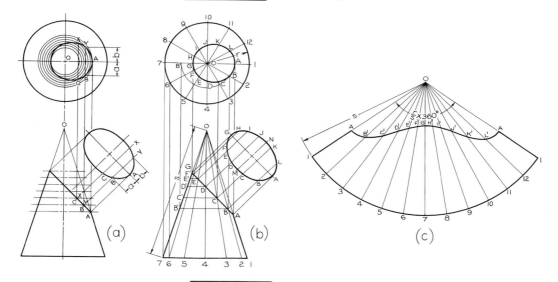

FIGURE 8.31 Plane and Cone.

DEVELOPING A PLANE AND AN OBLIQUE CYLINDER The intersection of a plane and an oblique cylinder are developed similarly, as shown in Figure 8.29.

THE INTERSECTION OF A PLANE AND A PYRAMID AND THE RESULTING DEVELOPMENT The intersection of a plane and a pyramid is a trapezoid, as shown in Figure 8.30.

THE INTERSECTION AND DEVELOPMENT OF A PLANE AND A CONE The intersection of a plane and a cone is an ellipse, as shown in

Figure 8.31. If a series of horizontal cutting planes is passed perpendicular to the axis, as shown, each plane will cut a circle from the cone that will shown in true size and shape in the top view. Points in which these circles intersect the original cutting plane are points on the ellipse. Since the cutting plane is shown on edge in the front view, all of these piercing points can be projected from there to the others, as shown.

For the development of the lateral surface of a cone, you can consider the cone as a pyramid having an infinite number of edges. The development is similar to that for a pyramid.

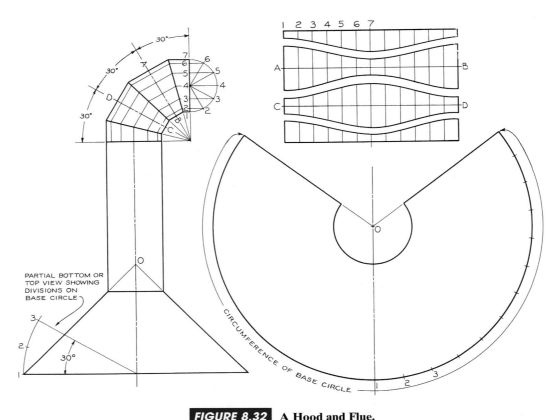

FIGURE 8.32   A Hood and Flue.

THE DEVELOPMENT OF A HOOD AND FLUE   The development of a hood and flue is shown in Figure 8.32. Since the hood is a conical surface, it may be developed as shown above. The two end sections of the elbow are cylindrical surfaces. The two middle sections of the elbow are cylindrical surfaces, but their bases are not perpendicular to the axes, so they will not develop into straight lines. Develop them similar to an oblique cylinder. Make auxiliary planes AB and DC perpendicular to the axes so they cut right sections from the cylinders, which will develop into the straight lines AB and CD in the developments. By arranging the developments as shown, the elbow can be constructed from a rectangular sheet of metal without wasting material. The patterns are shown separated after cutting.

■ 8.30 TRANSITION PIECES

A transition piece is one that connects two differently shaped, differently sized, or skewed-position openings. In most cases, transition pieces are composed of plane surfaces and conical surfaces, as shown in Figure 8.33. You will learn about developing conical surfaces by triangulation next. Triangulation can also be used to develop, approximately, certain warped surfaces. Transition pieces are used extensively in air conditioning, heating, ventilating, and similar construction.

■ 8.31 TRIANGULATION

Triangulation is simply a method of dividing a surface into a number of triangles and transferring them to the development. To find the development of an oblique cone by triangulation, divide the base of the cone in the top view into any number of equal parts and draw an element at each division point, as shown in Figure 8.34. Find the true length of each element. If the divisions of the base are comparatively small, the lengths of the chords may be used in the development to represent the lengths of the respective arcs. Since the development is symmetrical, it is necessary to lay out only half the development, as shown.

GRAPHICS SPOTLIGHT

High Technology Is First Mate In the Race for America's Cup*

A klaxon sounds and the sleek hull of the racing yacht begins knifing through cool, calm waters at increasing speed. The bow heaves as it pushes through the flat water, raising a wave as it moves faster and sending ripples fanning from the waterline. People riding along feel the breeze in their faces as the hull glides at speeds of up to 14 knots, but the motion and sensations last only a few seconds. This dash by a 25-foot-long scale model covers only about 800 feet and takes place in a huge water tank, not the open ocean. But it nonetheless signals a beginning of the next race for the America's Cup, the world's most prestigious sailing competition.

Challengers hoping to capture the America's Cup from New Zealand in 2000 have already begun extensive research and testing directed toward designing and buiding the fastest sailboats in their class. Months, even years, before construction begins on the boats, teams of naval architects, designers, computer scientists, model builders and engineers engage in a technological competition to create machines that can complete a two-hour race a few minutes or seconds ahead of rivals. It is this competition that brings yacht builders to the David Taylor Model Basin at the Naval Surface Warfare Center. Here, where the Navy tows and tests models of its future destroyers, frigates and other warships in the world's largest towing tanks, John K. Marshall, president of the New York Yacht Club, watches as fiberglass models of different racing hulls go through their paces.

"This is our surface warfare, so perharps it's appropriate that we are here," said Mr. Marshall, director of the Young America campaign, a $40 million effort by the New York Yacht Club to build the boat that wins the right to challenge Team New Zealand for the oldest trophy in international sport.

"Sailing in a sport, an athletic competition for which people must train and develop their skills," Mr. Marshall said. "But the America's Cup is also a technology competition and it's always been that way".

"If your boat is as fast, you can win sailing skill," said Bob Billingham of America One, the group organizing the entry for the St. Francis Yacht Club of San Francisco. "But you can't win with a slower boat.

So team worldwide labor to refine their designs, jealously guarding studies of hull shapes, sails and even rigging hardware to deny the smallest secret to competitors. Yachts in this class are so closely matched that no advantage is insignificant. Veterans remember when an Australian challenger in 1983 snatched the cup from the United States for the first time with help from a radical innovation, a winged keel. The Australians kept the keel secret until the last moment to prevent competitors from trying to copy it.

Getting the fastest boat requires working within the strictures of tight rules governing the overall design of this class of vessels. From 1958 to 1987, sailors competed for the America's Cup in 12-Meter Class yachts. But in 1989, a multinational group of yacht designers developed rules for a new International America's Cup Class, which first appeared in the 1992 competition.

The new America's Cup boats are lighter, faster, narrower and longer and carry more sail than their predecessors, with canoe-like bodies made of carbon fiber material instead of aluminum. The design is based on a mathematical formula that balances a boat's waterline length, sail area and displacement so that adding significantly to one dimension requires decreasing others. Generally, an America's Cup contender in about 75 feet long, supports a mast that stands 115 feet off the water, has a keel 14 feet deep and weighs 45,000 to 48,000 pounds. More than 40,000 pounds are in a lead ballast bulb at the base of the keel.

Registering to challenge Team New Zealand, who gained the prize in 1995 sailing Black Magic, 16 yacht clubs and syndicates from 10 nations, including 5 from the United States, have so far paid their $200,000 entry fees. Experts estimate that 10 or 12 of these groups will raise enough money to build at least one boat and that perhaps 4 or 5 of the competitors will muster the talent and expertise to produce first-rate vessels with a chance of winning. The competitors are to assemble in the harbor off Auckland in October 1999 and begin a series of match races to determine the best boat to be named the official challenger for the America's Cup. The winning boat will then race Team New Zealand's best new yacht in a best-of-seven series held in February and March of 2000 in the Hauraki Gulf, northeast of Auckland.

With such intense competition, any weight reduction or change in hull shape, sail design, or the placement of components, like the keel or rudder, that resluts in even a 1 percent increase in performance is significant.

Increasingly, the teams rely on computer simulation and the ability to test many design ideas in the cyberseas of a mathematical model before building and trying them in the real word. "Engineers use sophistcated software, known as *computational fluid dynamics programs*", said John Kuhn, a naval architect at Science Applications International Corporation in San Diego, a technical firm supporting the San Francisco group. The programs simulate the fluid flow around hulls and appendages, like rudders, keels and ballast bulbs, or the movement of air around masts or sails. Results from the programs, which calculate pressure and drag, give engineers the information they need to design components that are then tested in tanks or wind tunnels.

Information from these tests the *fluid dynamics programs* then go into a larger computer simulation called a *velocity prediction program*, or VPP. This program combines design specifications with environmental variables like wind, wave and temperature to make predictions on how fast a boat will sail in specific conditions. "A VPP integrates the work of different people working on different parts of a boat and predicts how an overall design will perform on water," said Mr. Kuhn, technical coordinator for the San Francisco group. "These programs are not perfect, but they help tell you how basic elements contribute to a design".

With each competition, said Duncan MacLane, technology project manager for the New York group, the designers are seeing better matches between computer predictions and actual performance. Still, advice from naval architects and other experts like Bruce Farr, the principal designer for the New York group, remains crucial.

"There is still a lot of art in the design process," Mr. MacLane said, "with many of the improvements we are considering coming from the intuition of designers. There is still a designer in front of the computer screen dealing with the nuances, making very subtle alignments in the design that produces a winner".

Tom Schnackenberg, who heads the design team for the New Zealand group, said that although computational and model testing made significant contributions to producing a boat, only full-scale testing and analysis of a real vessel could confirm its design performance. "In the real world", Mr. Scknackenberg said, "we often find full-scale results at variance with predictions". Team New Zealand plans to build at least one boat and use its older championship yachts in its preparations, said Alan Sefton, a spokesman for the group.

Many of the teams preparing new boats are paying extra attention to the sails and riggings, partly because of differences in these areas that appeared in the last cup race, said Mr. Kuhn of the San Francisco group. The mast and rigging on Black Magic, the New Zealand boat, was set farther back than those on other races, and the New Zealanders displayed sails with unusual shapes.

"Everyone in looking at sails and rigging because sail aerodynamics is one of the least understood elements of design," Mr. Kuhn said. "This is where we may be able to squeeze out some more performance". He and other experts said any group that wished to seek the America's Cup but had yet to start this kind of research and planning was probably out of luck. "The race," Mr. Kuhn said, "has already begun".

*Adapted from "High Technology is First Mate in the Race for America's Cup," by Warren E. Leary, *New York Times*, July 21, 1998.

■ 8.32 THE DEVELOPMENT OF A TRANSITION PIECE CONNECTING RECTANGULAR PIPES ON THE SAME AXIS

The transition piece can be a frustum of a pyramid that connects rectangular pipes on the same axis, as shown in Figure 8.35. As a check on the development, lines parallel on the surface must also be parallel on the development.

■ 8.33 FINDING THE INTERSECTION OF A PLANE AND A SPHERE AND FINDING THE APPROXIMATE DEVELOPMENT OF THE SPHERE

The intersection of a plane and a sphere is a circle, with the diameter of the circle depending on where the plane is located. Any circle cut by a plane through the center of the sphere is called a *great circle*. If a plane passes

through the center and is perpendicular to the axis, the resulting great circle is called the *equator*. If a plane contains the axis, it will cut a great circle called a *meridian*.

The surface of a sphere is double curved and is not developable. The surface may be developed approximately by dividing it into a series of zones and substituting for each zone a portion of a right-circular cone. If the conical surfaces are inscribed within the sphere, the development will be smaller than the spherical surface, but if the conical surfaces are circumscribed about the sphere, the development will be larger. If the conical surfaces are partly inside and partly outside the sphere, the resulting development is closely approximate to the spherical surface. This method of developing a spherical surface is the *polyconic* method and is shown in Figure 8.36a. It is used on government maps of the United States.

FIGURE 8.33 Transition Pieces.

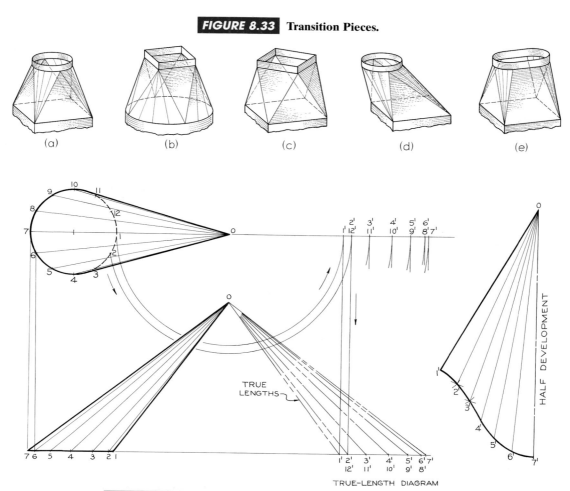

(a) (b) (c) (d) (e)

TRUE LENGTHS

TRUE-LENGTH DIAGRAM

HALF DEVELOPMENT

FIGURE 8.34 Development of an Oblique Cone by Triangulation.

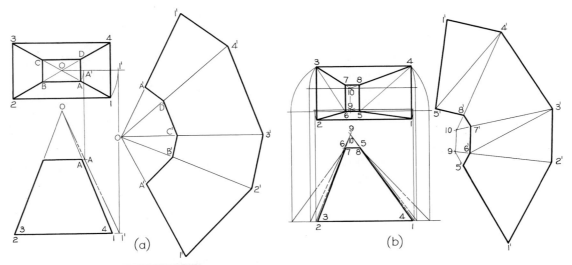

FIGURE 8.35 Development of a Transition Piece. Connecting Rectangular Pipes on the Same Axis.

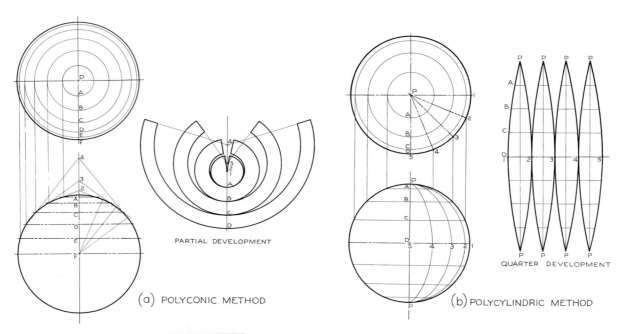

FIGURE 8.36 Approximate Development of a Sphere.

Another method of making an approximate development of the double-curved surface of a sphere is to divide the surface into equal sections with meridian planes and substitute cylindrical surfaces for the spherical sections. The cylindrical surfaces may be inscribed within the sphere or circumscribed about it or located partly inside and partially outside. The development of the series of cylindrical surfaces is an approximate development of the spherical surface. This method is the *polycylindric* method, sometimes called the gore method, as shown in Figure 8.36b.

■ 8.34 COMPUTER GRAPHICS

Using 3-D CAD, any view can be generated in one or two steps, eliminating the need to project auxiliary views manually. It is still very important to have a clear understanding of which line of sight will produce a true-size view or a view which shows a true dihedral angle. When measuring or dimensioning a view from a CAD screen, if the surface or angle is not true size, the automatic dimension from the CAD system will be that of the apparent, or projected, distance. Incorrectly dimensioned dihedral angles can be a common error in CAD drawings created by inexperienced operators. Solid modeling techniques can be used to create accurate intersections between various solids. Some CAD programs have commands which will create transition pieces which blend solids of two differing shapes—for example, a sweep/join operation.

■ KEY WORDS

auxiliary view	dihedral angle	double-curved surface	conic section
true size	development	double-curved surface of revolution	transition piece
primary auxiliary view	stretchout line		triangulation
secondary auxiliary view	intersection	torus	great circle
folding line	ruled surface	ellipsoid	equator
reference plane	generatrix	hyperboloid	meridian
second previous view	element	developable surface	polyconic
partial auxiliary view	plane	polyhedra	polycylindric
auxiliary section	single-curved surface	regular polyhedra	
third auxiliary view	warped surface	solid of revolution	

■ CHAPTER SUMMARY

- An auxiliary view can be used to create a projection that shows the true length of a line or true size of a plane.
- An auxiliary view can be directly produced using CAD if the original object was drawn as a 3-D model.
- Folding lines or reference lines represent the edge views of projection planes.
- Points are projected between views parallel to the line of sight and perpendicular to the reference lines or folding lines.
- A common use of auxiliary views is to show dihedral angles in true size.
- Curves are projected to auxiliary views by plotting them as points.
- A secondary auxiliary view can be constructed from a previously drawn (primary) auxiliary view.

- The technique for creating the development of solids is determined by the basic geometric shape. Prisms, pyramids, cylinders, and cones each have a particular development technique.
- The intersection of two solids is determined by plotting the intersection of each surface and transferring the intersection points to each development.
- Cones and pyramids use radial development. Prisms and cylinders use parallel development.
- Truncated solids, cones, and pyramids are created by developing the whole solid and then plotting the truncated endpoints on each radial element.
- Transition pieces are developed by creating triangular surfaces that approximate the transition from rectangular to circular. The smaller the triangular surfaces, the more accurate the development.

REVIEW QUESTIONS

1. What is meant by true length? By true size?

2. Why is a true-length line always parallel to an adjacent reference line?

3. If an auxiliary view is drawn from the front view, its depth dimensions would be the same as in what other views?

4. Describe one method for transferring depth between views.

5. What is the difference between a complete auxiliary view and a partial auxiliary view?

6. How many auxiliary views are necessary to draw the true size of an inclined plane? Of an oblique plane?

7. What is the angle between the reference-plane (or folding line) and the direction-of-sight lines?

8. How is the development of a pyramid similar to the development of a cone?

9. When developing a truncated cone or pyramid, why is the complete solid developed first?

10. What descriptive geometry techniques are used to determine the intersection points between two solids?

11. What is a transition piece?

12. What is a stretchout line?

13. Which parts of a development are true size and true shape?

14. What building trades use developments and intersections?

AUXILIARY VIEW PROJECTS

The projects in Figures 8.37–8.69 are to be drawn with CAD or freehand. If partial auxiliary views are not assigned, the auxiliary views are to be complete views of the entire object, including all necessary hidden lines.

It is often difficult to space the views of an auxiliary-view sketch. Make sure to provide enough space for the auxiliary view by lightly blocking in the overall dimensions first and by blocking in the overall dimensions of the auxiliary view. Add more detail after you have established the basic layout of the sketch. If metric or decimal dimensions are to be included, refer to the chapter on dimensioning.

A wide selection of intersection and development projects is provided in Figures 8.70–8.75. These projects are designed to fit size B (11 x 17-inch) or A3 (297 x 420-mm) sheets. Because developments are used to create patterns, they should be drawn accurately or dimensioned. They can also be solved on most CAD systems, using either 2-D or solid modeling.

DESIGN PROJECT

Packaging design for personal care products such as toothpaste, hand soap, and shampoo can be a determining factor in the product's sales success. Toothpaste containers vary from traditional squeeze tubes to elaborate pump dispensers that produce multicolored toothpaste mixtures.

Design a new or improved dispenser package for toothpaste or a similar product. Consider ease of use and the suitability of your design for the container's ultimate function. For example, a dispenser used by arthritis sufferers might be particularly easy to open. A dispenser used by children might incorporate an element of fun. Dispensers also have to keep the product clean and fresh.

Make your design a sensible candidate for mass production, striving for a low consumer price and the conservation of raw materials. Should your dispenser be disposable, reusable, or refillable? Use the graphic communication skills you have learned so far to represent your dispenser design clearly.

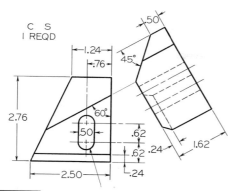

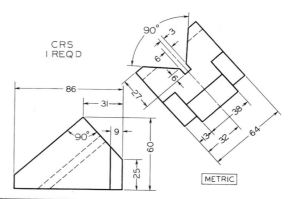

FIGURE 8.37 RH Finger. Given: Front and auxiliary views. Required: Complete front, auxiliary, left-side, and top views (Layout A–3 or A4–3 adjusted).

FIGURE 8.38 V-Block. Given: Front and auxiliary views. Required: Complete front, top, and auxiliary views (Layout A–3 or A4–3 adjusted).

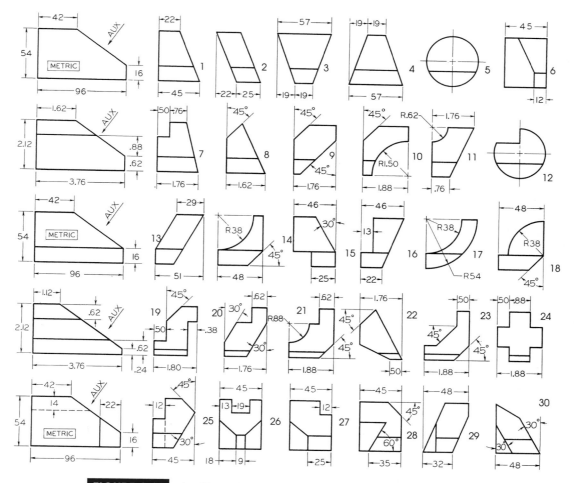

FIGURE 8.39 Auxiliary View Problems. Make freehand sketch or CAD drawing of selected problem as assigned. Draw given front and right-side views, and add incomplete auxiliary view, including all hidden lines (Layout A–3 or A4–3 adjusted). If assigned, design your own right-side view consistent with given front view, and then add complete auxiliary view.

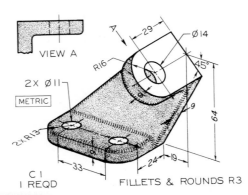

FIGURE 8.40   Anchor Bracket. Draw necessary views or partial views (Layout A–3 or A4–3 adjusted).

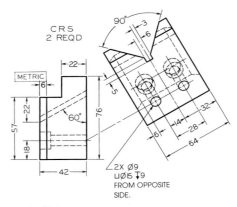

FIGURE 8.43   Guide Block. Given: Right-side and auxiliary views. Required: Right-side, auxiliary, plus front and top views—all complete (Layout B–3 or A3–3).

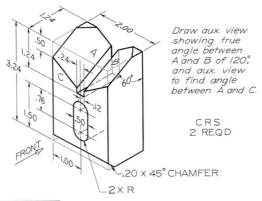

FIGURE 8.41   Centering Block. Draw complete front, top, and right-side views, plus indicated auxiliary views (Layout B–3 or A3–3).

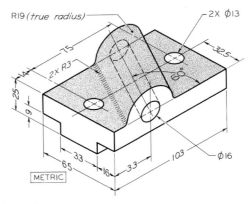

FIGURE 8.44   Angle Bearing. Draw necessary views, including a complete auxiliary view (Layout A–3 or A4–3 adjusted).

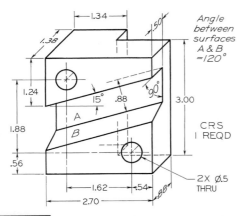

FIGURE 8.42   Clamp Slide. Draw necessary views completely (Layout B–3 or A3–3).

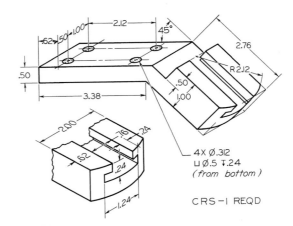

FIGURE 8.45   Guide Bracket. Draw necessary views or partial views (Layout B–3 or A3–3).

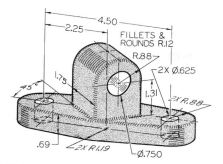

FIGURE 8.46   Rod Guide. Draw necessary views, including complete auxiliary view showing true shape of upper rounded portion (Layout B–4 or A3–4 adjusted).

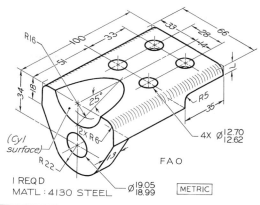

FIGURE 8.49   Angle Guide. Draw necessary views, including a partial auxiliary view of cylindrical recess (Layout B–4 or A3–4 adjusted).

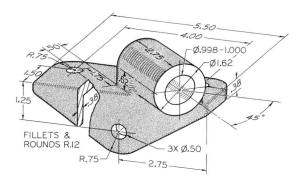

FIGURE 8.47   Brace Anchor. Draw necessary views, including partial auxiliary view showing true shape of cylindrical portion (Layout B–4 or A3–4 adjusted).

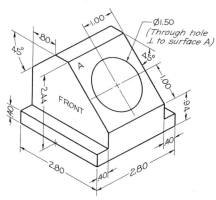

FIGURE 8.50   Holder Block. Draw front and right-side views (2.800 apart) and complete auxiliary view of entire object showing true shape of surface A and all hidden lines (Layout A–3 or A4–3 adjusted).

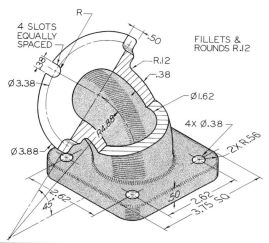

FIGURE 8.48   458 Elbow. Draw necessary views, including a broken section and two half views of flanges (Layout B–4 or A3–4 adjusted).

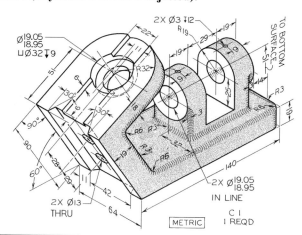

FIGURE 8.51   Control Bracket. Draw necessary views, including partial auxiliary views and regular views (Layout C–4 or A2–4).

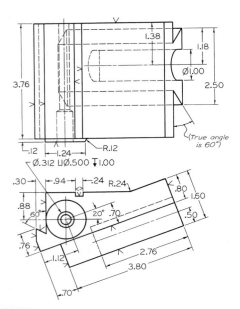

FIGURE 8.52 Tool Holder Slide. Draw given views, and add complete auxiliary view showing true curvature of slot on bottom (Layout B–4 or A3–4 adjusted).

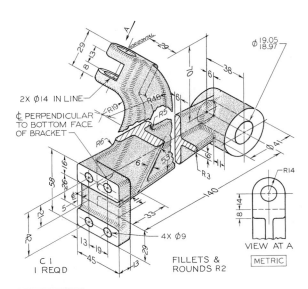

FIGURE 8.54 Guide Bearing. Draw necessary views and partial views, including two partial auxiliary views (Layout C–4 or A2–4).

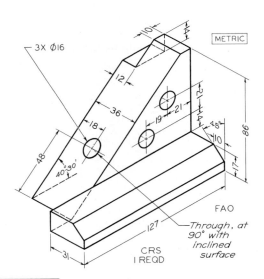

FIGURE 8.53 Adjuster Block. Draw necessary views, including complete auxiliary view showing true shape of inclined surface (Layout B–4 or A3–4 adjusted).

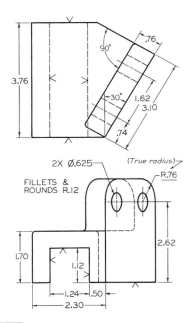

FIGURE 8.55 Drill Press Bracket. Draw given views and add complete auxiliary view showing true shape of inclined face (Layout B–4 or A3–4 adjusted).

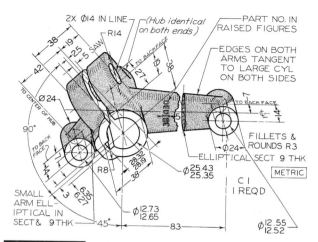

FIGURE 8.56   Brake Control Lever. Draw necessary views and partial views (Layout B–4 or A3–4 adjusted).

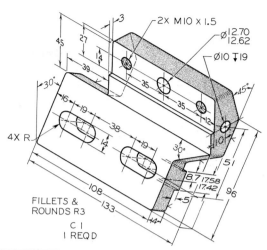

FIGURE 8.58   Cam Bracket. Draw necessary views or partial views as needed. For threads, see §§11.12 and 11.13 (Layout B–4 or A3–4 adjusted).

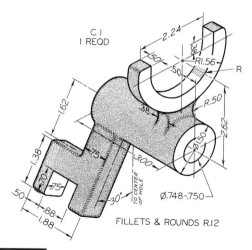

FIGURE 8.57   Shifter Fork. Draw necessary views, including partial auxiliary view showing true shape of inclined arm (Layout B–4 or A3–4 adjusted).

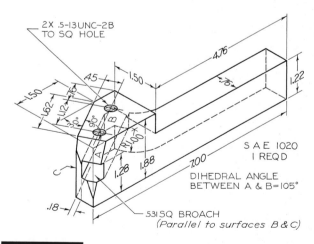

FIGURE 8.59   RH Tool Holder. Draw necessary views, including partial auxiliary views showing 105° angle and square hole true size. For threads, see §§11.12 and 11.13 (Layout B–4 or A3–4 adjusted).

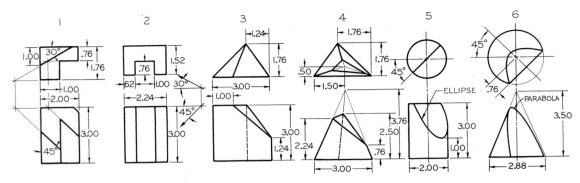

FIGURE 8.60 Draw secondary auxiliary views, complete, which (except Proj. 2) will show the true sizes of the inclined surfaces. In Proj. 2 draw secondary auxiliary view as seen in direction of arrow (Layout B–3 or A3–3).

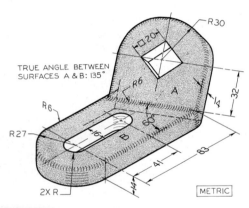

FIGURE 8.61 Control Bracket. Draw necessary views including primary and secondary auxiliary views so that the latter shows true shape of oblique surface A (Layout B–4 or A3–4 adjusted).

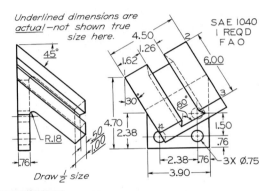

FIGURE 8.63 Dovetail Slide. Draw complete given views and auxiliary views, including view showing true size of surface 1–2–3–4 (Layout B–4 or A3–4 adjusted).

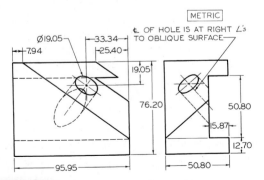

FIGURE 8.62 Holder Block. Draw given views and primary and secondary auxiliary views so that the latter shows true shape of oblique surface (Layout B–4 or A3–4 adjusted).

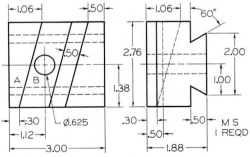

Draw primary aux. view showing angle between planes A and B; then secondary auxiliary view showing true size of surface A.

FIGURE 8.64 Dovetail Guide. Draw given views plus complete auxiliary views as indicated (Layout B–4 or A3–4 adjusted).

a

A fully rendered and
assembled turbine display.
(All courtesy of Autodesk, Inc.)

b

An exploded wireframe
display.

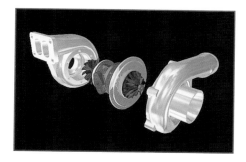

c

A surface-rendered
exploded display.

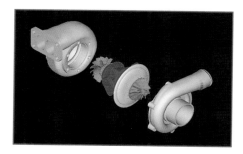

d

A surface-rendered,
color-coded display.

VRML images found at the Dynojet Research Incorporated web site. When viewed with a VRML browser, these images have a 3D quality. *(All courtesy of Dynojet Research, Inc.)*

a, b

Use of CAD speeds the design process and reduces costs. These models of a robot welder and a Zip drive let manufacturers simulate their products before they were even built. *(Reproduced by permission of SDRC, Inc.)*

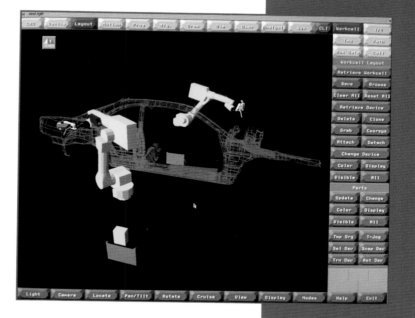

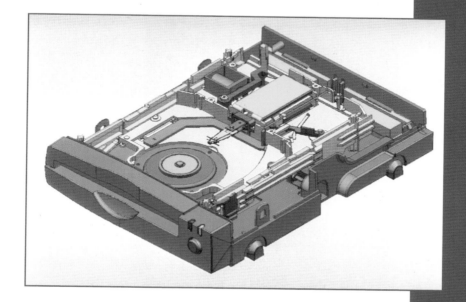

PLATE 4

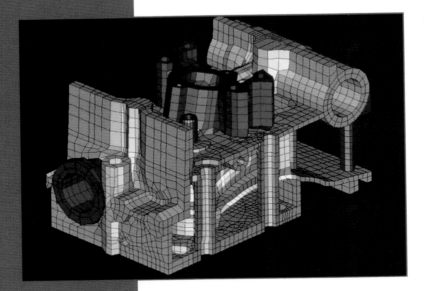

a

A Finite Element Model of diesel engine components created with I-DEAS software. *(Courtesy of SDRC, Inc.)*

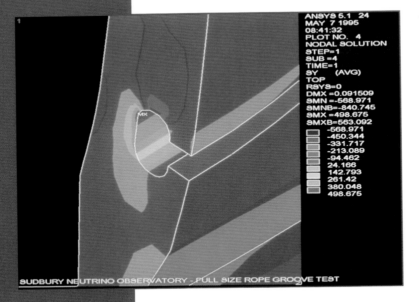

b

A structural stress analysis of the Sudbury Neutrino Observatory created with Auto FEA by ANSYS. *(Courtesy of ANSYS)*

a

An engineer using
virtual reality to help
in autmobile design.
A special helmet, goggles,
and glove allow the
engineer to simulate
the driving experience
as seen below. *(Courtesy of
Ford Motor Company)*

b

CAD software and
virtual reality equip-
ment allow users to
experience "virtual
rooms" and decide
whether they like
the design before
construction. *(Courtesy
of Truevision)*

PLATE 6

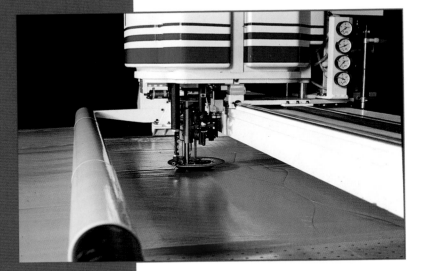

An automated cutting system uses patterns designed with CAD software to produce precise designs. *(Courtesy of Gerber Scientific, Inc.)*

b

Modern manufacturing depends heavily on computer controlled robots. At this Chrysler plant, 66 robots apply spot welds. *(Courtesy of Chrysler Motor Corp.)*

a, b, c

Examples of solid models created with I-DEAS software. Solid modelling can help bring designs to life. *(All courtesy of SDRC, Inc.)*

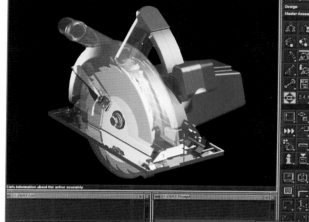

a, b, c

CAD software is very useful as a tool to visualize architectural designs. All these images were created in AutoCAD. *(All courtesy of Autodesk, Inc.)*

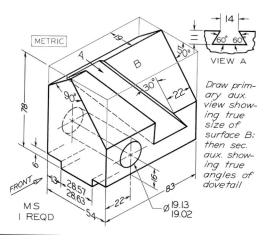

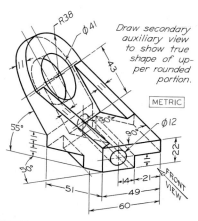

FIGURE 8.65 Adjustable Stop. Draw complete front and auxiliary views plus partial right-side view. Show all hidden lines (Layout C–4 or A2–4).

FIGURE 8.66 Tool Holder. Draw complete front view, and primary and secondary auxiliary views as indicated (Layout B–4 or A3–4 adjusted).

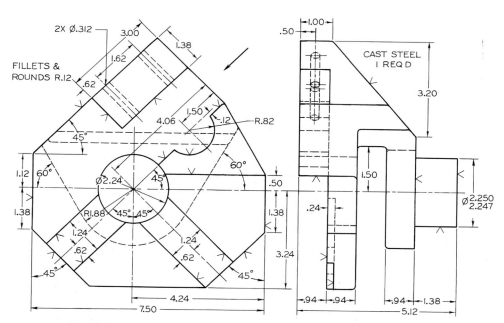

FIGURE 8.67 Box Tool Holder for Turret Lathe. Given: Front and right-side views. Required: Front and left-side views, and complete auxiliary view as indicated by arrow (Layout C–4 or A2–4).

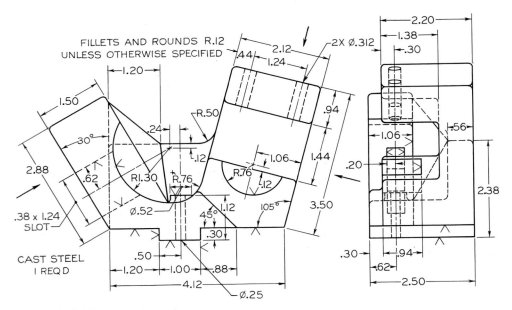

FIGURE 8.68 Pointing Tool Holder for Automatic Screw Machine. Given: Front and right-side views. Required: Front view and three partial auxiliary views (Layout C–4 or A2–4).

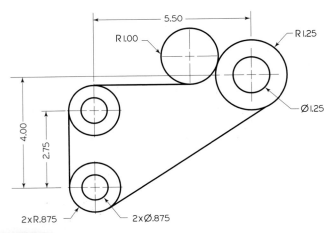

FIGURE 8.69 Print Roller. Given: Right-side view. Design your own front and auxiliary view (Use Layout A–3 or A4–3 adjusted). If assigned, use CAD to create a partial auxiliary view.

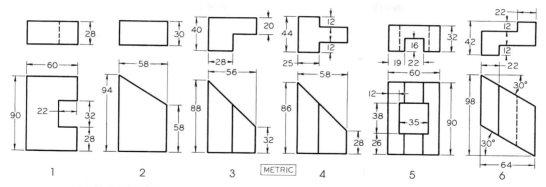

FIGURE 8.70   Draw given views and develop lateral surface (Layout A3–3 or B–3).

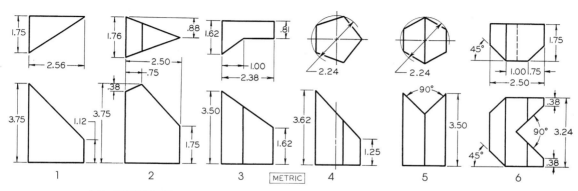

FIGURE 8.71   Draw given views and develop lateral surface (Layout A3–3 or B–3).

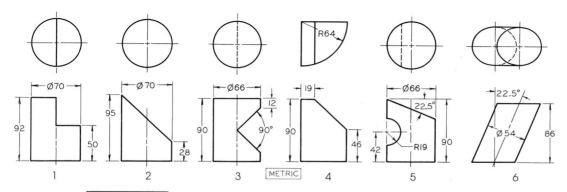

FIGURE 8.72   Draw given views and develop lateral surface (Layout A3–3 or B–3).

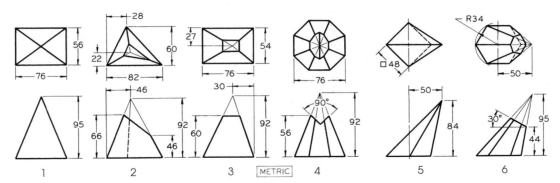

FIGURE 8.73   Draw given views and develop lateral surface (Layout A3–3 or B–3).

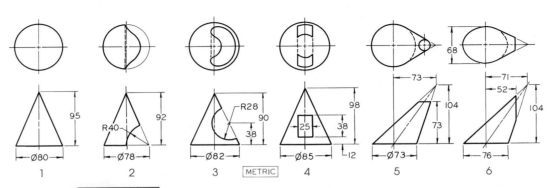

FIGURE 8.74   Draw given views and develop lateral surface (Layout A3–3 or B–3).

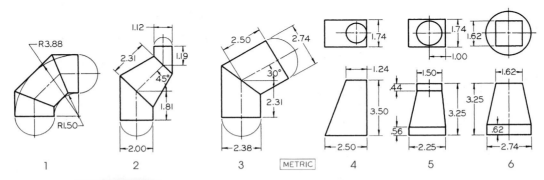

FIGURE 8.75   Draw given views of the forms and develop lateral surfaces (Layout A3–3 or B–3).

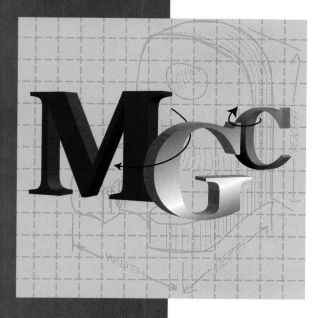

C H A P T E R   9

DIMENSIONING

OBJECTIVES

After studying the material in this chapter, you should be able to:

1. Use conventional dimensioning techniques to describe size and shape accurately on an engineering drawing.

2. Create and read a drawing at a specified scale.

3. Correctly place dimension lines, extension lines, angles, and notes.

4. Use aligned and unidirectional dimensioning systems.

5. Dimension circles, arcs, and inclined surfaces

6. Identify precision ranges for typical manufacturing operations.

7. Apply finish symbols and notes to a drawing.

OVERVIEW

We have all heard of the "rule of thumb." Actually, at one time an inch was defined as the width of a thumb, and a foot was simply the length of a man's foot. In old England, an inch used to be "three barley corns, round and dry." In the time of Noah and the ark, the cubit was the length of a man's forearm, or about 18 inches.

In 1791, France adopted the **meter** (1 meter = 39.37 inches; 1 inch = 25.4 mm), from which the metric system evolved. In the meantime, England was setting up a more accurate measurement for the **yard**, which was legally defined in 1824 by act of Parliament. A foot was one-third of a yard, and an inch was one-thirty-sixth of a yard. From these specifications, graduated rulers, scales, and many types of measuring devices have been developed, achieving even more accuracy of measurement and inspection.

Until this century, common fractions were considered adequate for dimensions. Then, as designs became more complicated and interchangeable parts became necessary to support mass production, more accurate specifications were required, leading to the decimal-inch system, or the SI system.

So far you have been learning to create drawings to describe the shape and position of objects you design. Dimensions and notes define the size, finish, and other manufacturing processes so that the drawing fully defines what you want manufactured. Dimensions describe the size and location of features of an object. Standards organizations prescribe exactly how dimensions should appear and the general rules for their selection and placement in the drawing, but it takes skill to dimension drawings so that their interpretation is clear and unambiguous.

The ability of CAD systems to automatically dimension drawings has improved dramatically, and CAD systems excel at drawing dimensions exactly to standard, but they are not good at selecting which dimension to show or where to place it in a drawing. This takes a level of intelligence that is not part of most CAD systems.

■ 9.1 INTERNATIONAL SYSTEM OF UNITS

The current rapid growth of worldwide science and commerce has fostered an international system of units (*SI system*) suitable for measurements in the physical and biological sciences and in engineering. The seven basic units of measurement are the meter (length), kilogram (mass), second (time), ampere (electric current), kelvin (thermodynamic temperature), mole (amount of substance), and candela (luminous intensity).

The SI system is gradually coming into use in the United States, especially by the many multinational companies in the chemical, electronics, and mechanical industries. A tremendous effort is now under way to convert all standards of the American National Standards Institute (ANSI) to the SI units in conformity with the International Standards Organization (ISO) standards.

■ 9.2 SIZE DESCRIPTION

You have been learning to completely describe an object's shape. The need for interchangeability of parts is the basis for modern *dimensioning* (see ANSI/ASME Y14.5M–1994). Drawings today must be dimensioned so that production personnel all over the world can make mating parts that will fit properly when assembled or when used to replace parts.

The increasing need for precision manufacturing and interchangeability has shifted responsibility for size control to the designing engineer. The production worker no longer must assume responsibility to ensure that parts fit, but only to properly interpret the instructions given on the drawings. You should be familiar with materials and methods of construction and with production requirements in order to create drawings that define exactly what you want to have manufactured.

A drawing submitted to production should show the object in its completed condition and should contain all necessary information specifying the final part. When you are dimensioning a drawing, keep in mind the finished piece, the production processes required, and above all, the function of the part in the total assembly. Whenever possible, give dimensions that are convenient for producing the part. Give sufficient dimensions so that it will not be necessary to scale from the drawing or assume any dimensions. Don't dimension to points or surfaces that are not accessible to the worker. Dimensions should not be duplicated or superfluous. Only those needed to produce and inspect the part against the design specifications should be given. Keep in mind that the dimensions you use to *make the drawing* are not necessarily the dimensions required to easily *make and inspect the part*. Provide functional dimensions that can be interpreted to make the part as you want it built.

■ 9.3 SCALE OF DRAWING

Drawings are usually made to scale, which is indicated in the title block. It helps to visualize the object if you

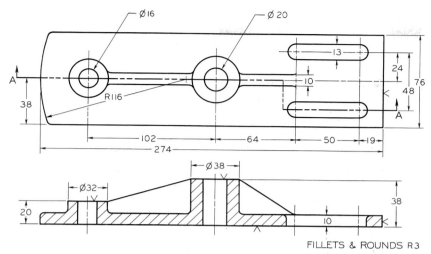

FIGURE 9.1 **Dimensioning Technique. Dimensions in Millimeters.**

have an approximate idea of its size, even though you should never scale the drawing for a needed dimension. Many standard title blocks include a note such as Do not scale drawing for dimensions.

A heavy straight line should be drawn under any dimension that is not to scale, or the abbreviation NTS (not to scale) should be used. When a change made in a drawing is not important enough to justify correcting the drawing, the practice is to just change the dimension. If a dimension does not match the appearance in the drawing, the part is made as dimensioned, not as pictured. Many manufacturers will confirm that the drawing is correct if there seems to be an error, however, it is your responsibility to specify exactly what you want built. When a drawing is prepared on a CAD system, make sure to define dimensions according to the proper standards. Because it is easy to edit CAD drawings, you should generally fix the drawing geometry and not merely change dimension values when making changes.

■ 9.4 LEARNING TO DIMENSION

Dimensions are given in the form of distances, angles, and notes regardless of the dimensioning units being used. For both CAD and sketching, the ability to dimension properly in millimeters, decimal inch, or fractional inch requires the following:

1. *Technique of dimensioning:* The standard for appearance of lines, the spacing of dimensions, the size of arrowheads, and so on allow others to interpret your drawing. A typical dimensioned drawing is shown in Figure 9.1. Note the strong contrast between the visible lines of the object and the thin lines used for the dimensions.

2. *Placement of dimensions:* Use logical placement for dimensions according to standard practices so that they are legible, easy to find, and easy to interpret.

3. *Choice of dimensions:* The dimensions you choose to show affect how your design is manufactured. In the past, manufacturing processes were considered the governing factor in dimensioning. Now function is considered first and manufacturing processes second. Dimension first for function and then review the dimensioning to see if you can make improvements for the purposes of production without adversely affecting the final result. Using a "geometric breakdown" will help you select dimensions. Generally, dimensions determined from a geometric breakdown will be defined by the part's function, but you should logically analyze the functional requirements of the part in the assembly.

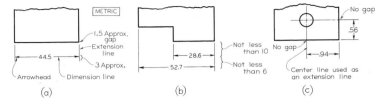

FIGURE 9.2   Dimensioning Technique.

■ 9.5 TOLERANCE

When a finished part is measured, it may vary slightly from the exact dimension specified. *Tolerance* is the total amount that the feature on the actual part is allowed to vary from what is specified by the dimension. You will learn a number of ways to specify tolerances in the next chapter. A good understanding of tolerance is important to understanding dimensioning, especially when choosing which dimensions to show. For now, keep in mind that tolerance can be specified generally by giving a note on the drawing such as All tolerances ±.01 inch unless otherwise noted.

■ 9.6 LINES USED IN DIMENSIONING

A *dimension line* is a thin, dark, solid line terminated by arrowheads, indicating the direction and extent of a dimension. In machine drawing, the dimension line is usually broken near the middle to place the dimension value in the line. In structural and architectural drawing, the dimension figure is placed above an unbroken dimension line.

As shown in Figure 9.2b, the dimension line nearest the object outline should be spaced at least 10 mm (3/8 inch) away. All other parallel dimension lines should be at least 6 mm (1/4 inch) apart, and more if space is available. *The spacing of dimension lines should be uniform throughout the drawing.*

An *extension line* is a thin, dark, solid line that extends from a point on the drawing to which a

dimension refers. The dimension line meets the extension lines at right angles, except in special cases. A gap of about 1.5 mm (1/16 inch) should be left where the extension line would join the object outline. The extension line should extend about 3 mm (1/8 inch) beyond the outermost arrowhead.

The foregoing dimensions for lettering height, spacing, and so on should be increased approximately 50 percent for drawings that are to be reduced to one-half size for the working print. Otherwise the lettering and dimensioning often are not legible.

A *centerline* is a thin, dark line alternating long and short dashes. Centerlines are commonly used as extension lines in locating holes and other symmetrical features. When extended for dimensioning, centerlines cross over other lines of the drawing without gaps. Always end centerlines with a long dash. Refer to Figure 9.2 for examples of lines used in dimensioning.

■ 9.7 ARROWHEADS

Arrowheads, shown in Figure 9.3, indicate the extent of dimensions. They should be uniform in size and style throughout the drawing, not varied according to the size of the drawing or the length of dimensions. Sketch arrowheads freehand so that the length and width have a ratio of 3:1. The arrowhead's length should be equal to the height of the dimension values (about 3 mm or 1/8 inch long). For best appearance, fill in the arrowhead, as in Figure 9.3d.

FIGURE 9.3   Arrowheads.

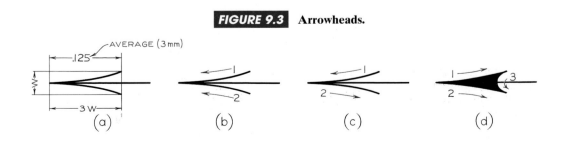

Practical Tips
Arrowheads

When using the arrowhead method in which both strokes are directed toward the point, it is easier to make the strokes toward yourself.

■ 9.8 LEADERS

A *leader* is a thin, solid line directing attention to a note or dimension and starting with an arrowhead or dot. Use an arrowhead to start the leader when you can point to a line in the drawing, such as the edge of a hole; use a dot to start the leader when locating something within the outline of the object. A leader should be an inclined straight line drawn at a large angle, except for the short horizontal shoulder (about 6 mm or 1/4 inch) extending from the center of the first or last line of lettering for the note. A leader to a circle should be a radial line, which is a line that would pass through the center of the circle if extended. See Figure 9.4 for examples.

For the best appearance, make leaders—

- near each other and parallel
- across as few lines as possible

Don't make leaders—

- parallel to nearby lines of the drawing
- through a corner of the view
- across each other
- longer than needed
- horizontal or vertical

■ 9.9 DIRECTION OF DIMENSION FIGURES

Figure 9.5 shows the two systems of reading direction for dimension values. In the preferred *unidirectional system*, approved by ANSI, all dimension figures and notes are lettered horizontally and are read from the bottom of the sheet. It is easier to use and read, especially on large drawings. In the *aligned system*, all dimension figures are aligned with the dimension lines so that they may be read from the bottom or right side of the sheet. Dimension lines in this system should not run in the directions included in the shaded area of Figure 9.6 if avoidable. (Figures 9.5 and 9.6 appear on page 284.)

In both systems, dimensions and notes shown with leaders are aligned with the bottom of the drawing. Notes without leaders should also be aligned with the bottom of the drawing.

■ 9.10 FRACTIONAL, DECIMAL, AND METRIC DIMENSIONS

In the early days of machine manufacturing in the United States, workers would scale the undimensioned design drawing to find needed dimensions. It was their responsibility to see that the parts fit together properly. Workers were skilled and very accurate, and excellent fits were obtained. Hand-built machines were often beautiful examples of precision craftsmanship.

The system of units and common fractions is still used in architectural and structural work, in which close accuracy is relatively unimportant and the steel tape or framing square is used to make measurements. Architectural and structural drawings are often dimensioned in this manner, and commodities, such as pipe and lumber, are identified by standard nominal sizes that are close to the actual dimensions.

As industry has progressed, there has been greater and greater demand for more accurate specifications of the important functional dimensions—more accurate than the 1/64 inch permitted using an engineers',

FIGURE 9.4 Leaders.

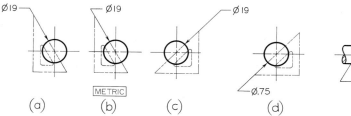

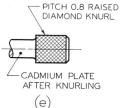

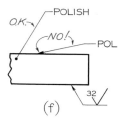

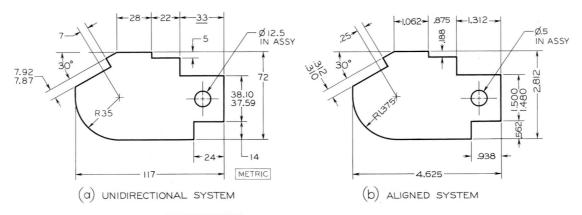

(a) UNIDIRECTIONAL SYSTEM   (b) ALIGNED SYSTEM

FIGURE 9.5   **Directions of Dimension Figures.**

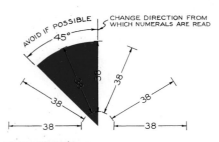

FIGURE 9.6   **Directions of Dimensions.**

architects', and machinists' scale. Since it was cumbersome to use fractions smaller than 1/64 inch, it became the practice to give decimal dimensions, such as 4.2340 and 3.815, for the dimensions requiring greater accuracy. However, some dimensions, such as standard nominal sizes of materials, punched holes, drilled holes, threads, keyways, and other features are still expressed in whole numbers and common fractions.

Drawings may be dimensioned entirely with whole numbers and common fractions, or entirely with decimals, or with a combination of the two. However, more recent practice is to use the decimal-inch system and the metric system as recommended by ANSI. Millimeters and inches in decimal form can be added, subtracted, multiplied, and divided more easily than can fractions. For inch-millimeter equivalents of decimal and common fractions, see the inside back cover of this book.

■ 9.11 DECIMAL SYSTEMS

A decimal system based on the decimal inch or the millimeter has many advantages and is compatible with most measuring devices and machine tools. The millimeter is the commonly used unit for most metric engineering drawings. To facilitate the changeover to metric dimensions, many drawings are dual-dimensioned in millimeters and decimal inches.

Complete decimal dimensioning uses decimals for all dimensions except where certain commodities, such as pipe and lumber, are identified by standardized nominal designations. In these systems, two-place inch or one-place millimeter decimals are used when a common fraction has been regarded as sufficiently accurate. *Combination dimensioning* uses decimals for all dimensions except for the nominal sizes of parts or features, such as bolts, screw threads, keyseats, or other items that use standard fractional designations (ANSI/ASME Y14.5M–1994).

One-place millimeter decimals are used when tolerance limits of ±0.1 mm or more can be permitted. Two (or more)–place millimeter decimals are used for tolerance limits less than ±0.1 mm. Fractions are considered to have the same tolerance as two-place decimal-inch dimensions when determining the number of places to retain in the conversion to millimeters. Keep in mind that 0.1 mm is approximately equal to .004 inch.

Two-place inch decimals are used when tolerance limits of ±.010 or more can be permitted. Three or more decimal places are used for tolerance limits less than ±.010. In two-place decimals, the second place preferably should be an even digit (for example, .02, .04, and .06 are preferred to .01, .03, or .05) so that when the dimension is divided by 2 (for example, when determining the radius from a diameter), the result will still be a two-place decimal. However, odd two-place decimals are used when required for design purposes, such as in dimensioning points on a smooth curve or when strength or clearance is a factor.

Hands On 9.1
Dimensioning technique

Sketch lines, arrowheads, leaders, dimension values, and gaps and sizes similar to the examples shown at left.

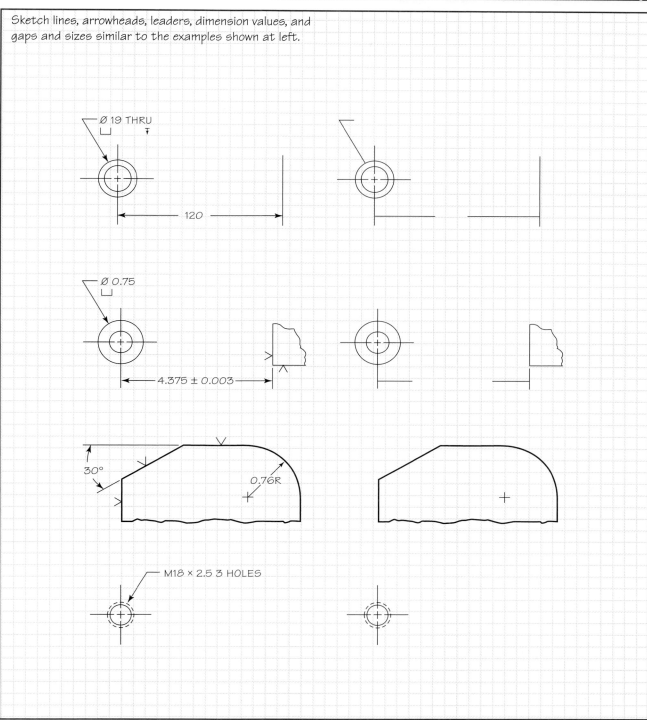

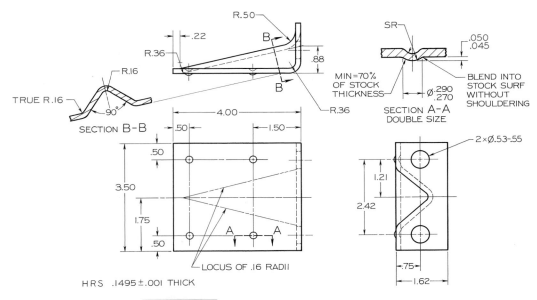

FIGURE 9.7 Complete Decimal Dimensioning.

A typical example of the use of the complete decimal-inch system is shown in Figure 9.7. The use of the preferred decimal-millimeter system is shown in Figure 9.8.

Use the following rule when rounding a decimal value to fewer places, regardless of whether it's decimal inch or metric:

- If the number following the rounding position is a 5, round to an even number.
- If the number following the rounding position is less than 5, make no change.
- If the number following the rounding position is more than 5, round up.

Here are some examples:

- 3.46325 becomes 3.463 when rounded to three places.
- 8.37652 becomes 8.377 when rounded to three places.
- 4.365 becomes 4.36 when rounded to two places.
- 4.366 becomes 4.36 when rounded to two places.

■ 9.12 DIMENSION VALUES

Good hand-lettering is important for dimension values on sketches. The shop produces according to the directions on the drawing, and to save time and prevent costly mistakes, all lettering should be perfectly legible.

Do not crowd dimension figures into limited spaces, making them illegible. There are techniques for showing dimension values outside extension lines or in combination with leaders. Use the methods shown in Figure 9.9, on page 289, when there is not enough room for both the figure and the dimension line inside the extension lines. If necessary, a removed partial view (or detail) may be drawn to an enlarged scale to provide the space needed for clear dimensioning.

Make all decimal points bold, allowing ample space. Where the metric dimension is a whole number, do not show either a decimal point or a zero. Where the metric dimension is less than 1 mm, a zero precedes the decimal point. Where the dimension exceeds a whole number by a fraction of 1 mm, the last digit to the right of the decimal point is not followed by a zero except when expressing tolerances. Figures 9.10a–d, on page 289, shows examples of correct metric dimension values.

Where the decimal-inch dimension is used on drawings, a zero is not used before the decimal point of values less than 1 inch. The decimal-inch dimension is expressed to the same number of decimal places as its tolerance. Zeros are added to the right of the decimal point as necessary. Correct decimal dimension values are shown in Figures 9.10e–j on page 289.

Never letter a dimension value over any line on the drawing; if necessary, break the line. Place dimension

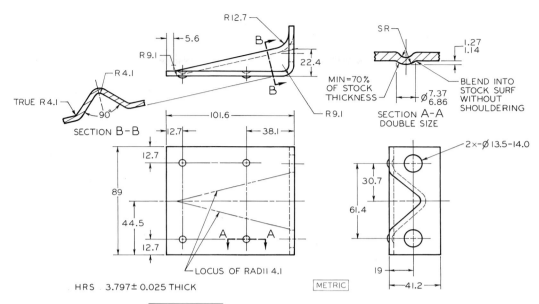

FIGURE 9.8 Complete Metric Dimensioning.

values outside sectioned areas, if possible. When a dimension must be placed on a sectioned area, leave an opening in the section lining for the dimension figure. Refer to Figure 9.11 on page 289 for showing dimensions in sectioned views.

In a group of parallel dimension lines, the numerals should be staggered, as in Figure 9.12a, and not stacked up one above the other, as in Figure 9.12b. (See page 289.)

DUAL DIMENSIONING

Dual dimensioning is used to show metric and decimal-inch dimensions on the same drawing. Two methods of displaying the dual dimensions are shown below:

POSITION METHOD The millimeter dimension is placed above the inch dimension, and the two are separated by a dimension line or by an added line when the unidirectional system of dimensioning is used. An

Hands On 9.2
Rounding dimensional values

Using the rounding rules you have learned, practice rounding the numbers below:

Number	Round to Two Decimals	Round to Three Decimals
4.2885		
76.4935		
23.2456		
11.7852		
9.0348		

alternative arrangement is the millimeter dimension to the left of the inch dimension, with the two separated by a slash line, or virgule. Placement of the inch dimension above or to the left of the millimeter dimension is also acceptable. Each drawing should illustrate the dimension identification as $\frac{\text{MILLIMETER}}{\text{INCH}}$ or MILLIMETER/INCH.

EXAMPLES

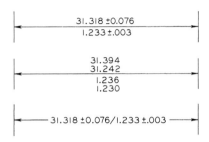

BRACKET METHOD   In this method, the millimeter dimension is enclosed in square brackets,[ ]. The location of this dimension is optional but should be uniform on any drawing—that is, above or below or to the left or the right of the inch dimension. Each drawing should include a note to identify the dimension values, such as DIMENSIONS IN [] ARE MILLIMETERS.

EXAMPLES

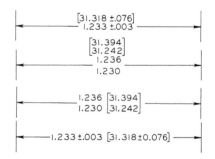

◼ 9.13 MILLIMETERS AND INCHES

Millimeters are indicated by the lowercase letters mm placed one space to the right of the numeral, as in 12.5 mm. *Meters* are indicated by the lowercase m placed similarly, as in 50.6 m. *Inches* are indicated by the symbol ″ placed slightly above and to the right of the numeral: 2-1/2″. *Feet* are indicated by the symbol ′ similarly placed: 3′– 0, 5′– 6, 10′– 0-1/4. It is customary in such expressions to omit the inch mark.

It is standard practice to omit millimeter designations and inch marks on a drawing except when there is a possibility of misunderstanding. For example,

1 VALVE should be 1″ VALVE. Where some inch dimensions are shown on a millimeter-dimensioned drawing, the abbreviation in. follows the inch values.

In some industries, all dimensions, regardless of size, are given in inches; in others, dimensions up to and including 72 inches are given in inches, and dimensions greater than 72 inches are given in feet and inches. In structural and architectural drafting, all dimensions of 1 foot or over are usually expressed in feet and inches.

If suitable, the drawing should contain a note stating Unless otherwise specified, all dimensions are in millimeters (or in inches, as applicable).

◼ 9.14 PLACEMENT OF DIMENSION AND EXTENSION LINES

The correct placement of dimension lines and extension lines is shown in Figure 9.13a on page 290. Rules for the placement of dimensions help you to dimension your drawings so that they are clear and readable. They also help locate dimensions in standard places so that someone manufacturing the part doesn't have to hunt all over a complicated drawing to find where a dimension might be placed. In addition, the rules can help you to avoid mistakes by using general practices for good dimensioning. You cannot always follow every rule to the letter, so keep in mind that the ultimate goal is to dimension the drawing clearly so that the parts are built to your specifications. These general rules will help you place dimensions properly:

- Dimension lines should not cross extension lines as in Figure 9.13b on page 290. It is perfectly OK for extension lines to cross one another, but they should not be shortened like those shown in Figure 9.13c.
- Place shorter dimensions nearest to the outline.
- Dimension lines should not coincide with or continue any line of the drawing, as in Figure 9.13d.
- Avoid crossing dimension lines wherever possible.
- Dimensions should be lined up and grouped together as much as possible, as in Figure 9.14 on page 290.
- Place dimensions between views when possible, but only attached to a single view. This way it is clear that the dimension relates to the feature, which can be seen in more than one view.
- Extension lines and centerlines can cross visible lines of the object to locate dimensions of interior

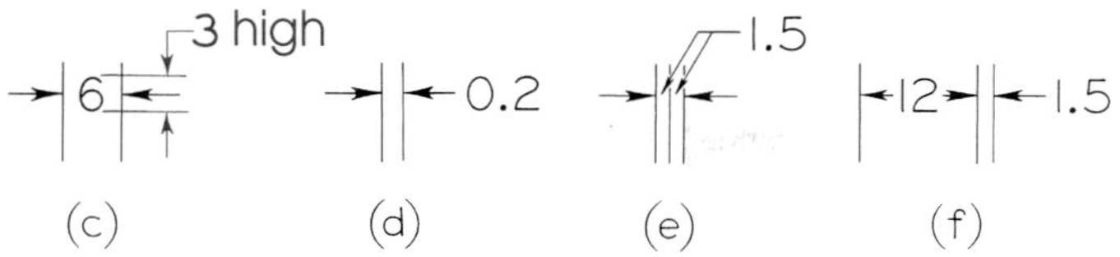

3 high

6

0.2

1.5

12   1.5

(c)   (d)   (e)   (f)

FIGURE 9.9   Dimension Figures. Metric Dimensions (c)–(f)

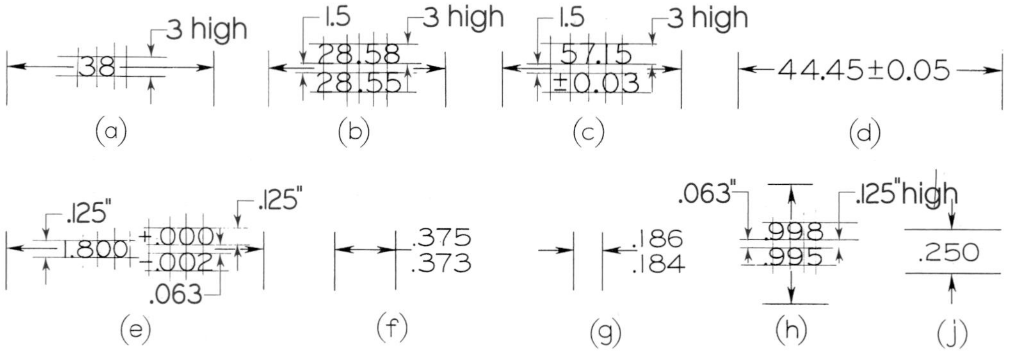

3 high

38

1.5
28.58
28.55
3 high

1.5
57.15
±0.03
3 high

44.45±0.05

(a)   (b)   (c)   (d)

.125"
1.800
+.000
−.002
.063

.125"

.375
.373

.186
.184

.063"
998
995

.125"high

.250

(e)   (f)   (g)   (h)   (j)

FIGURE 9.10   Decimal Dimension figures. Metric Dmensions (a)–(d).

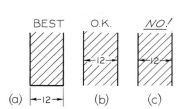

BEST   O.K.   NO!

12   12

(a) 12   (b)   (c)

FIGURE 9.11   Dimensions and Section Lines. Metric.

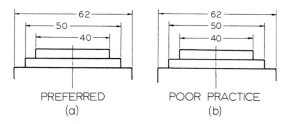

62
50
40
PREFERRED
(a)

62
50
40
POOR PRACTICE
(b)

FIGURE 9.12   Staggered Numerals. Metric.

features. Do not leave a gap in either line when crossing object lines, as shown in Figure 9.15b on page 290. To fit dimensions into a crowded area, you may leave gaps in extension lines near arrowheads so the dimensions show clearly, as shown in Figure 9.16 on page 290.

- Dimension lines are usually drawn at right angles to extension lines, unless showing them otherwise improves clarity, as in Figure 9.17 on page 290.

- Avoid dimensioning to hidden lines. (See Figure 9.18 on page 291.)

- *Dimensions should not be placed on a view unless doing so promotes the clarity of the drawing,* as shown in Figure 9.19 on page 291. In complicated drawings it is often necessary to place dimensions on a view.

- When a dimension must be placed in a hatched area or on the view, leave an opening in the hatching or a break in the lines for the dimension values, as shown in Figures 9.19b and 9.19c.

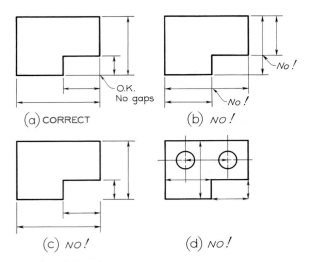

FIGURE 9.13 Dimension and Extension Lines.

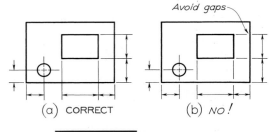

FIGURE 9.14 Grouped Dimensions.

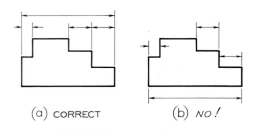

FIGURE 9.15 Crossing Lines.

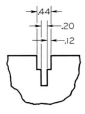

FIGURE 9.16 Placement of Dimensions.

- Give dimensions where the shapes are shown— where the contours of the object are defined—as is shown in Figure 9.20 on page 291. Do not attach dimensions to visible lines where the meaning is not clear, such as the dimension 20 shown in Figure 9.20b on page 291.

- Notes for holes are usually placed where you see the circular shape of the hole, as in Figure 9.20a, but give the diameter of an external cylindrical shape where it appears rectangular. This way it is near the dimension for the length of the cylinder.

- Locate holes in the view that shows the shape of the hole clearly.

■ 9.15 DIMENSIONING ANGLES

You should dimension angles by specifying the angle in degrees and a linear dimension as shown in Figure 9.21a on page 292. You can also give coordinate dimensions for two legs of a right triangle, as shown in Figure 9.21b. The coordinate method is better when a high degree of accuracy is required. Variations in degrees of angle are hard to control because the amount of variation increases with the distance from the vertex of the angle. Methods of indicating angles are shown in Figure 9.21. The tolerancing of angles is discussed in Chapter 10.

In civil engineering drawings, *slope* represents the angle with the horizontal, whereas *batter* is the angle referred to the vertical. Both are expressed by making one member of the ratio equal to 1, as shown in Figure 9.22 on page 289. *Grade*, as of a highway, is similar to slope but is expressed in percentage of rise per 100 feet of run. Thus a 20-foot rise in a 100-foot run is a grade of 20 percent. In structural drawings, angular measurements are made by giving the ratio of run to rise, with the larger size being 12 inches. These right triangles are referred to as *bevels*.

■ 9.16 DIMENSIONING ARCS

A circular arc is dimensioned in the view where you see its true shape by giving the value for its radius preceded by the abbreviation R. You may mark the centers with small crosses to clarify the drawing, but not for small or unimportant radii or undimensioned arcs. When there is room enough, both the radius value and the arrowhead

FIGURE 9.17 Placement of Dimensions.

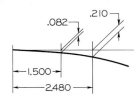

are placed inside the arc. If not, the arrowhead is left inside but the value is moved outside, or both the arrowhead and value are moved outside. When section lines or other lines are in the way, you can use a leader and place the value and leader outside of the section-lined or crowded area. For a long radius, when the center falls outside the available space, the dimension line is drawn toward the actual center; but a false center may be indicated and the dimension line "jogged" to it.

■ 9.17 FILLETS AND ROUNDS

Individual fillets and rounds are dimensioned like other arcs. If there are only a few and they are obviously the same size, giving one typical radius is preferred. However, fillets and rounds are often numerous on a drawing, and they usually are some standard size, such as metric R3 and R6, or R.125 and R.250 when decimal-inch. In this case, give a general note in the lower portion of the drawing, such as:

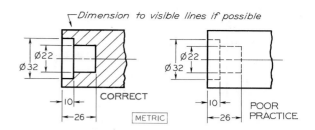

FIGURE 9.18 Placement of Dimensions.

FILLETS R6 AND ROUNDS R3 UNLESS OTHERWISE SPECIFIED

or

ALL CASTING RADII R6 UNLESS NOTED

or simply

ALL FILLETS AND ROUNDS R6

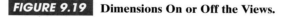

FIGURE 9.19 Dimensions On or Off the Views.

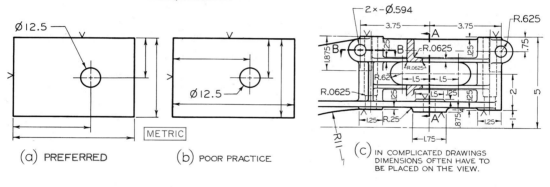

FIGURE 9.20 Contour Dimensioning.

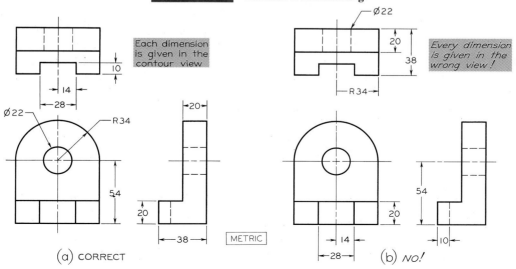

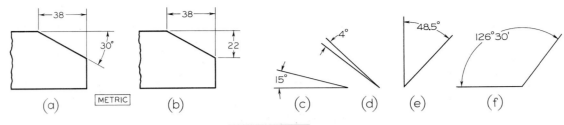

FIGURE 9.21 Angles.

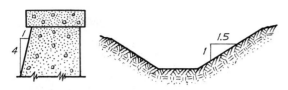

FIGURE 9.22 Angles in Civil Engineering Projects.

■ 9.18 GEOMETRIC BREAKDOWN

Engineering structures are composed largely of simple geometric shapes, such as the prism, cylinder, pyramid, cone, and sphere, as shown in Step by Step 9.1. They may be exterior (positive) or interior (negative) forms. For example, a steel shaft is a positive cylinder, and a round hole is a negative cylinder.

These shapes result directly from design necessity—keeping forms as simple as possible—and from the requirements of the fundamental manufacturing operations. Forms having plane surfaces are produced by planing, shaping, milling, and so forth, while forms having cylindrical, conical, or spherical surfaces are produced by turning, drilling, reaming, boring, countersinking, and other rotary operations, which you will learn about later in this chapter.

The dimensioning of engineering structures involves two basic steps:

1. Give the dimensions showing the sizes of the simple geometric shapes, called *size dimensions*.
2. Give the dimensions locating these elements with respect to each other, called *location dimensions.* Note that a location dimension locates a 3-D geometric element and not just a surface; otherwise, all dimensions would have to be classified as location dimensions.

This process of geometric analysis helps you determine the features of the object and their relationships to one another, but it is not enough just to dimension geometry. You must also consider the function of the part in the assembly and the manufacturing requirements in the shop.

■ 9.19 SIZE DIMENSIONS: PRISMS

The right rectangular prism is probably the most common geometric shape. Front and top views are dimensioned as shown in Figures 9.23a and 9.23b. The height and width are usually given in the front view, and the depth in the top view. The vertical dimensions can be placed on the left or right, usually in-line. Place the horizontal dimension between views as shown and not above the top or below the front view. Front and side views should be dimensioned as in Figures 9.23c and 9.23d. An example of size dimensions for a machine part made entirely of rectangular prisms is shown in Figure 9.24.

■ 9.20 SIZE DIMENSIONS: CYLINDERS

The right circular cylinder is the next most common geometric shape and is commonly seen as a shaft or a hole. Cylinders are usually dimensioned by giving the diameter and length where the cylinder appears as a rectangle. If the cylinder is drawn vertically, give the length at the right or left, as in Figure 9.25. If the cylinder is drawn horizontally give the length above or below the rectangular view, as in Figure 9.25.

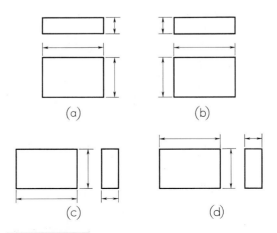

FIGURE 9.23 Dimensioning Rectangular Prisms.

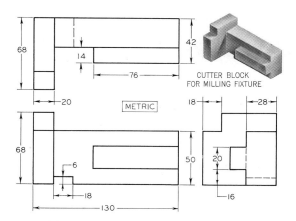

FIGURE 9.24 Dimensioning a Machine Part Composed of Prismatic Shapes.

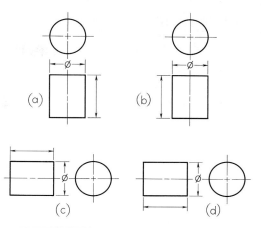

FIGURE 9.25 Dimensioning Cylinders.

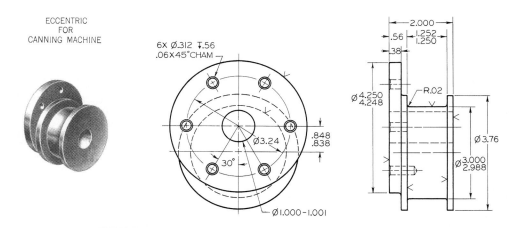

FIGURE 9.26 Dimensioning a Machine Part That Is Composed of Cylindrical Shapes.

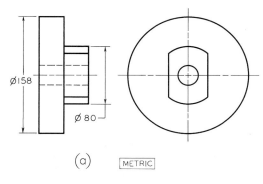

USE OF "Ø" TO INDICATE CIRCULAR SHAPE

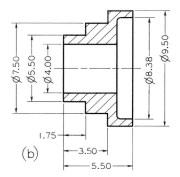

USE OF "Ø" TO OMIT CIRCULAR VIEW

FIGURE 9.27 Use of Ø in Dimensioning Cylinders.

Step by Step 9.1
Dimensioning by geometric breakdown

To dimension the object shown in isometric at right, use geometric breakdown as follows:

1. Consider the geometric features of the part.

In this case the features to be dimensioned include:

- two positive prisms
- one positive cylinder
- one negative cone
- five negative cylinders

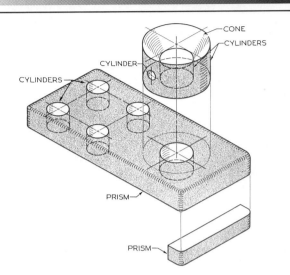

2. Specify the size dimensions for each feature by lettering the dimension values as indicated. (In this illustration, the word "size" indicates the various dimension values.) Note that the four cylinders of the same size can be specified with one dimension.

3. Finally, locate the geometric features with respect to each other. (Actual values would replace the word "location" in this illustration.) Always check to see that the object is fully dimensioned.

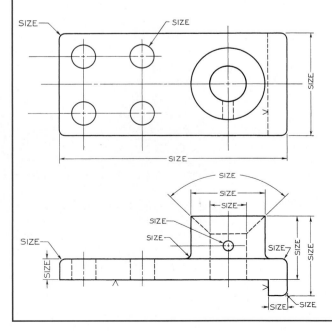

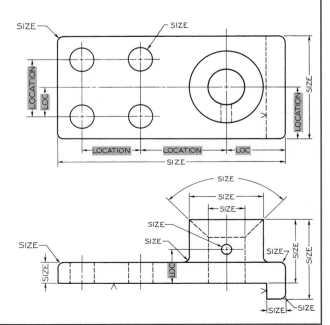

Examples are shown in Figure 9.26 on page 293. Do not use a diagonal diameter inside the circular view, except when clarity is improved. Using several diagonal diameters on the same center becomes very confusing.

The radius of a cylinder should never be given because measuring tools, such as the micrometer caliper, are designed to check diameters. Holes are usually dimensioned by means of notes specifying the diameter and the depth, as shown in Figure 9.26, with or without manufacturing operations.

The diameter symbol Ø should be given before all diameter dimensions, as in Figure 9.27a on page 293 (ANSI/ASME Y14.5M–1994). In some cases, the symbol Ø may be used to eliminate the circular view, as shown in Figure 9.27b. The abbreviation DIA following the numerical value will be found on older decimal-inch drawings.

■ 9.21 SIZE DIMENSIONING OF HOLES

Figure 9.28, on page 296, shows standard symbols used in dimensioning. For example, countersunk, counterbored, and tapped holes are usually specified by standard symbols or abbreviations, as shown in Figures 9.29 and 9.31 on pages 296 and 297. The order of items in a note corresponds to the order of procedure in the shop in producing the hole. The leader of a note should point to the circular view of the hole, if possible. When the circular view of the hole has two or more concentric circles, as for counterbored, countersunk, or tapped holes, the arrowhead should touch the outer circle. Examples are shown in Figure 9.31. Two or more holes can be dimensioned by a single note and by specifying the number of holes, as shown at the top of Figure 9.29.

It is widely acceptable to use decimal fractions for both metric or inch drill sizes, as shown in Figure 9.29b. For numbered or letter-size drills (listed in Appendix 16), specify the decimal size or give the number or letter designation followed by the decimal size in parentheses—for example #28(.1405) or "P" (.3230). Metric drills are all in decimal sizes and are not designated by number or letter.

Specify only the dimensions of the holes, without a note listing whether the holes are to be drilled, reamed, or punched, as shown in Figures 9.29c and 9.29d. The manufacturing technician or engineer is usually better suited to determine the least expensive process to use that will achieve the tolerance required.

■ 9.22 LOCATION DIMENSIONS

After you have specified the sizes of the geometric shapes composing the structure, give location dimensions to show the relative positions of these geometric shapes. Figure 9.30a, on page 296, shows rectangular shapes located by their faces. In Figure 9.30b, cylindrical or conical holes or bosses, or other symmetrical shapes, are located by their centerlines. Location dimensions for holes are preferably given where the holes appear circular, as shown in Figure 9.32. on page 298.

In general, location dimensions should be built from a finished surface or from an important center or centerline. Location dimensions should lead to finished surfaces wherever possible because rough castings and forgings vary in size, and unfinished surfaces cannot be relied on for accurate measurements. The *starting dimension*, used in locating the first machined surface on a rough casting or forging, must necessarily lead from a rough surface or from a center or a centerline of the rough piece.

When several cylindrical surfaces have the same centerline, as in Figure 9.27b, you do not need location dimensions to show they are concentric; the centerline is enough. Holes equally spaced about a common center may be dimensioned by giving the diameter of the *circle of centers*, or **bolt circle**. Use a note such as .750 X 3 to indicate repetitive features or dimensions, where the X means times and the 3 indicates the number of repeated features. Put a space between the letter X and the dimension as shown. Unequally spaced holes are located by means of the bolt circle diameter plus angular measurements with reference to *only one* of the centerlines. Examples are shown in Figure 9.33 on page 298.

Where greater accuracy is required, coordinate dimensions should be given, as shown in Figure 9.33c. In this case, the diameter of the bolt circle is enclosed in parentheses to indicate that it is to be used only as a **reference dimension**. Reference dimensions are given for information only. They are not intended to be measured and do not govern the manufacturing operations. They represent calculated dimensions and are often useful in showing the intended design sizes.

When several nonprecision holes are located on a common arc, they are dimensioned by giving the radius and the angular measurements from a **baseline**, as shown in Figure 9.34a on page 299. In this case, the baseline is the horizontal centerline.

In Figure 9.34b, the three holes are on a common centerline. One dimension locates one small hole from the center; the other gives the distances between the small holes. Note the dimension at X is left off. This method is used when the distance between the small holes is the important consideration. If the relation between the center hole and each of the small holes is more important, then include the distance at X and make the overall dimension a reference dimension.

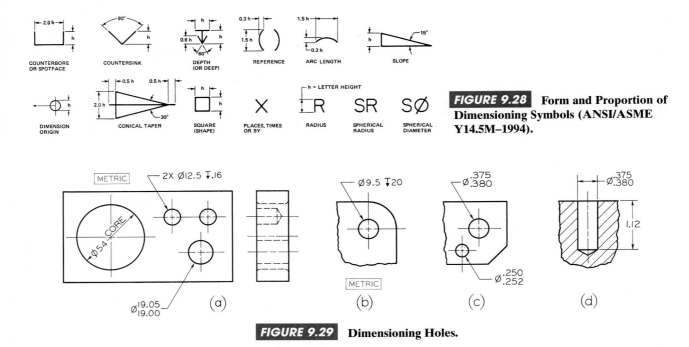

FIGURE 9.28 Form and Proportion of Dimensioning Symbols (ANSI/ASME Y14.5M–1994).

FIGURE 9.29 Dimensioning Holes.

Figure 9.34c shows another example of coordinate dimensioning. The three small holes are on a bolt circle whose diameter is given for reference purposes only. From the main center, the small holes are located in two mutually perpendicular directions.

Another example of locating holes by means of linear measurements is shown in Figure 9.34d. In this case, one measurement is made at an angle to the coordinate dimensions because of the direct functional relationship of the two holes.

In Figure 9.34e, the holes are located from two baselines, or *datums*. When all holes are located from a common datum, the sequence of measuring and machining operations is controlled, overall tolerance accumulations are avoided, and proper functioning of the finished part is assured. The datum surfaces selected must be more accurate than any measurement made from them, must be accessible during manufacture, and must be arranged to facilitate tool and fixture design. It may be necessary to specify accuracy of the datum surfaces in terms of straightness, roundness, flatness, and so on, which you will learn about in the next chapter.

Figure 9.34f shows a method of giving, in a single line, all the dimensions from a common datum. Each dimension except the first has a single arrowhead and is accumulative in value. The overall dimension is separate.

These methods of locating holes are applicable to locating pins or other symmetrical features.

Practice dimensioning using Worksheet 9.1–9.3.

■ 9.23 SYMBOLS AND SIZE DIMENSIONS: MISCELLANEOUS SHAPES

A variety of dimensioning symbols were introduced by ANSI/ASME (Y14.5M–1994) to replace traditional terms or abbreviations. These symbols are given with

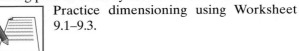

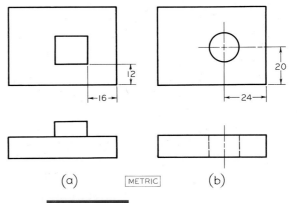

FIGURE 9.30 Location Dimensions.

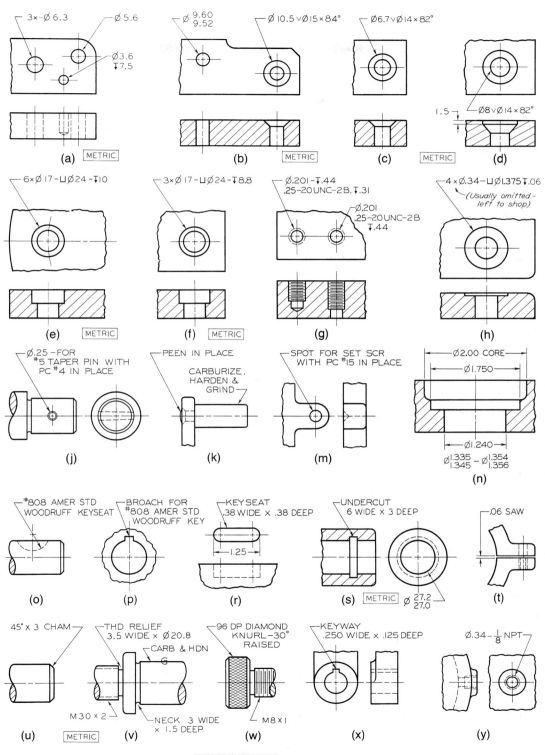

FIGURE 9.31 Local Notes.

construction details in Figure 9.28. Traditional terms and abbreviations are suitable for use where the symbols are not desired. Examples of some of these symbols are given in Figure 9.35 on page 299.

A triangular prism is dimensioned by giving the height, width, and displacement of the top edge in the front view and the depth in the top view, as shown in Figure 9.36a on page 299.

A rectangular pyramid is dimensioned by giving the heights in the front view and the dimensions of the base and the centering of the vertex in the top view, as shown in Figure 9.36b. If the base is square, it is necessary to give the dimensions for only one side of the base, provided it is labeled SQ or preceded by the square symbol, as shown in Figure 9.36c.

A cone is dimensioned by giving its altitude and the diameter of the base in the triangular view, as shown in Figure 9.36d. A frustum of a cone may be dimensioned by giving the vertical angle and the diameter of one of the bases, as shown in Figure 9.36e. Another method is to give the length and the diameters of both ends in the front view. Still another is to give the diameter at one end and the amount of taper per foot in a note.

Figure 9.36f shows a two-view drawing of a plastic knob. Overall, it is spherical and is dimensioned by giving its diameter preceded by the abbreviation and symbol for spherical diameter, SØ, or you may see the older notation where it is followed by the abbreviation SPHER. A bead around the knob is in the shape of a torus and is dimensioned by giving the thickness of the ring and the outside diameter, as shown. In Figure 9.36g, a spherical end is dimensioned by a radius preceded by the abbreviation SR. Internal shapes corresponding to the external shapes in Figure 9.36 would be dimensioned in a similar manner.

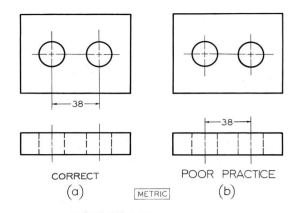

FIGURE 9.32 Locating Holes.

■ 9.24 MATING DIMENSIONS

In dimensioning a single part, its relation to mating parts must be taken into consideration. For example, in Figure 9.37a (on page 301) a guide block fits into a slot in a base. Those dimensions common to both parts are *mating dimensions*, as indicated.

These mating dimensions should be given on the multiview drawings in the corresponding locations, as shown in Figures 9.37b and 9.37c. Other dimensions are not mating dimensions since they do not control the accurate fitting together of two parts. The actual values of two corresponding mating dimensions may not be exactly the same. For example, the width of the slot in Figure 9.37b may be dimensioned 1/32 inch (0.8 mm) or several thousandths of an inch larger than the width of the block in Figure 9.37c, but these are mating dimensions figured from a single basic width. Mating dimensions need to be specified in the corresponding locations on the two parts and toleranced to ensure proper fitting of the parts.

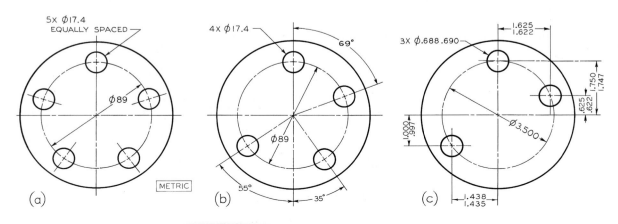

FIGURE 9.33 Locating Holes about a Center.

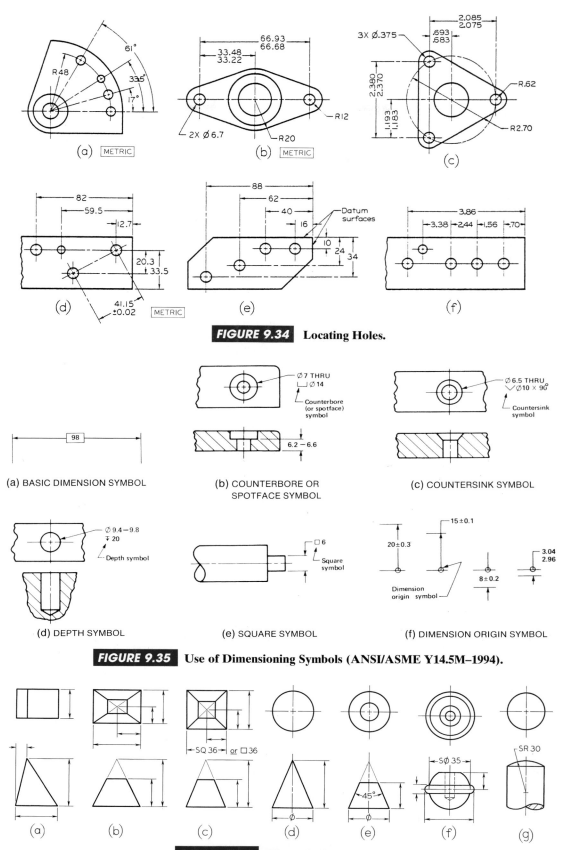

FIGURE 9.34 Locating Holes.

(a) BASIC DIMENSION SYMBOL

(b) COUNTERBORE OR SPOTFACE SYMBOL

(c) COUNTERSINK SYMBOL

(d) DEPTH SYMBOL

(e) SQUARE SYMBOL

(f) DIMENSION ORIGIN SYMBOL

FIGURE 9.35 Use of Dimensioning Symbols (ANSI/ASME Y14.5M–1994).

FIGURE 9.36 Dimensioning Various Shapes.

In Figure 9.38a the dimension A is a necessary mating dimension and should appear on both the drawings of the bracket and of the frame. In Figure 9.38b, which shows a redesign of the bracket into two parts, dimension A is not used on either part because it is not necessary to control closely the distance between the cap screws. But dimensions F are now essential mating dimensions and should appear on the drawings of both parts. The remaining dimensions E, D, B, and C are not considered to be mating dimensions since they do not directly affect the mating of the parts.

■ 9.25 MACHINE, PATTERN, AND FORGING DIMENSIONS

In Figure 9.37a, the base is machined from a rough casting; the pattern maker needs certain dimensions to make the pattern, and the machinist needs certain dimensions for the machining. In some cases one dimension will be used by both. Again, in most cases, these dimensions will be the same as those resulting from a geometric breakdown, but it is important to identify them to assign values to them.

Figure 9.39 on page 302 shows the same part as in Figure 9.37, with the machine dimensions and pattern dimensions identified by the letters M and P. The pattern maker is interested only in the dimensions required to make the pattern, and the machinist, in general, is concerned only with the dimensions needed to machine the part. Frequently, a dimension that is convenient for the machinist is not convenient for the pattern maker, or vice versa. Since the pattern maker uses the drawing only once, while making the pattern, and the machinist refers to it continuously, the dimensions should be given primarily for the convenience of the machinist.

If the part is large and complicated, two separate drawings are sometimes made—one showing the pattern dimensions and the other the machine dimensions. The usual practice, however, is to prepare one drawing for both the pattern maker and the machinist.

For forgings, it is common practice to make separate forging drawings and machining drawings. A forging drawing of a connecting rod, showing only the dimensions needed in the forge shop, is shown in Figure 9.40 on page 302. A machining drawing of the same part, but containing only the dimensions needed in the machine shop, is shown in Figure 9.41 on page 303.

Unless a decimal system is used, the pattern dimensions are nominal, usually to the nearest 1/16 inch, and given in whole numbers and common fractions. If a machine dimension is given in whole numbers and common fractions, the machinist is usually allowed a tolerance of ±1/64 inch. Some companies specify a tolerance of ±.010 inch on all common fractions. If greater accuracy is required, the dimensions are given in decimal form.

■ 9.26 DIMENSIONING OF CURVES

Curved shapes may be dimensioned by giving a group of radii, as shown in Figure 9.42a on page 303. Note that in dimensioning the R126 arc, whose center is inaccessible, the center may be moved inward along a centerline and a jog made in the dimension line. Another method is to dimension the outline envelope of a curved shape so that the various radii are self-locating from "floating centers," as shown in Figure 9.42b. Either a circular or a noncircular curve may be dimensioned by means of coordinate dimensions, or datums, as in Figure 9.42c.

■ 9.27 DIMENSIONING OF ROUNDED-END SHAPES

The method used for dimensioning rounded-end shapes depends on the degree of accuracy required. When precision is not necessary, the methods used are those that are convenient for manufacturing, as in Figures 9.43a to 9.43c on page 304.

In Figure 9.43a, the link to be cast or to be cut from sheet metal or plate is dimensioned as it would be laid out for manufacture, by giving the center-to-center distance and the radii of the ends. Only one such radius dimension is necessary, but that the number of places may be included with the size dimension.

In Figure 9.43 the pad on a casting, with a milled slot, is dimensioned from center to center for the convenience of both the pattern maker and the machinist in layout. An additional reason for the center-to-center distance is that it gives the total travel of the milling cutter, which can be easily controlled by the machinist. The width dimension indicates the diameter of the milling cutter, so it is better to give the diameter of a machined slot. On the other hand, a cored slot should be dimensioned by radius in conformity with the pattern maker's layout procedure.

In Figure 9.43c the semicircular pad is laid out in a similar manner to the pad in Figure 9.43b, except that angular dimensions are used. Angular tolerances can be used if necessary.

When accuracy is required, the methods shown in Figures 9.43d–g are recommended. Overall lengths of rounded-end shapes are given in each case, and radii are indicated, but without specific values. The center-to-center distance may be required for accurate location of some holes.

In Figure 9.43g the hole location is more critical than the location of the radius, so the two are located independently, as shown.

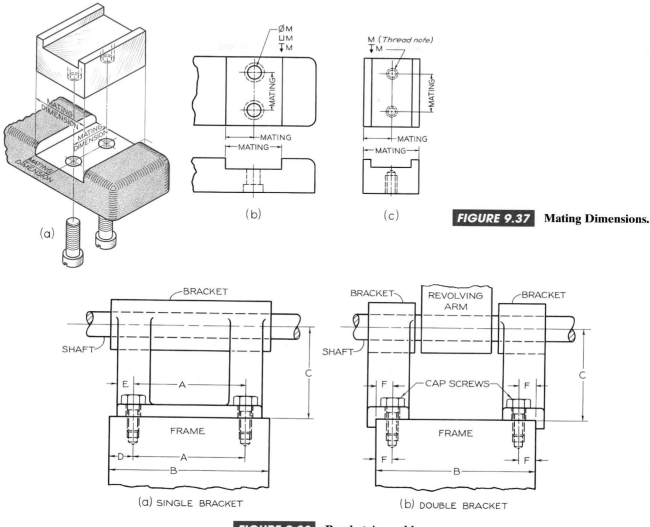

FIGURE 9.37 Mating Dimensions.

(a) SINGLE BRACKET

(b) DOUBLE BRACKET

FIGURE 9.38 Bracket Assembly.

9.28 SUPERFLUOUS DIMENSIONS

All necessary dimensions must be shown, but avoid giving unnecessary or superfluous dimensions, as shown in Figures 9.44a–l on page 306. Do not repeat dimensions on the same view or on different views, or give the same information in two different ways.

As Figure 9.44b shows, it can be impossible to determine how the designer intended to apply the tolerance when a dimension is given two different ways. When chaining dimensions, one dimension of the chain should be left out, as shown, so that the machinist works from one surface only. This is particularly important where an accumulation of tolerances can cause problems with how parts fit or function.

Do not omit dimensions, such as those at the right in Figure 9.44b, thinking that the holes are symmetrical and will be understood to be centered. One of the

two location dimensions should be given. As the creator of the drawing, you should specify exactly how the part is to be built and inspected.

As shown in Figure 9.44e, when one dimension clearly applies to several identical features, or a uniform thickness, it need not be repeated, but the number of places should be indicated. Dimensions for fillets and rounds and other noncritical features need not be repeated, nor need the number of places be specified. For example, the radii of the rounded ends in Figures 9.43a–f need not be repeated.

9.29 FINISH MARKS

A *finish mark* is used to indicate that a surface is to be machined, or finished, as on a rough casting or forging. To the patternmaker or diemaker, a finish mark means that allowance of extra metal in the rough workpiece

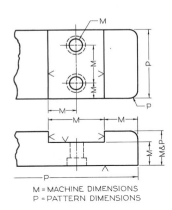

M = MACHINE DIMENSIONS
P = PATTERN DIMENSIONS

FIGURE 9.39   Machine and Pattern Dimensions.

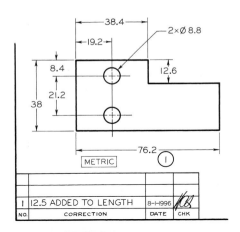

FIGURE 9.40   Revisions.

must be provided for the machining. On drawings of parts to be machined from rolled stock, finish marks are generally unnecessary, because it is obvious that the surfaces are finished. Similarly, it is not necessary to show finish marks when the dimension implies a finished surface, such as Ø6.22–6.35 (metric) or Ø2.45–2.50 (decimal-inch).

As shown in Figure 9.45 (on page 307), three styles of finish marks, the general ∨ symbol, the new basic √ symbol, and the old ✗ symbol, are used to indicate an ordinary smooth machined surface. The ∨ symbol is like a capital V, made about 3 mm (1/8″) high in conformity with the height of dimensioning lettering. The extended √ symbol, preferred by ANSI, is like a larger capital with the right leg extended. The short leg is made about 5 mm (3/16″) high and the height of the long leg is about 10 mm (3/8″). The basic symbol may be altered for more elaborate surface texture specifications.

The point of the ∨ symbol should be directed inward toward the body of metal in a manner similar to that of a tool bit. The √ symbol is not shown upside down (see Figure 9.46 on page 308).

Figure 9.45c shows a simple casting having several finished surfaces. In Figure 9.45d, two views of the same casting show how the finish marks are indicated on a drawing. *The finish mark is shown only on the edge view of a finished surface and is repeated in any other view in which the surface appears as a line, even if the line is a hidden line.*

If a part is to be finished all over, finish marks should be omitted, and a general note, such as FINISH ALL OVER or FAO, should be lettered on the lower portion of the sheet.

The several kinds of finishes are detailed in machine shop practice manuals. The following terms are among the most commonly used: *finish all over, rough finish, file finish, sand blast, pickle, scrape, lap, hone, grind, polish, burnish, buff, chip, spotface, countersink, counterbore, core, drill, ream, bore, tap, broach,* and *knurl.* When it is necessary to control the surface texture of finished surfaces beyond that of an ordinary machine finish, the √ symbol is used as a base for the more elaborate surface quality symbols.

■ 9.30   SURFACE ROUGHNESS, WAVINESS, AND LAY

The modern demands of the automobile, the airplane, and other modern machines that can stand heavier loads and higher speeds with less friction and wear have increased the need for accurate control of surface quality by the designer regardless of the size of the feature. Simple finish marks are not adequate to specify surface finish on such parts.

Surface finish is intimately related to the functioning of a surface, and proper specification of finish of such surfaces as bearings and seals is necessary. Surface quality specifications should be used only where needed, since the cost of producing a finished surface becomes greater as the quality of the surface called for is increased. Generally, the ideal surface finish is the roughest that will do the job satisfactorily.

The system of surface texture symbols recommended by ANSI/ASME (Y14.36M–1996) for use on drawings, regardless of the system of measurement used, is now broadly accepted by American industry. These symbols are used to define **surface texture**,

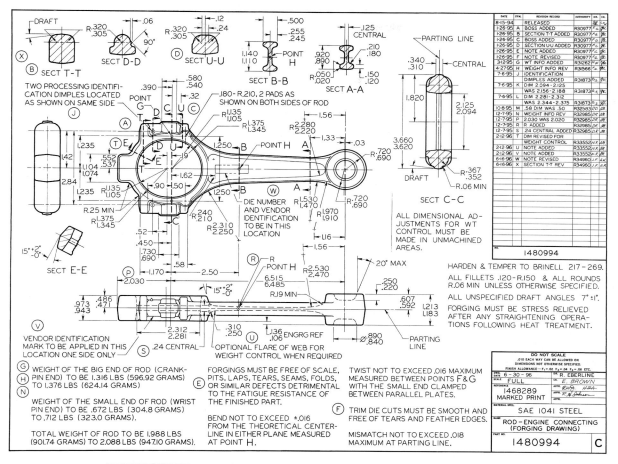

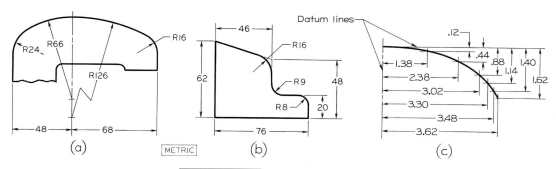

FIGURE 9.41 Forging Drawing of Connecting Rod. *Courtesy of Cadillac Motor Car Division.*

FIGURE 9.42 Dimensioning Curves.

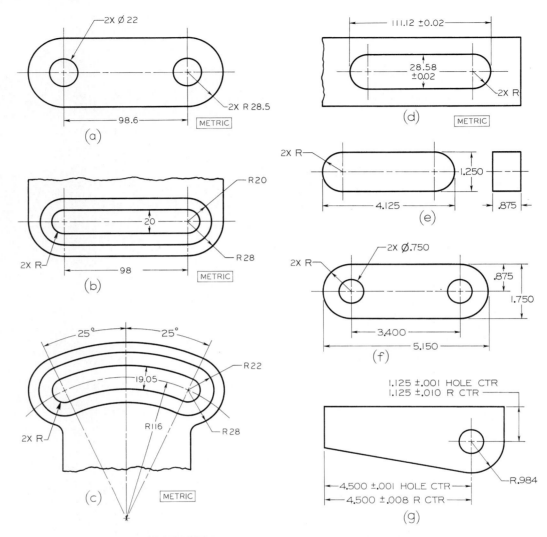

FIGURE 9.43 Dimensioning Rounded-End Shapes.

roughness, and *lay*. See Figure 9.47 on page 309 for the meaning and construction of these symbols. The basic surface texture symbol in Figure 9.47a indicates a finished or machined surface by any method, just as does the general V symbol. Modifications to the basic surface texture symbol, shown in Figures 9.47b–d, define restrictions on material removal for the finished surface. Where surface texture values other than roughness average (R_a) are specified, the symbol must be drawn with the horizontal extension, as shown in Figure 9.47e. Construction details for the symbols are given in Figure 9.47f.

Applications of the surface texture symbols are given in Figure 9.48a on page 309. Note that the

symbols read from the bottom and/or the right side of the drawing and that they are not drawn at any angle or upside down.

Measurements for roughness and waviness, unless otherwise specified, apply in the direction that gives the maximum reading, usually across the lay, as shown in Figure 9.48b. The recommended roughness height values are given in Table 9.1.

When it is necessary to indicate the roughness-width cutoff values, the standard values used are listed in Table 9.2. If no value is specified, the 0.80 value is assumed.

When maximum waviness height values are required, the recommended values to be used are as given in Table 9.3.

GRAPHICS SPOTLIGHT

Semiautomatic Dimensioning Using CAD

DIMENSIONING CONTROLS PART

Dimensioning is an important skill because the dimensions given in the drawing control how the part will be constructed and how tolerance values will be applied. Even when a drawing or model database is exported for direct machining, the machinist must know which fits and dimensions are critical, and where the part can vary. No parts are created exactly to the same size as the dimensions specified, so the designer must make it clear what allowances are possible. Dimensioning CAD drawings is accomplished using a suite of dimensioning tools provided by the software. Programs like AutoCAD 2000 call their tools semiautomatic dimensioning because the dimension lines, values, arrowheads, and extension lines are created for you automatically, but you must still choose where you will place dimensions in the drawing.

DIMENSION STYLES

AutoCAD 2000 lets you create different families of dimension appearances, called dimension styles. You use this to change the appearance of dimensions for different types of drawings. For example, architectural drawings have a different standard for their appearance than mechanical drawings, and civil drawings may still have yet another appearance. To create dimension styles and set their appearance in AutoCAD 2000 you use the Dimension Styles dialog box. You can quickly pick it from the Dimensioning Toolbar shown in Figure A.

PARENT & CHILD STYLES

AutoCAD 2000 uses child styles to let you change the appearance of dimension types within the style; for example radial dimensions can have a different appearance than linear dimensions, or ordinate dimensions. You can have a different appearance for each of these types of dimensions: linear, radial, angular, diameter, ordinate, and leader. You can think of child styles like this. If you have a child, they generally resemble you; have brown eyes if you do, etc. But the child may decide to dye

their hair. After that, no amount of you dying your hair will change the appearance of the child's hair. This is essentially how child dimension styles work. You can set the child style for a type of dimension so that it looks different from the parent style. Once a characteristic of a child dimension style is set differently than the parent style, changing the parent no longer changes the child. You can use these styles to manage the appearance of the dimensions in your drawing so that you do not have to tweak individual dimensions. Dimension styles also allow you to have a consistent approach to controlling the appearance of the dimensions in the drawing so that you know how the dimensions will update if you make a change. Figure B shows the dialogue box you can use.

(B)

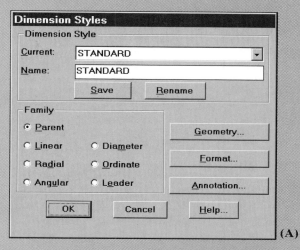

(A)

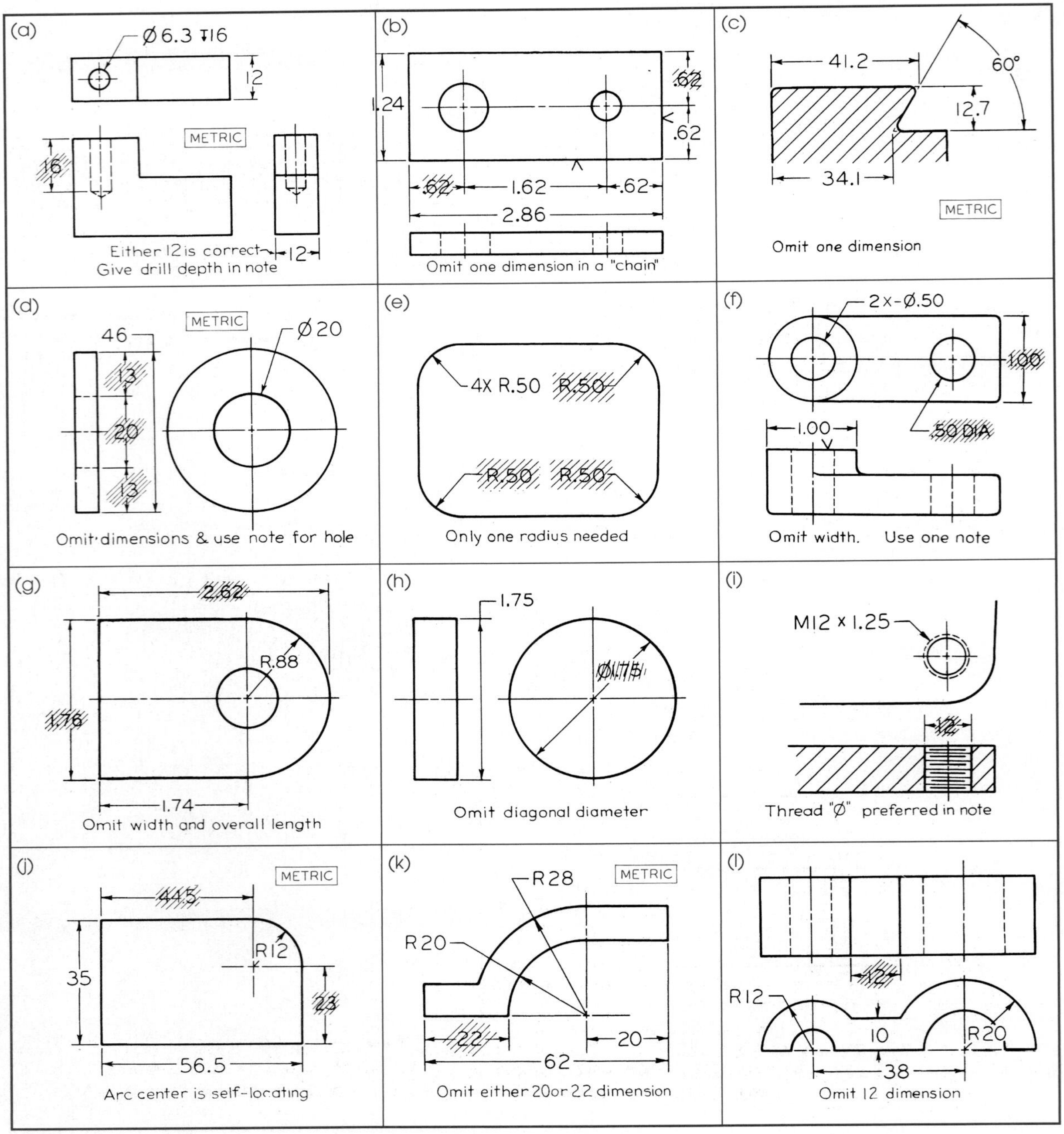

FIGURE 9.44 Superfluous Dimensions.

When it is desired to indicate lay, the lay symbols in Figure 9.49 (on page 311) are added to the surface texture symbols as per the examples given. Selected applications of the surface texture values to the symbols are given and explained in Figure 9.50 on page 311.

A typical range of surface roughness values that may be obtained from various production methods is shown in Figure 9.51 on page 312. Preferred roughness-height values are shown at the top of the chart.

■ 9.31 NOTES

It is usually necessary to supplement the direct dimensions with notes. Notes should be brief and carefully worded to allow only one interpretation. Notes should always be lettered horizontally on the sheet and arranged systematically. They should not be crowded and should not be placed between views, if possible. Notes are classified as *general notes* when they apply to an entire drawing and as **local notes** when they apply to specific items.

GENERAL NOTES   General notes should be lettered in the lower right-hand corner of the drawing, above or to the left of the title block or in a central position below the view to which they apply.

EXAMPLES

FINISH ALL OVER (FAO)
BREAK SHARP EDGES TO R0.8
G33106 ALLOY STEEL–BRINELL 340–380
ALL DRAFT ANGLES 3° UNLESS OTHERWISE SPECIFIED
DIMENSIONS APPLY AFTER PLATING

In machine drawings, the title strip or title block will carry many general notes, including those for materials, general tolerances, heat treatments, and patterns.

LOCAL NOTES   Local notes apply to specific operations only and are connected by a leader to the point at which such operations are performed, as shown in Figure 9.52 on page 313. The leader should be

attached at the front of the first word of a note, or just after the last word, and not at any intermediate place.

Use common abbreviations in notes, such as THD, DIA, MAX. Less common abbreviations should be avoided. All abbreviations should conform to ANSI Y1.1–1989. See Appendix 4 for ANSI abbreviations.

In general, leaders and notes should not be placed on the drawing until the dimensioning is substantially completed. Notes and lettering should not touch lines of the drawing or title block. If notes are lettered first, they may be in the way of necessary dimensions and will have to be moved.

When using CAD to add text for drawing notes, keep in mind the final scale to which the drawing will be plotted. You may need to enlarge the text in order for it to be legible when plotted to a smaller scale.

■ 9.32 DIMENSIONING OF THREADS

Local notes are used to specify dimensions of threads. For tapped holes, the notes should, if possible, be attached to the circular views of the holes, as shown in Figure 9.52g. For external threads, the notes are usually placed in the longitudinal views, where the threads are more easily recognized, as in Figures 9.52v and 9.52w. For a detailed discussion of thread notes, see Chapter 11.

■ 9.33 DIMENSIONING OF TAPERS

A *taper* is a conical surface on a shaft or in a hole. The usual method of dimensioning a taper is to give the amount of taper in a note, such as TAPER 0.167 ON DIA (with TO GAGE often added), and then give the diameter at one end with the length or give the diameter at both ends and omit the length. Taper on diameter means the difference in diameter per unit of length.

Standard machine tapers are used on machine spindles, shanks of tools, or pins and are described in "Machine Tapers" in ANSI/ASME B5.10–1994. Such standard tapers are dimensioned on a drawing by giving the diameter (usually at the large end), the length, and a

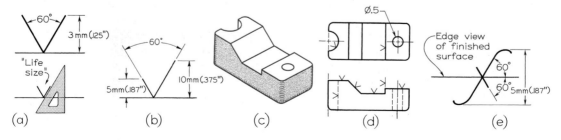

FIGURE 9.45   Finish Marks.

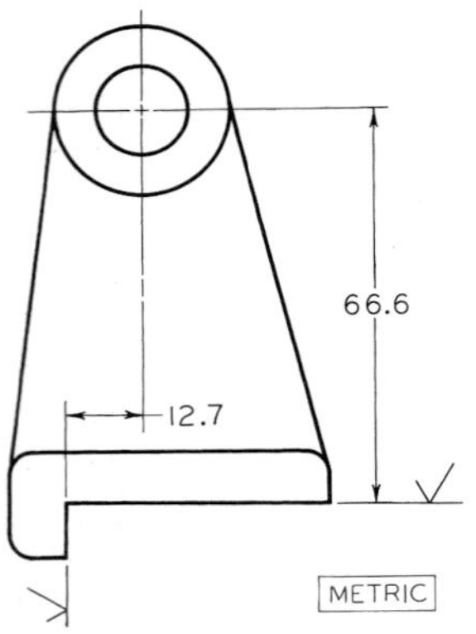

FIGURE 9.46 Dimensions to Finished Surfaces.

note, such as NO. 4 AMERICAN NATIONAL STANDARD TAPER, as shown in Figure 9.53a on page 314.

For not-too-critical requirements, a taper may be dimensioned by giving the diameter at the large end, the length, and the included angle, all with proper tolerances, as shown in Figure 9.53b. Or the diameters of both ends, plus the length, may be given with necessary tolerances.

For close-fitting tapers, the amount of **taper per unit on diameter** is indicated as shown in Figures 9.53c and 9.53d. A gage line is selected and located by a comparatively generous tolerance, while other dimensions are given appropriate tolerances as required.

■ 9.34 DIMENSIONING OF CHAMFERS

A **chamfer** is a beveled or sloping edge. It is dimensioned by giving the length of the offset and the angle, as in Figure 9.54a on page 314. A 45-degree chamfer also may be dimensioned in a manner similar to that shown in Figure 9.54a, but usually it is dimensioned by note, with or without the word CHAM, as in Figure 9.54b.

■ 9.35 SHAFT CENTERS

Shaft centers are required on shafts, spindles, and other conical or cylindrical parts for turning, grinding, and other operations. Such a center may be dimensioned, as shown in Figure 9.55 on page 314. Normally the centers are produced by a combined drill and countersink.

■ 9.36 DIMENSIONING KEYWAYS

Methods of dimensioning keyways for Woodruff keys and stock keys are shown in Figure 9.56 on page 314. Note, in both cases, the use of a dimension to center the keyway in the shaft or collar. The preferred method of dimensioning the depth of a keyway is to give the dimension from the bottom of the keyway to the opposite side of the shaft or hole, as shown. The

method of computing such a dimension is shown in Figure 9.56d. Values for A may be found in machinists' handbooks.

For general information about keys and keyways see Appendix 21.

■ 9.37 DIMENSIONING OF KNURLS

A **knurl** is a roughened surface to provide a better handgrip or to be used for a press fit between two parts. For handgrip purposes, it is necessary only to give the pitch of the knurl, the type of knurling, and the length of the knurled area, as shown in Figures 9.57a and 9.57b on page 315. To dimension a knurl for a press fit, the toleranced diameter before knurling should be given, as shown in Figure 9.57c. A note should be added that gives the pitch and type of knurl and the minimum diameter after knurling (see ANSI/ASME B94.6–1984 (R1995)).

■ 9.38 DIMENSIONING ALONG CURVED SURFACES

When angular measurements are unsatisfactory, chordal dimensions, as shown in Figure 9.58a on page 315, or linear dimensions on the curved surfaces, as shown in Figure 9.58b, may be given.

Micro-meters[a] (μm)	Micro-inches (μin.)	Micro-meters[a] (μm)	Micro-inches (μin.)
0.012	0.5	1.25	50
0.025	1	1.60	63
0.050	2	2.0	80
0.075	3	2.5	100
0.10	4	3.2	125
0.125	5	4.0	180
0.15	6	5.0	200
0.20	8	6.3	250
0.25	10	8.0	320
0.32	13	10.0	400
0.40	16	12.5	500
0.50	20	15	600
0.63	25	20	800
0.80	32	25	1000
1.00	40		

[a] Micrometers are the same as thousandths of a millimeter(1μm = 0.001 mm)

TABLE 9.1 *Preferred Series Roughness Average Values (Ra) (ANSI/ASME Y14.36–1996). Recommended values are in color.*

	Symbol	Meaning
(a)	√	Basic Surface Texture Symbol. Surface may be produced by any method except when the bar or circle, (b) or (d), is specified.
(b)	▽	Material Removal By Machining Is Required. The horizontal bar indicates that material removal by machining is required to produce the surface and that material must be provided for that purpose.
(c)	3.5 ▽	Material Removal Allowance. The number indicates the amount of stock to be removed by machining in millimeters (or inches). Tolerances may be added to the basic value shown or in a general note.
(d)	ᵩ	Material Removal Prohibited. The circle in the vee indicates that the surface must be produced by processes such as casting, forging, hot finishing, cold finishing, die casting, powder metallurgy or injection molding without subsequent removal of material.
(e)	√‾	Surface Texture Symbol. To be used when any surface characteristics are specified above the horizontal line or to the right of the symbol. Surface may be produced by any method except when the bar or circle, (b) or (d), is specified.

(f)

3 X
1.5 X
60°
3 X APPROX
60°
0.00
1.5 X
3 X
LETTER HEIGHT = X

FIGURE 9.47 Surface Texture Symbols and Construction (ANSI/ASME Y14.36M–1996).

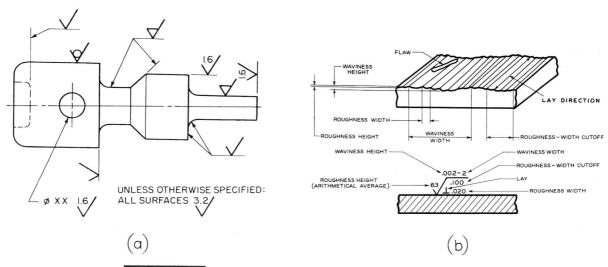

UNLESS OTHERWISE SPECIFIED:
ALL SURFACES 3.2/

ø X X 1.6/

(a)

FLAW
WAVINESS HEIGHT
ROUGHNESS WIDTH
ROUGHNESS HEIGHT
WAVINESS HEIGHT
ROUGHNESS HEIGHT (ARITHMETICAL AVERAGE)
WAVINESS WIDTH
LAY DIRECTION
ROUGHNESS–WIDTH CUTOFF
WAVINESS WIDTH
ROUGHNESS–WIDTH CUTOFF
LAY
ROUGHNESS WIDTH
.002-2
63 .100
.020

(b)

FIGURE 9.48 Application of Surface Texture Symbols and Surface Characteristics (ANSI/ASME Y14.36M–1996).

■ 9.39 SHEET-METAL BENDS

In sheet-metal dimensioning, allowance must be made for bends. The intersection of the plane surfaces adjacent to a bend is called the *mold line*, and this line, rather than the center of the arc, is used to determine dimensions, as shown in Figure 9.59 on page 315. The following procedure for calculating bends is typical. If the two inner plane surfaces of an angle are extended, their line of intersection is called the IML or *inside mold line*, as shown in Figures 9.60a–c on page 316. Similarly, if the two outer plane surfaces are extended, they produce the OML or *outside mold line*. The *centerline of bend* (⌀ B) refers primarily to the machine on which the bend is made and is at the center of the bend radius.

The length, or *stretchout*, of the pattern equals the sum of the flat sides of the angle plus the distance around the bend measured along the *neutral axis*. The distance around the bend is called the *bend allowance*. When metal bends, it compresses on the inside and stretches on the outside. At a certain zone in between, the metal is neither compressed not stretched, and this is called the neutral axis, as shown in Figure 9.60d. The neutral axis is usually assumed to be 0.44 of the thickness from the inside surface of the metal.

The developed length of material, or bend allowance (BA), to make the bend is computed from the empirical formula

$$BA = (0.017453R + 0.0078T)N,$$

where R = radius of bend, T = metal thickness, and N = number of degrees of bend as in Figure 9.60c.

■ 9.40 TABULAR DIMENSIONS

A series of objects having like features but varying in dimensions may be represented by one drawing, as shown in Figure 9.61 on page 316. Letters are substituted for dimension figures on the drawing, and the varying dimensions are given in tabular form. The dimensions of many standard parts are given in this manner in catalogs and handbooks.

■ 9.41 STANDARDS

Dimensions should be given, wherever possible, to make use of readily available materials, tools, parts, and gages. The dimensions for many commonly used machine elements—such as bolts, screws, nails, keys, tapers, wire, pipes, sheet metal, chains, belts, ropes, pins, and rolled metal shapes—have been standardized,

Millimeters (mm)	Inches (in.)	Millimeters (mm)	Inches (in.)
0.08	.003	2.5	.1
0.25	.010	8.0	.3
0.80	.030	25.0	1.0

TABLE 9.2 *Standard Roughness Sampling Length (Cutoff) Values (ANSI/ASME Y14.36–1996).*

Millimeters (mm)	Inches (in.)	Millimeters (mm)	Inches (in.)
0.0005	.00002	0.025	.001
0.0008	.00003	0.05	.002
0.0012	.00005	0.08	.003
0.0020	.00008	0.12	.005
0.0025	.0001	0.20	.008
0.005	.0002	0.25	.010
0.008	.0003	0.38	.015
0.012	.0005	0.50	.020
0.020	.0008	0.80	.030

TABLE 9.3 *Preferred Series Maximum Waviness Height Values (ANSI/ASME Y14.36–1996).*

and the drafter must obtain these sizes from company standards manuals, from published handbooks, from ANSI standards, or from manufacturers' catalogs. Tables of some of the more common items are given in the Appendix of this text.

Such standard parts are not delineated on detail drawings unless they are to be altered for use; they are conventionally drawn on assembly drawings and are listed in parts lists. Common fractions are often used to indicate the nominal sizes of standard parts or tools. If the complete decimal-inch system is used, all such sizes are ordinarily expressed by decimals—for example, .250 DRILL instead of 1/4 DRILL. If the all-metric system of dimensioning is used, then the *preferred* metric drill of the approximate same size (.24800) will be indicated as a 6.30.

■ 9.42 COORDINATE DIMENSIONING

Basic coordinate dimensioning practices are generally compatible with the data requirements for tape or computer-controlled automatic production machines.

LAY SYMBOLS

SYM	DESIGNATION	EXAMPLE	SYM	DESIGNATION	EXAMPLE
=	Lay parallel to the line representing the surface to which the symbol is applied.	DIRECTION OF TOOL MARKS	X	Lay angular in both directions to line representing the surface to which symbol is applied.	DIRECTION OF TOOL MARKS
⊥	Lay perpendicular to the line representing the surface to which the symbol is applied.	DIRECTION OF TOOL MARKS	M	Lay multidirectional	
C	Lay approximately circular relative to the center of the surface to which the symbol is applied.	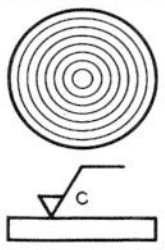	R	Lay approximately radial relative to the center of the surface to which the symbol is applied.	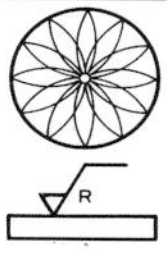

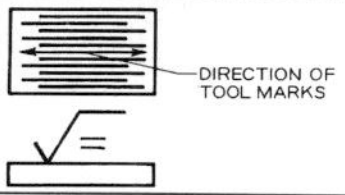

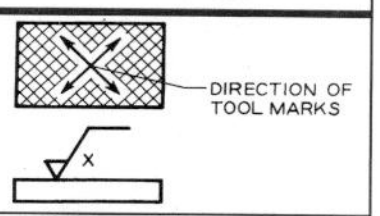

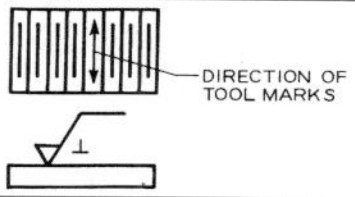

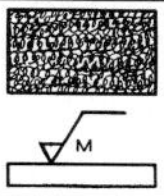

FIGURE 9.49 Lay Symbols (ANSI/ASME Y14.36M–1996).

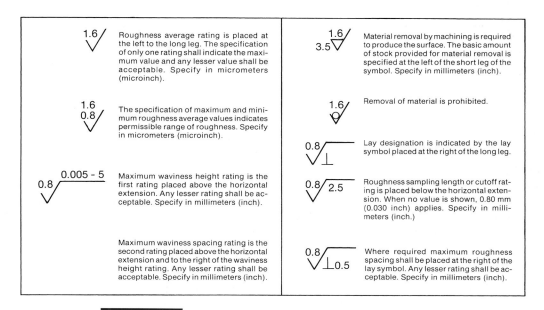

FIGURE 9.50 Application of Surface Texture Values to Symbol (ANSI/ASME Y14.36M–1996).

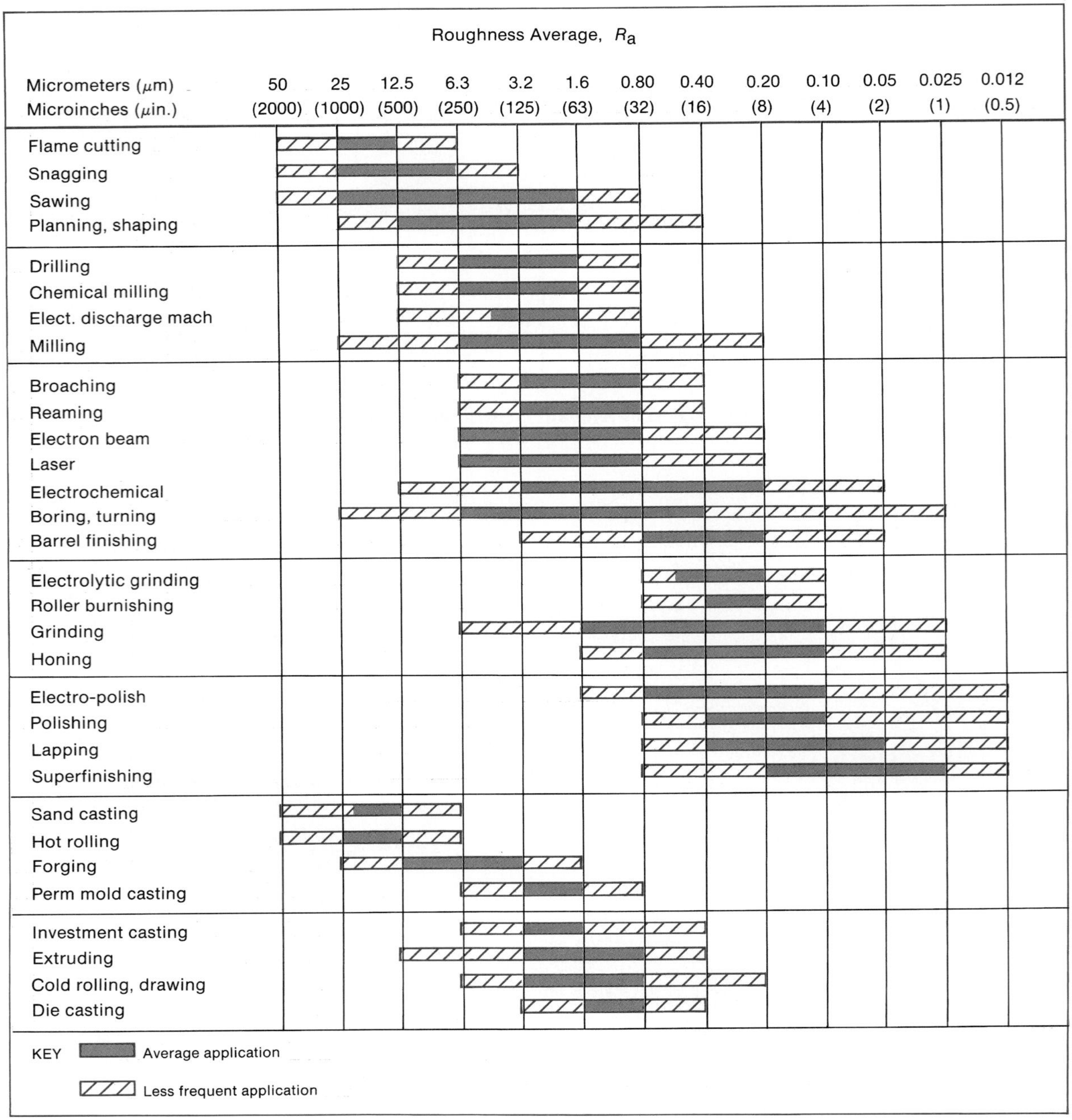

Roughness Average, R_a

| Micrometers (μm) | 50 | 25 | 12.5 | 6.3 | 3.2 | 1.6 | 0.80 | 0.40 | 0.20 | 0.10 | 0.05 | 0.025 | 0.012 |
Microinches (μin.)	(2000)	(1000)	(500)	(250)	(125)	(63)	(32)	(16)	(8)	(4)	(2)	(1)	(0.5)
Flame cutting													
Snagging													
Sawing													
Planning, shaping													
Drilling													
Chemical milling													
Elect. discharge mach													
Milling													
Broaching													
Reaming													
Electron beam													
Laser													
Electrochemical													
Boring, turning													
Barrel finishing													
Electrolytic grinding													
Roller burnishing													
Grinding													
Honing													
Electro-polish													
Polishing													
Lapping													
Superfinishing													
Sand casting													
Hot rolling													
Forging													
Perm mold casting													
Investment casting													
Extruding													
Cold rolling, drawing													
Die casting													

KEY ▇ Average application

▨ Less frequent application

FIGURE 9.51 Surface Roughness Produced by Common Production Methods (ANSI/ASME B46.1–1985). The ranges shown are typical of the processes listed. Higher or lower values may be obtained under special conditions.

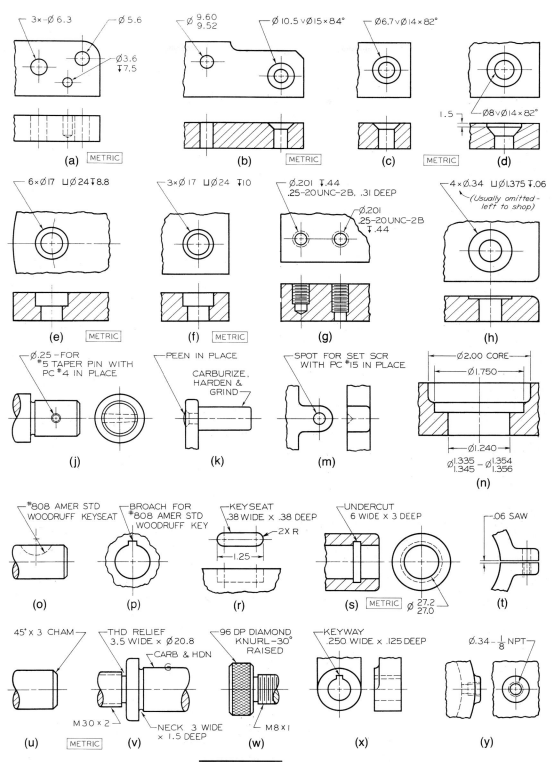

FIGURE 9.52 Local Notes.

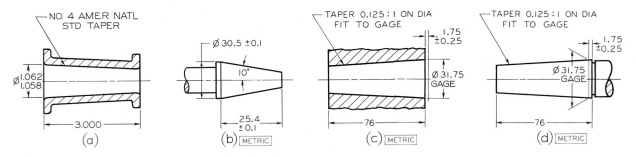

FIGURE 9.53   Dimensioning Tapers.

However, to design for automated production, you should consult the manufacturing machine manuals before making production drawings. Here are basic guidelines for coordinate dimensioning:

1. A set of three mutually perpendicular datum or reference planes is usually required for coordinate dimensioning. These planes either must be obvious or must be clearly identified, as shown in Figure 9.62 on page 317.

2. The designer selects as origins for dimensions those surfaces or features most important to the functioning of the part. Enough of these features are selected to position the part in relation to the set of mutually perpendicular planes. All related dimensions are then made from these planes. Rectangular coordinate dimensioning without dimension lines is shown in Figure 9.63 on page 317.

3. All dimensions should be in decimals.

4. Angles should be given, where possible, in degrees and decimal parts of degrees.

5. Standard tools, such as drills, reamers, and taps, should be specified when required.

6. All tolerances should be determined by the design requirements of the part, not by the capability of the manufacturing machine.

■ 9.43 DO'S AND DON'TS OF DIMENSIONING

The following checklist summarizes briefly most of the situations in which a beginning designer is likely to make a mistake in dimensioning. Students should check the drawing by this list before submitting it to the instructor.

1. Each dimension should be given clearly so that it can be interpreted in only one way.

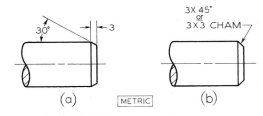

FIGURE 9.54   Dimensioning Chamfers.

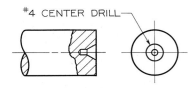

FIGURE 9.55   Shaft Center.

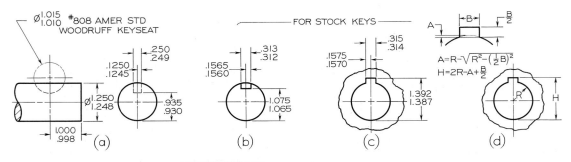

FIGURE 9.56   Dimensioning Keyways.

2. Dimensions should not be duplicated, nor should the same information be given in two different ways—except for dual dimensioning—and no dimensions should be given except those needed to produce or inspect the part.

3. Dimensions should be given between points or surfaces that have a functional relation to each other or that control the location of mating parts.

4. Dimensions should be given to finished surfaces or important centerlines, in preference to rough surfaces, wherever possible.

5. Dimensions should be given so that it will not be necessary for the machinist to calculate, scale, or assume any dimension.

6. Dimension features in the view where the feature's shape is best shown.

7. Dimensions should be placed in the views where the features dimensioned are shown true shape.

8. Dimensioning to hidden lines should be avoided wherever possible.

9. Dimensions should not be placed on a view unless clarity is promoted and long extension lines are avoided.

10. Dimensions applying to two adjacent views should be placed between views, unless clarity is promoted by placing some of them outside.

11. The longer dimensions should be placed outside all intermediate dimensions so that dimension lines will not cross extension lines.

12. In machine drawing, all unit marks should be omitted, except when necessary for clarity—for example, 1″ VALVE.

13. Don't expect production personnel to assume that a feature is centered (as a hole on a plate), but give a location dimension from one side. However, if a hole is to be centered on a symmetrical rough casting, mark the centerline and omit the locating dimension from the centerline.

14. A dimension should be attached to only one view, not to extension lines connecting two views.

15. Detail dimensions should line up in chain fashion.

16. A complete chain of detail dimensions should be avoided; it is better to omit one. Otherwise reference overall dimension by enclosing it within parentheses.

17. A dimension line should never be drawn through a dimension figure. A figure should never be lettered over any line of the drawing. The line can be broken if necessary.

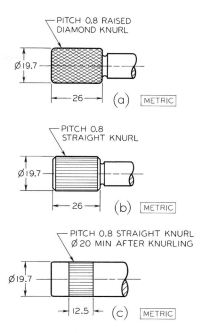

FIGURE 9.57 Dimensioning Knurls.

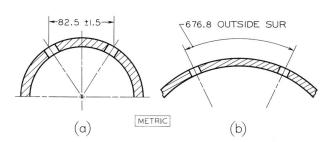

FIGURE 9.58 Dimensioning along Curved Surfaces

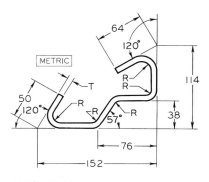

FIGURE 9.59 Profile Dimensioning.

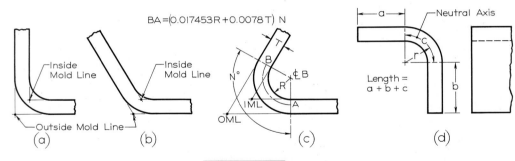

$$BA = (0.017453R + 0.0078T) N$$

FIGURE 9.60　Bends.

18. Dimension lines should be spaced uniformly throughout the drawing. They should be at least 10 mm (.38 inch) from the object outline and 6 mm (.25 inch) apart.

19. No line of the drawing should be used as a dimension line or coincide with a dimension line.

20. A dimension line should never be joined end to end with any line of the drawing.

21. Dimension lines should not cross, if avoidable.

22. Dimension lines and extension lines should not cross, if avoidable. (Extension lines may cross each other.)

23. When extension lines cross extension lines or visible lines, no break in either line should be made.

24. A centerline may be extended and used as an extension line, in which case it is still drawn like a centerline.

25. Centerlines should not extend from view to view.

26. Leaders for notes should be straight, not curved, and point to the center of circular views of holes wherever possible.

27. Leaders should slope at 45, 30, or 60 degrees with horizontal, but may be made at any convenient angle except vertical or horizontal.

28. Leaders should extend from the beginning or the end of a note, with the horizontal "shoulder" extending from mid-height of the lettering.

29. Dimension figures should be approximately centered between the arrowheads, except in a stack of dimensions, where they should be staggered.

30. Dimension figures should be about 3 mm (.13 inch) high for whole numbers and 6 mm (.25 inch) high for fractions.

31. Dimension figures should never be crowded or in any way made difficult to read.

32. Dimension figures should not be lettered over lines or sectioned areas unless necessary, in which case a clear space should be reserved for the dimension figures.

33. Dimension figures for angles should generally be lettered horizontally.

FIGURE 9.61　Tabular Dimensioning.

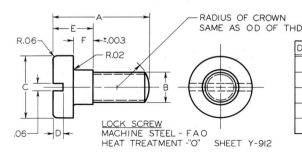

DETAIL	A	B	C	D	E	F	UNC THD	STOCK	LBS
1	.62	.38	.62	.06	.25	.135	.312–18	Ø.75	.09
2	.88	.38	.62	.09	.38	.197	.312–18	Ø.75	.12
3	1.00	.44	.75	.12	.38	.197	.375–16	Ø.875	.19
4	1.25	.50	.88	.12	.50	.260	.437–14	Ø.1	.30
5	1.50	.56	1.00	.16	.62	.323	.5–13	Ø1.125	.46

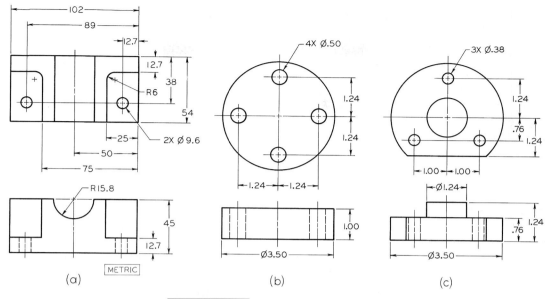

FIGURE 9.62 Coordinate Dimensioning.

34. Fraction bars should never be inclined except in confined areas, such as in tables.

35. The numerator and denominator of a fraction should never touch the fraction bar.

36. Notes should always be lettered horizontally on the sheet.

37. Notes should be brief and clear, and the wording should be standard in form.

38. Finish marks should be placed on the edge views of all finished surfaces, including hidden edges and the contour and circular views of cylindrical surfaces.

39. Finish marks should be omitted on holes or other features where a note specifies a machining operation.

40. Finish marks should be omitted on parts made from rolled stock.

41. If a part is finished all over, all finish marks should be omitted and the general note FINISH ALL OVER or FAO should be used.

42. A cylinder is dimensioned by giving both its diameter and length in the rectangular view, except when notes are used for holes. A diagonal diameter in the circular view may be used in cases where it increases clarity.

FIGURE 9.63 Rectangular Coordinate Dimensioning Without Dimension Lines (ANSI/ASME Y14.5M–1994).

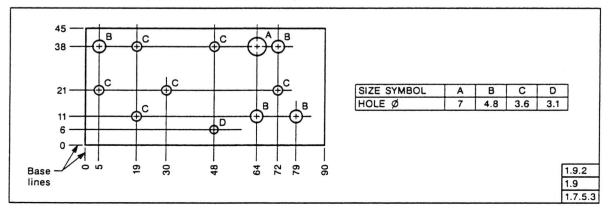

43. Manufacturing processes are generally determined by the tolerances specified, rather than specifically noted in the drawing. When the manufacturing process must be noted for some reason—such as for dimension holes to be bored, drilled, and reamed—use leaders which preferably point toward the center of the circular views of the holes. Give the manufacturing processes in the order they would be performed.

44. Drill sizes should be expressed in decimals, giving the diameter. For drills designated by number or letter, the decimal size must also be given.

45. In general, a circle is dimensioned by its diameter, an arc by its radius.

46. Diagonal diameters should be avoided, except for very large holes and for circles of centers. They may be used on positive cylinders when clarity is gained.

47. A diameter dimension value should always be preceded by the symbol ∅.

48. A radius dimension should always be preceded by the letter R. The radial dimension line should have only one arrowhead, and it should pass through or point through the arc center and touch the arc.

49. Cylinders should be located by their centerlines.

50. Cylinders should be located in the circular views, if possible.

51. Cylinders should be located by coordinate dimensions in preference to angular dimensions where accuracy is important.

52. When there are several rough, noncritical features obviously the same size (fillets, rounds, ribs, etc.), it is necessary to give only typical (abbreviation TYP) dimensions or to use a note.

53. When a dimension is not to scale, it should be underscored with a heavy straight line or marked NTS or NOT TO SCALE.

54. Mating dimensions should be given correspondingly on both drawings of mating parts.

55. Pattern dimensions should be given in two-place decimals or in common whole numbers and fractions to the nearest zero.

56. Decimal dimensions should be used for all machining dimensions.

57. Cumulative tolerances should be avoided where they affect the fit of mating parts.

■ 9.44 DO'S AND DON'TS OF PRACTICAL DESIGN

Figures 9.64 and 9.65 (on pages 321 and 322) contain a number of examples in which knowledge of manufacturing processes and limitations is essential for good design.

Many difficulties in producing good castings result from abrupt changes in section or thickness. In Figure 9.64a, rib thicknesses are uniform so that the metal will flow easily to all parts. Fillet radii are equal to the rib thickness—a good general rule to follow. When it is necessary to join a thin member to a thicker member, the thin member should be thickened as it approaches the intersection, as shown in Figure 9.64b.

In Figures 9.64c, 9.64g, and 9.64h, coring is used to produce walls with more-uniform sections. In Figure 9.64d, an abrupt change in sections is avoided by making thinner walls and leaving a collar, as shown.

Figures 9.64e and 9.64f show examples in which the preferred design tends to allow the castings to cool without introducing internal stresses. The less desirable design is more likely to crack as it cools, since there is no give in the design. Curved spokes are preferable to straight spokes, and an odd number of spokes is better than an even number because direct stresses along opposite spokes are avoided.

The design of a part may cause unnecessary trouble and expense for the pattern shop and foundry without any gain in the usefulness of the design. For example, in the poor designs in Figures 9.64j and 9.64k, one-piece patterns would not withdraw from the sand, and two-piece patterns would be necessary. In the preferred examples, the design is just as useful and is conductive to economical work in the pattern shop and foundry.

As shown in Figure 9.65a, a narrower piece of stock sheet metal can be used for certain designs that can be linked or overlapped. In this case, the stampings may be overlapped if dimension W is increased slightly, as shown. By such an arrangement, great savings in scrap metal can often be effected.

The maximum hardness that can be obtained in the heat treatment of steel depends on the carbon content of the steel. To get this hardness, it is necessary to cool rapidly, or quench, after heating to the temperature required. In practice it is often impossible to quench uniformly because of the design. In the design in Figure 9.65b, the piece is solid and will harden well on the outside, but will remain soft and relatively weak on the inside. As shown in the preferred example, a hollow piece can be quenched from both the outside

Hands On 9.4
Dimensioning true or false self quiz

Some of the following statements are wrong.

R   e
a   d

True False   D i m e n s i o n s   s h o u l d   n o t
b   e
d   u   p   l   i   c   a   t   e   d   ,   n   o   r
s   h   o   u   l   d   t   h   e
s   a   m   e   i   n   f   o   r-
m   a   t   i   o   n   b   e
g   i   v   e   n   i   n
t   w   o   d   i   f   f   e   r   e   n   t
w   a   y   s
.

True False   D i m e n s i o n s   s h o u l d
b   e
p   l   a   c   e   d
w   h   e   r   e
f   e   a   t   u   r   e   s
a   r   e   s   h   o   w   n
t   r   u   e   s   h   a   p   e   .

True   False   P r o d u c t i o n
p   e   r   s   o   n   n   e   l
w   i   l   l

---

and inside. Thus, it is possible for a hardened hollow shaft to be stronger than a hardened solid shaft.

As shown in Figure 9.65c, the addition of a rounded groove, called a *neck*, around a shaft next to a shoulder will eliminate a practical difficulty in precision grinding. It is not only more expensive to grind a sharp internal corner, but such sharp corners often lead to cracking and failure.

The design at the right in Figure 9.65d eliminates a costly reinforced weld, which would be required by the design on the left. The strong virgin metal with a generous radius is present at the point at which the stress is likely to be most severe. It is possible to make the design on the left as strong as the design on the right, but it is more expensive and requires expert skill and special equipment.

It is difficult to drill into a slanting surface, as shown at the left in Figure 9.65e. The drilling is greatly facilitated if a boss is provided, as shown at the right.

In Figure 9.65f, the design at the left requires accurate boring or reaming of a blind hole all the way to a flat bottom, which is difficult and expensive. It is better to drill deeper than the hole is to be finished, as shown at the right, to provide room for tool clearance and for chips.

In the upper example in Figure 9.65g, the drill and counterbore cannot be used for the hole in the center piece because of the raised portion at the right end. In the approved example, the end is redesigned to provide access for the drill and counterbore.

In the top design in Figure 9.65h, the ends are not the same height. As a result, each flat surface must be machined separately. In the design below, the ends are the same height, the surfaces are in-line horizontally, and only two machining operations are necessary. It is always good design to simplify and limit the machining as much as possible.

The design at the left in Figure 9.65j requires that the housing be bored for the entire length to receive a pressed bushing. If the cored recess is made as shown, machining time can be decreased. This assumes that average loads would be applied in use.

In Figure 9.65k, the lower bolt shown is encircled by a rounded groove no deeper than the root of the thread. This makes a gentle transition from the small diameter at the root of the threads and the large diameter of the body of the bolt, producing less stress concentration and a stronger bolt. In general, sharp internal corners should be avoided because these are points of stress concentration and possible failure.

In Figure 9.65m, a .2500 steel plate is being pulled, as shown by the arrows. Increasing the radius of the inside corners increases the strength of the plate by distributing the load over a greater area.

■ KEY WORDS

meter	combination dimensioning	bolt circle	taper per unit on diameter
yard	millimeters	reference dimension	chamfer
SI system	meters	baseline	knurl
dimensioning	inches	datums	mold line
dimension line	feet	mating dimensions	inside mold line
extension line	slope	finish mark	outside mold line
centerline	batter	surface texture	centerline of bend
leader	grade	roughness	stretchout
unidirectional system	bevels	lay	neutral axis
aligned system	size dimensions	local notes	bend allowance
complete decimal dimensioning	location dimensions	taper	neck
		standard machine tapers	

■ CHAPTER SUMMARY

• To increase clarity, dimensions and notes are added to a drawing to precisely describe size, location, and manufacturing process.

• Drawings are scaled to fit on a standard sheet of paper. Drawings created by hand are drawn to scale. CAD drawings are drawn full size and scaled when they are printed.

• The three types of scales are metric, engineers', and architects'.

• Dimensions and notes are placed on drawings according to prescribed standards.

• Dimensions that are incorrectly placed on a drawing are considered just as wrong as if the numbers in the dimension were incorrect.

■ REVIEW QUESTIONS

1. Which type of line is never crossed by any other line when dimensioning an object?

2. How is geometric analysis used in dimensioning?

3. What is the difference between a size dimension and a location dimension?

4. Which dimension system allows dimensions to be read from the bottom and from the right? When can a dimension be read from the left?

5. Sketch an example of dimensioning an angle.

6. When are finish marks used? Draw two types.

7. How are negative and positive cylinders dimensioned? Draw examples.

8. How are holes and arcs dimensioned? Draw examples.

9. What are notes and leaders used for?

10. Why is it important to avoid including superfluous dimensions?

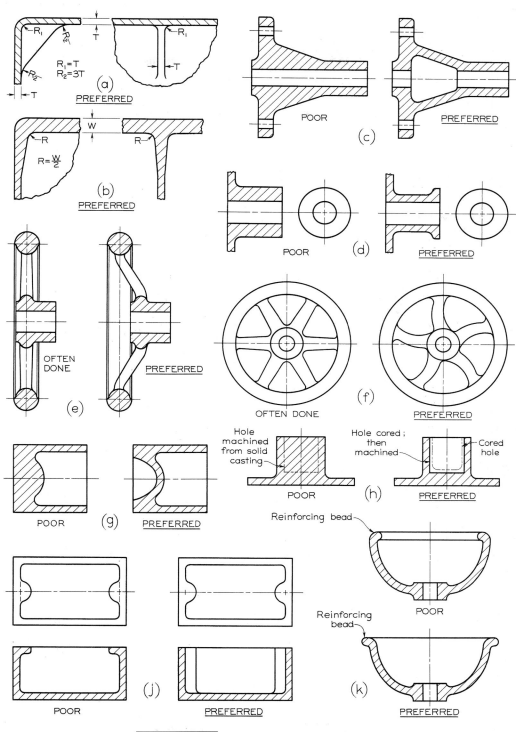

FIGURE 9.64 Casting Design Do's and Don'ts.

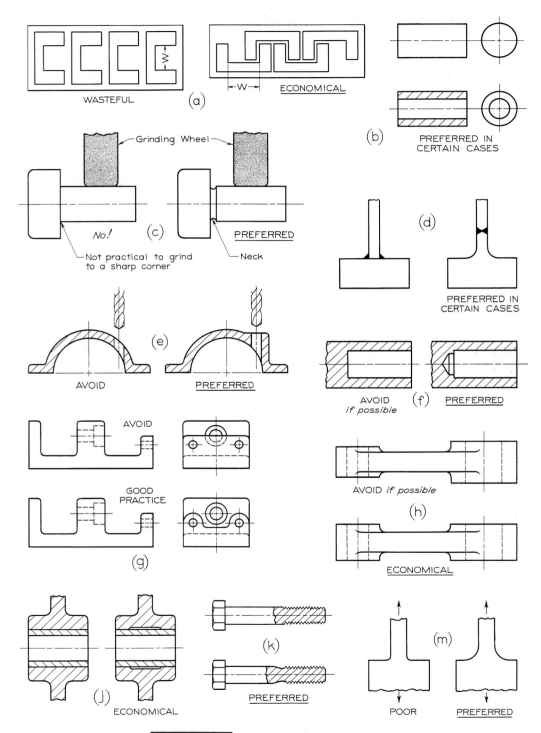

FIGURE 9.65 Casting Design Do's and Don'ts.

■ DIMENSIONING PROJECTS

You will practice dimensioning more using working drawings assigned from other chapters. A number of special dimensioning projects are available here in Figures 9.66 through 9.68. The problems are designed for Layout A–3 (8.5 x 11 inches) and are to be sketched and dimensioned to a full-size sale. Layout A4–3 (297 x 420 mm) may be used with appropriate adjustments in the title strip layout.

■ DESIGN PROJECT

Design a hand held brain teaser puzzle with parts that can be assembled and disassembled, or that can fit together in a variety of ways. Rubic's Cube is one example of a brain teaser, but yours may be simpler or more complex. Consider what materials and processes should be used to effectively mass produce your puzzle. Sketch pictorial and orthographic views to a full size scale, specifying dimensions according to the practices you learned in this chapter. Use a general note specifying the tolerance, material and finishes necessary to manufacture the parts.

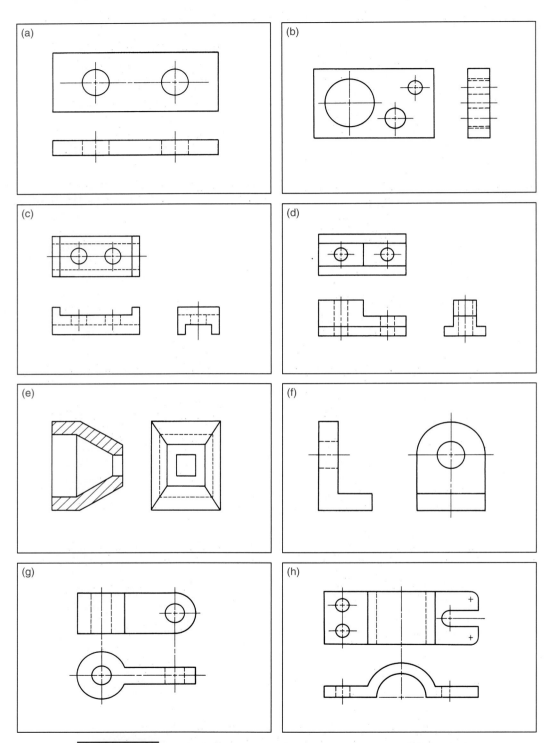

FIGURE 9.66   Using Layout A–3 or A4–3 (adjusted), sketch or use CAD to draw the parts shown. To determine sizes, make measurements from the figure. Assume that the parts are shown here at one-half size. Dimension drawing completely in one-place millimeters or two-place inches as assigned, full size. See inside back cover for decimal-inch and millimeter equivalents.

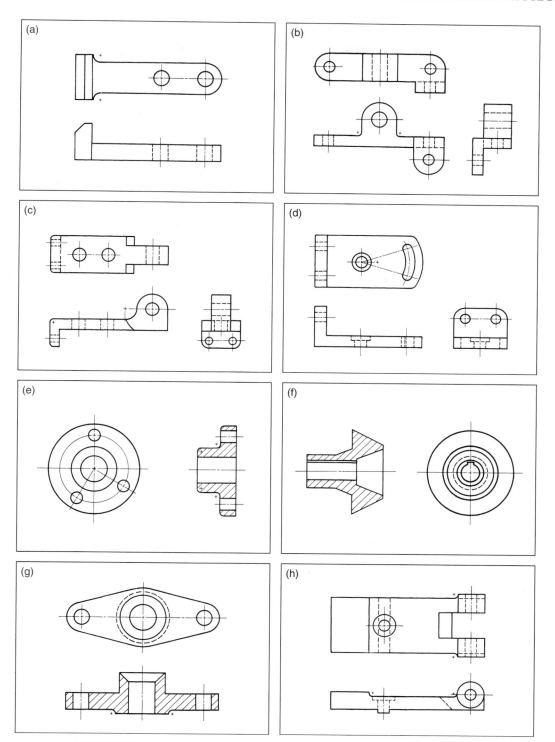

FIGURE 9.67   Using Layout A–3 or A4–3 (adjusted), sketch or use CAD to draw the parts shown. To determine sizes, make measurements from the figure. Assume that the parts are shown here at one-half size. Dimension drawing completely in one-place millimeters or two-place inches as assigned, full size. See inside back cover for decimal-inch and millimeter equivalents.

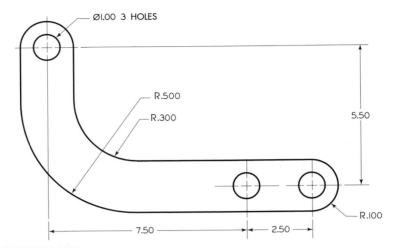

Ø1.00 3 HOLES

R.500

R.300

5.50

R.100

7.50

2.50

FIGURE 9.68 Using Sheet Layout A–3 of A4–3 (adjusted), draw the OML and IML.

C H A P T E R   1 0

DESIGN, PROCESSES, AND DRAWING

OBJECTIVES

After studying the material in this chapter, you should be able to:

1. Define *manufacturing* as we know it today.
2. Discuss the importance of the initial design stages in product development and manufacture.
3. Define *concurrent engineering* and show how this systematic approach integrates the design and manufacturing processes.
4. Explain the importance of computer-aided design, engineering, and manufacture in product development.
5. Explain the importance of prototypes and rapid prototyping in modern manufacturing processes.
6. Define *design for manufacture* and *design for manufacture and assembly.*
7. Explain the importance of proper material selection based on their properties, cost and availability, appearance, service life, and recycle potentials.
8. List the manufacturing processes typically used today.
9. Explain the importance of dimensional accuracy and surface finish.
10. List the typical measuring devices used in manufacturing.
11. Discuss the benefits of computer-integrated manufacturing.

OVERVIEW

Take a few moments and inspect various objects around you: your pen, watch, calculator, telephone, chair, and light fixtures. You will soon realize that all these objects had a different shape at one time. You could not find them in nature as they appear in your room. They have been transformed from various raw materials and assembled into the shapes that you now see.

Some objects are made of a single part, such as nails, bolts, wire or plastic coat hangers, metal brackets, and forks. However, most objects such as aircraft jet engines (invented in 1939), ball-point pens (1938), toasters (1926), washing machines (1910), air conditioners (1928), refrigerators (1931), photocopiers (1949), all types of machines and thousands of other products are made of an assembly of several parts made from a variety of materials. All are made by various processes that we call "manufacturing."

Manufacturing, in its broadest sense, is the process of converting raw materials into products. It encompasses (1) product design, (2) selection of raw materials, and (3) selection of processes by which manufacturing of goods take place, using various production methods and techniques.

Manufacturing is the backbone of any industrialized nation. Its importance is emphasized by the fact that, as an economic activity, it comprises approximately 20% to 30% of the value of all goods and services produced. The level of manufacturing activity is directly related to the economic health of a country. Generally, the higher the level of manufacturing activity in a country, the higher is the standard of living of its people.

Manufacturing also involves activities in which the manufactured product is itself used to make other products. Examples are large presses to shape sheet metal for car bodies, machinery to make bolts and nuts, and sewing machines for making clothing. An equally important aspect of manufacturing activities is servicing and maintaining this machinery during its useful life.

Engineering drawings, whether created with drawing instruments or CAD, are detailed instructions for manufacturing the described objects. The drawings must give information regarding shape, size, materials, finish, and, sometimes, the manufacturing process required. This chapter provides engineers with information about terms and processes used in manufacturing; information that will assist them with their drawings.

■ 10.1 "MANUFACTURING" DEFINED

The word "manufacturing" is derived from the Latin *manu factus*, meaning "made by hand." In the modern sense, manufacturing involves making products from raw materials by various processes, machinery, and operations, following a well-organized plan for each activity required. The word product means something that is produced, and the words product and production first appeared sometime during the 15th century. The word production is often used interchangeably with the word manufacturing. Whereas manufacturing engineering is the term used widely in the United States to describe this area of industrial activity, the equivalent term in other countries is production engineering.

Because a manufactured item has undergone a number of processes in which a piece of raw material has become a useful product, it has a value—defined as monetary worth or marketable price. For example, as the raw material for ceramics, clay has a certain value as mined. When the clay is used to make a ceramic cutting tool or electrical insulator, value is added to the clay. Similarly, a wire coat hanger or a nail has a value over and above the cost of a piece of wire from which it is made. Thus manufacturing has the important function of adding value.

Manufacturing may produce discrete products, meaning individual parts, or continuous products. Nails, gears, balls for bearings, beverage cans, and engine blocks are examples of discrete parts, even though they are mass produced at high production rates. On the other hand, a spool of wire, metal or plastic sheet, tubes, hose, and pipe are continuous products, which may be cut into individual lengths and thus become discrete parts.

Manufacturing is generally a complex activity involving a wide variety of resources and activities such as:

- Product design
- Purchasing
- Marketing
- Machinery and tooling
- Manufacturing
- Sales
- Process planning
- Production control
- Shipping

- Materials
- Support services
- Customer service

Manufacturing activities must be responsive to several demands and trends:

1. A product must fully meet design requirements and product specifications and standards.

2. A product must be manufactured by the most environmentally friendly and economical methods.

3. Quality must be built into the product at each stage, from design to assembly, rather than relying on quality testing after the product is made. Furthermore, quality should be appropriate to the product's use.

4. In a highly competitive environment, production methods must be sufficiently flexible so as to respond to changing market demands, types of products, production rates, production quantities, and on-time delivery to the customer.

5. New developments in materials, production methods, and computer integration of both technological and managerial activities in a manufacturing organization must constantly be evaluated with a view to their appropriate, timely, and economic implementation.

6. Manufacturing activities must be viewed as a large system, each part of which is interrelated to others. Such systems can be now modeled in order to study the effect of factors such as changes in market demands, product design, and materials. Various other factors and production methods affect product quality and cost.

7. A manufacturing organization must constantly strive for higher levels of quality and productivity (defined as the optimum use of all its resources: materials, machines, energy, capital, labor, and technology). Output per employee per hour in all phases must be maximized. Zero-based part rejection and waste are also an integral aspect of productivity.

■ 10.2 THE DESIGN PROCESS AND CONCURRENT ENGINEERING

The *design process* for a product first requires a clear understanding of the functions and the performance expected of that product. The product may be new, or it may be a revised version of an existing product. We

all have observed, for example, how the design and style of radios, toasters, watches, automobiles, and washing machines have changed. The market for a product and its anticipated uses must be defined clearly, with the assistance of sales personnel, market analysts, and others in the organization. Product design is a critical activity because it has been estimated that 70% to 80% of the cost of product development and manufacture is determined at the initial design stages.

Traditionally, design and manufacturing activities have taken place sequentially rather than concurrently or simultaneously (Figure 10.1a). Designers would spend considerable effort and time in analyzing components and preparing detailed part drawings; these drawings would then be forwarded to other departments in the organization, such as materials departments where, for example, particular alloys and vendor sources would be identified. The specifications would then be sent to a manufacturing department where the detailed drawings would be reviewed and processes selected for efficient production. While this approach seems logical and straightforward, in practice it has been found to be extremely wasteful or resources.

In theory, a product can flow from one department in an organization to another and directly to the marketplace, but in practice there are usually difficulties encountered. For example, a manufacturing engineer may wish to taper the flange on a part to improve its castability, or a different alloy may be desirable, thus necessitating a repeat of the design analysis stage to ensure that the product will still function satisfactorily. These iterations, also shown in Figure 10.1a, are certainly wasteful of resources but, more importantly, of time.

There is a great desire, originally driven by the consumer electronics industry, to bring products to market as quickly as possible. The rationale is that products introduced early enjoy a greater percentage of the market and hence profits, and have a longer life before obsolescence (clearly a concern with consumer electronics). For these reasons, concurrent engineering, also called simultaneous engineering, has come to the fore.

A more modern product development approach is shown in Figure 10.1b. While there is a general product flow from market analysis to design to manufacturing, there are recognized iterations which occur in the process. The main difference to the more modern approach is that all disciplines are involved in

(a)

(b)

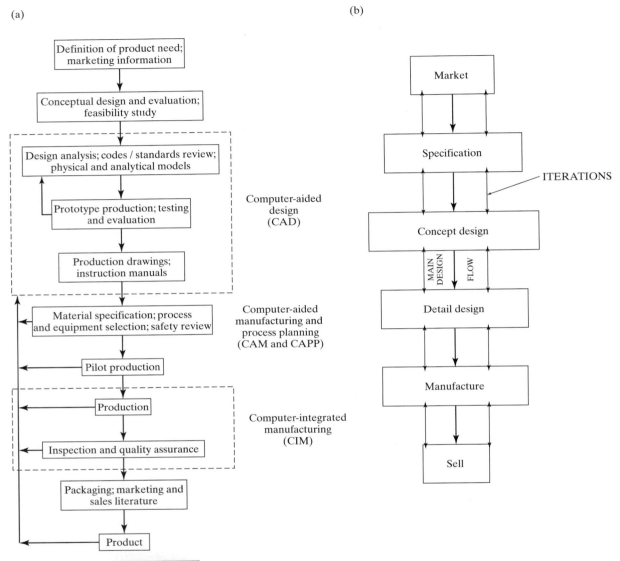

FIGURE 10.1 **Design and manufacturing activities traditionally have taken place sequentially rather than concurrently. With CAM, the design process can work simultaneously with the manufacturing process.**

the early design stages, so that the iterations which naturally occur result in less wasted effort and lost time. A key to the approach is the now well-recognized importance of communication between and within disciplines. That is, while there must be communication between engineering and marketing and service functions, so too must there be avenues of interactions between engineering sub-disciplines, for example, design for manufacture, design recyclability and design for safety.

The design process begins with the development of an original product concept. An innovative approach

to design is highly desirable—and even essential —at this stage for the product to be successful in the marketplace. Innovative approaches can also lead to major savings in material and production costs. The design engineer or product designer must be knowledgeable of the interrelationships among materials, design, and manufacturing, as well as the overall economics of the operation.

Concurrent engineering is a systematic approach integrating the design and manufacture of products with the view of optimizing all elements involved in the life cycle of the product. *Life cycle* means that all aspects of a product (such as design, development, production, distribution, use, and its ultimate disposal and recycling) are considered simultaneously. The basic goals of concurrent engineering are to minimize product design and engineering changes and the time and costs involved in taking the product from design concept to production and introduction of the product into the marketplace.

The philosophy of *life cycle engineering* requires that the entire life of a product be considered in the design stage (i.e., the design, production, distribution, use and disposal/recycling must be considered simultaneously). Thus a well-designed product is functional (design stage), well manufactured (production), packaged so that it safely arrives to the end user or customer (distribution), functions effectively for its intended life and has components which can be easily replaced for maintenance or repair (use), and can be disassembled so that components can be recycled (disposal).

Although the concept of concurrent engineering appears to be logical and efficient, its implementation can take considerable time and effort when those using it either do not work as a team or fail to appreciate its real benefits. It is apparent that for concurrent engineering to succeed it must:

a. Have the full support of the upper management,

b. Have multifunctional and interactive teamwork, including support groups, and

c. Utilize all available technologies.

There are numerous examples of the benefits of concurrent engineering. An automotive company, for example, has reduced the number of components in an engine by 30%, and as a result has decreased its weight by 25% and cut manufacturing time by 50%. The concurrent engineering concept can be implemented not only in large organizations but in smaller companies as well. This is particularly noteworthy in view of the fact that 98% of U.S. manufacturing establishments have fewer than 500 employees.

For both large and small companies, product design often involves preparing analytical and physical models of the product, as an aid to studying factors such as forces, stresses, deflections, and optimal part shape. The necessity for such models depends on product complexity. Today, constructing and studying analytical models is simplified through the use of computer-aided design (CAD), engineering (CAE), and manufacturing (CAM) techniques.

■ 10.3 COMPUTER-AIDED DESIGN AND PRODUCT DEVELOPMENT

Computer-aided design (CAD) allows the designer to conceptualize objects more easily without having to make costly illustrations, models, or prototypes. These systems are now capable of rapidly and completely analyzing designs, from a simple bracket to complex structures. For example, the two-engine Boeing 777 passenger airplane was designed completely by computer (paperless design) with AutoCAD 2000 workstations linked to eight computers (Figure 10.2). The airplane was constructed directly from the CAD/CAM software developed and no prototypes or mockups were built, unlike previous models.

■ 10.4 COMPUTER-AIDED ENGINEERING ALLOWS FOR FUTURE MODIFICATION

Using *computer-aided engineering*, the performance of structures subjected to static or fluctuating loads and various temperatures can now be simulated, analyzed, and tested efficiently, accurately, and more quickly than ever. The information developed can be stored, retrieved, displayed, printed, and transferred anywhere in the organization. Designs can be optimized and modifications can be made directly and easily at any time.

FIGURE 10.2   **Boeing 777 in Flight.** *Courtesy of Boeing Co.*

■ 10.5   COMPUTER-AIDED ENGINEERING LINKS ALL PHASES OF MANUFACTURING

Computer-aided manufacturing (CAM) involves all phases of manufacturing by utilizing and processing further the large amount of information on materials and processes collected and stored in the organization's database. Computers now assist manufacturing engineers and others in organizing tasks such as programming numerical control of machines; programming robots for material handling and assembly; designing tools, dies, and fixtures; and maintaining quality control.

On the basis of the models developed using the above mentioned techniques, the product designer selects and specifies the final shape and dimensions of the product, its dimensional accuracy and surface finish, and the materials to be used. The selection of materials is often made with the advice and cooperation of materials engineers, unless the design engineer is also experienced and qualified in this area. An important design consideration is how a particular component is to be assembled into the final product. Lift the hood of your car and observe how hundreds of components are put together in a limited space.

■ 10.6   THE ROLE OF PROTOTYPES AND RAPID PROTOTYPING IN PRODUCT DEVELOPMENT

The next step in the production process is to make and test a *prototype*, that is, an original working model of the product. An important development is *rapid prototyping*, which relies on CAD/CAM and various manufacturing techniques (using metallic or nonmetallic materials) to quickly produce prototypes in the form of a solid physical model of a part and at low cost. For example, prototyping new automotive components by traditional methods of shaping, forming, machining, etc. costs hundreds of millions of dollars a year; some components may take a year to produce. Rapid prototyping can cut these costs as well as development times significantly. These techniques are being advanced further so that they can be used for low-volume economical production of actual parts.

Tests of prototypes must be designed to simulate as closely as possible the conditions under which the product is to be used. These include environmental conditions such as temperature and humidity, as well as the effects of vibration and repeated use and misuse of the product. Computer-aided engineering techniques are now capable of comprehensively and rapidly performing such simulations. During this stage, modifications in the original design, materials selected, or production methods may be necessary. After this phase has been completed, appropriate process plans, manufacturing methods (Table 10.1), equipment, and tooling are selected with the cooperation of manufacturing engineers, process planners, and others involved in production.

■ 10.7   DESIGN FOR MANUFACTURE, ASSEMBLY, DISASSEMBLY, AND SERVICE

As we have seen, design and manufacturing must be intimately interrelated; they should never be viewed as separate disciplines or activities. Each part or component of a product must be designed so that it not only meets design requirements and specifications, but also can be manufactured economically and efficiently. This approach improves productivity and allows a manufacturer to remain competitive.

This broad view has become recognized as the area of *design for manufacture* (DFM). It is a comprehensive approach to production of goods and integrates the design process with materials, manufacturing methods, process planning, assembly, testing, and quality assurance. Effectively implementing design for manufacture requires that designers have a fundamental understanding of the characteristics, capabilities, and limitations of materials, manufacturing processes, and related operations, machinery, and equipment. This knowledge includes characteristics such as variability in machine performance, and dimensional accuracy and surface finish of the workpiece, processing time, and the effect of processing method on part quality.

Designers and product engineers must be able to assess the impact of design modifications on manufacturing process selection, assembly, inspection, tools and dies, and product cost. Establishing quantitative relationships is essential in order to optimize the design for ease of manufacturing and assembly at minimum product cost (also called producibility). Computer-aided design, engineering, manufacturing, and process planning techniques, using powerful computer programs, have become indispensable to those conducting such analysis. New developments include expert systems, which have optimization capabilities, thus expediting the traditional iterative process in design optimization.

After individual parts have been manufactured, they have to be assembled into a product. Assembly is an important phase of the overall manufacturing

Shape of Feature	Production Method
Flat surfaces	Rolling, planing, broaching, milling, shaping, grinding
Parts with cavities	End milling, electrical-discharge machining, electrochemical machining, ultrasonic machining, casting
Parts with sharp features	Permanent mold casting, machining, grinding, fabricating, powder metallurgy
Thin hollow shapes	Slush casting, electroforming, fabricating
Tubular shapes	Extrusion, drawing, roll forming, spinning, centrifugal casting
Tubular parts	Rubber forming, expanding with hydraulic pressure, explosive forming, spinning
Curvature on thin sheets	Stretch forming, peen forming, fabricating, assembly
Opening in thin sheets	Blanking, chemical blanking, photochemical blanking
Cross-sections	Drawing, extruding, shaving, turning, centerless grinding
Square edges	Fine blanking, machining, shaving, belt grinding
Small holes	Laser, electrical discharge machining, electrochemical machining
Surface textures	Knurling, wire brushing, grinding, belt grinding, shot blasting, etching, deposition
Detailed surface feature	Coining, investment casting, permanent-mold casting, machining
Threaded parts	Thread cutting, thread rolling, thread grinding, chasing
Very large parts	Casting, forging, fabricating, assembly
Very small parts	Investment casting, machining, etching, powder metallurgy, nanofabrication, micromachining

TABLE 10.1   **Shapes and Some Common Methods of Production.**

operation and requires consideration of the ease, speed, and cost of putting parts together. Also, many products must be designed so that disassembly is possible, enabling the products to be taken apart for maintenance, servicing, or recycling of their components. Because assembly operations can contribute significantly to product cost, *design for assembly* (DFA) as well as design for disassembly are now recognized as important aspects of manufacturing. Typically, a product that is easy to assemble is also easy to disassemble. The latest trend now includes design for service, ensuring that individual parts or sub-assemblies in a product are easy to reach and service.

Methodologies and computer software (CAD) have been developed for DFA utilizing 3-D conceptual designs and solid models. In this way, subassembly and assembly times and costs are minimized while maintaining product integrity and performance; the system also improves the product's ease of disassembly. The trend now is to combine design for manufacture and design for assembly into the more comprehensive *design for manufacture and assembly* (DFMA) which recognizes the inherent in-terrelationships between design and manufacturing.

There are several methods of assembly, such as using fasteners, adhesives, or by welding, soldering, and brazing, each with its own characteristics and requiring different operations. The use of a bolt and nut, for example, requires preparation of holes that must match in location and size. Hole generation requires operations such as drilling or punching, which take additional time, require separate operations, and produce scrap. On the other hand, products assembled with bolts and nuts can be taken apart and reassembled with relative ease.

Parts can also be assembled with adhesives. This method, which is being used extensively in aircraft and automobile production, does not require holes. However, surfaces to be assembled must match properly and be clean because joint strength is adversely affected by the presence of contaminants such as dirt, dust, oil, and moisture. Unlike mechanical fastening, adhesively joined components, as well as those that are welded, are not usually designed to be taken apart and reassembled, hence are not suitable for the important purposes of recycling individual parts in the product.

Parts may be assembled by hand or by automatic equipment and robots. The choice depends on factors such as the complexity of the product, the number of parts to be assembled, the protection required to prevent damage or scratching of finished surfaces of the parts, and the relative costs of labor and machinery required for automated assembly.

■ 10.8 MATERIAL SELECTION

An ever-increasing variety of materials is now available, each having its own characteristics, applications, advantages, and limitations. The following are the general types of materials used in manufacturing today either individually or in combination.

- Ferrous metals: carbon, alloy, stainless, and tool and die steels.
- Nonferrous metals: aluminum, magnesium, copper, nickel, titanium, superalloys, refractory metals, beryllium, zirconium, low-melting alloys, and precious metals.
- Plastics: thermoplastics, thermosets, and elastomers.
- Ceramics, glass ceramics, glasses, graphite, diamond, and diamond-like materials.
- Composite materials: reinforced plastics, metal-matrix and ceramic-matrix composites. These are also known as engineered materials.
- *Nanomaterials*, shape-memory alloys, amorphous alloys, superconductors, and various other materials with unique properties.

As new materials are developed, the selection of appropriate materials becomes even more challenging. Aerospace structures, as well as products such as sporting goods, have been at the forefront of new material usage. The trend has been to use more titanium and composites for the airframe of commercial aircraft, with a gradual decline of the use of aluminum and steel. There are constantly shifting trends in the usage of materials in all products, driven principally by economic considerations as well as other considerations.

■ 10.9 PROPERTIES OF MATERIALS

When selecting materials for products, we first consider their mechanical properties: strength, toughness, ductility, hardness, elasticity, fatigue, and creep. The strength-to-weight and stiffness-to-weight ratios of material are also important, particularly for aerospace and automotive applications. Aluminum, titanium, and reinforced plastics, for example, have higher ratios than steels and cast irons. The mechanical properties specified for a product and its components should, of course, be for the conditions under which the product is expected to function. We then consider the physical properties of density, specific heat, thermal expansion and conductivity, melting point, and electrical and magnetic properties.

Chemical properties also play a significant role in hostile as well as normal environments. Oxidation, corrosion, general degradation of properties, toxicity, and flammability of materials are among the important factors to be considered. In some commercial airline disasters, for example, many deaths have been caused by toxic fumes from burning nonmetallic materials in the aircraft cabin.

Manufacturing properties of materials determine whether they can be cast, formed, machined, welded, and heat treated with relative ease (Table 10.2). The method(s) used to process materials to the desired shapes can adversely affect the product's final properties, service life, and its cost.

■ 10.10 COST AND AVAILABILITY OF MATERIALS

Cost and availability of raw and processed materials and manufactured components are major concerns in manufacturing. Competitively, the economic aspects of material selection are as important as the technological considerations of properties and characteristics of materials.

If raw or processed materials or manufactured components are not available in the desired shapes, dimensions and quantities, substitutes and/or additional processing will be required, which can contribute significantly to product cost. For example, if we need a round bar of a certain diameter and it is not available in standard form, then we have to purchase a larger rod and reduce its diameter by some means, such as machining, drawing through a die, or grinding. It should be noted, however, that a product design can be modified to take advantage of standard dimensions of raw materials, thus avoiding additional manufacturing costs.

Reliability of supply, as well as demand, affects cost. Most countries import numerous raw materials that are essential for production. The United States, for example, imports the majority of raw materials such as natural rubber, diamond, cobalt, titanium, chromium, aluminum, and nickel from other countries. The broad political implications of such reliance on other countries are self-evident.

Alloy	Castability	Weldability	Machinability
Aluminum	E	F	G-E
Copper	F-G	F	F-G
Gray cast iron	E	D	G
White cast iron	G	VP	VP
Nickel	F	F	F
Steels	F	E	F
Zinc	E	D	E

E, excellent; G, good; F, fair; D, difficult; VP, very poor

TABLE 10.2 General Manufacturing Characteristics of Various Alloys.

Different costs are involved in processing materials by different methods. Some methods require expensive machinery, others require extensive labor, and still others require personnel with special skills, a high level of education, or specialized training.

■ 10.11 APPEARANCE, SERVICE LIFE, AND RECYLING

The appearance of materials after they have been manufactured into products influences their appeal to the consumer. Color, feel, and surface texture are characteristics that we all consider when making a decision about purchasing a product.

Time- and service-dependent phenomena such as wear, fatigue, creep, and dimensional stability are important. These phenomena can significantly affect a product's performance and, if not controlled, can lead to total failure of the product. Similarly, compatibility of materials used in a product is important. Friction and wear, corrosion, and other phenomena can shorten a product's life or cause it to fail prematurely. An example is galvanic corrosion between mating parts made of dissimilar metals.

Recycling or proper disposal of materials at the end of their useful service lives has become increasingly important in an age when we are more conscious of preserving resources and maintaining a clean and healthy environment. Note, for example, the use of biodegradable packaging materials or recyclable glass bottles and aluminum beverage cans. The proper treatment and disposal of toxic wastes and materials are also a crucial consideration.

■ 10.12 MANUFACTURING PROCESSES

Before preparing a drawing for the production of a part, the drafter/designer should consider what manufacturing processes are to be used. These processes will determine the representation of the detailed feature s of the part, the choice of dimensions, and the machining of processing accuracy. Many processes are used to produce parts and shapes (Table 10.1) and there is usually more than one method of manufacturing a part from a given material. The broad categories of processing methods for materials are:

a. *Casting*: Expendable molds (i.e., sand casting) and permanent molds (Figure 10.3).

b. *Forming and shaping*: Rolling, forging, extrusion, drawing, sheet forming, powder metallurgy, and molding (Figure 10.4a-d).

c. *Machining*: Turning, boring, drilling, milling, planing, shaping, broaching, grinding, ultrasonic machining; chemical, electrical, and electrochemical machining; and high-energy beam machining (Figure 10.5a-g).

d. *Joining*: Welding, brazing, soldering, diffusion bonding, adhesive bonding, and mechanical joining (Figure 10.6a-b).

e. *Finishing*: Honing, lapping, polishing, burnishing, deburring, surface treating, coating, and plating.

Selection of a particular manufacturing process, or a series of processes, depends not only on the shape to be produced but also on many other factors pertaining to material properties (Table 10.2). Brittle and hard materials, for example, cannot be shaped easily, whereas they can be cast or machined by several methods. The manufacturing process usually alters the properties of materials. Metals that are formed at room temperature, for example, become stronger, harder, and less ductile than they were before processing.

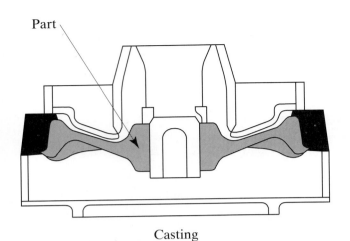

Part

Casting

FIGURE 10.3 **This casting mold is an example of a permanent mold.**

Two steel mounting brackets are shown in Figure 10.7, one made by casting, and the other by stamping of sheet metal. Note that there are some differences in the designs, although the parts are basically alike. Each of these two manufacturing processes has its own advantages and limitations, as well as production rates and manufacturing cost.

Manufacturing engineers are constantly being challenged to find new solutions to manufacturing problems and cost reduction. For a long time, for example, sheet metal parts were cut and fabricated by traditional tools, punches, and dies. Although they are still widely used, some of these operations are now being replaced by laser cutting techniques (Figure 10.8). With advances in computer technology, we can automatically control the path of the laser, thus increasing the capability for producing a wide variety of shapes accurately, repeatedly, and economically.

FIGURE 10.4 **Examples of forming and shaping methods.**

(a)

(b)

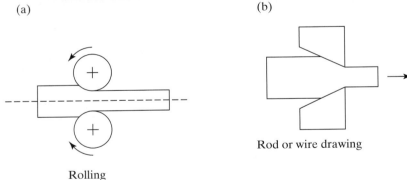

Rolling

Rod or wire drawing

(c)

(d)

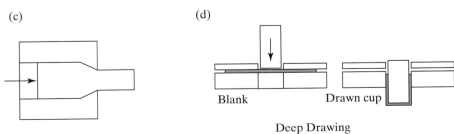

Blank

Drawn cup

Deep Drawing

Extrusion

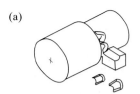

Turning

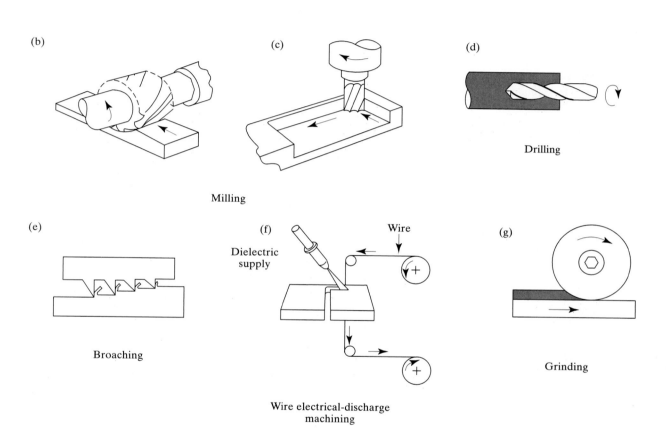

Milling

Drilling

Broaching

Wire electrical-discharge
machining

Grinding

FIGURE 10.5   **Examples of different types of machining.**

Joining

FIGURE 10.6   **Example of joining method.**

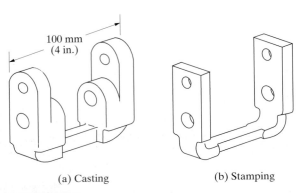

(a) Casting   (b) Stamping

FIGURE 10.7   Two steel mounting brackets (a) made by casting, (b) made by stamping.

■ 10.13 DIMENSIONAL ACCURACY AND SURFACE FINISH

Size, thickness, and shape complexity of the part have a major bearing on the manufacturing process selected to produce it. Flat parts with thin cross-sections, for example, cannot be cast properly. Complex parts cannot be formed easily and economically, whereas they may be cast or fabricated from individual pieces.

Tolerances and surface finish obtained in hot-working operations cannot be as good as those obtained in cold-working (room temperature) operations because dimensional changes, warpage, and surface oxidation occur during processing at elevated temperatures. Some casting processes produce a better surface finish than others because of the different types of mold materials used and their surface finish.

The size and shape of manufactured products vary widely. For example, the main landing gear for a twin-engine, 400-passenger Boeing 777 jetliner is 4.3 m (14 ft) high, with three axles and six wheels, made by forging and machining processes (Figure 10.2). At the other extreme is the generation of a 0.05-mm (0.002-in.) diameter hole at one end of a 0.35-mm (0.014-in.) diameter needle (Figure 10.9), using a process called electrical-discharge machining. The hole is burr-free and has a location accuracy of ±0.003 mm (0.0001 in.).

Another small-scale manufacturing example is given in Figure 10.10, which shows microscopic gears as small as 100 m (0.004 in.) in diameter. These gears have possible applications such as powering *microrobots* to repair human cells, *microknives* in surgery, and camera shutters for precise photography. The gears are made by a special electroplating and x-ray etching technique of metal plates coated with a polymer film. The center hole in these gears is so small that a human hair cannot pass through it. Such small-scale operations

FIGURE 10.8   Cutting sheet metal with a laser beam.
Courtesy of Rofin-Sinar, Inc., and Manufacturing Engineering Magazine, Society of Manufacturing Engineers.

are called *nanotechnology* and *nanofabrication* ("nano" meaning one billionth).

Ultraprecision manufacturing techniques and machinery are now being developed and are coming into more common use. For machining mirrorlike surfaces, for example, the cutting tool is a very sharp diamond tip and the equipment has very high stiffness and must be operated in a room where the temperature is controlled within 1 °C. Highly sophisticated techniques such as molecular-beam epitaxy and scanning-tunneling engineering are being implemented to obtain accuracies on the order of the atomic lattice (0.1 nm; 10^{-8} in.).

■ 10.14 MEASURING DEVICES USED IN MANUFACTURING

Although the machinist uses various measuring devices depending on the kind of dimensions (fractional, decimal, or metric) shown on the drawing, it is evident that to dimension correctly, the engineering designer must have at least a working knowledge of the common measuring tools. The machinists' *steel rule*, or *scale*, is a commonly used measuring tool in the shop (Figure 10.11a). The smallest division on one scale of this rule is 0, and such a scale is used for common fractional dimensions. Also, many machinists' rules have a decimal scale with the smallest division of .010, which is used for dimensions given on the drawing by the decimal system. For

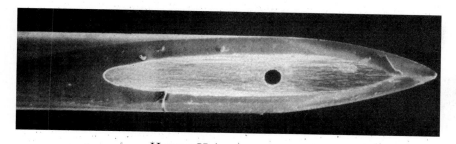

FIGURE 10.9 A 0.05-mm hole produced in a needle, using the electrical-discharge machining process. *Courtesy of Derata Corporation.*

Human Hair

FIGURE 10.10 Microscopic gear with a diameter on the order of 100μm, made by a special etching process. *Courtesy of Wisconsin Center for Applied Microelectronics, University of Wisconsin–Madison.*

checking the nominal size of outside diameters, the *outside spring caliper* and steel scale are used, as shown in Figs. 10.11b and 10.11c. Likewise, the *inside spring caliper* is used for checking nominal dimensions, as shown in Figs. 10.11d and 10.11e. Another use for the outside caliper (Figure 10.11f) is to check the nominal distance between holes (center to center). The *combination square* may be used for checking height (Figure 10.11g) and for a variety of other measurements. Measuring devices are also available that have metric scales.

For dimensions that require more precise measurements, the *vernier caliper* (Figs. 10.11h and 10.11j) or the *micrometer caliper* (Figure 10.11k) may be used. It is common practice to check measurements to 0.025 mm (.0010) with these instruments, and in some instances they are used to measure directly to 0.0025 mm (.00010).

Many of the measuring devices discussed here have been supplemented with newer, more sophisticated tools. Computerized measuring devices have broadened the range of accuracy previously attainable. Figure 10.12 illustrates an ultraprecision electronic digital readout

micrometer and caliper that contain integral microprocessors. In addition to the hand-held printer/recorder providing a hard-copy output of measurements, the printer also calculates and lists statistical mean, minimum, and maximum values, as well as standard deviation.

Most measuring devices in manufacturing are adjustable so they can each be employed to measure any size within their range of designed usage. There is also a need for measuring devices designed to be used for only one particular dimension. These are called *fixed gages* because their setting is fixed and cannot be changed.

A common type of fixed gage consists of two carefully finished rounds. One might think of each of these rounds as being 25.4 mm (1.000) in diameter and 38 mm (1.5000) long. Let one of these diameters be slightly larger than the other. One can see that, for a certain range of hole sizes, the smaller round will enter the hole but the larger will not. If the larger round diameter is made slightly greater than the largest acceptable hole diameter and if the diameter of the smaller round is made slightly less than the smallest acceptable hole diameter, then the large round *will never go* into any acceptable hole but the small round *will go* into any acceptable hole. A fixed gage consisting of two such rounds is called a "go–no go" gage. There are, of course, many kinds of "go–no go" gages.

The subject of gages and gaging is a specialized field and involves so many technical considerations that many large companies employ highly trained workers to attend to nothing but this one feature of their operations.

■ 10.15 OPERATIONAL AND MANUFACTURING COSTS

The design and cost of tooling, the lead time required to begin production, and the effect of workpiece material on tool and die life are major considerations. Depending on its size, shape, and expected life, the cost of tooling can be substantial. For example, a set of steel dies for stamping sheet-metal fenders for automobiles may cost about $2 million.

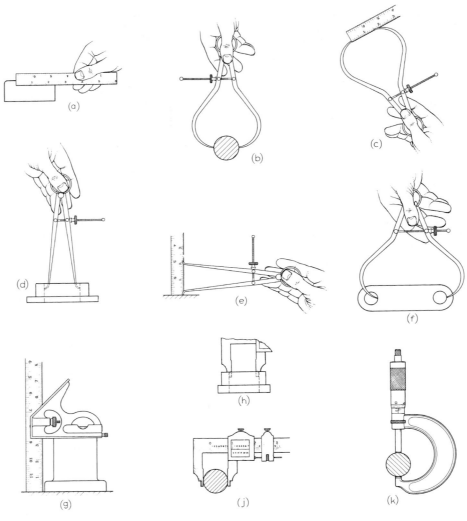

FIGURE 10.11 Measuring Devices Used by the Machinist.

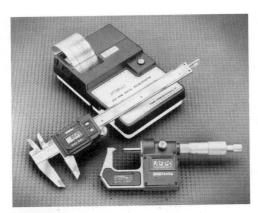

FIGURE 10.12 Computerized Measurement System.

Courtesy of Fred V. Fowler Co., Inc.

For parts made from expensive materials, the lower the scrap rate, the more economical the production process will be; thus, every attempt should be made for *zero-base waste*. Because it generates chips, machining may not be more economical than forming operations, all other factors being the same.

Availability of machines and equipment and operating experience within the manufacturing facility are also important cost factors. If they are not available, some parts may have to be manufactured by outside firms. Automakers, for example, purchase many parts from outside vendors, or have them made by outside firms according to their specifications.

The number of parts required (quantity) and the required production rate (pieces per hour) are important in determining the processes to be used and the

Hands On 10.1
Tolerances

You can get a feeling for the concept of tolerance by considering examples. What would be the most reasonable tolerance in the cases listed below?

Case:	A window opening in a building		
Tolerance:	±1 ft.	±1 in.	±.125 in.

Case:	A model out of soap		
Tolerance:	±.03 mm	±.3 mm	±3 mm

Case:	Socket for a light bulb		
Tolerance:	±6 in.	±.6 in.	±.06 in.

Case:	Gears for a bicycle		
Tolerance:	±.5 in.	±.05 in.	±.005 in.

Case:	A fitting for chimney pipe		
Tolerance:	±.1 mm	±1 mm	±10 mm

Case:	The hinge for a car door		
Tolerance:	±2 in.	±.2 in.	±.02 in.

Case:	A shovel handle		
Tolerance:	±30mm.	±3mm	±.3mm

economics of production. Beverage cans or transistors, for example, are consumed in numbers and at rates much higher than telescopes and propellers for ships.

The operation of machinery has significant environmental and safety implications. Depending on the type of operation, some processes adversely affect the environment, such as the use of oil-base lubricants in hot metalworking processes. Unless properly controlled, such processes can cause air, water, and noise pollution. The safe use of machinery is another important consideration, requiring precautions to eliminate hazards in the workplace.

■ 10.16 CONSEQUENCES OF IMPROPER SELECTION OF MATERIALS AND PROCESSES

Numerous examples of product failure can be traced to improper selection of material or manufacturing processes or improper control of process variables. A component or a product is generally considered to have failed when:

- It stops functioning (broken shaft, gear, bolt, cable, or turbine blade).

- It does not function properly or perform within required specification limits (worn bearings, gears, tools, and dies).

- It becomes unreliable or unsafe for further use (frayed cable in a winch, crack in a shaft, poor connection in a printed-circuit board, or delamination of a reinforced plastic component).

■ 10.17 NET-SHAPE MANUFACTURING

Since not all manufacturing operations produce finished parts, additional operations may be necessary. For example, a forged part may not have the desired dimensions or surface finish; thus additional operations such as machining or grinding may be necessary. Likewise, it may be difficult, impossible, or economically undesirable to produce a part with holes using just one manufacturing process, thus necessitating additional processes such as drilling. Also, the holes produced by a particular manufacturing process may not have the proper roundness, dimensional accuracy, or surface finish, thus creating a need for additional operations such as honing.

Finishing operations can contribute significantly to the cost of a product. Consequently, the trend has been for *net-shape* or *near-net-shape manufacturing*, in which the part is made as close to the final desired dimensions, tolerances, surface finish, and specifications as possible. Typical examples of such manufacturing methods are near-net-shape forging and casting of parts, stamped sheet-metal parts, injection molding of plastics, and components made by powder-metallurgy techniques.

■ 10.18 COMPUTER-INTEGRATED MANUFACTURING

The major goals of automation in manufacturing facilities are to integrate various operations to improve productivity, increase product quality and uniformity, minimize cycle times, and reduce labor costs. Beginning in the 1940s, automation has accelerated because of rapid advances in control systems for machines and in computer technology.

Few developments in the history of manufacturing have had a more significant impact than computers. Computers are now used in a very broad range of applications, including control and optimization of manufacturing processes, material handling, assembly, automated inspection and testing of products, as well as inventory control and numerous management activities. Beginning with computer graphics and computer-aided design and manufacturing, the use of computers has been extended to *computer-integrated manufacturing* (CIM). Computer-integrated manufacturing is particularly effective because of its capability for:

- Responsiveness to rapid changes in market demand and product modification.
- Better use of materials, machinery and personnel, and reduced inventory.
- Better control of production and management of the total manufacturing operation.
- High-quality products at low cost.

The major applications of computers in manufacturing are:

a. *Computer numerical control (CNC)* is a method of controlling the movements of machine components by direct insertion of coded instructions in the form of numerical data. Numerical control was first implemented in the early 1950s and was a major advance in automation of machines.

b. *Adaptive control (AC).* The parameters in a manufacturing process are adjusted automatically to optimize production rate and product quality, and to minimize cost. Parameters such as forces, temperatures, surface finish, and dimensions of the part are monitored constantly. If they move outside the acceptable range, the system adjusts the process variables until the parameters again fall within the acceptable range.

c. *Industrial robots.* Introduced in the early 1960s, industrial robots (Figs. 10.13 and 10.14) have been replacing humans in operations that are repetitive, boring, and dangerous, thus reducing the possibility of human error, decreasing variability in product quality, and improving productivity. Robots with sensory perception capabilities are being developed (*intelligent robots*), with movements that simulate those of humans.

d. *Automated handling of materials.* Computers have allowed highly efficient handling of materials and products in various stages of completion (*work in*

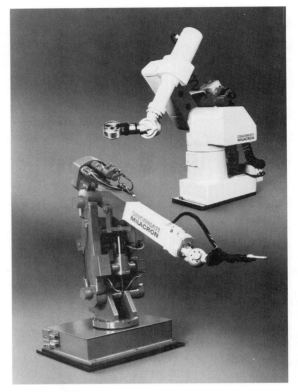

FIGURE 10.13 **Industrial Robots.** *Courtesy of Cincinnati Milacron.*

progress), such as from storage to machines, from machine to machine, and at the points of inspection, inventory, and shipment.

e. *Automated and robotic assembly systems* are replacing costly assembly by operators. Products are designed or redesigned so that they can be assembled more easily by machine (Figure 10.15).

f. *Computer-aided process planning (CAPP)* is capable of improving productivity in a plant by optimizing process plans, reducing planning costs, and improving the consistency of product quality and reliability. Functions such as cost estimating and work standards (time required to perform a certain operation) can also be incorporated into the system.

g. *Group technology (GT).* The concept of group technology is that parts can be grouped and produced by classifying them into families, according to similarities in design and similarities in manufacturing processes to produce the part. In this way, part designs and process plans can be standardized and families of parts can be produced efficiently and economically.

FIGURE 10.14   Robotic Welding on Ford Automobile Assembly Line.
Courtesy of Ford Motor Co.

FIGURE 10.15   Automated Manufacturing System. *Courtesy of Cargill Detroit.*

Digital Polish for Factory Floors

Without question, this has been the decade in which software moved irrevocably into the industrial designer's world. In corporate offices across the country, engineers have been booting up programs that let them tinker, in three dimensions, with every permutation and combination of a product's design. Be the item as lowly as a dinner plate or as complex as a Boeing 777 jet, the goal has always been the same: save time and money in getting products to market.

It was just a matter of time before engineers would aim their computers at designing and refining the assembly lines on which those products are made. Apparently, that time has come:

- These days, before it approves a design for a new car or van, the Ford Motor Company checks the plan against computer models of its factory floor. Often, through subtle changes like relocating a few seams or shaving a few millimeters from the length of a fender, Ford can lop weeks off the time it takes to prepare an old plant to make a new car.

- The Dow Chemical Company now uses computers to simulate its methods for making plastics, running what-if scenarios to fine-tune the temperatures, pressures and rates at which it feeds in raw materials. Dow can now switch production among 15 different grades of plastics in minutes, with almost no wasted material. Before computer modeling, the process took two hours and yielded lots of useless byproducts.

- The International Business Machines Corporation, the world's largest computer maker and an early convert to factory-simulation software, has learned that its assembly lines can accommodate diverse products of similar size, as long as IBM tweaked its conveyor belts to deliver different parts and products to different work areas. One early result: with almost no retooling, IBM expanded a plant in Charlotte, NC, that had made only banking systems to include voice-messaging systems, bar-code readers and devices to program pacemakers.

"No one wants to build a new assembly line, if they can re-use the one they have," said Frank Lerchenmuller, an IBM vice president for engineering technology solutions. Of course not. But until the advent of software that can simulate assembly lines and the movements of the people who run them, no one really knew a good way to gauge what they could salvage from an older factory. Now, with software from companies like Tecnomatix Technologies Inc. of Israel, Dassault Systemes of France and Aspen Technology in the United States, it is as feasible to design or remodel an entire plant as it is to reconfigure a car.

Production engineers in industries as diverse as chemicals, automobiles and aluminum smelting are manipulating virtual pictures of their plants and processes to see whether moving a clamp or adding a new ingredient will make existing equipment more productive, or will enable the same assembly line to skip freely from product to product. Some are even testing out a new virtual reality program that enables engineers wearing special goggles to detect problems by "walking through and around" a three-dimensional model of their factory designs. The entire relationship between product design and production engineering is being turned on its ear. No longer is it enough for designers to create products that can be made and maintained efficiently. Increasingly, management is asking them whether the products can be manufactured with a minimum of retooling or work disruption—and if not, whether it is worth giving up a particular feature to wring time and money from the manufacturing process.

Of course, there is little allure to the time-honored alternative to computer modeling—setting up an actual assembly line and trouble-shooting it piece by piece. "It's too risky and expensive to try new tools or methods if you have to build prototype hardware to test every change," said Rani Agarwal Finstad, director of manufacturing math-based systems for GM Powertrain, a manufacturer of car components that recently used simulation software to squeeze a month out of the process of programming robots to make new crankshafts.

The market for simulation software is growing, if slowly. Bernard Charles, Dassault's president, calculates that companies spent $300 million globally for software that simulates manufacturing operations last year. That is nothing compared with the $4 billion he estimates that industry spent on product-design software, but "there's a growing understanding that computer simulations can replace physical mockups of plants the same way they've replaced mockups of products," Mr. Charles said. Apparently, IBM agrees; it is marketing Dassault's factory-simulation software in the United States.

Testimonials to factory modeling are easy to find. The engineers who design the automation systems sold by Rockwell International have used simulation to design assembly lines that can handle different-sized items, enabling factories to produce small batches of products cost-effectively. "We're headed to where modeling will let us design plants that can efficiently build a single-lot size of one item," said Randall L. Freeman, vice president for global marketing at Rockwell Automation.

Computer simulations of the tread-etching process has enabled tire makers like Goodyear Tire and Rubber to switch production from one type of tire to another in about an hour—a process that previously took an entire work shift. And simulations have shown cookie companies like the Nabisco unit of RJR Nabisco Holdings how to use the same packaging machines to make five-pound bags for price clubs, one-pound bags for grocery stores and six-cookie packs for vending machines.

Various forces are driving the trend toward computer modeling. For one thing, computer technology has finally caught up with manufacturing pipe dreams. "Only recently have computers been powerful enough to quickly simulate what happens if you change something in a chemical reactor," said David E. Waite, Dow's manufacturing manager for information technology.

Economic and marketplace forces are at work, too. Companies that spent much of the 1990's paring ancillary product lines and work forces are now trimming capital investment, lest shareholders think they have lost their cost-cutting touch. Consumers, meanwhile, have grown increasingly picky and expect to be able to choose among myriad colors, sizes and shapes for almost any product. That means that manufacturers must mix and match parts as the orders come in. And that, in turn, means having tools that can respond to electronic commands to switch paint wells, move clamps or change packaging and labels.

Adapted from "Digital Polish for Factory Floors; Software Simulations Head to Better Assembly Lines," by Claudia H. Deutsch, *The New York Times,* March 22, 1999.

h. *Just-in-time production (JIT).* The principal of JIT is that supplies are delivered just in time to be used, parts are produced just in time to be made into subassemblies and assemblies, and products are finished just in time to be delivered to the customer. In this way, inventory-carrying costs are low, part defects are detected right away, productivity is increased, and high-quality products are made at low cost.

i. *Cellular manufacturing.* Cellular manufacturing involves workstations, which are *manufacturing cells* usually containing several machines and with a central robot, each performing a different operation on the part.

j. *Flexible manufacturing systems (FMS)* integrate manufacturing cells into a large unit, all interfaced with a central computer. Flexible manufacturing systems have the highest level of efficiency, sophistication, and productivity in manufacturing. Although costly, they are capable of producing parts randomly and changing manufacturing sequences on different parts quickly, thus they can meet rapid changes in market demand for various types of products.

k. *Expert systems,* which are basically intelligent computer programs, are being developed rapidly with capabilities to perform tasks and solve difficult real-life problems as human experts would.

l. *Artificial intelligence (AI)* involves the use of machines and computers to replace human intelligence. Computer-controlled systems are becoming capable of learning from experience and making decisions that optimize operations and minimize costs. *Artificial neural networks,* which are designed to simulate the thought processes of the human brain, have the capability of modeling and simulating production facilities, monitoring and controlling manufacturing processes, diagnosing problems in machine performance, conducting financial planning, and managing a company's manufacturing strategy.

■ 10.19 SHARED MANUFACTURING

Although large corporations can afford to implement modern technology and take risks, smaller companies generally have difficulty in doing so with their limited personnel, resources, and capital. More recently, the concept of *shared manufacturing* has been proposed. This consists of a regional or nationwide network of manufacturing facilities with state-of-the-art equipment for training, prototype development and small-scale production runs, and is available to help small companies develop products that compete in the global marketplace.

In view of these advances and their potential, some experts have envisaged the factory of the future. Although highly controversial and viewed as unrealistic by some,

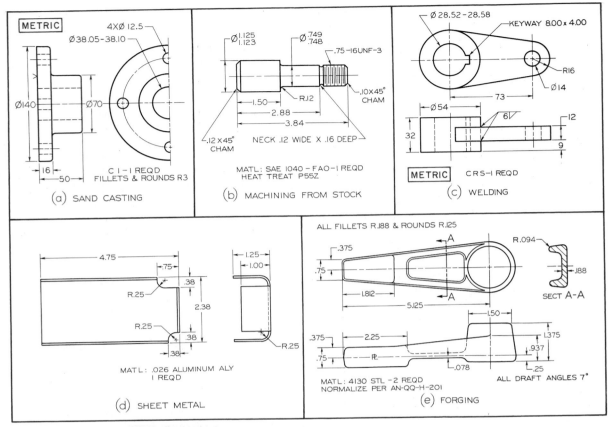

FIGURE 10.16 Comparison of Drawings for Different Manufacturing Processes.

this is a system in which production will take place with little or no direct human intervention. The human role is expected to be confined to supervision, maintenance, and upgrading of machines, computers, and software.

■ 10.20 MANUFACTURING METHODS AND THE DRAWING

In designing a part, consider what materials and manufacturing processes are to be used. These processes will determine the representation of the detailed features of the part, the choice of dimensions, and the machining or processing accuracy. Principal types of metal forming are (1) casting, (2) machining from standard stock, (3) welding, (4) forming from sheet stock, and (5) forging. A knowledge of these processes, along with a thorough understanding of the intended use of the part, will help determine some basic manufacturing processes. Drawings that reflect these manufacturing methods are shown in Figure 10.16.

In sand casting, for example, as shown in Figure 10.16a, all cast surfaces remain rough textured, with all corners filleted or rounded. Sharp corners indicate that

at least one of the surfaces is finished (i.e., machined further usually to produce a flat surface), and finish marks are shown on the edge view of the finished surface.

In drawings of parts machined from standard stock, as shown in Figure 10.16b, most surfaces are represented as machined. In some cases, as on shafting, the surface existing on the raw stock is often accurate enough without further finishing. Corners are usually sharp, but fillets and rounds are machined when necessary. For example, an interior corner may be machined with a radius to provide greater strength.

On welding drawings, as shown in Figure 10.16c, the several pieces are cut to size, brought together, and then welded. Welding symbols (listed in Appendix 32) indicate the welds required. Generally there are no fillets and rounds except those generated during the welding process itself. Certain surfaces may be machined after welding or, in some cases, before welding. Notice that lines are shown where the separate pieces are joined.

On sheet-metal drawings, as shown in Figure 10.16d, the thickness of material is uniform and is usually given in the material specification note rather than by a dimension on the drawing. Bend radii and bend reliefs at corners are specified according to standard practice. For dimensions, either the decimal-inch or metric dimensioning systems may be used. Allowances of extra material for joints may be required when the flat blank size is being determined.

For forged parts, separate drawings are usually made for the die maker and for the machinist. Thus, a forging drawing, as shown in Figure 10.16e, provides only the information to produce the forging, and the dimensions given are those needed by the die maker. All corners are rounded and filleted and are so shown on the drawing. The draft is drawn to scale and is usually specified by degrees in a note.

■ KEY WORDS

adaptive control	computer-integrated manufacturing	forming	net-shape manufacturing
artificial intelligence		group technology	producibility
artificial neural networks	computer numerical control	industrial robots	product cost
automated assembly	concurrent engineering	joining	protoytpes
automated materials handling	design for manufacture, assembly, disassembly and service	just-in-time production	rapid prototyping
casting		life cycle engineering	robotic assembly
cellular manufacturing	dimensional accuracy	machining	scale
computer-aided design, engineering, and manufacturing	expert systems	manufacturing engineering	shaping
	finishing	microknives	shared manufacturing
	fixed gages	microrobots	steel rule
computer-aided process planning	flexible manufacturing systems	nanofabrication	ultraprecision manufacturing
		nanotechnology	

■ CHAPTER SUMMARY

- Modern manufacturing involves product design, selection of materials, and selection of processes. The process of transforming raw materials into a finished product is called the manufacturing process.
- The design process requires a clear understanding of the functions and performance expected of that product.
- Concurrent engineering integrates the design process with production to optimize the life cycle of the product.
- Computer-aided design, engineering, and manufacturing are used to construct and study models (prototypes) allowing the designer to conceptualize objects more easily and more cost efficiently.

- The selection of appropriate materials is key to successful product development.
- Manufacturing processing methods have changed dramatically over the last few decades. More cost and time efficient processes can be implemented using computer-integrated manufacturing.
- Special dimensioning techniques are used for surfaces that have been mchined by one of the manufacturing processes.
- Drawing conventions combined with detailed notes represent arious manufacturing processes. These drawings and notes are the manufacturing instructions that the shop technician uses to creae the desired object.

■ REVIEW QUESTIONS

1. List the three important phases in the manufacturing process.
2. Define concurrent engineering and explain how it can be used to enhance the design and manufacturing process.
3. Explain the benefits of rapid prototyping.
4. List four types of materials used in manufacturing today.

5. List the five broad categories of manufacturing processing.
6. Give at least two examples of nanotechnology.
7. List four types of measuring devices.
8. Give three consequences of improper selection of materials and processes.
9. List four applications of computer-integrated manufacturing.

C H A P T E R 1 1

TOLERANCING

OBJECTIVES

After studying the material in this chapter, you should be able to:

1. Describe the nominal size, tolerance, limits, and allowance of two mating parts.
2. Identify a clearance fit, interference fit, and transition fit.
3. Describe the basic hole and basic shaft systems.
4. Dimension mating parts using limit dimensions, unilateral tolerances, and bilateral tolerances.
5. Describe the classes of fit and give examples of each.
6. Draw geometric tolerancing symbols.
7. Specify position and geometric tolerances.

OVERVIEW

Interchangeable manufacturing allows parts made in different places to still fit together when assembled. It is essential to mass production that parts all fit together properly, and this interchangeability requires effective size control by the engineer.

For example, an automobile manufacturer subcontracts parts manufacturing to other companies—both parts for new automobiles and replacement parts for repairs. All parts must be enough alike so that each could fit properly in any assembly. Parts can be made to very close dimensions, even to a few millionths of an inch or thousandths of a millimeter—as in gage blocks—but highly accurate parts are extremely expensive and there will still be some variation between the exact dimension and the actual size of the part.

Fortunately, exact sizes are not needed. The accuracy needed in a part depends on its function. A manufacturer of children's tricycles would soon go out of business if the parts were made with jet-engine accuracy—no one would be willing to pay the price. Providing a tolerance with a dimension allows it to be specified with whatever degree of accuracy required.

Quality in manufacturing is primarily a factor of machining tolerances. Products with small variations in shape and size are considered high quality and can command higher prices. Waste results when the manufacturing process cannot maintain shape and size within prescribed limits. By monitoring the manufacturing processes and reducing waste, a company can improve profits. This direct relationship to profits is why tolerancing is critical to manufacturing success.

Tolerancing is an extension of dimensioning. It provides additional information about the shape, size, and position of every feature of a product. It communicates how to manufacture a product. CAD programs often provide features for dimensioning, tolerancing, and checking fits and interferences that can assist in the tolerancing process.

■ 11.1 TOLERANCE DIMENSIONING

Tolerance is the total amount a specific dimension can vary (ANSI/ASME Y14.5M–1994). For example, a dimension given as 1.625 ±.002 means that the manufactured part may be 1.6270 or 1.6230 or anywhere between these *limit dimensions*. The tolerance is .0040. Since greater accuracy costs more money, specify as generous a tolerance as possible which will still permit satisfactory functioning of the part.

Tolerances are assigned so that any two mating parts will fit together, as shown in Figure 11.1a. In this case, the actual hole may not be less than 1.250 inch and not more than 1.251 inch; these are the limits for the dimension and the difference between them (.001 inch) is the tolerance. Likewise, the shaft must be between limits 1.248 inch and 1.247 inch; the tolerance for the shaft is .001 inch. A metric version is shown in Figure 11.1b.

An illustration of the dimensions in Figure 11.1a is shown in Figure 11.2a. The maximum shaft is shown solid, and the minimum shaft is shown phantom. The difference, .001 inch, is the tolerance for the shaft. Similarly, the tolerance for the hole is the difference between the two limits shown, or .001 inch. The loosest fit, or maximum clearance, occurs when the smallest shaft is in the largest hole, as shown in Figure 11.2b. The tightest fit, or minimum clearance, occurs when the largest shaft is in the smallest hole, as shown in Figure 11.2c. The difference between the largest allowable shaft size and the smallest allowable hole size (.002 inch in this case) is called the ***allowance***. The average clearance is .003 inch, so any shaft will fit inside any hole interchangeably.

In metric dimensions, the limits for the hole are 31.75 mm and 31.78 mm; their difference, 0.03 mm, is the tolerance. Similarly, the limits for the shaft are 31.70 mm and 31.67 mm and the tolerance is 0.03 mm.

When parts are required to fit properly in assembly but not to be interchangeable, parts are not always toleranced but sometimes just indicated to be made to fit at assembly, as shown in Figure 11.3.

■ 11.2 SIZE DESIGNATIONS

You should become familiar with the definitions of terms that apply in tolerancing (ANSI/ASME Y14.5M–1994). *Nominal size* is used for general identification and is usually expressed in common fractions. In Figure 11.1 the nominal size of both hole and shaft, which is 1-1/4 inch, would be 1.25 inch or 31.75 mm.

Basic size, or *basic dimension*, is the theoretically exact size from which limits of size are determined by applying allowances and tolerances. It is the size from which limits are determined for the size, shape, or location of a feature. In Figure 11.1a the basic size is the same as the nominal size, 1-1/4 inch, or 1.250 inch (31.75 mm in Figure 11.1b).

Hands On 11.1
Determining minimum and maximum clearances

Determine the hole tolerance, shaft tolerance, allowance (minimum clearance), and maximum clearance for the parts shown. Write your answers in the spaces provided below.

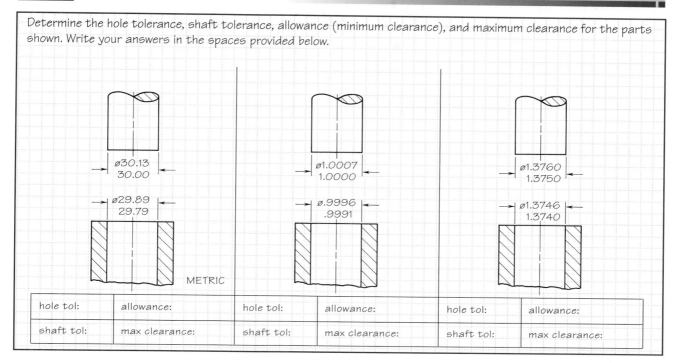

ø30.13
30.00

ø29.89
29.79

METRIC

ø1.0007
1.0000

ø.9996
.9991

ø1.3760
1.3750

ø1.3746
1.3740

hole tol:	allowance:	hole tol:	allowance:	hole tol:	allowance:
shaft tol:	max clearance:	shaft tol:	max clearance:	shaft tol:	max clearance:

FIGURE 11.1 Fits Between Mating Parts.

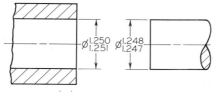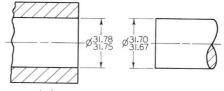

ø1.250
1.251   ø1.248
1.247

ø31.78
31.75   ø31.70
31.67

(a) LIMIT DIMENSIONS

(b) LIMIT DIMENSIONS – METRIC

FIGURE 11.2 Limit Dimensions.

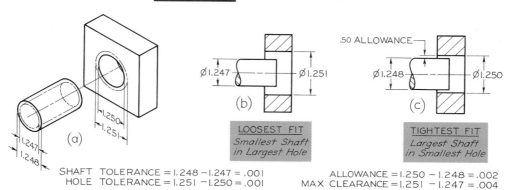

1.250
1.251

1.247
1.248

(a)

ø1.247   ø1.251

(b)

.50 ALLOWANCE

ø1.248   ø1.250

(c)

LOOSEST FIT
Smallest Shaft
in Largest Hole

TIGHTEST FIT
Largest Shaft
in Smallest Hole

SHAFT TOLERANCE = 1.248 – 1.247 = .001
HOLE TOLERANCE = 1.251 – 1.250 = .001

ALLOWANCE = 1.250 – 1.248 = .002
MAX CLEARANCE = 1.251 – 1.247 = .004

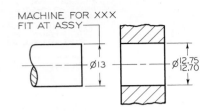

FIGURE 11.3   Noninterchangeable Fit.

Actual size is the measured size of the finished part.

Allowance is the minimum clearance space (or maximum interference) between mating parts. In Figure 11.2c, the allowance is the difference between the size of the smallest hole, 1.250 inch, and the size of the largest shaft, 1.248 inch—or .002 inch. Allowance represents the tightest permissible fit. For clearance fits this difference will be positive, but for interference fits it will be negative.

■ 11.3 FITS BETWEEN MATING PARTS

Fit is used to signify the range of tightness or looseness from a combination of allowances and tolerances in mating parts (ANSI B4.1–1967 (R1994) and ANSI B4.2–1978 (R1994)). There are four general types of fits between parts:

1. *Clearance fit*, where an internal member fits into an external member (as a shaft in a hole), always has space or clearance between the parts. In Figure 11.2c, the largest shaft is 1.248 inch and the smallest hole is 1.250 inch, giving a minimum air space (allowance) of .002 inch between the parts. In a clearance fit the allowance is always positive.

2. *Interference fit*, where the internal member is always larger than the external member, requires that the parts be forced together. In Figure 11.4a the smallest shaft is 1.2513 inch and the largest hole is 1.2506 inch, so the interference of metal between parts amounts to at least .00070. For the largest size shaft and smallest hole, the interference would be .0019 inch. An interference fit always has a negative allowance.

3. *Transition fit* results in either a clearance or interference condition. In Figure 11.4b the smallest shaft, 1.2503 inch, will fit into the largest hole, 1.2506 inch. But the largest shaft, 1.2509 inch, will have to be forced into the smallest hole, 1.2500 inch.

4. *Line fit* is where the limits are specified so that a clearance or surface contact results when mating parts are assembled.

■ 11.4 SELECTIVE ASSEMBLY

If allowances and tolerances are specified properly, mating parts are completely interchangeable. But for close fits, it is necessary to specify very small allowances and tolerances, and the cost may be very high. To avoid this expense, either manual or computer-controlled selective assembly is often used. In *selective assembly*, all parts are inspected and classified into several grades according to actual sizes, so that "small" shafts can be matched with "small" holes, "medium" shafts with "medium" holes, and so on. In this way, acceptable fits may be obtained at less expense than by machining all mating parts to very accurate dimensions. Selective assembly is generally better than interchangeable assembly for transition fits, since either clearance or interference is allowed.

FIGURE 11.4   Fits Between Parts.

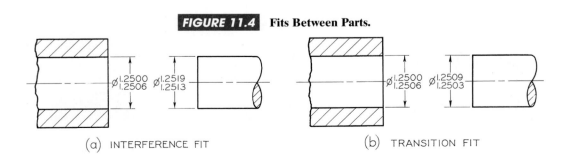

(a) INTERFERENCE FIT
(b) TRANSITION FIT

■ 11.5 BASIC HOLE SYSTEM

Reamers, broaches, and other standard tools are often used to produce holes, and standard plug gages are used to check the actual sizes. On the other hand, shafts are easily machined down to any size desired. Therefore, toleranced dimensions are commonly figured on the *basic hole system*, in which the minimum hole is taken as the basic size. Then the allowance is determined, and tolerances are applied.

■ 11.6 BASIC SHAFT SYSTEM

In some industries, such as textile machinery manufacturing, which use a great deal of cold-finished shafting, the *basic shaft system* is often used. It is advantageous when several parts having different fits, when they are required on a single shaft, or when the shaft for some reason can't be machined to size easily. This system should be used only when there is a reason for it. In this system, the maximum shaft is taken as the basic size, an allowance for each mating part is assigned, and tolerances are applied.

In Figure 11.5, the maximum size of the shaft, .500 inch, is the basic size. For a clearance fit, an allowance of .002 inch is decided upon, giving the minimum hole size of .502 inch. Tolerances of .003 inch and .001 inch, respectively, are applied to the hole and the shaft to obtain the maximum hole, .505 inch, and the minimum shaft, .499 inch. The minimum clearance is the difference between the smallest hole and the largest shaft (.502 inch − .500 inch = .002 inch), and the maximum clearance is the difference between the largest hole and the smallest shaft (.505 inch − .499 inch = .006 inch).

In the case of an interference fit, the minimum hole size would be found by subtracting the desired allowance from the basic shaft size.

Step by Step 11.1
Using the basic hole system

1. Determine where mating parts fit. Since the hole will be machined with a standard-size tool, its size will be used to determine the fit. In the figure shown, the minimum size of the hole, .500 inch, is used as the basic size.

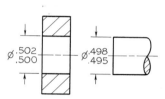

BASIC HOLE FIT

2. Determine the type of fit and apply the allowance to the basic size. For a clearance fit, an allowance of .002 inch is subtracted from the basic hole size, making the maximum shaft size .498 inch since it is easier to machine the shaft down to a smaller size than to apply the allowance to the hole.

3. Apply the tolerance. Tolerances of .002 inch and .003 inch, respectively, are applied to the hole and the shaft to obtain the maximum hole of .502 inch and the minimum shaft of .495 inch. Thus, the minimum clearance is the difference between the smallest hole and the largest shaft (.500 inch − .4980 inch = .002 inch), and the maximum clearance is the difference between the largest hole and the smallest shaft (502 inch − .495 inch = .007 inch).

Interference Fit

In the case of an interference fit, the maximum shaft size would be found by adding the desired allowance (the maximum interference) to the basic hole size.

In Figure 11.4a the basic size is 1.2500 inch. The maximum interference decided upon was .0019 inch, which when added to the basic size gives 1.2519 inch, the largest shaft size.

FIGURE 11.5 **Basic-Shaft System.**

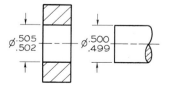

BASIC SHAFT FIT

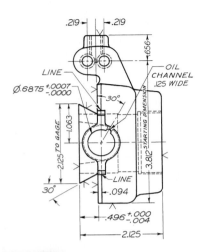

FIGURE 11.6 **Toleranced Decimal Dimensions.**

■ 11.7 SPECIFICATION OF TOLERANCES

A tolerance of a decimal dimension must be given in decimal form, as shown in Figure 11.6.

General tolerances on decimal dimensions in which tolerances are not given may also be covered in a printed note, such as

DECIMAL DIMENSION TO BE HELD TO ±001.

If a dimension of 3.250 is given, the worker would machine between the limits 3.249 and 3.251 (see Figure 11.11).

Tolerances for metric dimensions may be covered in a note, such as

METRIC DIMENSIONS TO BE HELD TO ±0.08.

So that when the given dimension of 3.25 is converted to millimeters, the worker machines between the limits of 82.63 mm and 82.74 mm.

Every dimension on a drawing should have a tolerance, either direct or by general tolerance note, except that commercial material is often assumed to have the tolerances set by commercial standards.

It is customary to indicate an overall general tolerance for all common fraction dimensions by means

of a printed note in or just above the title block, as shown in Figure 11.7.

EXAMPLE

ALL FRACTIONAL DIMENSIONS ±1/64″
UNLESS OTHERWISE SPECIFIED.

General angular tolerances also may be given as

ANGULAR TOLERANCE ±1°.

Here are several methods of expressing tolerances in dimensions that are approved by ANSI (ANSI/ASME Y14.5M–1994):

1.   Limit dimensioning. In this preferred method, the maximum and minimum limits are specified, as shown in Figure 11.8. The maximum value is placed above the minimum value, as shown in Figure 11.8a. In single-line note form, the low limit precedes the high limit separated by a dash, as shown in Figure 11.8b.

2.   Plus-or-minus dimensioning. In this method the basic size is followed by a plus-or-minus expression for the tolerance. The result can be either unilateral, where the tolerance only applies in one direction so that one value is zero, or bilateral, where either the same or different values are added and subtracted, as shown in Figure 11.9. If two unequal tolerance numbers are given—one plus and one minus—the plus is placed above the minus. One of the numbers may be zero. If the plus value and minus value are the same, a single value is given, preceded by the plus-or-minus symbol (±), as shown in Figure 11.10.

The *unilateral system of tolerances* allows variations in only one direction from the basic size. This method is advantageous when a critical size is approached as material is removed during manufacture, as in the case of close-fitting holes and shafts. In Figure 11.9a the basic size is 1.878 inch (47.70 mm). The tolerance of .002 inch (0.05 mm) is all in one direction—toward the smaller size. If the dimension is for a shaft diameter, the basic size of 1.878 inch (47.70 mm) is nearer the critical size, so the tolerance is taken away from the critical size. A unilateral tolerance is always all plus or all minus, but the zeros for the other tolerance value should be shown as in Figure 11.9a.

The *bilateral system of tolerances* allows variations in both directions from the basic size. Bilateral tolerances are usually given for location dimensions or any dimensions that can be allowed to vary in either direction. In Figure 11.9b, the basic size is 1.876 inch (47.65 mm), and the actual size may be larger by .002 inch (0.05 mm) or smaller by .001 inch (0.03 mm). If

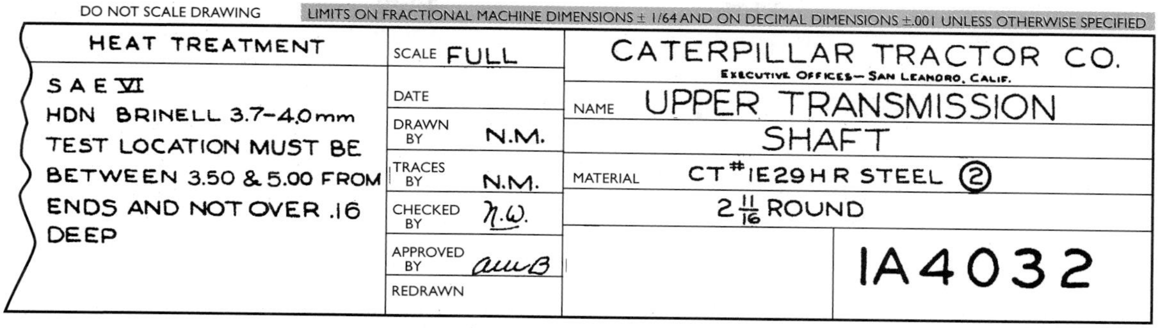

DO NOT SCALE DRAWING   LIMITS ON FRACTIONAL MACHINE DIMENSIONS ± 1/64 AND ON DECIMAL DIMENSIONS ±.001 UNLESS OTHERWISE SPECIFIED

HEAT TREATMENT		CATERPILLAR TRACTOR CO.
SAE Ⅵ HDN BRINELL 3.7–4.0mm TEST LOCATION MUST BE BETWEEN 3.50 & 5.00 FROM ENDS AND NOT OVER .16 DEEP	SCALE FULL	EXECUTIVE OFFICES— SAN LEANDRO, CALIF.
	DATE	NAME UPPER TRANSMISSION SHAFT
	DRAWN BY N.M.	
	TRACES BY N.M.	MATERIAL CT#1E29 HR STEEL ②
	CHECKED BY 𝓷.𝓌.	2 11/16 ROUND
	APPROVED BY 𝒶𝓊𝓌𝓑	IA4032
	REDRAWN	

FIGURE 11.7   General Tolerance Notes.

equal variation in both directions is allowed, the plus-or-minus symbol (±) is used, as shown in Figure 11.10.

A typical example of limit dimensioning is given in Figure 11.11.

3. *Single-limit dimensioning*. It is not always necessary to specify both limits. MIN or MAX is often placed after a number to indicate minimum or maximum dimensions desired where other elements of design determine the other unspecified limit. For example, a thread length may be dimensioned as MINFULLTHD or a radius dimensioned as R .05 MAX . Other applications include depths of holes, chamfers, and so on.

4. *Angular tolerances* are usually bilateral and in terms of degrees, minutes, and seconds:

$$25° ± 1°, 25° 0' ± 0° 15', OR 25° ± 0.25°.$$

■ 11.8 AMERICAN NATIONAL STANDARD LIMITS AND FITS

The American National Standards Institute has issued ANSI B4.1–1967 (R1994), "Preferred Limits and Fits for Cylindrical Parts," defining terms and recommending preferred standard sizes, allowances, tolerances, and fits in terms of the decimal inch. This standard gives a series of standard classes of fits on a unilateral-hole basis so that the fit produced by mating parts of a class of fit will produce approximately similar performance throughout the range of sizes. These tables give standard allowances for any given size or type of fit; they also prescribe the standard limits for the mating parts which will produce the fit.

The tables are designed for the basic hole system (see Appendixes 5–9). For coverage of the metric system of tolerances and fits, see Appendixes 11–14.

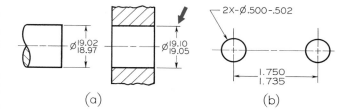

FIGURE 11.8   Method of Giving Limits.

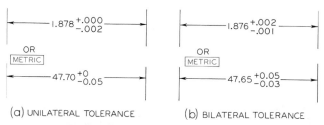

FIGURE 11.9   Tolerance Expression.

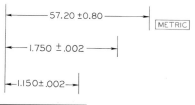

FIGURE 11.10   Bilateral Tolerances.

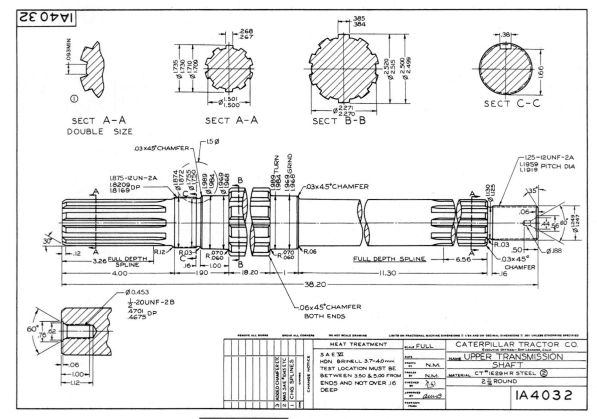

FIGURE 11.11   **Limit Dimensions.**

Table 11.1 gives the three general types of fits, the five subtypes, their letter symbols and descriptions.

In the tables for each class of fit, the range of nominal sizes of shafts or holes is given in inches. To simplify the tables and reduce the space required to present them, the other values are given in *thousandths of an inch.* Minimum and maximum limits of clearance are given; the top number is the least clearance, or the allowance, and the lower number the maximum clearance, or loosest fit. Then, under the heading "Standard Limits" are the limits for the hole and for the shaft that are to be applied to the basic size to obtain the limits of size for the parts, using the basic hole system.

■ 11.9   ACCUMULATION OF TOLERANCES

In tolerance dimensioning, it is very important to consider the effect of one tolerance on another. When the location of a surface is affected by more than one tolerance value, the tolerances are *cumulative.* For example, in Figure 11.12a, if dimension Z is omitted, surface

A will be controlled by both dimensions X and Y, and there can be a total variation of .010 inch instead of the variation of .005 inch permitted by dimension Y. If the part is made to the minimum tolerances of X, Y, and Z, the total variation in the length of the part will be .015 inch, and the part can be as short as 2.985 inch. However, the tolerance on the overall dimension W is only .005 inch, permitting the part to be only as short as 2.995 inch. The part is controlled in too many different ways—it is over-dimensioned.

In some cases, for functional reasons, it may be desirable to hold all three dimensions (such as X, Y, and Z shown in Figure 11.12a) closely without regard to the overall width of the part. In such cases the overall dimension should be made a *reference dimension* placed inside parentheses. In other cases it may be desired to hold two dimensions, (such as X and Y in Figure 11.12a), and the overall width of the part closely. In that case, a dimension such as Z shown in Figure 11.12a should be omitted or given as a reference dimension only.

General fit types and subtypes

Fit Type	Symbol	Subtype	Description
CLEARANCE	RC	Running or sliding fits	Running-and-sliding fits, (Appendix 5), are intended to provide a similar running performance, with suitable lubrication allowance, throughout the range of sizes. The clearances for the first two classes, used chiefly as slide fits, increase more slowly with diameter than the other classes, so that accurate location is maintained even at the expense of free relative motion.
LOCATIONAL	LC	Clearance fits	Locational fits (Appendixes 6–8) are fits intended to determine only the location of the mating parts; they may provide rigid or accurate location, as with interference fits, or provide some freedom of location, as with clearance fits. Accordingly, they are divided into three groups: clearance fits, transition fits, and interference fits.
	LT	Transition clearance or interference fits	
	LN	Locational interference fits	
INTERFERENCE	FN	Force or shrink fits	Force or shrink fits (Appendix 9) constitute a special type of interference fit, normally characterized by maintenance of constant bore pressures throughout the range of sizes. The interference therefore varies almost directly with diameter, and the difference between its minimum and maximum value is small to maintain the resulting pressures within reasonable limits.

TABLE 11.1 *Fit Types.*

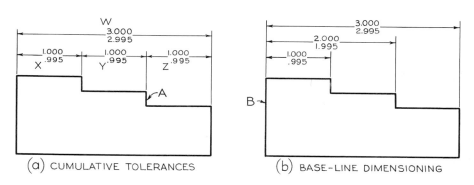

(a) CUMULATIVE TOLERANCES   (b) BASE-LINE DIMENSIONING

FIGURE 11.12 Cumulative Tolerances.

Hands On 11.2
Sketching dimensions with tolerances

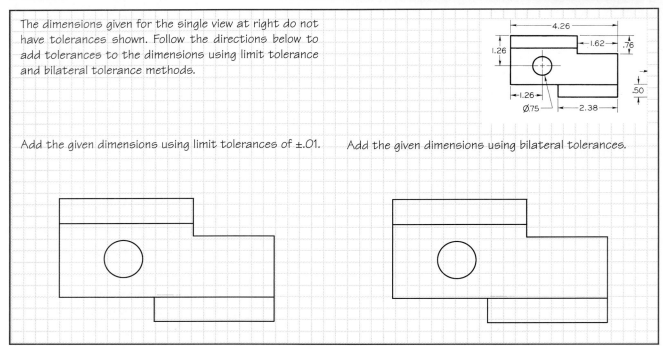

The dimensions given for the single view at right do not have tolerances shown. Follow the directions below to add tolerances to the dimensions using limit tolerance and bilateral tolerance methods.

Add the given dimensions using limit tolerances of ±.01.

Add the given dimensions using bilateral tolerances.

As a rule, it is best to dimension each surface so that it is affected by only one dimension. This can be done by referring all dimensions to a single datum surface, such as B, as shown in Figure 11.12b.

■ 11.10 TOLERANCES AND MACHINING PROCESSES

Tolerances should be as generous as possible and still permit satisfactory use of the part. The tighter the tolerance, the more expensive it is to manufacture the part. Great savings can be gained from the use of less expensive tools, from lower labor and inspection costs, and from reduced scrapping of material.

Figure 11.13 shows a chart, to be used as a general guide, with the tolerances achievable by the indicated machining processes. You can convert these to metric values by multiplying by 25.4 and rounding to one less decimal place.

■ 11.11 METRIC SYSTEM OF TOLERANCES AND FITS

The preceding material on limits and fits between mating parts applies for both systems of measurement. A system of preferred metric limits and fits by

the International Organization for Standardization (ISO) is in the ANSI B4.2 standard. The system is specified for holes, cylinders, and shafts, but it is also adaptable to fits between parallel surfaces of such features as keys and slots. The following terms for metric fits, illustrated in Figure 11.14, are somewhat similar to those for decimal-inch fits:

1. Basic size is the size from which limits or deviations are assigned. Basic sizes, usually diameters, should be selected from a table of preferred sizes, as shown in Figure 11.15.

2. Deviation is the difference between the basic size and the hole or shaft size. This is equivalent to the tolerance in the decimal-inch system.

3. Upper deviation is the difference between the basic size and the permitted maximum size of the part. This is comparable to maximum tolerance in the decimal-inch system.

4. Lower deviation is the difference between the basic size and the minimum permitted size of the part. This is comparable to minimum tolerance in the decimal-inch system.

Range of Sizes From	Range of Sizes To and Including	Tolerances								
.000	.599	.00015	.0002	.0003	.0005	.0008	.0012	.002	.003	.005
.600	.999	.00015	.00025	.0004	.0006	.001	.0015	.0025	.004	.006
1.000	1.499	.0002	.0003	.0005	.0008	.0012	.002	.003	.005	.008
1.500	2.799	.00025	.0004	.0006	.001	.0015	.0025	.004	.006	.010
2.800	4.499	.0003	.0005	.0008	.0012	.002	.003	.005	.008	.012
4.500	7.799	.0004	.0006	.001	.0015	.0025	.004	.006	.010	.015
7.800	13.599	.0005	.0008	.0012	.002	.003	.005	.008	.012·	.020
13.600	20.999	.0006	.001	.0015	.0025	.004	.006	.010	.015	.025

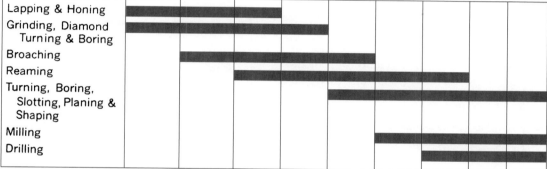

Lapping & Honing
Grinding, Diamond Turning & Boring
Broaching
Reaming
Turning, Boring, Slotting, Planing & Shaping
Milling
Drilling

FIGURE 11.13 Tolerances Related to Machining Processes.

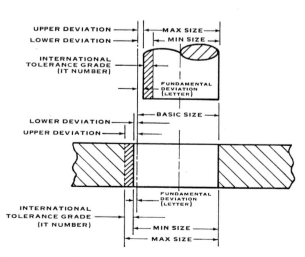

FIGURE 11.14 Terms Related to Metric Limits and Fits [ANSI B4.2– 1978 (R1994)].

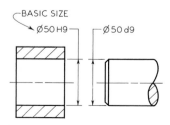

FIGURE 11.15 Methods of Specifying Tolerances with Symbols for Mating Parts.

5. Fundamental deviation is the deviation closest to the basic size. This is comparable to minimum allowance in the decimal-inch system.

6. Tolerance is the difference between the permitted minimum and maximum sizes of a part.

7. International tolerance grade (IT) is a set of tolerances that varies according to the basic size and provides a uniform level of accuracy within the grade. For example, in the dimension 50H8 for a close-running fit, the IT grade is indicated by the numeral 8. (The letter H indicates that the tolerance is on the hole for the 50-mm dimension.) In all, there are 18 IT grades—IT01, IT0, and IT1 through IT16 (shown in Figures 11.16 and 11.17)—for IT grades related to machining processes and for the practical use of the IT grades (see also Appendix 10).

8. Tolerance zone refers to the relationship of the tolerance to basic size. It is established by a combination of the fundamental deviation indicated by a letter and the IT grade number. In the dimension 50H8, for the close-running fit, the H8 specifies the tolerance zone, as shown in Figure 11.17.

9. The *hole-basis system of preferred fits* is a system in which the basic diameter is the minimum size. For the generally preferred hole-basis system, shown in Figure 11.15a, the fundamental deviation is specified by the uppercase letter H.

10. The *shaft-basis system of preferred fits* is a system in which the basic diameter is the maximum size of the shaft. The fundamental deviation is given by the lowercase letter f, as shown in Figure 11.18b.

11. An *interference fit* results in an interference between two mating parts under *all* tolerance conditions.

12. A *transition fit* results in either a clearance or an interference condition between two assembled parts.

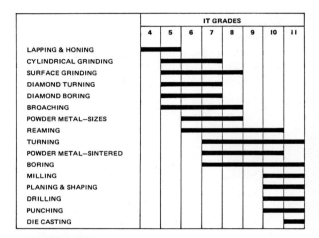

FIGURE 11.16 International Tolerance Grades Related to Machining Processes [ANSI B4.2–1978 (R1994)].

FIGURE 11.17 Practical Use of International Tolerance Grades.

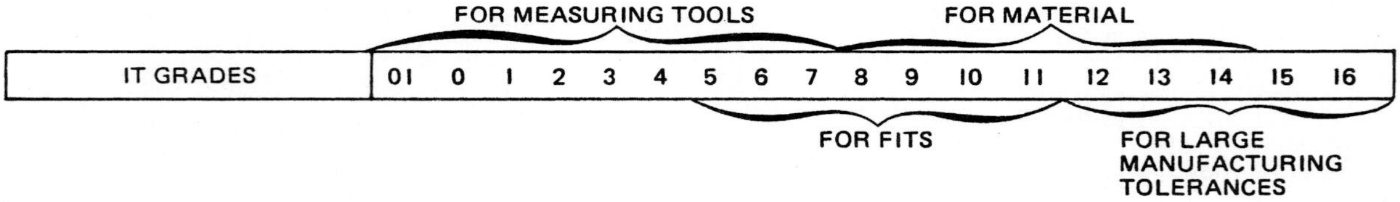

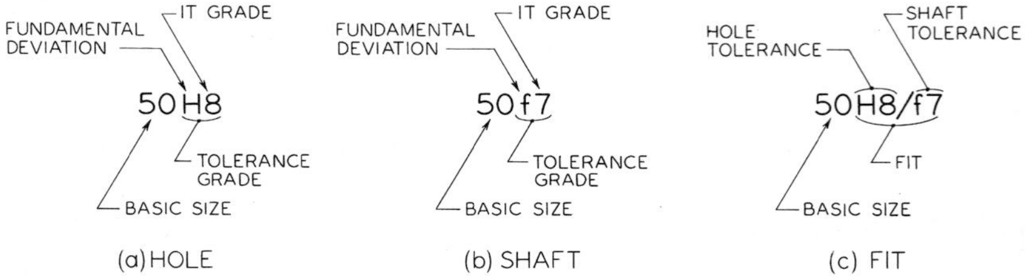

FIGURE 11.18 Application of Definitions and Symbols to Holes and Shafts [ANSI B4.2–1978 (R1994)].

(a) HOLE

(b) SHAFT

(c) FIT

13. *Tolerance symbols* are used to specify the tolerances and fits for mating parts, as shown in Figure 11.18c. For the hole-basis system, the 50 indicates the diameter in millimeters, the capital letter H indicates the fundamental deviation for the hole, and the lowercase letter f indicates the deviation for the shaft. The numbers following the letters indicate the IT grade. Note that the symbols for the hole and shaft are separated by a slash. Tolerance symbols for a 50-mm-diameter hole may be given in several acceptable forms, as shown in Figure 11.19. The values in parentheses are for reference only and may be omitted. The upper and lower limit values may be found in Appendix 11.

■ 11.12 PREFERRED SIZES

The preferred basic sizes for computing tolerances are given in Table 11.2. Basic diameters should be selected from the first-choice column since these are readily available stock sizes for round, square, and hexagonal products.

■ 11.13 PREFERRED FITS

The symbols for either the hole-basis or shaft-basis preferred fits (clearance, transition, and interference) are given in Table 11.2. Fits should be selected from this table for mating parts where possible.

The values corresponding to the fits are found in Appendixes 11–14. Although second- and third-choice basic-size diameters are possible, they must be calculated from tables not included in this text. For the generally preferred hole-basis system, note that the ISO symbols range from H11/c11 (loose running) to H7/u6 (force fit). For the shaft-basis system, the preferred symbols range from C11/h11 (loose fit) to U7/h6 (force fit).

Suppose that you want to use the symbols to specify the dimensions for a free-running (hole-basis) fit for a proposed diameter of 48 mm. Since 48 mm is not listed as a preferred size in Table 11.2, the design is altered to use the acceptable 50-mm diameter. From the preferred fits descriptions in Table 11.3, the free-running (hole-basis) fit is H9/d9. To determine the upper and lower deviation limits of the hole as given in the preferred hole-basis table (Appendix 11), follow across from the basic size of 50 to H9 under "Free Running." The limits for the hole are 50.000 and 50.062 mm. Then the upper and lower limits of deviation for the shaft are found in the d9 column under "Free Running." They

50 H8   50H8$\left(\begin{smallmatrix}50.039\\50.000\end{smallmatrix}\right)$ $\begin{smallmatrix}50.039\\50.000\end{smallmatrix}$(50H8)

(a) PREFERRED   (b)   (c)

FIGURE 11.19 Acceptable Methods of Giving Tolerance Symbols (ANSI/ASME Y14.5M–1994).

Basic Size mm		Basic Size mm		Basic Size mm	
First Choice	Second Choice	First Choice	Second Choice	First Choice	Second Choice
1		10		100	
	1.1		11		110
1.2		12		120	
	1.4		14		140
1.6		16		160	
	1.8		18		180
2		20		200	
	2.2		22		220
2.5		25		250	
	2.8		28		280
3		30		300	
	3.5		35		350
4		40		400	
	4.5		45		450
5		50		500	
	5.5		55		550
6		60		600	
	7		70		700
8		80		800	
	9		90		900
				1000	

TABLE 11.2 *Preferred Sizes [ANSI B4.2–1978 (R1994)].*

are 49.920 and 49.858 mm, respectively. Limits for other fits are established in a similar manner.

Limits for the shaft-basis dimensioning are determined similarly from the preferred shaft-basis table in Appendix 13. See Figures 11.19 and 11.20 for acceptable methods of specifying tolerances by symbols on drawings. A single note for the mating parts (free-running fit, hole-basis) would be Ø50 H9/d9, as shown in Figure 11.20.

■ 11.14 GEOMETRIC TOLERANCING

Geometric tolerances state the maximum allowable variations of a form or its position from the perfect geometry implied on the drawing. The term geometric

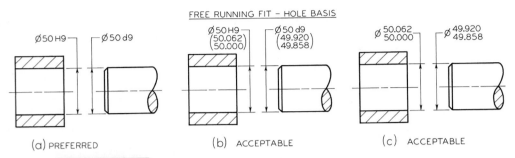

FIGURE 11.20   Methods of Specifying Tolerances with Symbols for Mating Parts.

ISO Symbol		Description
Hole Basis	**Shaft[a] Basis**	
Clearance Fits		
H11/c11	C11/h11	***Loose-running*** fit for wide commercial tolerances or allowances on external members.
H9/d9	D9/h9	***Free-running*** fit not for use where accuracy is essential, but good for large temperature variations, high running speeds, or heavy journal pressures.
H8/f7	F8/h7	***Close-running*** fit for running on accurate machines and for accurate location at moderate speeds and journal pressures.
H7/g6	G7/h6	***Sliding*** fit not intended to run freely, but to move and turn freely and locate accurately.
Transition Fits		
H7/h6	H7/h6	***Locational clearance*** fit provides snug fit for locating stationary parts; but can be freely assembled and disassembled.
H7/k6	K7/h6	***Locational transition*** fit for accurate location, a compromise between clearance and interference.
H7/n6	N7/h6	***Locational transition*** fit for more accurate location where greater interference is permissible.
Interference Fits		
H7/p6	P7/h6	***Locational interference*** fit for parts requiring rigidity and alignment with prime accuracy of location but without special bore pressure requirements.
H7/s6	S7/h6	***Medium drive*** fit for ordinary steel parts or shrink fits on light sections, the tightest fit usable with cast iron.
H7/u6	U7/h6	***Force*** fit suitable for parts which can be highly stressed or for shrink fits where the heavy pressing forces required are impractical.

More clearance →

More interference →

[a] The transition and interference shaft-basis fits shown do not convert to exactly the same hole-basis fit conditions for basic sizes in the range from Q through 3 mm. Interference fit P7/h6 converts to a transition fit H7/p6 in the above size range.

TABLE 11.3   *Preferred Fits [ANSI B4.2–1978 (R1994)].*

refers to various forms, such as a plane, a cylinder, a cone, a square, or a hexagon. Theoretically these are perfect forms, but because it is impossible to produce perfect forms, it may be necessary to specify the amount of variation permitted. Geometric tolerances specify either the diameter or the width of a tolerance zone within which a surface or the axis of a cylinder or a hole must be if the part is to meet the required accuracy for proper function and fit. When tolerances of form are not given on a drawing, it is customary to assume that, regardless of form variations, the part will fit and function satisfactorily.

Tolerances of form and position (or location) control such characteristics as straightness, flatness, parallelism, perpendicularity (squareness), concentricity, roundness, angular displacement, and so on.

Methods of indicating geometric tolerances by means of **geometric characteristic symbols**, rather than by traditional notes, are recommended. See the latest Dimensioning and Tolerancing Standard (ANSI/ASME Y14.5M–1994) for more complete coverage.

■ 11.15 SYMBOLS FOR TOLERANCES OF POSITION AND FORM

Since traditional notes for specifying tolerances of position (location) and form (shape) may be confusing or unclear, may require too much space, and may not be understood internationally, most multinational companies have adopted symbols for such specifications (ANSI/ASME Y14.5M–1994). These ANSI symbols, shown in Table 11.4, provide an accurate and concise means of specifying geometric characteristics and tolerances in a minimum of space. The symbols may be supplemented by notes if the precise geometric requirements cannot be conveyed by the symbols. For construction details of the geometric tolerancing symbols, see Appendix 37.

Geometric characteristic symbols				Modifying symbols	
	Type of Tolerance	**Characteristic**	**Symbol**	**Term**	**Symbol**
For individual features	Form	Straightness	—	At maximum material condition	Ⓜ
		Flatness	▱	At least material condition	Ⓛ
		Circularity (roundness)	○	Projected tolerance zone	Ⓟ
		Cylindricity	⌀⃫	Free state	Ⓕ
For individual or related features	Profile	Profile of a line	⌒	Tangent plane	Ⓣ
		Profile of a surface	⌓	Diameter	⌀
For related features	Orientation	Angularity	∠	Spherical diameter	S⌀
		Perpendicularity	⊥	Radius	R
		Parallelism	//	Spherical radius	SR
				Controlled radius	CR
	Location	Position	⌖	Reference	( )
		Concentricity	◎	Arc length	⌒
		Symmetry	＝	Statistical tolerance	⟨ST⟩
	Runout	Circular runout	↗ *	Between	↔
		Total runout	↗↗ *		

* ARROWHEADS MAY BE FILLED OR NOT FILLED

TABLE 11.4 *Geometric Characteristic and Modifying Symbols (ASME Y14.5M–1994).*

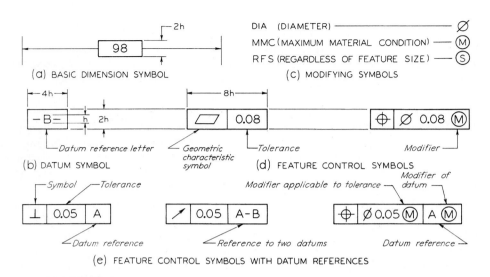

FIGURE 11.21   Use of Symbols for Tolerance of Position and Form (ASME Y14.5M–1994).

Combinations of the various symbols and their meanings are given in Figure 11.21. Application of the symbols to drawing are illustrated in Figure 11.22. The geometric characteristic symbols plus the supplementary symbols are explained and illustrated below with material adapted from ANSI/ASME Y14.5M–1994:

1. The ***basic dimension symbol*** is identified by the enclosing frame symbol, as shown in Figure 11.21a. The basic dimension, or size, is the value used to describe the theoretically exact size, shape, or location of a feature. It is the basis from which permissible variations are established either by specifying tolerances on other dimensions, by tolerances given in notes or by using feature control frames.

2. The ***datum identifying symbol*** consists of a capital letter in a square frame and a leader line extending from the frame to the concerned feature and terminating with a triangle. The triangle may be filled or not filled, as shown in Figure 11.21b. Letters of the alphabet (except I, O, and Q) are used as datum-identifying letters. A point, line, plane, cylinder, or other geometric form assumed to be exact for purposes of computation may serve as a datum from which the location or geometric relationship of features of a part may be established, as shown in Figure 11.23.

3. ***Supplementary symbols*** include the symbols for MMC (maximum material condition—or minimum hole diameter, maximum shaft diameter) and LMC (least material condition—or maximum hole diameter, minimum shaft diameter), as shown in Figure 11.21c. The abbreviations MMC and LMC are also used in notes (see also Table 11.4).

The symbol for diameter precedes the specified tolerance in a feature control symbol, as shown in Figure 11.21d. This symbol for diameter should precede the dimension. For narrative notes, you can use the abbreviation DIA.

4. ***Combined symbols*** are found when individual symbols, datum reference letters, and needed tolerances are combined in a single frame, as shown in Figure 11.21e.

A form tolerance is given by a feature control symbol made up of a frame around the appropriate geometric characteristic symbol plus the allowable tolerance. A vertical line separates the symbol and the tolerance, as shown in Figure 11.21d. Where needed, the tolerance should be preceded by the symbol for the diameter and followed by the symbol for MMC or LMC.

Reference to a datum is indicated in the feature control symbol by placing the datum reference letter after either the geometric characteristic symbol or the tolerance. Vertical lines separate the entries, and where applicable the datum reference letter entry includes the symbol for MMC or LMC, as shown in Figure 11.21.

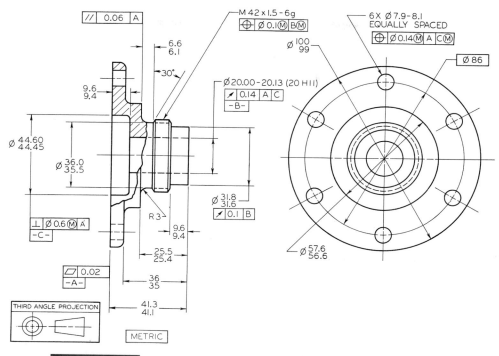

FIGURE 11.22 Application of Symbols to Position and Form Tolerance Dimensions (ASME Y14.5M–1994).

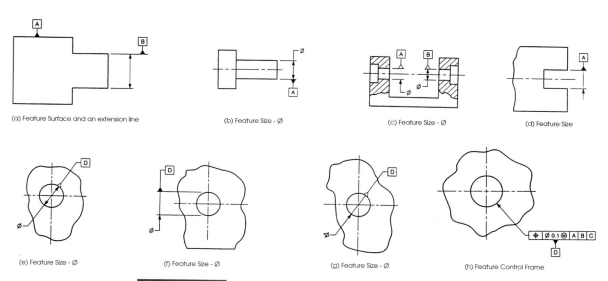

(a) Feature Surface and an extension line

(b) Feature Size - Ø

(c) Feature Size - Ø

(d) Feature Size

(e) Feature Size - Ø

(f) Feature Size - Ø

(g) Feature Size - Ø

(h) Feature Control Frame

FIGURE 11.23 Placement of Datum Feature Symbol (ASME Y14.5M–1994).

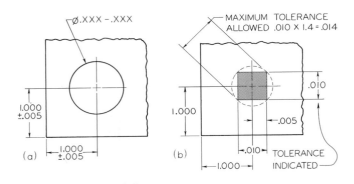

FIGURE 11.24   Tolerance Zones.

■ 11.16 POSITIONAL TOLERANCES

Figure 11.24a shows a hole located from two surfaces at right angles to each other. In Figure 11.24, the center may lie anywhere within a square tolerance zone, the sides of which are equal to the tolerances. The total variation along either diagonal of the square by the coordinate method of dimensioning will be 1.4 times greater than the indicated tolerance. When the location of the hole is off in a diagonal direction, the area of the square tolerance zone is increased by 57 percent without exceeding the tolerance permitted.

Features located by toleranced angular and radial dimensions will have a wedge-shaped tolerance zone.

If four holes are dimensioned with rectangular coordinates, as in Figure 11.25a, specifying a tolerance describes a square zones in which the center of the hole must be located as shown in Figures 11.25b and 11.25c. Because of the shape of the square zone, the tolerance for the location of the center of the hole is greater in the diagonal direction than the indicated tolerance.

In Figure 11.25a, hole A is selected as a datum, and the other three are located from it. The square tolerance zone for hole A results from the tolerances on the two rectangular coordinate dimensions locating hole A. The sizes of the tolerance zones for the other three holes result from the tolerances between the holes, while their locations will vary according to the actual location of the datum hole A. Two of the many possible zone patterns are shown in Figures 11.25b and 11.25c.

With the dimensions shown in Figure 11.25a, it is difficult to say whether the resulting parts will actually fit the mating parts satisfactorily, even though they conform to the tolerances shown on the drawing.

Geometric tolerancing provides a method to accurately tolerance features based on their geometry so that these tolerancing problems don't occur. Geometric tolerancing uses feature control frames to define the specific geometric tolerance. (See §11.21.) This is called **true-position dimensioning**. Using it, the tolerance zone for each hole will be a circle, with the size of the circle depending on the amount of variation permitted from true position.

Feature control symbols are related to the feature by one of several methods illustrated in Figure 11.22.

FIGURE 11.25   Tolerance Zones.

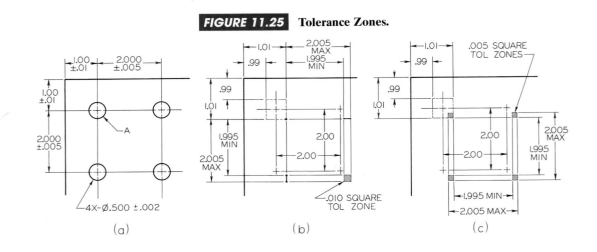

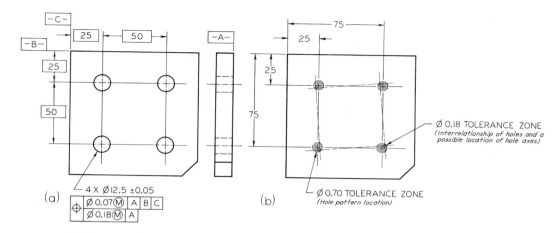

FIGURE 11.26 True-Position Dimensioning [ANSI Y14.5M–1982 (R1988)].

The following methods are preferred:

1. Adding the symbol to a note or dimension pertaining to the feature
2. Running a leader from the symbol to the feature
3. Attaching the side, end, or corner of the symbol frame to an extension line from the feature
4. Attaching a side or end of the symbol frame to the dimension line pertaining to the feature

A true-position dimension specifies the theoretically exact position of a feature. The location of each feature, such as a hole, slot, or stud, is given by untoleranced *basic dimensions* identified by the enclosing frame or symbol. To prevent misunderstandings, true position should be established with respect to a datum. In simple arrangements, the choice of a datum may be obvious and not require identification.

Positional tolerances are indicated in a feature control frame attached to a feature on the object. Positional tolerances describe a cylindrical zone for the tolerance as shown in Figure 11.26. This cylindrical tolerance zone has a diameter equal to the positional tolerance, and its length is equal to the length of the feature unless otherwise specified. Its axis must be within this cylinder, as shown in Figure 11.27.

The centerline of the hole may coincide with the centerline of the cylindrical tolerance zone, as shown in Figure 11.27a. It may be parallel to it but displaced so that it remains within the tolerance cylinder, as shown in Figure 11.27b. Or it may be inclined while remaining within the tolerance cylinder, as shown in Figure 11.27c. In this last case the positional tolerance also defines the limits of squareness variation.

FIGURE 11.27 Cylindrical Tolerance Zone (ASME Y14.5M–1994).

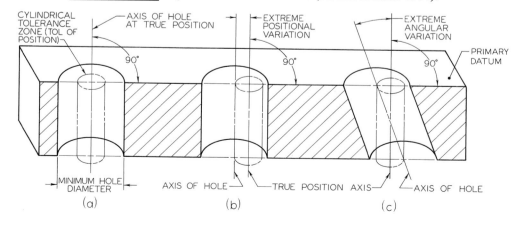

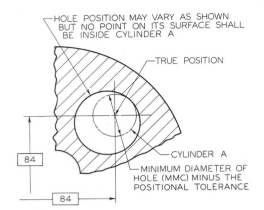

HOLE POSITION MAY VARY AS SHOWN
BUT NO POINT ON ITS SURFACE SHALL
BE INSIDE CYLINDER A

TRUE POSITION

CYLINDER A

MINIMUM DIAMETER OF
HOLE (MMC) MINUS THE
POSITIONAL TOLERANCE

84

84

FIGURE 11.28 **True Position Interpretation (ASME Y14.5M–1994).**

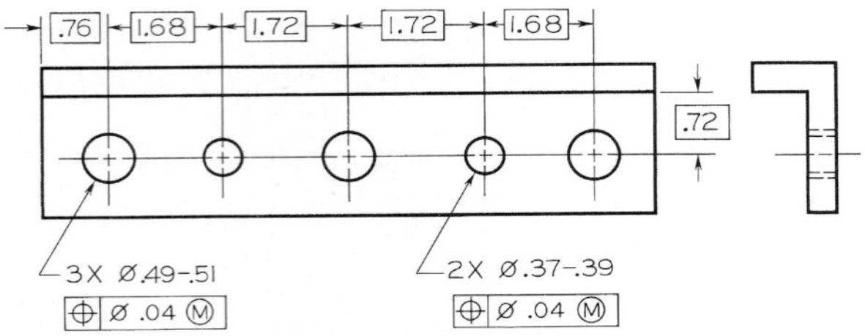

.76 1.68 1.72 1.72 1.68

.72

3X Ø.49-.51
$\bigoplus$ Ø .04 $\textcircled{M}$

2X Ø.37-.39
$\bigoplus$ Ø .04 $\textcircled{M}$

FIGURE 11.29 **No Tolerance Accumulation.**

As shown in Figure 11.28, the positional tolerance specification indicates that all elements on the hole surface must be on or outside a cylinder whose diameter is equal to the minimum diameter or the maximum diameter of the hole minus the positional tolerance (diameter, or twice the radius), with the centerline of the cylinder located at true position.

The use of basic untoleranced dimensions to locate features at true position avoids the accumulation of tolerances even in a chain of dimensions, as shown in Figure 11.29.

While features, such as holes and bosses, may vary in any direction from the true-position axis, other features, such as slots, may vary on either side of a true-position plane, as shown in Figure 11.30.

Since the exact locations of the true positions are given by untoleranced dimensions, it is important to prevent the application of general tolerances to these. A note should be added to the drawing, such as

GENERAL TOLERANCES DO NOT APPLY TO BASIC DIMENSIONS.

■ 11.17 MAXIMUM MATERIAL CONDITION

Maximum material condition, or MMC, means that a feature of a finished product contains the maximum amount of material permitted by the toleranced dimensions shown for that feature. Holes, slots, or other internal features are at MMC when at minimum size. Shafts, pads, bosses, and other external features are at MMC when at their maximum size. A feature is at MMC for both mating parts when the largest shaft is in the smallest hole and there is the least clearance between the parts.

FIGURE 11.30 **Positional Tolerancing for Symmetry (ASME Y14.5M–1994).**

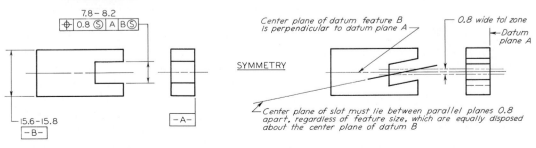

7.8 – 8.2
$\bigoplus$ 0.8 $\textcircled{S}$ A B $\textcircled{S}$

15.6-15.8
–B–

–A–

(a) THIS ON THE DRAWING . . .

Center plane of datum feature B
is perpendicular to datum plane A

0.8 wide tol zone

Datum plane A

SYMMETRY

Center plane of slot must lie between parallel planes 0.8
apart, regardless of feature size, which are equally disposed
about the center plane of datum B

(b) . . . MEANS THIS

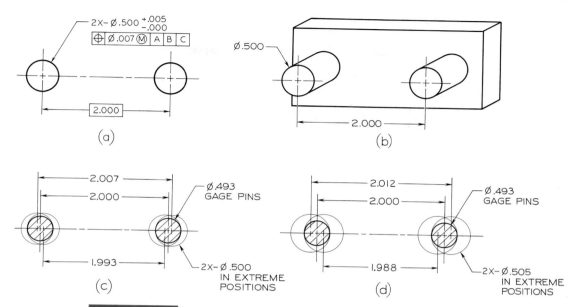

FIGURE 11.31 Maximum and Minimum Material Conditions—Two-Hole Pattern (ASME Y14.5M–1994).

In assigning positional tolerance to a hole, consider the size limits of the hole. If the hole is at MMC, or its smallest size, the positional tolerance is not affected, but if the hole is larger, the available positional tolerance is greater. In Figure 11.31a, two half-inch holes are shown. If they are exactly .500 inch in diameter (MMC, or smallest size) and are exactly 2.000 inch apart, a gage made of two round pins .500 inch in diameter fixed in a plate 2.000 inch apart, as shown in Figure 11.31b, should fit into them. However, the center-to-center distance between the holes may vary from 1.993 inch to 2.007 inch.

If the .500-inch-diameter holes are at their extreme positions, as in Figure 11.31c, the pins in the gage would have to be .007 inch smaller, or .493 inch in diameter, to fit into the holes. If the .500-inch-diameter holes are located at the maximum distance apart, the .493-inch-diameter gage pins would contact the inner sides of the holes; and if the holes are located at the minimum distance apart, the .493-inch-diameter pins would contact the outer surfaces of the holes, as shown. If gage-maker's tolerances are not considered, the gage pins would have to be .493 inch in diameter and exactly 2.000 inches apart if the holes are .500 inch in diameter, or MMC.

If the holes are .505 inch in diameter—that is, at maximum size—the same .493-inch-diameter gage

pins at 2.000 inches apart will fit them with the inner sides of the holes contacting the inner sides of the gage pins and the outer sides of the holes contacting the outer sides of the gage pins, as shown in Figure 11.31d. When the holes are larger, they may be further apart and still fit the pins. In this case they may be 2.012 inches apart, which is beyond the tolerance permitted for the center-to-center distance between the holes. Similarly, the holes may be as close together as 1.988 inches from center to center, which again is outside the specified positional tolerance.

So when holes are at maximum size a greater positional tolerance becomes available. Since all features may vary in size, it is necessary to make clear on the drawing at what basic dimension the true position applies. In all but a few exceptional cases, when the holes are larger, the additional positional tolerance is available without affecting function. They can still be freely assembled whether or not the holes or other features are within the specified positional tolerance. This practice has been recognized and used in manufacturing for years, as is evident from the use of fixed-pin gages, which have been commonly used to inspect parts and control the least favorable condition of assembly. Thus it has become common practice for both manufacturing and inspection to assume that positional tolerance applies to MMC and that greater positional tolerance becomes permissible when the part is not at MMC.

To avoid possible misinterpretation as to whether maximum material condition (MMC) applies, it should be clearly stated on the drawing by the addition of MMC symbols to each applicable tolerance or by suitable coverage in a document referenced on the drawing. When MMC is not specified on the drawing with respect to an individual tolerance, datum reference, or both, the following rules apply:

1. True-position tolerances and related datum references apply at MMC. For a tolerance of position, RFS (regardless of feature size) may be specified on the drawing with respect to the individual tolerance, datum reference, or both, as applicable.

2. All applicable geometric tolerances—such as angularity, parallelism, perpendicularity, concentricity, and symmetry tolerances, including related datum references apply at RFS, where no modifying symbol is specified. Circular runout, total runout, concentricity, and symmetry are applicable only on an RFS basis and cannot be modified to MMC or LMC. No element of the actual feature will extend beyond the envelope of the perfect form at MMC. MMC or LMC must be specified on the drawing where it is required.

■ 11.18 TOLERANCES OF ANGLES

Bilateral tolerances have traditionally been given on angles, as shown in Figure 11.32. Using bilateral tolerances, the wedge-shaped tolerance zone increases as the distance from the vertex of the angle increases. The use of angular tolerances may be avoided by using

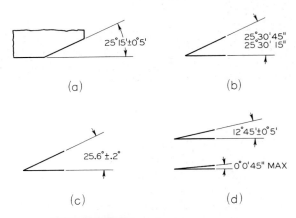

FIGURE 11.32   Tolerances of Angles.

gages. Taper turning is often handled by machining to fit a gage or by fitting to the mating part.

If an angular surface is located by a linear and an angular dimension, as shown in Figure 11.33a, the surface must lie within a tolerance zone, as shown in Figure 11.33b. The angular zone will be wider as the distance from the vertex increases. To avoid the accumulation of tolerance further out from the angle's vertex—the *basic angle tolerancing method*, shown in Figure 11.33c, is recommended (ASME Y14.5M–1994). The angle is indicated as a basic dimension, and no angular tolerance is specified. The tolerance zone is now defined by two parallel planes, resulting in improved angular control, as shown in Figure 11.33d.

FIGURE 11.33   Angular Tolerance Zones (ASME Y14.5M–1994).

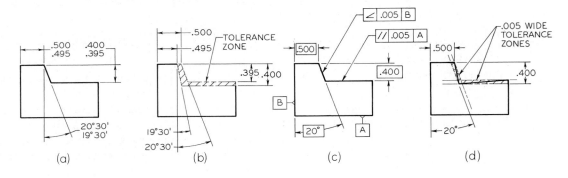

GRAPHICS SPOTLIGHT

Geometric Tolerances With AutoCAD 2000

AutoCAD 2000 has dialog boxes which allow you to create feature control frames for geometric dimensioning and tolerancing. When you pick the tolerance icon from AutoCAD's dimensioning toolbar, the dialog box shown in Fig. A appears on your screen and shows the standard tolerance symbols.

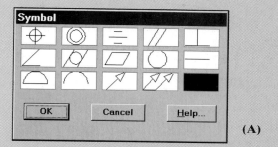

(A)

To begin creating a feature control frame, you double-click on the symbol you want to use. For example, you could double-click on the positional tolerance symbol shown in the upper left-hand corner of the dialog box. When you have selected a symbol, a new dialog box, shown in Fig. B will appear on the screen. You use it to create either a single feature control frame or a stacked feature control frame. You can also add diameter symbols, modifiers, datum references, or datum identifiers. The diameter symbol shown in the geometric tolerance dialog box was added by picking in the empty box below the heading "Dia." A diameter symbol appears automatically.

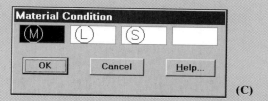

(C)

The area to the right of the diameter symbol is a text entry box that you use to type in the value you want to show for the tolerance.

To add a modifier symbol, pick in the next empty box to the right. The material condition dialog box (Fig. C) pops up on the screen. You can quickly pick the modifier you want to add from the dialog box.

Datum references can be created just as quickly by picking in the appropriate box and typing in the letter you want to use. Datum references can also have a modifier when used with certain types of tolerances. Again, just pick in the empty box below the modifier and use the dialog box that appears to add a symbol.

If you want stacked tolerances or a datum identifier, you continue on with the same basic procedure. When you pick OK at the end of the process, you will be prompted to pick a location to place the tolerance in the drawing. You can also use the leader command and select the option to place a tolerance at the end of a leader line. Using these dialog boxes, you can quickly add geometric tolerance symbols, as shown in Fig. D to your drawings. Creating the symbols is easy, but you must give careful consideration to what the placement of the symbols in the drawing means. Make sure to reflect the intent of the design and the tolerances that are required for the part to function correctly in the assembly. Specifying needlessly restrictive tolerances increases the cost of the part without adding more functionality to the design.

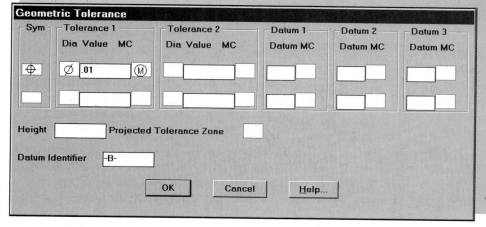

(B)   **(D)**

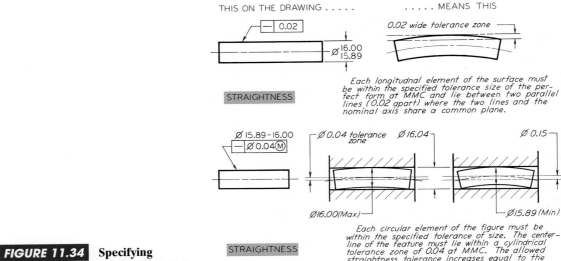

THIS ON THE DRAWING . . . . MEANS THIS

STRAIGHTNESS

Each longitudnal element of the surface must be within the specified tolerance size of the perfect form at MMC and lie between two parallel lines (0.02 apart) where the two lines and the nominal axis share a common plane.

STRAIGHTNESS

Each circular element of the figure must be within the specified tolerance of size. The centerline of the feature must lie within a cylindrical tolerance zone of 0.04 at MMC. The allowed straightness tolerance increases equal to the amount the feature departs from MMC.

FIGURE 11.34   **Specifying Straightness (ASME Y14.5M–1994).**

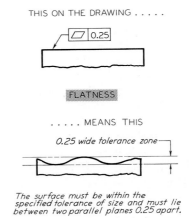

THIS ON THE DRAWING . . . . .

FLATNESS

. . . . . MEANS THIS

0.25 wide tolerance zone

The surface must be within the specified tolerance of size and must lie between two parallel planes 0.25 apart.

FIGURE 11.35   **Specifying Flatness (ASME Y14.5M– 1994).**

■ 11.19 FORM TOLERANCES FOR SINGLE FEATURES

Straightness, flatness, roundness, cylindricity, and in some instances, profile are form tolerances applicable to single features regardless of feature size.

1. *Straightness tolerance*, shown in Figure 11.34, specifies a tolerance zone within which an axis or all points of the considered element must lie. Straightness is a condition in which an element of a surface or an axis is a straight line.

2. *Flatness tolerance* specifies a tolerance zone defined by two parallel planes within which the surface must lie, as shown in Figure 11.35. Flatness is the condition of a surface having all elements in one plane.

3. *Roundness (circularity) tolerance*, shown in Figure 11.36, specifies a tolerance zone bounded by two concentric circles within which each circular element of the surface must lie. Roundness is a condition of a surface of revolution in which, for a cone

THIS ON THE DRAWING . . . . MEANS THIS

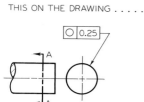

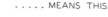

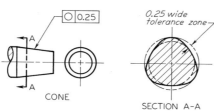

CYLINDER

CONE

SECTION A-A

ROUNDNESS

Each circular element of the surface in any plane perpendicular to a common axis must be within the specified tolerance of size and must lie between two concentric circles — one having a radius 0.25 larger than the other.

FIGURE 11.36 Specifying Roundness for a Cylinder or Cone (ASME Y14.5M–1994).

THIS ON THE DRAWING . . . . MEANS THIS

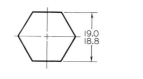

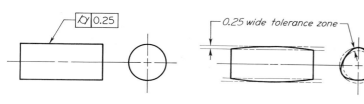

CYLINDRICITY

The cylindrical surface must be within the specified tolerance of size and must lie between two concentric cylinders — one having a radius 0.25 larger than the other.

FIGURE 11.37 Specifying Cylindricity (ASME Y14.5M–1994).

THIS ON THE DRAWING . . . . MEANS THIS

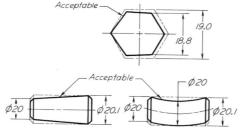

Tolerance zone or boundary within which forms may vary when no tolerance of form is given.

FIGURE 11.38 Acceptable Variations of Form—No Specified Tolerance of Form.

or cylinder, all points of the surface intersected by any plane perpendicular to a common axis are equidistant from that axis. For a sphere, all points of the surface intersected by any plane passing through a common center are equidistant from that center.

4. *Cylindricity tolerance* specifies a tolerance zone bounded by two concentric cylinders within which the surface must lie, as shown in Figure 11.37. This tolerance applies to both circular and longitudinal elements of the entire surface. Cylindricity is a condition of a surface of revolution in which all points of the surface are equidistant

from a common axis. When no tolerance of form is given, many possible shapes may exist within a tolerance zone, as shown in Figure 11.38.

5. *Profile tolerance* specifies a uniform boundary or zone along the true profile within which all elements of the surface must lie, as shown in Figures 11.39 and 11.40. A profile is the outline of an object in a given plane, or 2-D, figure. Profiles are formed by projecting a 3-D figure onto a plane or by taking cross sections through the figure, with the resulting profile composed of such elements as straight lines, arcs, or other curved lines.

THIS ON THE DRAWING . . . . .

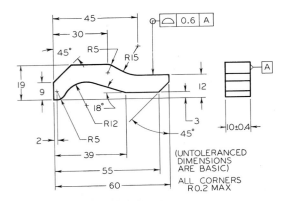

. . . . . MEANS THIS

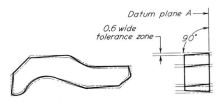

Surfaces all around must lie within two parallel
boundaries 0.6 apart equally disposed about
the true profile which are perpendicular to
datum plane A. Radii of part corners must
not exceed 0.2.

FIGURE 11.39   Specifying Profile of a Surface All Around (ASME
Y14.5M–1994).

THIS ON THE DRAWING . . . . .

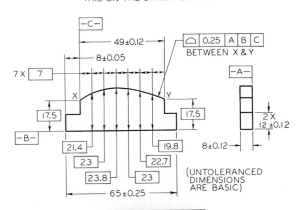

. . . . . MEANS THIS

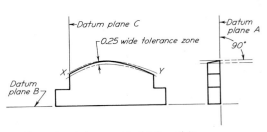

The surface between points X & Y must lie
between the two profile boundaries 0.25 apart,
equally disposed about the true profile, which
are perpendicular to datum plane A and positioned with respect to datum planes B & C.

FIGURE 11.40   Specifying Profile of a Surface between Points (ASME
Y14.5M–1994).

■ 11.20 FORM TOLERANCES FOR RELATED FEATURES

Angularity, parallelism, perpendicularity, and in some instances, profile are form tolerances applicable to related features. These tolerances control the attitude of features to one another (ASME Y14.5M–1994).

1. Angularity tolerance, shown in Figure 11.41, specifies a tolerance zone defined by two parallel planes at the specified basic angle (other than 90 degrees) from a datum plane or axis within which the surface or the axis of the feature must lie.

2. Parallelism tolerance, shown in Figures 11.42–11.44, specifies a tolerance zone defined by two parallel planes or lines parallel to a datum plane or axis, respectively, within which the surface or axis of the feature must lie. Also, parallelism tolerance may specify a cylindrical tolerance zone parallel to a datum axis within which the axis of the feature must lie.

3. Perpendicularity tolerance. Perpendicularity is the condition of a surface, median plane, or axis which is at 90 degrees to a datum plane or axis. A

THIS ON THE DRAWING . . . . .

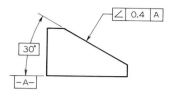

ANGULARITY

. . . . . MEANS THIS

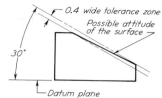

The surface must be within the specified tolerance of size and must lie between two parallel planes 0.4 apart which are inclined at 30° to the datum plane A.

FIGURE 11.41 Specifying Angularity for a Plane Surface (ASME Y14.5M–1994).

THIS ON THE DRAWING . . . . .

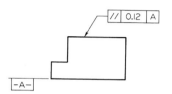

PARALLELISM

. . . . . MEANS THIS

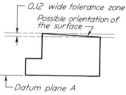

The surface must be within the specified tolerance of size and must lie between two planes 0.12 apart which are parallel to the datum plane A.

FIGURE 11.42 Specifying Parallelism for a Plane Surface (ASME Y14.5M–1994).

THIS ON THE DRAWING . . . . .

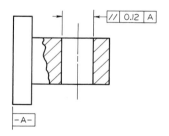

PARALLELISM

. . . . . MEANS THIS

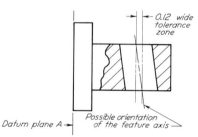

The feature axis must be within the specified tolerance of location and must lie between two planes 0.12 apart which are parallel to the datum plane, regardless of feature size.

FIGURE 11.43 Specifying Parallelism for an Axis Feature RFS (ASME Y14.5M–1994).

THIS ON THE DRAWING . . . . .

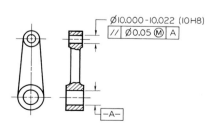

PARALLELISM

. . . . . MEANS THIS

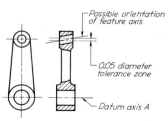

The feature axis must be within the specified tolerance of location. Where the feature is at maximum material condition (10.00), the maximum parallelism tolerance is 0.05 diameter. Where the feature departs from its MMC size, an increase in the parallelism tolerance is allowed which is equal to the amount of such departure.

FIGURE 11.44 Specifying Parallelism for an Axis Feature at MMC (ASME Y14.5M–1994).

perpendicularity tolerance specifies one of the following:

(a) A tolerance zone is defined by two parallel planes perpendicular to a datum plane, datum axis, or axis within which the surface of the feature must lie, as shown in Figure 11.45.

(b) A cylindrical tolerance zone perpendicular to a datum plane within which the axis of the feature must lie, as shown in Figure 11.46.

4. Concentricity tolerance. Concentricity is the condition in which the axes of all cross-sectional elements of a feature's surface of revolution are common to the axis of a datum feature. A concentricity tolerance specifies a cylindrical tolerance zone whose axis coincides with a datum axis and within which all cross-sectional axes of the feature being controlled must lie, as shown in Figure 11.47.

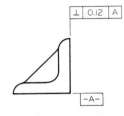

PERPENDICULARITY
FOR A PLANE SURFACE

(a)

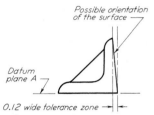

The surface must be within the specified tolerance of size and must lie between two parallel planes 0.12 apart which are perpendicular to the datum plane A.

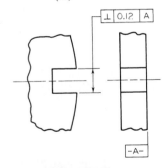

PERPENDICULARITY
FOR A MEDIAN PLANE

(b)

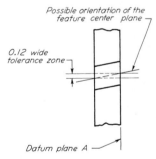

The feature center plane must be within the specified tolerance of location and must lie between two parallel planes 0.12 apart, regardless of feature size, which are perpendicular to the datum plane A.

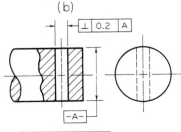

PERPENDICULARITY
FOR AN AXIS

(c)

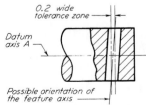

The feature axis must be within the specified tolerance of location and must lie between two planes 0.2 apart, regardless of feature size, which are perpendicular to the datum axis A.

FIGURE 11.45 Specifying Perpendicularity (ASME Y14.5M–1994).

■ 11.21 USING GEOMETRIC DIMENSIONING AND TOLERANCING

Geometric dimensioning and tolerancing (GDT) has evolved over the last forty years to become an indispensable tool for defining parts and features more accurately. GDT not only considers an individual part and its dimensions and tolerances, but views that part in relations to its related parts. This allows the designer more latitude in defining the parts feature's more accurately by not only considering the part's dimensions, but its tolerances at the initial design stage. GDT also simplifies the inspection process. This is accomplished through the use of ASME standards (ASME–Y14.5M), as we have discussed previously.

Individually manufactured parts and components must eventually be assembled in to products. We take for granted that each part of a lawnmower, for example, will mate properly with its other components when assembled. The wheels will slip into their axles, the pistons will fit properly into their cylinders, etc. Nothing should be too tight or too loose.

Geometric dimensioning and tolerancing, therefore, is important to both the design and manufacturing processes.

Applying GDT principles to the design process requires five steps:

Step 1: Define the part's functions. It is best to break the part down to its simplest functions. Be as specific as possible. For example, a lawnmower wheel's functions is to: (a). Give the product mobility: (b). Lift the mowing deck off the ground; (c) Add rigidity to the body, etc.

Step 2: List the functions by priority. Only one function should have top priority. This step can be difficult since many parts are designed to incorporate multiple functions. In our lawnmower wheels example, the function with top priority would be to give the product mobility.

Step 3: Define the datum reference frame. This step should be based on your list of priorities. This may mean creating several reference frames, each based on a priority on your list. The frame should be set up in either one, two, or three planes.

Step 4: Control selection. In most cases, several controls will be needed (e.g.,runout, position, concentricity, roughness, etc.). Begin with the simplest control. By "simplest" we mean least restrictive. Work from the least restrictive to the most restrictive set of controls.

Step 5: Calculate tolerances. Most tolerances are mathematically based. This step should be the easiest. Apply MMC, RFS, or LMC where indicated. Avoid completing this step first, it should always be your last. See Fig. 11.48 for a worksheet outlining these five steps.

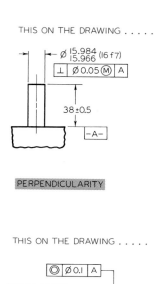

THIS ON THE DRAWING . . . . .

PERPENDICULARITY

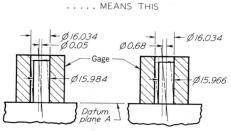

. . . . . MEANS THIS

The feature axis must be within the specified tolerance of location. Where the feature is at MMC (15.984) the maximum perpendicularity tolerance is 0.05 diameter. Where the feature departs from its MMC size, an increase in the perpendicularity tolerance is allowed which is equal to the amount of such departure.

FIGURE 11.46 Specifying Perpendicularity for an Axis, Pin, or Boss (ASME Y14.5M–1994).

THIS ON THE DRAWING . . . . .

CONCENTRICITY

. . . . . MEANS THIS

Extreme locational variation

Extreme attitude variation

0.1 diameter tolerance zone

Axis of of datum A

The feature axis must be within a cylindrical zone of 0.1 diameter, regardless of feature size, and whose axis coincides with the datum axis.

FIGURE 11.47 Specifying Concentricity (ASME Y14.5M–1994).

CAD programs generally allow the user to add tolerances to dimension values in the drawings. Geometric dimensioning and tolerancing symbols, finish marks, and other standard symbols are typically available as a part of the CAD program or as a symbol library.

Geometric dimensioning and tolerancing has become an essential part of today's manufacturing industry. To compete in today's marketplace, companies are required to develop and produce products of the highest quality, at lowest cost, and guarantee on-time delivery. Although considered by most to be a design

specification language, GDT is a manufacturing and inspection language as well, providing a means for uniform interpretation and understanding by these various groups. It provides both a national and international contract base for customers and suppliers.

■ 11.22 COMPUTER GRAPHICS

CAD programs generally allow the user to add tolerances to dimension values in the drawings. Geometric dimensioning and tolerancing symbols, finish marks, and other standard symbols are typically available as a part of the CAD program or as a symbol library.

1. DEFINE THE PART'S (FEATURE'S) FUNCTION:

Basic function _____

Additional function(s) _____

2. LIST THE FUNCTIONS IN ORDER OF PRIORITY

Function# _____   Function# _____   Function# _____

3. DEFINE DATUM REFERENCE FRAME

Part # _____   Function _____
Primary Datum Feature ____   Secondary Datum Feature ____

Part # _____   Function _____
Primary Datum Feature ____   Secondary Datum Feature ____

Part # _____   Function _____
Primary Datum Feature ____   Secondary Datum Feature ____

4. SELECT CONTROL TYPE

Part# _____   Part # _____
Control _____   Control _____

5. CALCULATE THE TOLERANCES

Part # _____   Part# _____   Part# _____

FIGURE 11.48 Geometric Dimensioning and Tolerancing Design Worksheet.

■ KEY WORDS

tolerance	limit dimensioning	international tolerance grade (IT)	supplementary symbols
limit dimensions	plus-or-minus dimensioning	tolerance zone	combined symbols
allowance	unilateral system of tolerance	hole-basis system of preferred fits	true-position dimensioning
nominal size	bilateral system of tolerances	shaft-basis system of preferred fits	maximum material condition
basic size	single-limit dimensioning	interference fit	basic angle tolerancing method
basic dimension	angular tolerances	transition fit	straightness tolerance,
actual size	reference dimension	tolerance symbols	flatness tolerance
allowance	basic size	geometric tolerances	roundness (circularity) tolerance
clearance fit	deviation	geometric dimensioning and tolerancing (GDT)	cylindricity tolerance
interference fit	upper deviation	geometric characteristic symbols	profile tolerance
transition fit	lower deviation		angularity tolerance
line fit	fundamental deviation	basic dimension symbol	parallelism tolerance
elective assembly	tolerance	datum identifying symbol	perpendicularity tolerance
basic hole system			concentricity tolerance
basic shaft system			
general tolerances			

■ CHAPTER SUMMARY

- Tolerance dimensioning describes the minimum and maximum limits for a size or location of a feature.
- There are several ways of dimensioning tolerances, including limit dimensions, unilateral tolerances, bilateral tolerances, and geometric tolerancing.
- Basic-hole tolerance systems are the most commonly used tolerance system because they assume the hole is nominal size and adjust the shaft to accommodate the tolerance.
- The amount of space between two mating parts at maximum material condition is called the allowance.
- Mating parts with large allowances are classified as having a clearance fit or a running-and-sliding fit.

- Mating parts with negative allowances are classified as having an interference fit or force fit.
- Mating parts are designed around a nominal size and class of fit. Other tolerances are calculated from these two values.
- High-quality parts are often dimensioned with geometric tolerancing to ensure that the size, shape, and relative geometric characteristics are properly defined.
- GDT has become an essential part of today's manufacturing industry. GDT is not only a design language, but an inspection language as well.

■ REVIEW QUESTIONS

1. What do the two numbers of a limit dimension mean?
2. Draw three different geometric tolerances that reference a datum. Label the information in each box.
3. Why is the basic-hole system more common than the basic-shaft system?
4. Give five examples of nominal sizes in everyday life. What is the purpose of a nominal size?

5. Give an example of two parts that would require a running-and-sliding fit. A force fit.
6. List five classes of fit.
7. Can one part have an allowance? Why?
8. Can two parts have a tolerance? Why?
9. Give an example of how GDT could be used as both a design and inspection tool.
10. List the five steps required to apply GDT to the design process.

■ DESIGN PROJECT

Design a system which will attach under the top of a standard wooden table or desk to contain a computer system keyboard. The system should allow the user to store the keyboard underneath the table surface when not in use. Provide a means to prevent the keyboard cable from becoming caught or tangled. Include detailed dimensioning and tolerancing.

■ TOLERANCING PROJECTS

Refer to Figures 11.49—11.53.

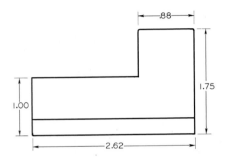

FIGURE 11.49   Proj. 11.1: Sketch the figure shown above. Use either limit dimensions, bilateral tolerances or geometric tolerancing to add a ø.375 hole to the left end of the part, located .50 inch from the bottom surface and 2 inches from the right end of the part. The location should be accurate to ±.005 and it's size accurate to within ±.002.

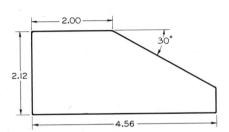

FIGURE 11.50   Proj 11.2: Add geometric dimensioning and tolerancing symbols to the drawing to so the following: a) Control the flatness of the bottom surface to a total tolerance of .001. b) Control perpendicularity of the left surface and bottom surface to .003. c) Control the tolerance for the 30° angle to .01.

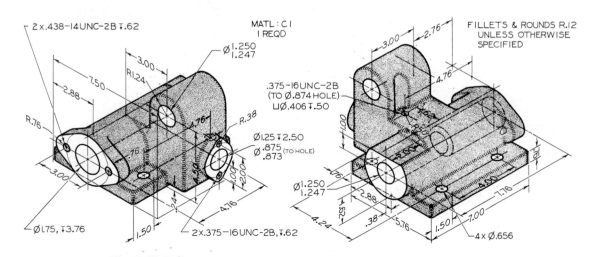

FIGURE 11.51   Automatic Stop Box.
Proj. 11.3: Create a detail drawing for the part shown in the two isometric views. Use standardized dimensioning and tolerancing symbols to replace notes as much as possible.

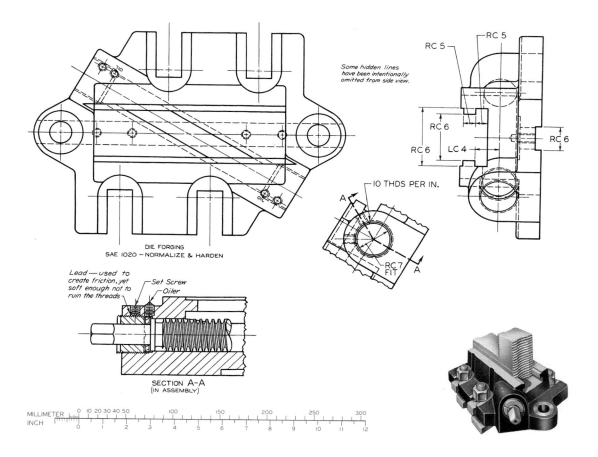

Some hidden lines
have been intentionally
omitted from side view.

RC 5

RC 5

RC 5

RC 6

RC 6

RC 6

LC 4

RC 6

DIE FORGING
SAE 1020 — NORMALIZE & HARDEN

10 THDS PER IN.

A

A

RC 7
FIT

Lead — used to
create friction, yet
soft enough not to
ruin the threads

Set Screw

Oiler

SECTION A–A
(IN ASSEMBLY)

MILLIMETER 0 10 20 30 40 50   100   150   200   250   300
INCH 0   1   2   3   4   5   6   7   8   9   10   11   12

FIGURE 11.52 Jaw Base for Chuck Jaw.
Proj. 11.4: Given: Top, right-side, and partial auxiliary views.
Required: Top, left-side (beside top), front, and partial auxiliary views
complete with dimensions, if assigned. Use metric or decimal-inch dimensions. Use American National Standard tables for indicated fits or convert
for metric values. See Appendixes 5–14.

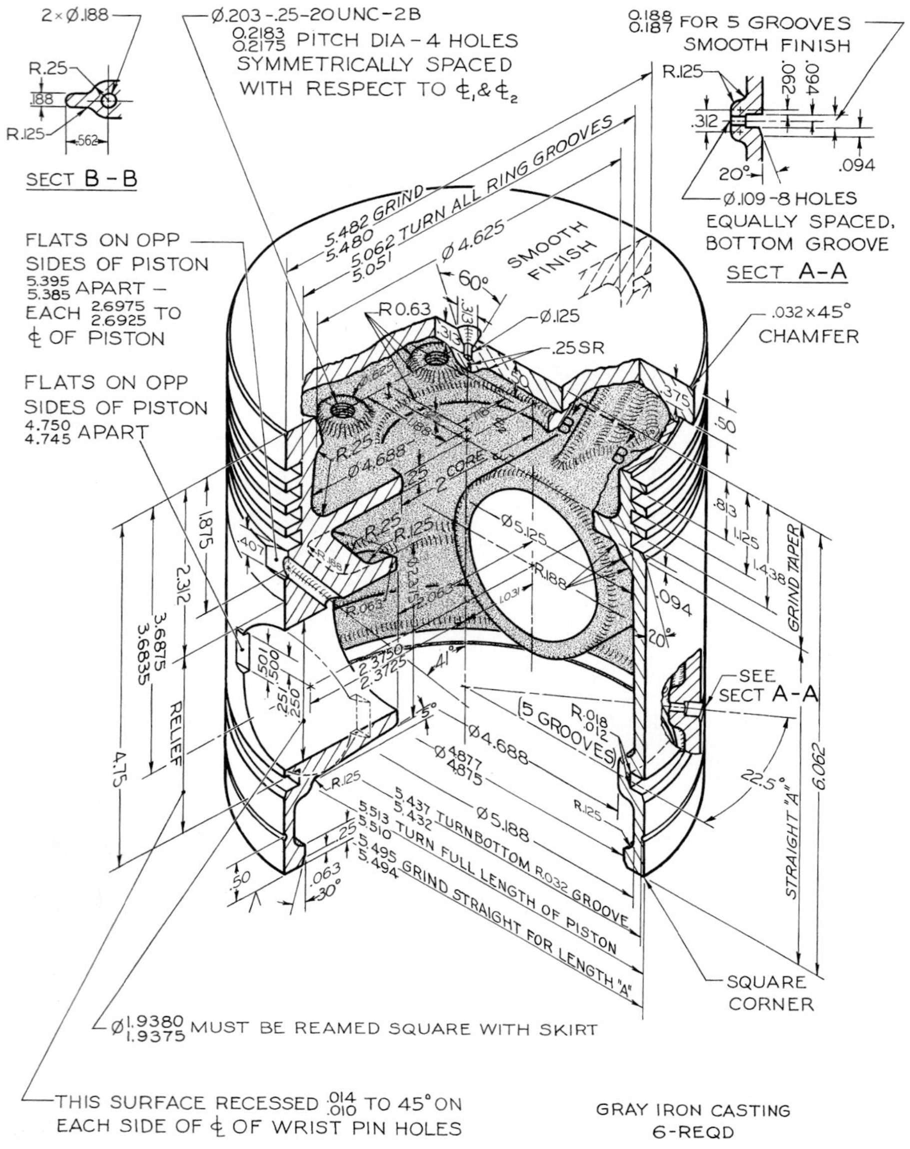

FIGURE 11.53   Caterpillar Tractor Piston.

Proj. 11.5: Make detail drawing full size on size C or A2 sheet. If assigned, use unidirectional decimal-inch system, converting all fractions to two-place decimal dimensions, or convert all dimensions to metric. Use standard symbols for dimensioning and tolerancing to replace notes.

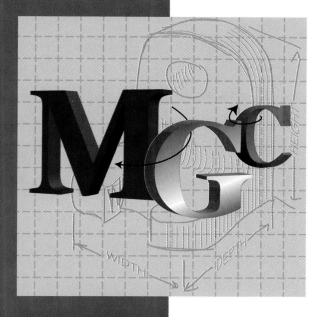

C H A P T E R   1 2

THREADS, FASTENERS, AND SPRINGS

OBJECTIVES

After studying the material in this chapter, you should be able to:

1. Define and label the parts of a screw thread.
2. Identify various screw thread forms.
3. Sketch detailed, schematic, and simplified threads.
4. Define typical thread specifications.
5. Identify various fasteners and describe their use.
6. Sketch various screw head types.
7. Sketch springs using break conventions.

OVERVIEW

The concept of the screw thread seems to have occurred first to Archimedes, the third-century-B.C. mathematician who wrote briefly on spirals and invented or designed several simple devices applying the screw principle. By the first century B.C., the screw was a familiar element but was crudely cut from wood or filed by hand on a metal shaft. Nothing more was heard of the screw thread until the 15th century.

Leonardo da Vinci understood the screw principle, and he created sketches showing how to cut screw threads by machine. In the 16th century, screws appeared in German watches and were used to fasten suits of armor. In 1569, the Frenchman Besson invented the screw-cutting lathe, but this method of screw production did not take hold for another century and a half; nuts and bolts continued to be made largely by hand. In the 18th century, during the Industrial Revolution, screw manufacturing started in England.

Threads and fasteners are the principal fastening devices used for assembling component parts. The shape of the helical thread is called the thread form. The metric thread form is the international standard, although the unified thread form is common in the United States. Other thread forms are used in specific applications. CAD drawing programs often use software that automatically depicts threads. The thread specification is a special leader note that defines the type of thread or fastener. This is an instruction for the shop technician so the correct type of thread is created during the manufacturing process.

To speed production time and reduce costs, many new types of fasteners are created every year. Existing fasteners are also modified to improve their insertion in mass production.

(a) Springs. (b) Screws and fasteners. *From* Machine Design: An Integrated Approach *by Robert Norton,* © *1996. Reprinted by permission of Prentice-Hall, Inc.*

■ 12.1 STANDARDIZED SCREW THREADS

In early times, there was no such thing as standardization. Nuts made by one manufacturer would not fit the bolts of another. In 1841 Sir Joseph Whitworth started crusading for a standard screw thread, and soon the Whitworth thread was accepted throughout England.

In 1864 the United States adopted a thread proposed by William Sellers of Philadelphia, but the Sellers nuts would not screw onto a Whitworth bolt or vice versa. In 1935 the American standard thread, with the same 60-degree V form of the old Sellers thread, was adopted in the United States. Still there was no standardization among countries. In peacetime it was a nuisance; in World War I it was a serious inconvenience; and in World War II the obstacle was so great that the Allies decided to do something about it. Talks began among the Americans, British, and Canadians, and in 1948 an agreement was reached on the unification of American and British screw threads. The new thread was called the *Unified screw thread*, and it represented a compromise between the American standard and Whitworth systems, allowing complete interchangeability of threads in three countries.

In 1946 an International Organization for Standardization (ISO) committee was formed to establish a single international system of metric screw threads. Consequently, through the cooperative efforts of the Industrial Fasteners Institute (IFI), several committees of the American National Standards Institute, and the ISO representatives, a metric fastener standard was prepared.[*]

Today screw threads are vital to our industrial life. They are designed for hundreds of different purposes. The three basic applications are as follows:

1. to hold parts together
2. to adjust parts with reference to each other
3. to transmit power.

[*] For a listing of ANSI standards for threads, fasteners, and springs, see Appendix 1.

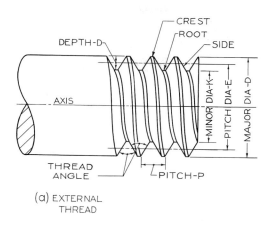

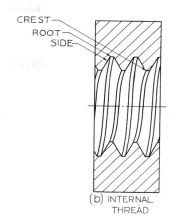

FIGURE 12.1 Screw Thread Nomenclature.

■ 12.2 SCREW THREAD TERMS

The following definitions apply to screw threads in general. Refer to Figure 12.1. For additional information regarding specific Unified and metric screw thread terms and definitions, refer to the following standards:

ANSI/ASME B1.1–1989
ANSI/ASME B1.7M–1984 (R1992)
ANSI/ASME B1.13M–1983 (R1989)
ANSI/ASME Y14.6–1978 (R1993)
ANSI/ASME Y14.6aM–1981 (R1993).

SCREW THREAD   A ridge of uniform section in the form of a helix on the external or internal surface of a cylinder.

EXTERNAL THREAD   A thread on the outside of a member, as on a shaft.

INTERNAL THREAD   A thread on the inside of a member, as in a hole.

MAJOR DIAMETER   The largest diameter of a screw thread (for both internal and external threads).

MINOR DIAMETER   The smallest diameter of a screw thread (for both internal and external threads).

PITCH   The distance from a point on a screw thread to a corresponding point on the next thread measured parallel to the axis. The pitch (P) is equal to 1 divided by the number of threads per inch.

PITCH DIAMETER   The diameter of an imaginary cylinder passing through threads where the widths of the threads and the widths of the spaces would be equal.

LEAD   The distance a screw thread advances axially in one turn.

ANGLE OF THREAD   The angle between the sides of the thread measured in a plane through the axis of the screw.

CREST   The top surface joining the two sides of a thread.

ROOT   The bottom surface joining the sides of two adjacent threads.

SIDE   The surface of the thread that connects the crest with the root.

AXIS OF SCREW   The longitudinal centerline through the screw.

DEPTH OF THREAD   The distance between the crest and the root of the thread measured normal to the axis.

FORM OF THREAD   The cross section of thread cut by a plane containing the axis.

SERIES OF THREAD   Standard number of threads per inch for various diameters.

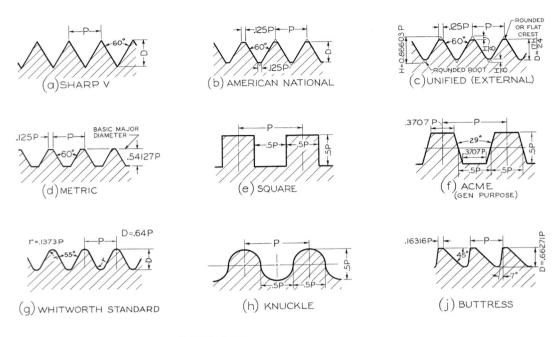

FIGURE 12.2 Screw Thread Forms.

■ 12.3 SCREW THREAD FORMS

The thread form is basically the shape of the thread. Various forms of threads are used for different purposes. The main uses for threads are to hold parts together, to adjust parts with reference to each other, and to transmit power. Figure 12.2 shows some of the typical thread forms:

Sharp-V thread (60 degrees) is useful for certain adjustments because of the increased friction resulting from the full thread face. It is also used on brass pipe work.

American national thread, with flattened roots and crests, is a stronger thread. This form replaced the sharp-V thread for general use.

Unified thread is the standard thread agreed upon by the United States, Canada, and Great Britain in 1948 and has replaced the American national form. The crest of the external thread may be flat or rounded, and the root is rounded; otherwise, the thread form is essentially the same as the American national. Some

earlier American national threads are still included in the new standard, which lists 11 different numbers of threads per inch for the various standard diameters, together with selected combinations of special diameters and pitches. The 11 series are the *coarse thread series* (UNC or NC), recommended for general use; the *fine thread series* (UNF or NF), for general use in automotive and aircraft work and in applications where a finer thread is required; the *extra fine series* (UNF or NF), which is the same as the SAE extra fine series, used particularly in aircraft and aeronautical equipment and generally for threads in thin walls; and the eight series of 4, 6, 8, 12, 16, 20, 28, and 32 threads with constant pitch. The 8UN or 8N, 12UN or 12N, and 16UN or 16N series are recommended for the uses corresponding to the old 8-, 12-, and 16-pitch American national threads. In addition, there are three special thread series—UNS, NS, and UN—that involve special combinations of diameter, pitch, and length of engagement.

The *unified extra fine thread* series (UNEF) has many more threads per inch for given diameters than any series of the American national or unified. The form of thread is the same as the American national. These small threads are used in thin metal where the length of thread engagement is small, in cases where close adjustment is required, and where vibration is great.

Metric thread is the standard screw thread agreed upon for international screw thread fasteners. The crest and root are flat, but the external thread is often rounded if formed by a rolling process. The form is similar to the American national and unified threads but with less depth of thread. The preferred metric thread for commercial purposes conforms to the ISO basic profile M for metric threads. This M profile design is comparable to the unified inch profile, but the two are not interchangeable. For commercial purposes, two series of metric threads are preferred—coarse (general purpose) and fine—much fewer than previously used.

Square thread is theoretically the ideal thread for power transmission, since its face is nearly at right angles to the axis, but due to the difficulty of cutting it with dies and because of other inherent disadvantages (such as the fact that split nuts will not readily disengage), square thread has been displaced to a large extent by the acme thread. Square thread is not standardized.

Acme thread is a modification of the square thread and has largely replaced it. It is stronger than the square thread, is easier to cut, and has the advantage of easy disengagement from a split nut, as on the lead screw of a lathe.

Standard worm thread (not shown) is similar to the acme thread but is deeper. It is used on shafts to carry power to worm wheels.

Whitworth thread was the British standard and has been replaced by the unified thread. The uses of Whitworth thread correspond to those of the American national thread.

Knuckle thread is usually rolled from sheet metal but is sometimes cast. In modified forms knuckle thread is used in electric bulbs and sockets, bottle tops, etc.

Buttress thread is designed to transmit power in one direction only. It is commonly used in large guns, in jacks, and in other mechanisms that have high-strength requirements.

Practical Tips
Chamfer on nuts and bolts

Nuts and bolts are usually chamfered (cut at an angle to form a beveled edge). This makes bolts easier to start into a threaded hole, and removes the sharp corners from nuts, making them easier to handle.

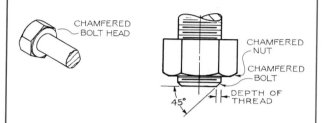

- The chamfer on bolts is usually at 45 degrees from the thread depth.

- The chamfer on nuts is shown as 30 degrees.

■ 12.4 THREAD SERIES

Five series of threads were used in the old ANSI standards:

1. *Coarse thread*—A general-purpose thread used for holding. Designated NC (national coarse).

2. *Fine thread*—A greater number of threads per inch; used extensively in automotive and aircraft construction. Designated NF (national fine).

3. *8-pitch thread*—All diameters have eight threads per inch. Used on bolts for high-pressure pipe flanges, cylinder-head studs, and similar fasteners. Designated 8N (national form, 8 threads per inch).

4. *12-pitch thread*—All diameters have 12 threads per inch; used in boiler work and for thin nuts on shafts and sleeves in machine construction. Designated 12N (national form, 12 threads per inch).

5. *16-pitch thread*—All diameters have 16 threads per inch; used where necessary to have a fine thread regardless of diameter, as on adjusting collars and bearing retaining nuts. Designated 16N (national form, 16 threads per inch).

■ 12.5 THREAD NOTES

Thread notes for metric, unified, and American national screw threads are shown in Figure 12.3. These same notes or symbols are used in correspondence, on shop and storeroom records, and in specifications for parts, taps, dies, tools, and gages.

Metric screw threads are designated basically by the letter M for metric profile followed by the nominal size (basic major diameter) and the pitch, both in millimeters and separated by the symbol ×. For example, the basic thread note M10 × 1.5 is adequate for most commercial purposes, as shown in Figure 12.3b. If needed, the class of fit and LH for left-hand designation is added to the note. (The absence of LH indicates an RH thread.)

If necessary, the length of the thread engagement—the letter S (short), N (normal), or L (long)—is added to the thread note. For example, the single note M10 × 1.5–6H/6g–N–LH combines the specifications for internal and external mating of left-hand metric threads of 10-mm diameter and 1.5-mm pitch with general-purpose tolerances and normal length of engagement.

If the thread is a multiple thread, the word DOUBLE, TRIPLE, or QUADRUPLE should precede the thread depth; otherwise, the thread is understood to be single.

A thread note for a blind tapped hole is shown in Figure 12.3a. A *tap drill* is sized to form a hole that will leave enough material for thread to be cut using a tap in order to form a threaded hole. In practice the tap drill size and depth are omitted and left up to the shop. In a complete note, the tap drill size and depth should be given. For tap drill sizes, see Appendix 15.

Thread notes for holes are preferably attached to the circular views of the holes. Thread notes for external threads are preferably given where the threaded shaft appears rectangular, as shown in Figures 12.3b–f. A sample special thread designation is 1–7N–LH.

General-purpose acme threads are indicated by the letter G, and centralizing acme threads by the letter C. Typical thread notes are 1–4 ACME–2G or 1–6 ACME–4C.

Thread notes for unified threads are shown in Figures 12.3j and 12.3k. Unified thread notes are distinguished from those for American national threads by the insertion of the letter U before the series letters, and by the letters A and B (for external or internal, respectively) after the numeral designating the class of fit. If the letters LH are omitted, the thread is understood to be RH. Some typical thread notes are:

–20 UNC–2A TRIPLE

–18 UNF–2B

1–16 UN–2A

■ 12.6 AMERICAN NATIONAL THREAD FITS

For general use, three classes of screw thread fits between mating threads (as between bolt and nut) have been established by ANSI.

These fits are produced by the application of tolerances listed in the standard and are as follows:

1. *Class 1 fit*—Recommended only for screw thread work where clearance between mating parts is essential for rapid assembly and where shake or play is not objectionable.

2. *Class 2 fit*—Represents a high quality of commercial thread product and is recommended for the great bulk of interchangeable screw thread work.

3. *Class 3 fit*—Represents an exceptionally high quality of commercially threaded product and is recommended only in cases where the high cost of precision tools and continual checking are warranted.

The standard for unified screw threads specifies tolerances and allowances defining the several classes of fit (degree of looseness or tightness) between mating threads. In the symbols for fit, the letter A refers to the external threads and B to internal threads. There are three classes of fit each for external threads (1A, 2A, 3A) and internal threads (1B, 2B, 3B). Classes 1A and 1B have generous tolerances, facilitating rapid assembly and disassembly. Classes 2A and 2B are used in the normal production of screws, bolts, and nuts, as well as in a variety of general applications. Classes 3A and 3B provide for applications needing highly accurate and close-fitting threads.

■ 12.7 METRIC AND UNIFIED THREAD FITS

Some specialized metric thread applications are specified by tolerance grade, tolerance position, class, and length of engagement. There are two general classes of metric thread fits. The first is for general-purpose applications and has a tolerance class of 6H for internal threads and a class of 6g for external threads. The second is used where closer fits are necessary and has a tolerance class of 6H for internal threads and a class of 5g6g for external threads. Metric thread tolerance classes of 6H/6g are generally assumed if not otherwise designated and are used in applications comparable to the 2A/2B inch classes of fits.

The single-tolerance designation of 6H refers to both the tolerance grade and position for the pitch

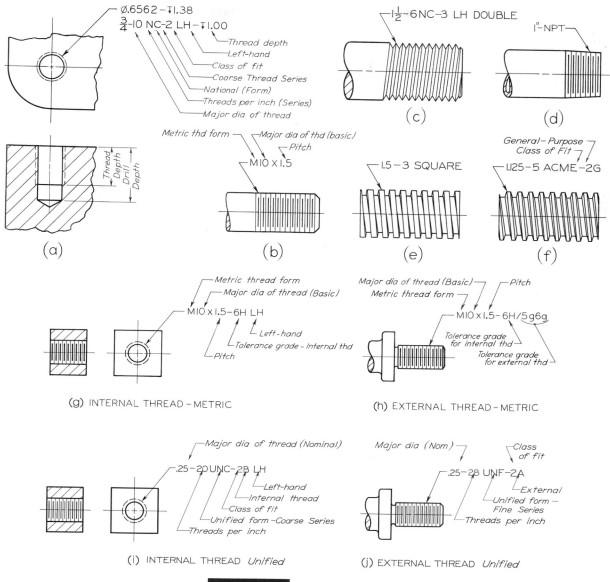

FIGURE 12.3 Thread Notes.

diameter and the minor diameter for an internal thread. The single-tolerance designation of 6g refers to both the tolerance grade and position for the pitch diameter and the major diameter of the external thread. A double designation of 5g6g indicates separate tolerance grades for the pitch diameter and for the major diameter of the external thread.

■ 12.8 THREAD PITCH

The *pitch* of any thread form is the distance parallel to the axis between corresponding points on adjacent threads, as shown in Figure 12.4.

For metric threads, this distance is specified in millimeters. The pitch for a metric thread that is included with the major diameter in the thread designation determines the size of the thread—for example, M10 x 1.5, as shown in Figure 12.4b.

For threads dimensioned in inches, the pitch is equal to 1 divided by the number of threads per inch. See Appendix 15 for thread tables giving more information on standard numbers of threads per inch for various thread series and diameters. For example, a unified coarse thread of 1-inch diameter has eight threads per inch, and the pitch (P) equals 1/8 inch (.125).

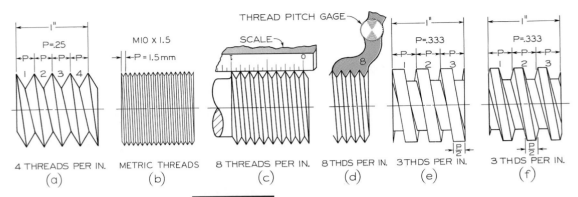

FIGURE 12.4 **Pitch of Threads.**

If a thread has only four threads per inch, the pitch and the threads themselves are quite large, as shown in Figure 12.4a. If there are 16 threads per inch, the pitch is only 1/16 inch (.063 inch), and the threads are relatively small, similar to those in Figure 12.4b.

The pitch or the number of threads per inch can be measured with a scale or with a *thread-pitch gage.*

■ 12.9 RIGHT-HAND AND LEFT-HAND THREADS

A *right-hand thread* is one that advances into a nut when turned clockwise, and a *left-hand thread* is one that advances into a nut when turned counterclockwise, as shown in Figure 12.5. A thread is always considered to be right-handed (RH) unless otherwise specified. A left-hand thread is always labeled LH on a drawing.

■ 12.10 SINGLE AND MULTIPLE THREADS

A *single thread*, as the name implies, is composed of one ridge, and the lead is therefore equal to the pitch. *Multiple threads* are composed of two or more ridges

running side by side. As shown in Figures 12.6a–c, the *slope line* is the hypotenuse of a right triangle whose short side equals .5P for single threads, P for double threads, 1.5P for triple threads, and so on. This applies to all forms of threads. In *double threads,* the lead is twice the pitch; in *triple threads,* the lead is three times the pitch, and so on. On a drawing of a single or triple thread, a root is opposite a crest; in the case of a double or quadruple thread, a root is drawn opposite a root. Therefore, in one turn, a double thread advances twice as far as a single thread, and a triple thread advances three times as far. RH double square and RH triple acme threads are shown in Figures 12.6d and 12.6e, respectively.

Multiple threads are used wherever quick motion, but not great power, is desired, as on ballpoint pens, toothpaste caps, valve stems, and so on. The threads on a valve stem are frequently multiple threads to impart quick action in opening and closing the valve. Multiple threads on a shaft can be recognized and counted by observing the number of thread endings on the end of the screw.

FIGURE 12.5 **Right-Hand and Left-Hand Threads.**

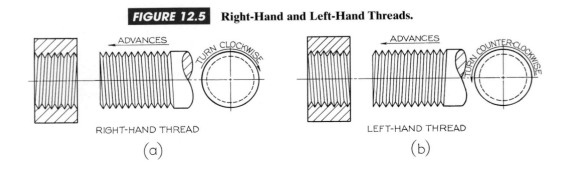

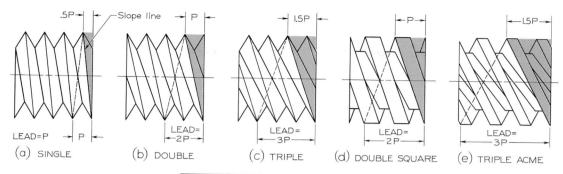

FIGURE 12.6 Multiple Threads.

■ 12.11 THREAD SYMBOLS

There are three methods of representing screw threads on drawings—the schematic, simplified, and detailed methods. *Schematic*, *simplified*, and *detailed* thread symbols may be combined on a single drawing.

Schematic and the more common simplified representations are used to show threads of small diameter, under approximately 1 inch or 25mm diameter on the plotted drawing. The symbols are the same for all forms of threads, such as metric, unified, square, and acme, but the thread specification identifies which is to be used.

Detailed representation is a closer approximation of the exact appearance of a screw thread, where the true profiles of the thread's form are drawn; but the helical curves are replaced by straight lines. The true projection of the helical curves of a screw thread takes too much time to draw, so it is rarely used in practice.

Do not use detailed representation unless the diameter of the thread on the drawing is more than 1 inch or 25 mm and then only to call attention to the thread when necessary. Schematic representation is much simpler to draw and still presents the appearance

of thread. Detailed representation is shown in Figure 12.7. Whether the crests or roots are flat or rounded, they are represented by single lines and not double lines, as in Figure 12.1; consequently, American national and unified threads are drawn the same way.

■ 12.12 EXTERNAL THREAD SYMBOLS

Simplified representation for external threads are shown in Figures 12.8a and 12.8b. The threaded portions are indicated by hidden lines parallel to the axis at the approximate depth of the thread, whether the cylinder appears rectangular or circular. The depth shown is not always the actual thread depth, just a representation of it. Use the table in Figure 12.10a for the general appearance of these lines.

When the schematic form is shown in section, as in Figure 12.8c, show the V's of the thread to make the thread obvious. It is not necessary to show the V's to scale or to the actual slope of the crest lines. To draw the V's, use the schematic thread depth, as shown in Figure 12.10a, and determine the pitch by drawing 60-degree V's.

FIGURE 12.7 Detailed Metric, American National, and Unified Threads.

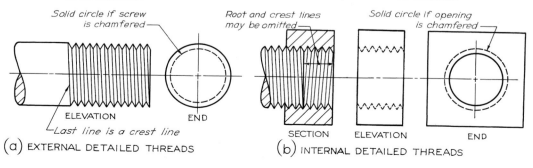

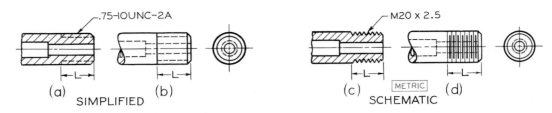

FIGURE 12.8 External Thread Symbols.

Schematic threads are indicated by alternate long and short lines, as shown in Figure 12.8d. The short lines representing the root lines are thicker than the long crest lines. Theoretically, the crest lines should be spaced according to actual pitch, but this would make them crowded and tedious to draw, defeating the purpose, which is to save time in sketching them. Space the crest lines carefully by eye, then add the heavy root lines halfway between the crest lines. Generally, lines closer together than about 1/16 inch are hard to distinguish. The spacing should be proportionate for all diameters. To help you gain a feeling for the correct proportions for the schematic symbols, they are given in Figure 12.10. You do not need to use these actual measurements in sketching schematic threads, just use them to get a feel for how far apart to make the lines.

■ 12.13 INTERNAL THREAD SYMBOLS

Internal thread symbols are shown in Figure 12.9. Note that the only differences between the schematic and simplified internal thread symbols occur in the sectional views. The representation of the schematic thread in section in Figures 12.9m, 12.9o, and 12.9p is exactly the same as the external representation shown

in Figure 12.8d. Hidden threads, by either method, are represented by pairs of hidden lines. The hidden dashes should be staggered, as shown.

In the case of blind tapped holes, the drill depth normally is drawn at least three schematic pitches beyond the thread length, as shown in Figures 12.9d, 12.9e, 12.9n, and 12.9o. The symbols in Figures 12.9f and 12.9p represent the use of a bottoming tap, when the length of thread is the same as the depth of drill. The thread length you sketch may be slightly longer than the actual given thread length. If the tap drill depth is known or given, draw the drill to that depth. If the thread note omits this information, as is often done in practice, sketch the hole three schematic thread pitches beyond the thread length. The tap drill diameter is represented approximately, not to actual size.

■ 12.14 DETAILED REPRESENTATION: METRIC, UNIFIED, AND AMERICAN NATIONAL THREADS

The detailed representation for metric, unified, and American national threads is the same, since the flats are disregarded.

FIGURE 12.9 Internal Thread Symbols.

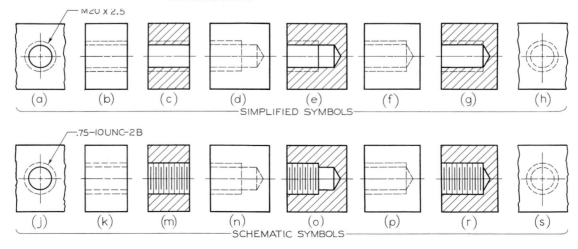

MAJOR DIAMETER	#5 (.125) TO #12 (.216)	.25	.3125	.375	.4375	.5	.5625	.625	.6875	.75	.8125	.875	.9375	1.
DEPTH, D	.03125	.03125	.03125	.0468	.0468	.0625	.0625	.0625	.0625	.0781	.0937	.0937	.0937	.0937
PITCH, P	.0468	.0625	.0625	.0625	.0625	.0937	.0937	.0937	.0937	.125	.125	.125	.125	.125

(For metric values: 1" = 25.4 mm or see inside front cover.)

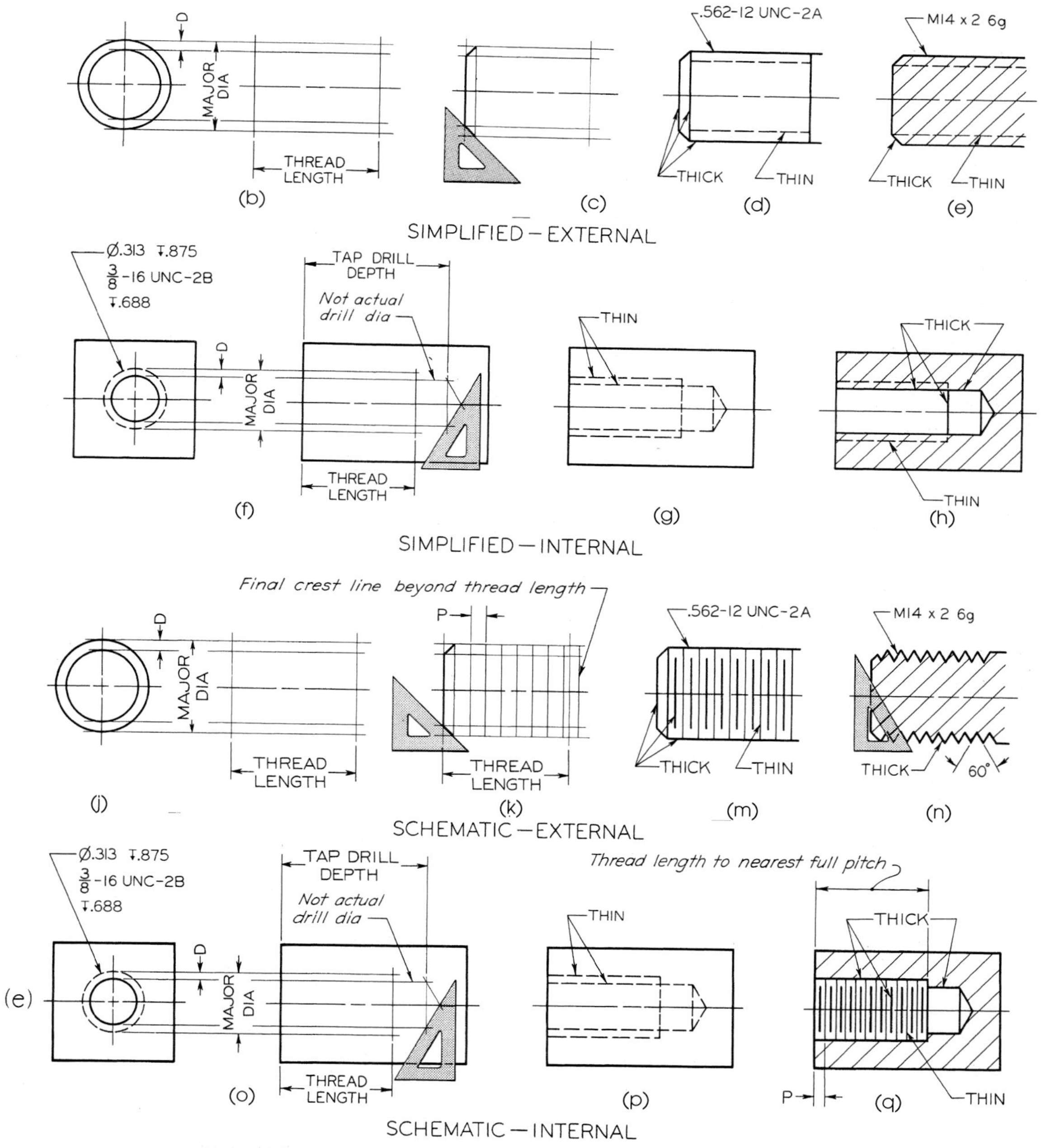

SIMPLIFIED – EXTERNAL

SIMPLIFIED – INTERNAL

SCHEMATIC – EXTERNAL

SCHEMATIC – INTERNAL

FIGURE 12.10   To Draw Thread Symbols—Simplified and Schematic.

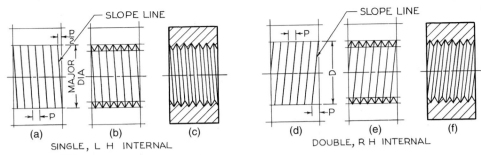

SINGLE, L H INTERNAL   DOUBLE, R H INTERNAL

FIGURE 12.11 Detailed Representation—Internal Metric, Unified, and American National Threads.

Internal detailed threads in section are drawn as shown in Figure 12.11. Notice that for LH threads the lines slope upward to the left (Figures. 12.11a to 12.11c), while for RH threads the lines slope upward to the right (Figs. 12.11d to 12.11f).

Figure 12.12 is an assembly drawing showing an external square thread partly screwed into a nut. The detail of the square thread at A is the same as in Step-by-Step 12.2. But when the external and internal threads are assembled, the thread in the nut overlaps and covers up half of the V, as shown at B.

The internal thread construction is the same as in Figure 12.13. Note that the thread lines representing the back half of the internal threads (since the thread is in section) slope in the opposite direction from those on the front side of the screw.

Steps in drawing a single internal square thread in section are shown in Figure 12.13. Note in Figure 12.13b that a crest is drawn opposite a root. This is the case for both single and triple threads. For double or quadruple threads, a crest is opposite a crest. Thus, the

construction in Figures 12.13a and 12.13b is the same for any multiple of thread. The differences appear in Figure 12.13c, where the threads and spaces are distinguished and outlined.

The same internal thread is shown in Figure 12.13e from an external view. The profiles of the threads are drawn in their normal position, but with hidden lines, and the sloping lines are omitted for simplicity. The end view of the same internal thread is shown in Figure 12.13f. Note that the hidden and solid circles are opposite those for the end view of the shaft.

◼ 12.15 DETAILED REPRESENTATION OF ACME THREAD

Detailed representation of acme threads is used only to call attention when details of the thread are important and the major diameter is larger than 1 inch or 25 mm on the drawing. The steps, shown in Figure 12.14, are as follows:

1. Make a centerline and lay out the length and major diameter of the thread, as shown in Figure 12.14a. Determine P by dividing 1 by the number of threads per inch (see Appendix 22). Make construction lines for the root diameter, making the thread depth P/2. Make construction lines halfway between crest and root guidelines.

2. On the intermediate construction lines, lay off spaces, as shown in Figure 12.14b.

3. Through alternate points, make construction lines for sides of threads at 15 degrees (instead of 14-1/2 degrees), as shown in Figure 12.14c.

4. Make construction lines for other sides of threads, as shown in Figure 12.14d. For single and triple threads, a crest is opposite a root, while for double and quadruple threads, a crest is opposite a crest. Finish tops and bottoms of threads.

5. Make parallel crest lines (Figure 12.14e).

FIGURE 12.12 Square Threads in Assembly.

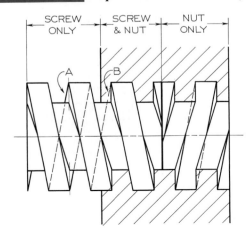

SCREW ONLY | SCREW & NUT | NUT ONLY

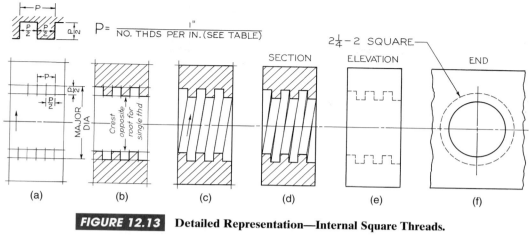

FIGURE 12.13 Detailed Representation—Internal Square Threads.

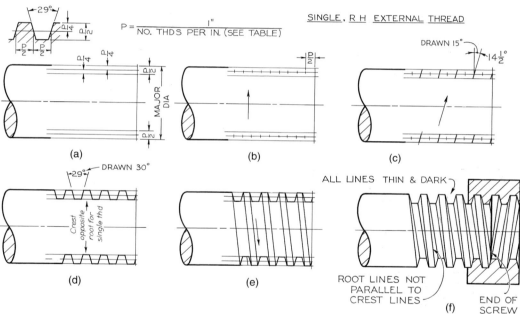

FIGURE 12.14 Detailed Representation—Acme Threads.

6. Make parallel root lines, and finish the thread profiles (Figure 12.14f). All lines should be thin and dark. The internal threads in the back of the nut will slope in the opposite direction to the external threads on the front side of the screw.

End views of acme threaded shafts and holes are drawn exactly like those for the square thread, as shown in Figures 12.12 and 12.13.

■ 12.16 USE OF PHANTOM LINES
Use phantom lines to save time when representing identical features, as shown in Figure 12.15. Threaded shafts and springs may be shortened without using conventional breaks, but must be correctly dimensioned. Phantom lines are not used much except in detail drawings.

■ 12.17 THREADS IN ASSEMBLY
Threads in an assembly drawing are shown in Figure 12.16. It is customary not to section a stud or a nut or any solid part unless necessary to show some internal shapes. When external and internal threads are sectioned in assembly, the V's are required to show the threaded connection.

Step by Step 12.1
Showing detailed thread

1. Make centerline and lay out length and major diameter as shown below.

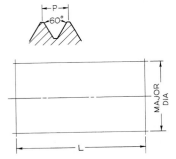

2. Find the number of threads per inch in Appendix 15 for American National and Unified Threads. This number depends on the major diameter of the thread and whether the thread is internal or external.

   Find P (pitch) by dividing 1 by the number of threads per inch. The pitch for metric threads is given directly in the thread designation. For example, the M14 x 2 thread has a pitch of 2 mm.

   Establish the slope of the thread by offsetting the slope line .5P for single threads, P for double threads, 1P for triple threads, and so on. For right-hand external threads, the slope line slopes upward to the left; for left-hand external threads, the slope line slopes upward to the right.

   By eye, mark off even spacing for the pitch. If using CAD, make a single thread and array the lines using the pitch as the spacing as shown below.

   P = PITCH *(See tables)*

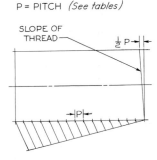

3. From the pitch points, make crest lines parallel to the slope line These should be dark, thin lines. Make two V's to establish the depth of thread, and sketch light guidelines for the root of thread, as shown.

   SINGLE, R H EXTERNAL THREAD

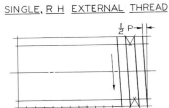

4. Finish the final 60-degree Vs. The Vs should be vertical; they should not lean with the thread.

   Make root lines as shown below. Root lines will not be parallel to crest lines, but should appear parallel to each other.

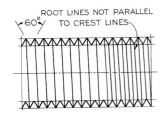

5. When the end is chamfered (usually 45 degrees with end of shaft, sometimes 30 degrees), the chamfer extends to the thread depth. The chamfer creates a new crest line, which you make between the two new crest points. It is not parallel to the other crest lines. When finished, all thread lines should be shown thin, but dark.

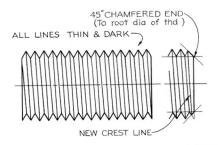

Step by Step 12.2
Detailed representation of square threads

Detailed representation of external square threads is only used when the major diameter is over about 1 inch or 25 mm, and it is important to show the detail of the thread on the finished sketch or plotted drawing. The steps to create detailed square thread are as follows:

1. Make a centerline and lay out the length and major diameter of the thread. Determine P by dividing 1 by the number of threads per inch (see Appendix 22). For a single RH thread, the lines slope upward to the left, and the slope line is offset as for all single threads of any form. On the upper line, use spacing equal to P/2, as shown.

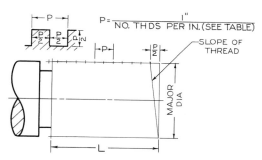

2. From the points on the upper line, draw guidelines for root of thread, making the depth as shown.

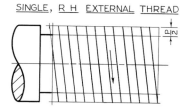

3. Make parallel visible back edges of threads.

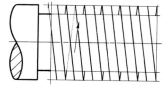

4. Make parallel visible root lines.

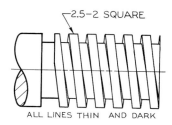

5. All lines should be thin and dark.

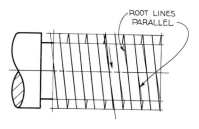

Practical Tip
End view of a shaft

The end view of the shaft illustrated in Step by Step 12.2 is shown below. Note that the root circle is hidden; no attempt is made to show the true projection.

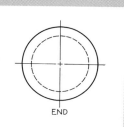

If the end of a shaft is chamfered, a solid circle would be drawn instead of the hidden circle.

Hands On 12.1

Sketching thread symbols

1. Sketch the proper threads on the sketches below using detailed representation.

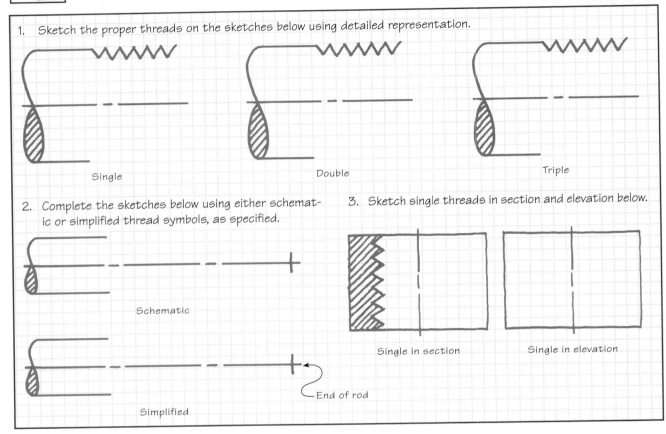

Single

Double

Triple

2. Complete the sketches below using either schematic or simplified thread symbols, as specified.

Schematic

Simplified

End of rod

3. Sketch single threads in section and elevation below.

Single in section

Single in elevation

FIGURE 12.15 Use of Phantom Lines.

■ 12.18 AMERICAN NATIONAL STANDARD PIPE THREADS

The American National Standard for pipe threads, originally known as the Briggs standard, was formulated by Robert Briggs in 1882. Two general types of pipe threads have been approved as American National Standard: *taper pipe threads* and *straight pipe threads*.

The profile of the tapered pipe thread is illustrated in Figure 12.17. The taper of the standard tapered pipe thread is 1 in 16, or .75″ per foot measured on the diameter and along the axis. The angle between the sides of the thread is 60°. The depth of the sharp V is

.8660p, and the basic maximum depth of the thread is .800p. The basic pitch diameters, E_0 and E_1, and the basic length of the effective external taper thread, L_2, are determined by the formulas

$$E_0 = D - (.050D + 1.1)1/n$$

$$E_1 = E + .0625L_1$$

$$L_2 = (.80D + 6.8)1/n$$

where D = basic outer diameter of pipe, E_0 = pitch diameter of thread at end of pipe, E_1 = pitch diameters of thread at large end of internal thread, L_1 = normal engagement by hand, and n = number of threads per inch.

ANSI also recommended two modified tapered pipe threads for (1) dry-seal pressure-tight joints (.880

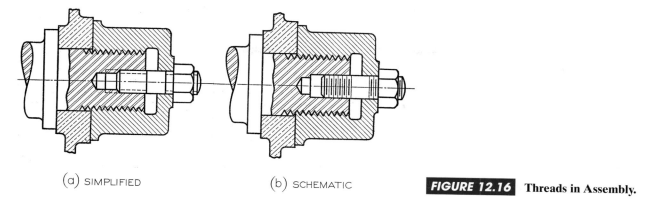

(a) SIMPLIFIED   (b) SCHEMATIC   **FIGURE 12.16**   **Threads in Assembly.**

per foot taper) and (2) rail-fitting joints. The former is used for metal-to-metal joints, eliminating the need for a sealer, and is used in refrigeration, marine, automotive, aircraft, and ordnance work. The latter is used to provide a rigid mechanical thread joint as is required in rail-fitting joints.

While tapered pipe threads are recommended for general use, there are certain types of joints in which straight pipe threads are used to advantage. The number of threads per inch, the angle, and the depth of thread are the same as on the tapered pipe thread, but the threads are cut parallel to the axis. Straight pipe threads are used for pressure-tight joints for pipe couplings, fuel and oil line fittings, drain plugs, free-fitting mechanical joints for fixtures, loose-fitting mechanical joints for locknuts, and loose-fitting mechanical joints for hose couplings.

Pipe threads are represented by detailed or symbolic methods in a manner similar to the representation of unified and American national threads. The symbolic representation (schematic or simplified) is recommended for general use regardless of the diameter, as shown in Figure 12.18. The detailed method is recommended only when the threads are large and when it is desired to show the profile of the thread, as for example, in a sectional view of an assembly.

As shown in Figure 12.18, it is not necessary to draw the taper on the threads unless there is some reason to emphasize it, since the thread note indicates whether the thread is straight or tapered. If it is desired to show the taper, it should be exaggerated, as shown in Figure 12.19, where the taper is drawn 1/16 inch per 1 inch *on radius* (or 6.75 inch per 1 foot on diameter) instead of the actual taper of 1/16 inch on diameter. American National Standard tapered pipe threads are indicated by a note giving the nominal diameter followed by the letters NPT (national pipe taper), as shown in Figure 12.19. When straight pipe threads are specified, the letters NPS (national pipe straight) are used. In practice, the tap drill size is normally not given in the thread note.

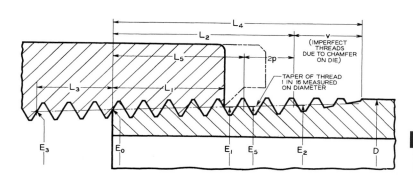

FIGURE 12.17 American National Standard Taper Pipe Thread [ANSI/ASME B1.20.1–1983 (R1992)].

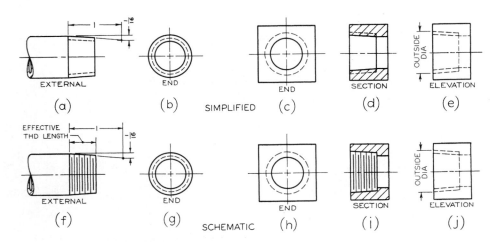

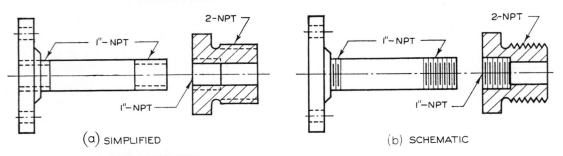

FIGURE 12.18 Conventional Pipe Thread Representation.

FIGURE 12.19 Conventional Representation of Pipe Threads.

■ 12.19 BOLTS, STUDS, AND SCREWS

The term **bolt** is generally used to denote a "through bolt" that has a head on one end, is passed through clearance holes in two or more aligned parts, and is threaded on the other end to receive a nut to tighten and hold the parts together, as shown in Figure 12.20a.

A hexagon-head **cap screw**, shown in Figure 12.20b, is similar to a bolt except it often has greater threaded length. It is often used when one of the parts being held together is threaded to act as a nut. The cap screw is screwed on with a wrench. Cap screws are not screwed into thin materials if strength is desired.

A **stud**, shown in Figure 12.20c, is a steel rod threaded on one or both ends. If threaded on both ends, it is screwed into place with a pipe wrench or with a stud driver. If threaded on one end, it is force fitted into place. As a rule, a stud is passed through a clearance hole in one member, is screwed into another member, and uses a nut on the free end, as shown.

A **machine screw** is similar to a slotted-head cap screw but usually smaller. It may be used with or without a nut. Figure 12.21 shows different screw head types.

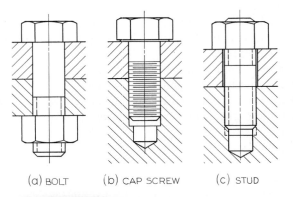

(a) BOLT (b) CAP SCREW (c) STUD

FIGURE 12.20 Bolt, Cap Screw, and Stud.

A **set screw** is a screw, with or without a head, that is screwed through one member and whose special point is forced against another member to prevent motion between the two parts.

Do not section bolts, nuts, screws, and similar parts when drawn in assembly because they do not have interior detail that needs to be shown.

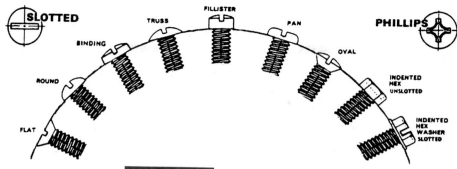

FIGURE 12.21 Types of Screwheads.

12.20 TAPPED HOLES

The bottom of a drilled hole, formed by the point of a twist drill, is cone-shaped, as shown in Figures 12.22a and 12.22b. When an ordinary drill is used to make holes that will be tapped, it is referred to as a tap drill. When drawing the drill point, use an angle of 30 degrees to approximate the actual 31-degree slope of the drill bit.

The thread length is the length of full or perfect threads. The tap drill depth does not include the cone point of the drill. In Figure 12.22c and 12.22d, the drill depth shown beyond the threads (labeled A) includes several imperfect threads produced by the chamfered end of the tap. This distance varies according to drill size and whether a plug tap or a bottoming tap is used to finish the hole.

A drawing of a tapped hole finished with a bottoming tap is shown in Figure 12.22e. Blind bottom-tapped holes are hard to form and should be avoided when possible. Instead, a relief with its diameter slightly greater than the major diameter of the thread is used, as shown in Figure 12.22f. Tap drill sizes for unified, American national, and metric threads are given in Appendix 15. Tap drill sizes and lengths may be given in the thread note, but are generally left to the manufacturer to determine. Since the tapped-thread length contains only full threads, it is necessary to make this length only one or two pitches beyond the end of the engaging screw. In simplified or schematic representation, don't show threads in the bottoms of tapped holes so that the ends of the screw show clearly, as in Figure 12.19b.

The thread length in a tapped hole depends on the major diameter and the material being tapped. The minimum engagement length (X), when both parts are steel, is equal to the diameter (D) of the thread. Table 12.1 shows different engagement lengths for different materials.

12.21 STANDARD BOLTS AND NUTS

American National Standard hexagon bolts and nuts[*] are made in both metric and inch sizes. Square bolts and nuts, shown in Figure 12.23, are only produced in inch sizes. Metric bolts, cap screws, and nuts also come in hexagon form. Square heads and nuts are chamfered at 30 degrees, and hexagon heads and nuts are chamfered at 15–30 degrees. Both are drawn at 30 degrees for simplicity.

BOLT TYPES   Bolts are grouped according to use: regular bolts for general use and heavy bolts for heavier use or easier wrenching. Square bolts come only in the regular type; hexagon bolts, screws, and nuts and square nuts are available in both regular and heavy.

Metric hexagon bolts are grouped according to use: regular and heavy bolts and nuts for general service and high-strength bolts and nuts for structural bolting.

---

[*] The ANSI standards cover several bolts and nuts. For complete details, see the standards.

FIGURE 12.22 Drilled and Tapped Holes.

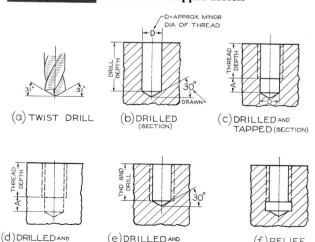

(a) TWIST DRILL

(b) DRILLED (SECTION)

(c) DRILLED AND TAPPED (SECTION)

(d) DRILLED AND TAPPED (ELEVATION)

(e) DRILLED AND BOTTOM TAPPED (SECTION)

(f) RELIEF

Practical Tips
Tapped holes

Preventing tap breakage

One of the chief causes of tap breakage is insufficient tap drill depth. When the depth is too short, the tap is forced against a bed of chips in the bottom of the hole. Don't specify a blind hole when a through hole of not much greater length can be used. When a blind hole is necessary, the tap drill depth should be generous.

Clearance holes

When a bolt or a screw passes through a clearance hole, the hole is often drilled 0.8 mm (1/32″) larger than the screw for screws of 3/8″ (10-mm) diameter and 1.5 mm (1/16″) larger for larger diameters. For more precise work, the clearance hole may be only 0.4 mm (1/64 inch) larger than the screw for diameters up to 3/8″ (10-mm) and 0.8 mm (1/32″) larger for larger diameters.

Closer fits may be specified for special conditions. The clearance spaces on each side of a screw or bolt need not be shown on a drawing unless it is necessary to show clearly that there is no thread engagement. When it is necessary to show that there is no thread engagement, the clearance spaces should be drawn about 1.2 mm (3/64″) wide.

HEXAGON BOLT　　　SQUARE BOLT
AND NUT　　　　　AND NUT
(a)　　　　　　　(b)

FIGURE 12.23　Standard Bolts and Nuts. Courtesy of Cordova Bolt Inc., Buena Park, CA.

Screw Material	Material of Parts	Thread Engagement
steel	steel	D
steel steel steel	cast iron brass bronze	1-1/2D
steel steel steel	aluminum zinc plastic	2D

TABLE 12.1　*Thread Engagement Lengths for Different Materials.*

FINISH　Square bolts and nuts, hexagon bolts, and hexagon flat nuts are *unfinished*. Unfinished bolts and nuts are not machined on any surface except for the threads. Hexagon cap screws, heavy hexagon screws, and all hexagon nuts, except hexagon flat nuts, are considered *finished* to some degree and have a "washer face" machined or otherwise formed on the bearing surface. The washer face is 1/64 inch thick (drawn 1/32 inch so that it will be visible on the plotted drawing), and its diameter is equal to 1.5 times the body diameter for the inch series.

For nuts, the bearing surface may also be a circular surface produced by chamfering. Hexagon screws and hexagon nuts have closer tolerances and a more finished appearance but are not completely machined. There is no difference in the drawing for the degree of finish on finished screws and nuts.

PROPORTIONS　Proportions for both inch and metric are based on the diameter (D) of the bolt body. These are shown in Figure 12.24.

For regular hexagon and square bolts and nuts, proportions are:

$$W = 1\text{-}1/2D \qquad H = 2/3\,D \qquad T = 7/8\,D$$

where W = width across flats, H = head height, and T = nut height.

For heavy hexagon bolts and nuts and square nuts, proportions are:

$$W = 1\text{-}1/2D + 1/8″ \,(\text{or} + 3\text{ mm})$$

$$H = 2/3\,D \qquad T = D$$

The washer face is always included in the head or nut height for finished hexagon screw heads and nuts.

THREADS　Square and hex bolts, hex cap screws, and finished nuts in the inch series are usually Class 2 and may have coarse, fine, or 8-pitch threads. Unfinished nuts have coarse threads and are Class 2B. For diameter and pitch specifications for metric threads, see Appendix 18.

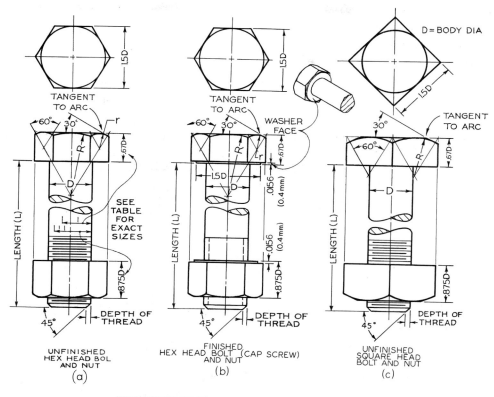

FIGURE 12.24 Bolt Proportions (Regular).

THREAD LENGTHS

For bolts or screws up to 6" (150 mm) in length:

Thread length = 2D + 1/4" (or + 6 mm)

For bolts or screws over 6" in length,

Thread length = 2D + 1/2" (or + 12 mm)

Fasteners too short for these formulas are threaded as close to the head as practical. For drawing purposes, use approximately three pitches. The threaded end may be rounded or chamfered, but it is usually drawn with a 45-degree chamfer from the thread depth, as shown in Figure 12.24.

BOLT LENGTHS Lengths of bolts have not been standardized because of the endless variety required by industry. Short bolts are typically available in standard length increments of 1/4 inch (6 mm), while long bolts come in increments of 1/2 inch to 1 inch (12 to 25 mm). For dimensions of standard bolts and nuts, see Appendix 18.

■ 12.22 DRAWING STANDARD BOLTS

Detail drawings show all of the necessary information defining the shape, size, material, and finish of a part.

Standard bolts and nuts do not usually require detail drawings unless they are to be altered (for example, by having a slot added through the end of a bolt) because they are usually stock parts that can easily be purchased. But you often need to show them on assembly drawings, which you will learn more about in Chapter 13.

Templates are available to help you add bolts quickly to sketches, or you can use the dimensions from Appendix 18 if accuracy is important, as in figuring clearances. In most cases a quick representation, where proportions are based on the body diameter, is sufficient. Three typical bolts illustrating the use of these proportions are shown in Figure 12.24.

Many CAD systems have fastener libraries that you can use to add a wide variety of nuts and bolts to your drawings. Often these symbols are based on a diameter of 1 inch so that you can quickly figure a scale at which to insert them. Other systems prompt for the diameter and lengths and create a symbol to your specifications. In 3-D models, when nuts and bolts are represented, the thread is rarely shown because it adds to the complexity and size of the drawing and is difficult to model. The thread specification is annotated in the drawing.

G R A P H I C S   S P O T L I G H T

Fastener Libraries

ENGINEERS SPEND TWENTY HOURS PER MONTH REDRAWING PARTS

Many engineers spend up to twenty hours each month redrawing standard parts or parts they purchase from outside vendors. They need to show how the parts fit in assembly or in order to specify which part to use. Using a library of these standard parts can save considerable time in creating engineering drawings and specifications. Some vendors are willing to provide free drawings for their parts in standard drawing format. Many resources are available on the World Wide Web. You can also purchase libraries of standard symbols.

THOUSANDS OF PARTS AVAILABLE IN PARTSPEC

One place you can go for standard fasteners and vendor part drawings is Autodesk's PartSpec software. PartSpec is an application that runs with AutoCAD

Manufacturer:	FASTENERS
Product:	(Selection Available)
Model:	

(Selection Available)
BINDING HEAD SCREW
BUTTON HEAD CAP SCREW
DOWEL PIN
FILLISTER HEAD SCREW
FLAT HEAD 100
FLAT HEAD 82
FLAT SOCKET
FLAT WASHER (A NARROW)
FLAT WASHER (A WIDE)
FLAT WASHER (B NARROW)
FLAT WASHER (B REGULAR)
FLAT WASHER (B WIDE)
HEX BOLT
HEX NUT
HEX NUT (HEAVY JAM)
HEX NUT (HEAVY)
HEX NUT (JAM)
HEX NUT (THICK)
HEX SCREW
HEX WASHER SCREW
LOCK WASHER (EXTRA DUTY)

(A)

Manufacturer:	FASTENERS
Product:	FILLISTER HEAD SCREW
Model:	ANSI-FRACTION

○ Drawing
○ Picture
○ Text

Sch   T
Back   L   F   R
Sect   B

Insert

Front View

SIZE: 1/4-20

Part Number:

Part Description: 1/4-20 FILLISTER HEAD SCREW

Previous

(B)

Release 12 or 13. You can use it to search through a part database containing thousands of parts on two CD-ROMs. The PartSpec window with standard fasteners selected is shown in Fig. A.

As you see in the figure above, you can select from various standard libraries and manufacturer databases. Once you have a library or manufacturer selected, you can choose from the list of products and the desired model for which drawings are available in the PartSpec database. Fig. B shows a fillister head screw, model ANSI-fraction, size 1/4-20, is shown as the selection. The front view drawing of the fillister head screw of that size is shown at the right of the dialog box. If you want to insert that view into your current AutoCAD drawing you can pick on the icon labeled Insert in the upper right. To select from other available views, you can pick on the T, F, B, L, R, Back, Sch, and Sect. buttons for the view desired. Drawings provided with PartSpec follow a set of standards to ensure their usability.

MANUFACTURER DRAWINGS AND ORDERING INFORMATION

Over twenty different manufactures are represented in the PartSpec database. You can select a manufacturer from the list available, and then pick a particular product and model or type in a part number or description used to search the database. You can also qualify the search by detailed information appropriate to the part. Once you have selected a part, you can get ordering information or manufacturer's specifications in text format. Fig. C shows Penn Engineering's Self-Clinching Flush Head Stud.

SAVE TIME SEARCHING FOR MATERIAL DATA WITH MATERIALSPEC

PartSpec has a counterpart for specifying materials, called MaterialSpec. It is a searchable text database on CD-ROM containing materials from five categories: plastics, metals, composites, ceramics, and military specifications (MIL5). You can choose materials by type, manufacturer, part name or number, description, property, or application. Add-on applications can provide a valuable resource for engineering. One of the biggest advantages of using CAD is that drawings can be re-used, re-scaled, or re-oriented for different purposes resulting in a valuable time savings. Remember the World Wide Web is also a valuable engineering resource for drawings and material information.

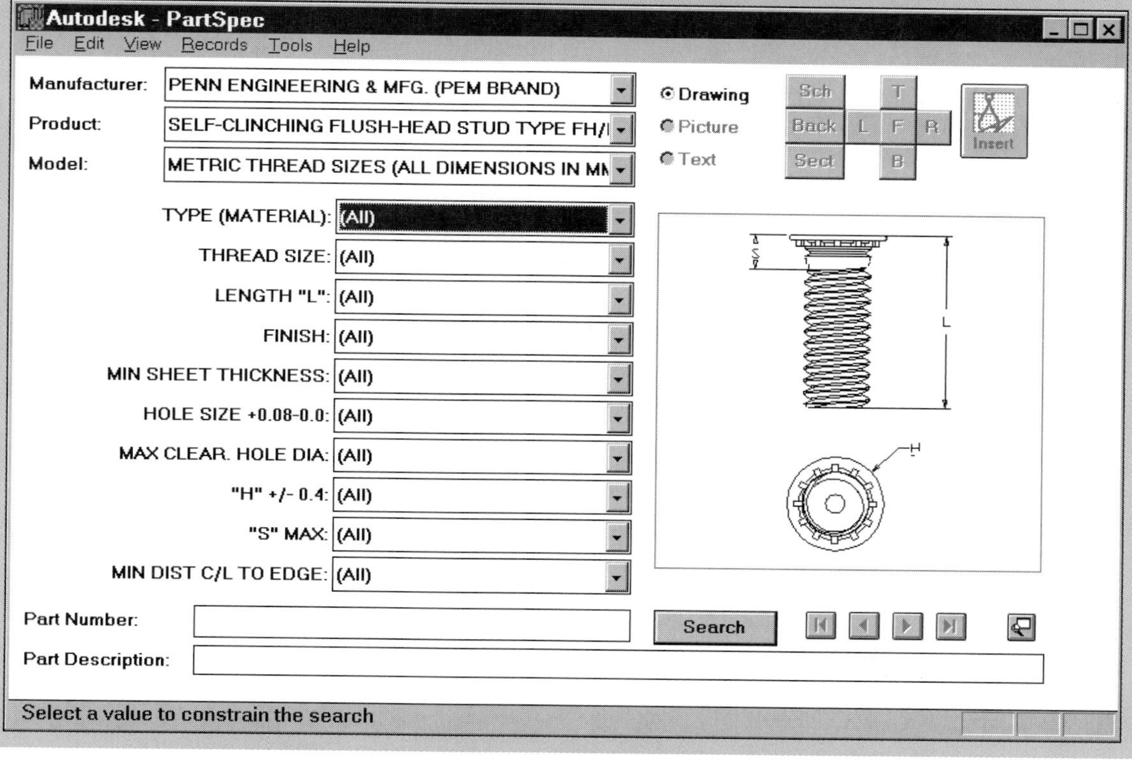

(C)

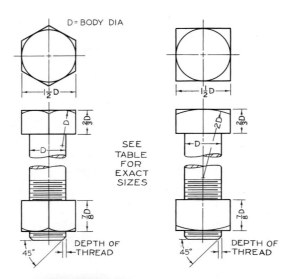

FIGURE 12.25 Bolts "Across Flats."

Generally, bolt heads and nuts should be drawn "across corners" in all views, regardless of projection. This conventional violation of projection is used to prevent confusion between the square and hexagon heads and nuts and to show actual clearances. Only when there is a special reason should bolt heads and nuts be drawn across flats, as shown in Figure 12.25.

■ 12.23 SPECIFICATIONS FOR BOLTS AND NUTS

In specifying bolts in parts lists, in correspondence, or elsewhere, the following information must be covered in order:

1. Nominal size of bolt body
2. Thread specification or thread note
3. Length of bolt
4. Finish of bolt
5. Style of head
6. Name

EXAMPLE (COMPLETE DECIMAL-INCH)

.75–10 UNC-2A X 2.50 HEXAGON CAP SCREW

EXAMPLE (ABBREVIATED DECIMAL-INCH)

.75 X 2.50 HEX CAP SCR

EXAMPLE (METRIC)

M8 X 1.25–40, HEX CAP SCR

Nuts may be specified as follows:

EXAMPLE (COMPLETE)

5/8–11 UNC-2B SQUARE NUT

EXAMPLE (ABBREVIATED)

5/8 SQ NUT

EXAMPLE (METRIC)

M8 X 1.25 HEX NUT

For either bolts or nuts, REGULAR or GENERAL PURPOSE are assumed if omitted from the specification. If the heavy series is intended, the word HEAVY should appear as the first word in the name of the fastener. Likewise, HIGH STRENGTH STRUCTURAL should be indicated for such metric fasteners. However, the number of the specific ISO standard is often included in the metric specifications—for example, HEXAGON NUT ISO 4032 M12 × 1.75. Finish need not be mentioned if the fastener or nut is correctly named.

■ 12.24 LOCKNUTS AND LOCKING DEVICES

Many types of special nuts and devices to prevent nuts from unscrewing are available, and some of the most common are shown in Figure 12.26. The American National Standard *jam nuts,* as shown in Figures 12.26a and 12.26b, are the same as the hexagon or hexagon flat nuts, except that they are thinner. The application shown in Figure 12.26b, where the larger nut is on top and is screwed on more tightly, is recommended. They are the same distance across flats as the corresponding hexagon nuts (1-1/2*D* or 1-1/2*D* + 1/8 inch). They are slightly over 1/2*D* in thickness but are drawn 1/2*D* for simplicity. They are available with or without the washer face in the regular and heavy types. The tops of all are flat and chamfered at 30 degrees, and the finished forms have either a washer face or a chamfered bearing surface.

The lock washer, shown in Figure 12.26c, and the cotter pin, shown in Figures 12.26e, 12.26g, and 12.26h, are very common (see Appendixes 27 and 30). The set screw, shown in Figure 12.26f, is often made to press against a plug of softer material, such as brass, which in turn presses against the threads without deforming them. For use with cotter pins (see Appendix 30), it is recommended to use a hex slotted nut (Figure 12.26g), a hex castle nut (Figure 12.26h), or a hex thick slotted nut or a heavy hex thick slotted nut.

Step by Step 12.3
Sketching hexagonal bolts, cap screws, and nuts

1. Determine the diameter of the bolt, the length (from the underside of the bearing surface to the tip), the style of head (square or hexagon), the type (regular or heavy), and the finish before starting to draw.

2. Lightly sketch the top view as shown, where D is the diameter of the bolt. Project the corners of the hexagon or square to the front view. Sketch the head and nut heights. Add the 1/64-inch (0.4-mm) washer face if needed. Its diameter is equal to the distance across flats of the bolt head or nut. Only the metric and finished hexagon screws or nuts have a washer face. The washer face is 1/64 inch thick, but is shown at about 1/32 inch (1 mm) for clearness. The head or nut height includes the washer face.

3. Represent the curves produced by the chamfer on the bolt heads and nuts as circular arcs, although they are actually hyperbolas. On drawings of small bolts or nuts under approximately 1/2 inch (12 mm) in diameter, where the chamfer is hardly noticeable, omit the chamfer in the rectangular view.

4. Chamfer the threaded end of the screw at 45 degrees from the schematic thread depth.

5. Show threads in simplified or schematic form for diameters of 1 inch (25 mm) or less on the drawing. Detailed representation is rarely used because it clutters the drawing and takes too much time.

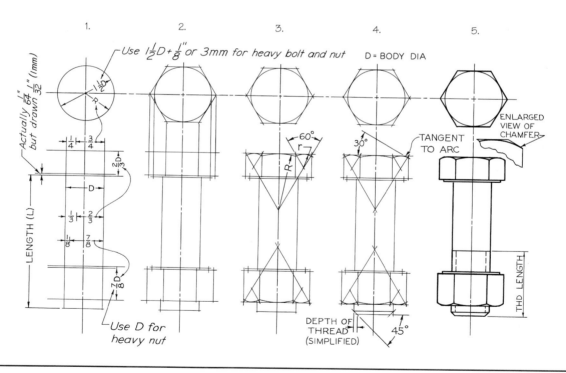

Similar metric locknuts and locking devices are available. See fastener catalogs for details.

Reid Tool is one company that has a free download of its catalog available as CAD files at http://www.reidtool.com/download.htm.

■ 12.25 STANDARD CAP SCREWS

Five types of American National Standard cap screws are shown in Figure 12.27. The first four of these have standard heads, while the socket head cap screws, as shown in Figure 12.27e, have several different shapes of round heads and sockets. Cap screws are normally finished and are used on machine tools and other

Step by Step 12.4
Sketching square bolts, cap screws, and nuts

1. Determine the diameter of the bolt, the length (from the underside of the bearing surface to the tip), the style of head (square or hexagon), the type (regular or heavy), and the finish before starting to draw.

2. Lightly sketch the top view as shown, where D is the diameter of the bolt. Project the corners of the hexagon or square to the front view. Sketch the head and nut heights. Add the 1/64-inch (0.4-mm) washer face if needed. Its diameter is equal to the distance across flats of the bolt head or nut. Only the metric and finished hexagon screws or nuts have a washer face. The washer face is 1/64 inch thick, but is shown at about 1/32 inch (1 mm) for clearness. The head or nut height includes the washer face.

3. Represent the curves produced by the chamfer on the bolt heads and nuts as circular arcs, although they are actually hyperbolas. On drawings of small bolts or nuts under approximately 1/2 inch (12 mm) in diameter, where the chamfer is hardly noticeable, omit the chamfer in the rectangular view.

4. Chamfer the threaded end of the screw at 45 degrees from the schematic thread depth.

5. Show threads in simplified or schematic form for diameters of 1 inch (25 mm) or less on the drawing. Detailed representation is rarely used because it clutters the drawing and takes too much time.

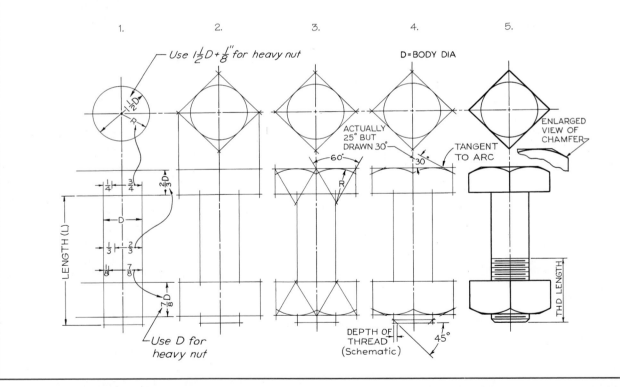

machines when accuracy and appearance are important. The ranges of sizes and exact dimensions are given in Appendixes 18 and 19. The hexagon head cap screw and hex socket head cap screw are also available in metric.

Cap screws ordinarily pass through a clearance hole in one member and screw into another. The clearance hole need not be shown on the drawing when the presence of the unthreaded clearance hole is obvious.

Cap screws are inferior to studs when frequent removal is necessary. They are used on machines requiring few adjustments. The slotted or socket-type heads are used for crowded conditions.

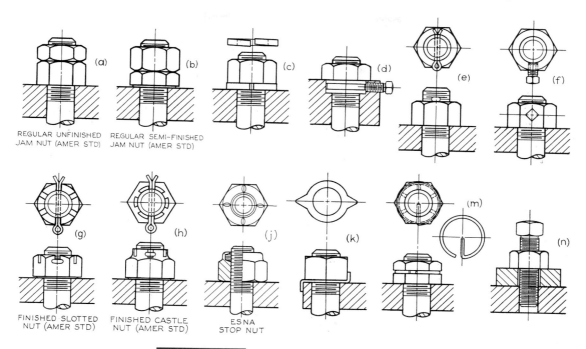

REGULAR UNFINISHED
JAM NUT (AMER STD)

REGULAR SEMI-FINISHED
JAM NUT (AMER STD)

FINISHED SLOTTED
NUT (AMER STD)

FINISHED CASTLE
NUT (AMER STD)

ESNA
STOP NUT

FIGURE 12.26 Locknuts and Locking Devices.

Actual dimensions may be used in drawing cap screws when exact sizes are necessary. Figure 12.27 shows the proportions in terms of body diameter (D) that are usually used. Hexagonal head cap screws are drawn similar to hex head bolts. The points are chamfered at 45 degrees from the schematic thread depth.

Note that screwdriver slots are drawn at 45 degrees in the circular views of the heads, without regard to true projection, and that threads in the bottom of the tapped holes are omitted so that the ends of the screws may be clearly seen. A typical cap screw note is as follows:

EXAMPLE (COMPLETE)

.375–16 UNC-2A X 2.5 HEXAGON HEAD CAP SCREW

EXAMPLE (ABBREVIATED)

.375 X 2.5 HEX HD CAP SCR

EXAMPLE (METRIC)

M20 X 2.5 X 80 HEX HD CAP SCR

■ 12.26 STANDARD MACHINE SCREWS

Machine screws are similar to cap screws but are usually smaller (.060-inch to .750-inch diameter) and the threads generally go all the way to the head. There are eight ANSI-approved forms of heads, which are shown in Appendix 20. The hexagonal head may be slotted if desired. All others are available in either slotted or recessed-head forms. Standard machine screws are produced with a naturally bright finish, not heat treated, and have plain-sheared ends, not chamfered. For similar metric machine screw forms and specifications, see Appendix 20.

Machine screws are used for screwing into thin materials, and all the smaller-numbered screws are threaded nearly to the head. They are used extensively in firearms, jigs, fixtures, and dies. Machine screw nuts are used mainly on the round-head, pan-head, and flat-head types and are usually hexagonal in form.

Exact dimensions of machine screws are given in Appendix 20, but they are seldom needed for drawing purposes. The four most common types of machine screws are shown in Figure 12.28, with proportions

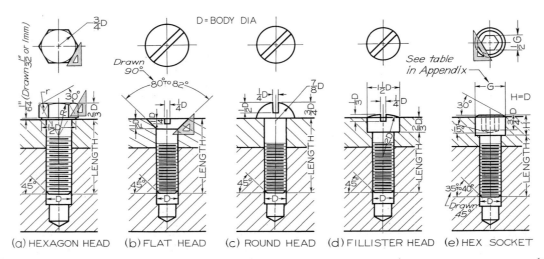

(a) HEXAGON HEAD   (b) FLAT HEAD   (c) ROUND HEAD   (d) FILLISTER HEAD   (e) HEX SOCKET

Hexagon Head Screws   Coarse, Fine, or 8-Thread Series, 2A. Thread length $= 2D + \frac{1}{4}''$ up to 6" long and $2D + \frac{1}{2}''$ if over 6" long. For screws too short for formula, threads extend to within $2\frac{1}{2}$ threads of the head for diameters up to 1". Screw lengths not standardized. For suggested lengths for metric Hexagon Head Screws, see Appendix 15.

Slotted Head Screws   Coarse, Fine, or 8-Thread Series, 2A. Thread length $= 2D + \frac{1}{4}''$. Screw lengths not standardized. For screws too short for formula, threads extend to within $2\frac{1}{2}$ threads of the head.

Hexagon Socket Screws   Coarse or Fine Threads, 3A. Coarse thread length $= 2D + \frac{1}{2}''$ where this would be over $\frac{1}{2}L$; otherwise thread length $= \frac{1}{2}L$. Fine thread length $= 1\frac{1}{2}D + \frac{1}{2}''$ where this would be over $\frac{3}{8}L$; otherwise thread length $= \frac{3}{8}L$. Increments in screw lengths $= \frac{1}{8}''$ for screws $\frac{1}{4}''$ to 1" long, $\frac{1}{4}''$ for screws 1" to 3" long, and $\frac{1}{2}''$ for screws $3\frac{1}{2}''$ to 6" long.

FIGURE 12.27   **Standard Cap Screws. See Appendixes 18 and 19.**

based on the diameter (D). Clearance holes and counterbores should be made slightly larger than the screws.

A typical machine screw note is as follows:

EXAMPLE (COMPLETE)

NO. 10 (.1900)–32 NF-3 X 5/8
FILLISTER HEAD MACHINE SCREW

EXAMPLE (ABBREVIATED)

NO. 10 (.1900) X 5/8 FILL HD MACH SCR

EXAMPLE (METRIC)

M8 X 1.25 X 30 SLOTTED PAN HEAD MACHINE SCREW

■ 12.27 STANDARD SET SCREWS

Set screws, shown in Figure 12.29, are used to prevent motion, usually rotary, between two parts, such as the movement of the hub of a pulley on a shaft. A set screw is screwed into one part so that its point bears

firmly against another part. If the point of the set screw is cupped, as shown in Figure 12.29e, or if a flat is milled on the shaft, as shown in Figure 12.29a, the screw will hold much more firmly. Obviously, set screws are not efficient when the load is heavy or when it is suddenly applied. Usually they are manufactured of steel and case hardened.

The American National Standard square-head set screw and slotted headless set screw are shown in Figures 12.29a and 12.29b. Two American National Standard socket set screws are illustrated in Figures 12.29c and 12.29d. American National Standard set screw points are shown in Figures 12.29e to 12.29k. Headless set screws have come into greater use because the projecting head of headed set screws has caused many industrial casualties; this has resulted in legislation prohibiting their use in many states.

Metric hexagon socket headless set screws with the full range of points are available. Nominal diameters of metric hex socket set screws are 1.6, 2, 2.5, 3, 4, 5, 6, 8, 10, 12, 16, 20, and 24 mm.

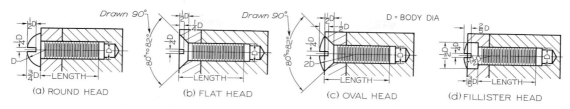

FIGURE 12.28 **Standard Machine Screws. See Appendix 20.**

Square-head set screws have coarse, fine, or 8-pitch threads and are Class 2A, but are usually furnished with coarse threads since the square-head set screw is generally used on the rougher grades of work. Slotted headless and socket set screws have coarse or fine threads and are Class 3A.

Nominal diameters of set screws range from number 0 up through 2 inches; set-screw lengths are standardized in increments of 1/32 inch to 1 inch depending on the overall length of the set screw.

Metric set-screw length increments range from 0.5 mm to 4 mm, again depending on overall screw length.

Set screws are specified as follows:

EXAMPLE (COMPLETE)

.375–16 UNC-2A × .75 SQUARE HEAD
FLAT POINT SET SCREW

EXAMPLE (ABBREVIATED)

.375 – × 1.25 SQ HD FL PT SS
.438 × .750 HEX SOC CUP PT SS
1/4 –20 UNC 2A × 1/2 SLOT. HDLS CONE PT SS

EXAMPLE (METRIC)

M10 × 1.5 × 12 HEX SOCKET HEAD SET SCREW

12.28 AMERICAN NATIONAL STANDARD WOOD SCREWS

Wood screws with three types of heads—flat, round, and oval—have been standardized. The approximate dimensions sufficient for drawing purposes are shown in Figure 12.30.

The Phillips-style recessed head is also available on several types of fasteners, as well as wood screws. Three styles of cross recesses have been standardized by ANSI. A special screwdriver is used, as shown in

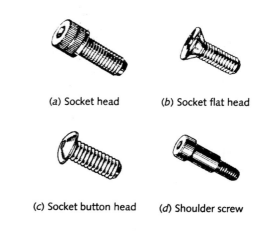

(a) Socket head   (b) Socket flat head

(c) Socket button head   (d) Shoulder screw

(e) Socket set screw

FIGURE 12.29 **American National Standard Set Screws.** *Courtesy of Cordova Bolt Inc., Buena Park, CA.*

Figure 12.31q, and this results in rapid assembly without damage to the head.

■ 12.29 MISCELLANEOUS FASTENERS

Many other types of fasteners have been devised for specialized uses. Some of the more common types are shown in Figure 12.31. A number of these are American National Standard round-head bolts, including carriage, button-head, step, and countersunk bolts.

Helical-coil-threaded inserts, as shown in Figure 12.31p, are shaped like a spring except that the cross section of the wire conforms to threads on the screw and in the hole. These are made of phosphor bronze or stainless steel, and they provide a hard, smooth protective lining for tapped threads in soft metals and plastics.

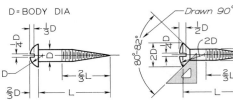

FIGURE 12.30 **American National Standard Wood Screws.**

■ 12.30 KEYS

Keys are used to prevent movement between shafts and wheels, couplings, cranks, and similar machine parts attached to or supported by shafts, as shown in Figure 12.32. A *keyseat* is in a shaft; a *keyway* is in the hub or surrounding part.

For heavy-duty functions, rectangular keys (flat or square) are used, and sometimes two rectangular keys are necessary for one connection. For even stronger connections, interlocking *splines* may be machined on the shaft and in the hole.

A *square key* is shown in Figure 12.32a, and a *flat key* in Figure 12.32b. The widths of keys are generally about one fourth the shaft diameter. In either case, one half the key is sunk into the shaft. The depth of the keyway or the keyseat is measured on the side—not the center—as shown in Figure 12.32a. Square and flat keys may have the top surface tapered 1/8 inch per foot, in which case they become square taper or flat taper keys.

A rectangular key that prevents rotary motion but permits relative longitudinal motion is a *feather key* and is usually provided with *gib heads*, or otherwise fastened so it cannot slip out of the keyway. A *gib head key*, shown in Figure 12.32c, is exactly the same as the square taper or flat taper key except that a gib head, which allows easy removal, is added. Square and flat keys are made from cold-finished stock and are not machined. For dimensions, see Appendix 21.

The *Pratt & Whitney key* (P&W key), shown in Figure 12.32d, is rectangular in shape with semicylindrical ends. Two-thirds of the height of the P&W key is sunk into the shaft keyseat (see Appendix 25).

The *Woodruff key* is semicircular in shape, as shown in Figure 12.33. The key fits into a semicircular key slot cut with a Woodruff cutter, as shown, and the top of the key fits into a plain rectangular keyway. Sizes of keys for given shaft diameters are not standardized, but for average conditions it is satisfactory to

FIGURE 12.31 **Miscellaneous Bolts and Screws.**

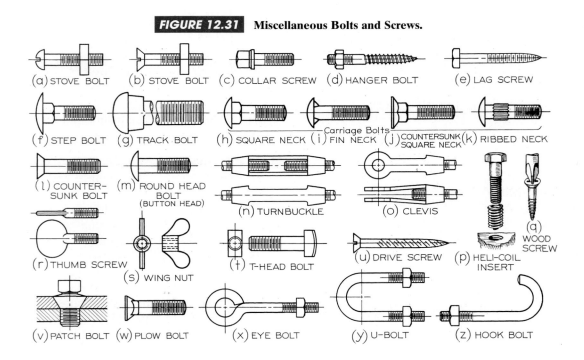

select a key whose diameter is approximately equal to the shaft diameter. For dimensions, see Appendix 23.

Typical specifications for keys are as follows:

.25 x 1.50 SQ KEY
No. 204 WOODRUFF KEY
1/4 x 1/16 x 1_ FLAT KEY
No. 10 P&W KEY

See manufacturers' catalogs for specifications for metric counterparts.

■ 12.31 MACHINE PINS

Machine pins include taper pins, straight pins, dowel pins, clevis pins, and cotter pins. For light work, the taper pin is effective for fastening hubs or collars to shafts, as shown in Figure 12.34, in which the hole through the collar and shaft is drilled and reamed when the parts are assembled. For slightly heavier duty, the taper pin may be used parallel to the shaft as for square keys (see Appendix 29).

Dowel pins are cylindrical or conical in shape and usually used to keep two parts in a fixed position or to preserve alignment. The dowel pin is most commonly used where accurate alignment is essential. Dowel pins are usually made of steel and are hardened and ground in a centerless grinder.

The clevis pin is used in a clevis and is held in place by a cotter pin. For the latter, see Appendix 30.

■ 12.32 RIVETS

Rivets are regarded as permanent fastenings as distinguished from removable fastenings, such as bolts and screws. Rivets are generally used to hold sheet metal or rolled steel shapes together and are made of wrought iron, carbon steel, copper, or occasionally other metals.

To fasten two pieces of metal together, holes are punched, drilled, or punched and then reamed, all slightly larger in diameter than the shank of the rivet. Rivet diameters are made from $d = 1.2 \sqrt{t}$ to $d = 1.4 \sqrt{t}$, where d is the rivet diameter and t is the metal thickness. The larger rivet diameter size is used for steel and single-riveted joints, and the smaller may be used for multiple-riveted joints. In structural work it is common practice to make the hole 1.6 mm (1/16 inch) larger than the rivet.

When the red-hot rivet is inserted, a "dolly bar" having a depression the shape of the driven head is held against the head. A riveting machine is then used to drive the rivet and to form the head on the plain end. This action causes the rivet to swell and fill the hole tightly.

FIGURE 12.32 Square and Flat Keys.

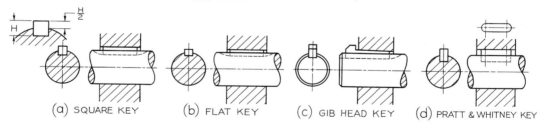

(a) SQUARE KEY   (b) FLAT KEY   (c) GIB HEAD KEY   (d) PRATT & WHITNEY KEY

FIGURE 12.33 Woodruff Keys and Key-Slot Cutter.

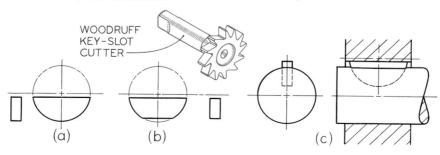

WOODRUFF KEY-SLOT CUTTER

(a)   (b)   (c)

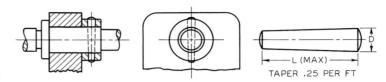

FIGURE 12.34 Taper Pin.

TAPER .25 PER FT

Large rivets or heavy hex structural bolts are often used in structural work of bridges and buildings and in ship and boiler construction, and they are shown in their exact formula proportions in Figure 12.35. The button head, as shown in Figure 12.35a, and countersunk head, as shown in Figure 12.35e, are the rivets most commonly used in structural work. The button head and cone head are commonly used in tank and boiler construction.

Typical riveted joints are illustrated in Figure 12.36. Notice that the rectangular view of each rivet shows the shank of the rivet with both heads made with circular arcs, and the circular view of each rivet is represented by only the outside circle of the head.

Since many engineering structures are too large to be built in the shop, they are built in the largest units possible and then are transported to the desired location. Trusses are common examples of this. The rivets driven in the shop are called *shop rivets*, and those driven on the job are called *field rivets*. However, heavy steel bolts are commonly used on the job for structural work. Solid black circles are used to represent field rivets, and other standard symbols are used to show other features, as shown in Figure 12.37.

For light work, small rivets are used. American National Standard small solid rivets are illustrated with dimensions showing their standard proportions in Figure 12.38 (ANSI/ASME B18.1.1–1972 (R1995)). Included in the same standard are tinners', coppers', and belt rivets. Metric rivets are also available.

Dimensions for large rivets can be found in ANSI/ASME B18.1.2–1972 (R1995). See manufacturers' catalogs for additional details.

Blind rivets, commonly known as Pop Rivets (Figure 12.39), are often used for fastening together thin sheet-metal assemblies. Blind rivets are hollow and are installed with manual or power-operated rivet guns which grip a center pin or mandrel, pulling the head into the body and expanding the rivet against the sheet metal. They are available in aluminum, steel, stainless steel, and plastic. As with any fastener, the designer should be careful to choose an appropriate material to avoid corrosive action between dissimilar metals.

■ 12.33 SPRINGS

A *spring* is a mechanical device designed to store energy when deflected and to return the equivalent amount of energy when released (ANSI Y14.13M–1981 (R1992)). Springs are commonly made of spring steel, which may be music wire, hard-drawn wire, or oil-tempered wire. Other materials used for compression springs include stainless steel, beryllium copper, and phosphor bronze. In addition, compression springs made of urethane plastic are used in applications where conventional springs would be affected by corrosion, vibration, or acoustic or magnetic forces. Springs are classified as *helical springs*, shown in Figure 12.40, or *flat springs*, shown in Figure 12.44. Helical springs are usually cylindrical but may also be conical.

FIGURE 12.35 Standard Large Rivets.

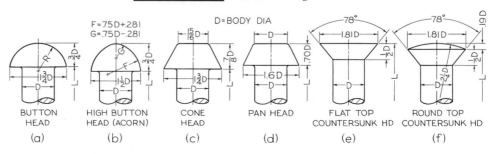

BUTTON HEAD	HIGH BUTTON HEAD (ACORN)	CONE HEAD	PAN HEAD	FLAT TOP COUNTERSUNK HD	ROUND TOP COUNTERSUNK HD
(a)	(b)	(c)	(d)	(e)	(f)

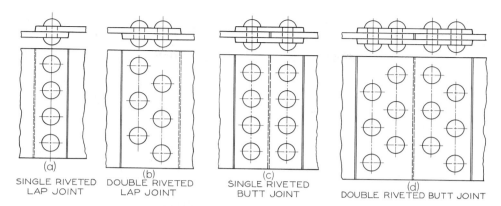

FIGURE 12.36   Common Riveted Joints.

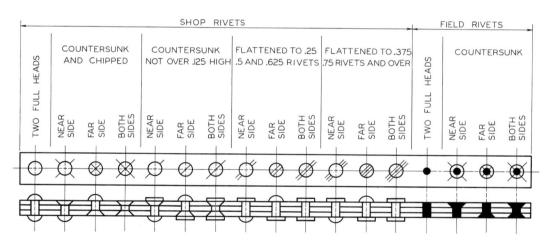

FIGURE 12.37   Conventional Rivet Symbols.

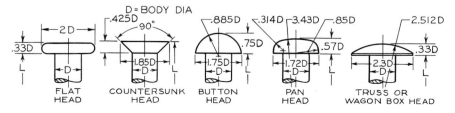

FIGURE 12.38   American National
Standard Small Solid Rivet
Proportions.

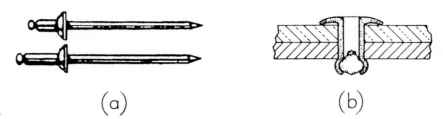

FIGURE 12.39 Blind Rivets (a) Before Installation, and (b) Installed.

(a)

(b)

There are three types of helical springs: *compression springs*, shown in Figure 12.41, which offer resistance to a compressive force; *extension springs*, shown in Figure 12.42, which offer resistance to a pulling force; and *torsion springs*, shown in Figure 12.43, which offer resistance to a torque or twisting force.

On working drawings, true projections of helical springs are never drawn because of the labor involved.

Instead, as in the drawing of screw threads, the detailed and schematic methods are used, where straight lines replace helical curves, as shown in Figure 12.40.

A square-wire spring is similar to the square thread with the core of the shaft removed, as shown in Figure 12.40b. Standard cross-hatching is used if the areas in section are large, as in Figures 12.40a and 12.40b. If these areas are small, the sectioned areas

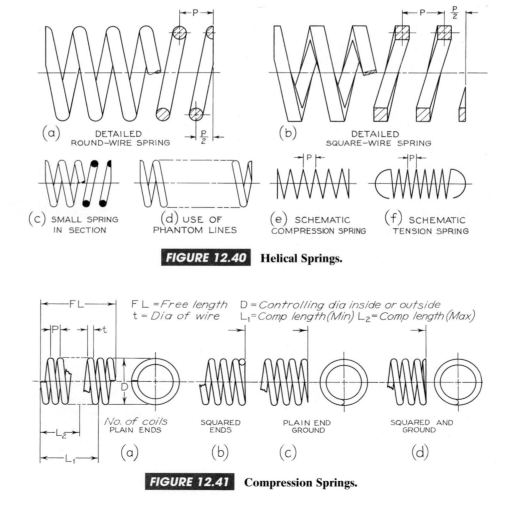

(a) DETAILED ROUND-WIRE SPRING

(b) DETAILED SQUARE-WIRE SPRING

(c) SMALL SPRING IN SECTION

(d) USE OF PHANTOM LINES

(e) SCHEMATIC COMPRESSION SPRING

(f) SCHEMATIC TENSION SPRING

FIGURE 12.40 Helical Springs.

FL = Free length   D = Controlling dia inside or outside
t = Dia of wire L_1= Comp length(Min) L_2= Comp length (Max)

No. of coils
PLAIN ENDS

SQUARED ENDS

PLAIN END GROUND

SQUARED AND GROUND

(a)

(b)

(c)

(d)

FIGURE 12.41   Compression Springs.

may be made solid black, as shown in Figure 12.40c. In cases where a complete picture of the spring is not necessary, use phantom lines to save time in drawing the coils, as shown in Figure 12.40d. If the drawing of the spring is too small to be represented by the outlines of the wire, use schematic representation, shown in Figures 12.40e and 12.40f.

Compression springs have plain ends, as shown in Figure 12.41a, or squared (closed) ends, as shown in Figure 12.41b. The ends may be ground, as shown in Figure 12.41c, or both squared and ground, as shown in Figure 12.41d. Required dimensions are indicated in the figure. When required, RH or LH is specified.

An extension spring may have any one of many types of ends, and it is therefore necessary to draw the spring or at least the ends and a few adjacent coils, as shown in Figure 12.42.

A typical torsion spring drawing is shown in Figure 12.43.

A typical flat spring drawing is shown in Figure 12.44. Other types of flat springs are *power springs* (or flat coil springs), *Belleville springs* (like spring washers), and *leaf springs* (commonly used in automobiles).

Many companies use a printed specification form to provide the necessary spring information, including data such as load at a specified deflected length, the load rate, finish, and type of service.

■ 12.34 DRAWING HELICAL SPRINGS

The construction for a schematic elevation view of a compression spring having six total coils is shown in Figure 12.45a. Since the ends are closed, or squared, two of the six coils are "dead" coils, leaving only four full pitches to be set off along the top of the spring.

If there are six total coils, as shown in Figure 12.45b, the spacings will be on opposite sides of the spring. The construction of an extension spring with six active coils and loop ends is shown in Figure 12.45c.

Figure 12.46 shows the steps in drawing a detailed section and elevation view of a compression spring. The given spring is shown pictorially in Figure 12.46a. Figure 12.46b shows the cutting plane through the centerline of the spring. In Figure 12.46c the cutting plane has been removed. Steps in constructing the sectional view are shown in Figures 12.46d–f. The corresponding elevation view is shown in Figure 12.46g.

If there is a fractional number of coils, such as the five coils in Figure 12.46h, the half-rounds of sectional wire are placed on opposite sides of the spring.

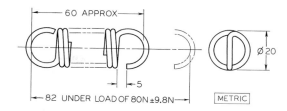

MATERIAL : 2.00 OIL TEMPERED SPRING STEEL WIRE
14.5 COILS RIGHT HAND
MACHINE LOOP AND HOOK IN LINE
SPRING MUST EXTEND TO 110 WITHOUT SET
FINISH : BLACK JAPAN

FIGURE 12.42 Extension Spring Drawing.

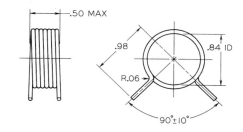

MATERIAL : .059 MUSIC WIRE
6.75 COILS RIGHT HAND NO INITIAL TENSION
TORQUE : 2.50 INCH LB AT 155° DEFLECTION SPRING MUST
DEFLECT 180° WITHOUT PERMANENT SET AND
MUST OPERATE FREELY ON .75 DIAMETER SHAFT
FINISH : CADMIUM OR ZINC PLATE

FIGURE 12.43 Torsion Spring Drawing.

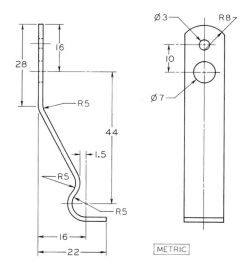

MATERIAL : 1.20 X 14.0 SPRING STEEL
HEAT TREAT : 44–48 C ROCKWELL
FINISH : BLACK OXIDE AND OIL

FIGURE 12.44 Flat Spring.

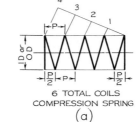
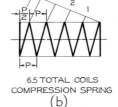
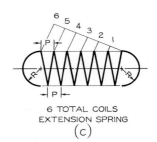

6 TOTAL COILS
COMPRESSION SPRING
(a)

6.5 TOTAL COILS
COMPRESSION SPRING
(b)

6 TOTAL COILS
EXTENSION SPRING
(c)

FIGURE 12.45   **Schematic Spring Representation. Courtesy of SDRC, Milford, OH.**

■ 12.35 COMPUTER GRAPHICS

Standard representations of threaded fasteners and springs, in both detailed and schematic forms, are available in CAD symbol libraries. Use of computer graphics frees the drafter from the need to draw time-consuming repetitive features by hand and also makes it easy to modify drawings if required.

In 3-D modeling, thread is not usually represented because it can be difficult to create and computer intensive to view and edit. Instead the nominal diameter of a threaded shaft or hole is usually created along with notation calling out the thread. Sometimes the depth of the thread is shown in the 3-D drawing to call attention to the thread and to help in determining fits and clearances.

FIGURE 12.46   **Steps in Detailed Representation of Spring.**

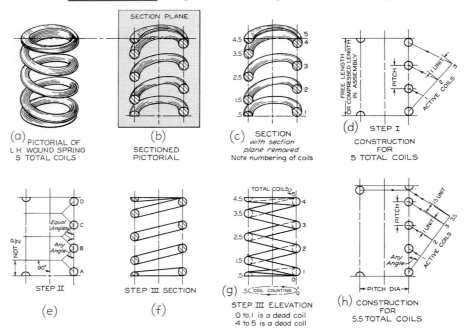

(a) PICTORIAL OF L H WOUND SPRING 5 TOTAL COILS

(b) SECTIONED PICTORIAL

(c) SECTION with section plane removed Note numbering of coils

(d) STEP I CONSTRUCTION FOR 5 TOTAL COILS

(e) STEP II

(f) STEP III SECTION

(g) STEP III ELEVATION 0 to.1 is a dead coil 4 to 5 is a dead coil

(h) CONSTRUCTION FOR 5.5 TOTAL COILS

■ KEY WORDS

unified screw thread	single thread	bolt	keys
tap drill	multiple threads	cap screw	keyseat
pitch	schematic	stud	keyway
right-hand thread	simplified	machine screw	rivet
left-hand thread	detailed	set screw	spring

■ CHAPTER SUMMARY

- There are many types of thread forms; however, metric and unified are the most common.
- The method of showing threads on a drawing is called the thread representation. The three types of thread representation are detailed, schematic, and simplified.
- The major diameter, pitch, and form are the most important parts of a thread specification.
- Thread specifications are dimensioned using a leader, usually pointing to the rectangular view of the threaded shaft or to the circular view of a threaded hole. The thread specification tells the manufacturing technician what kind of thread needs to be created.
- The nut and bolt is still the most common type of fastener. Many new types of fasteners are being created to streamline the production process.
- Keys and pins are special fasteners that attach a pulley to a shaft.
- The screw head determines what kind of tool will be necessary to install the fastener.

■ REVIEW QUESTIONS

1. Draw a typical screw thread using detailed representation, and label the parts of the thread.
2. Sketch a long spring and show how phantom lines are used to represent the middle part of the spring.
3. Draw several types of screw heads.
4. List five types of screws.
5. Why is the simplified thread representation the most commonly used drawing style?
6. List five fasteners that do not have any threads.
7. Write out a metric thread specification and a unified thread specification and label each part of the specification.
8. Which type of thread form is used on a light bulb?

■ THREAD AND FASTENER PROJECTS

Students are expected to make use of the information in this chapter and in various manufacturers' catalogs in connection with the working drawings at the end of the next chapter, where many different kinds of threads and fasteners are required. However, several projects are included here for specific assignment in this area (Figures 12.47 to 12.50).

■ DESIGN PROJECT

Design a system which uses thread to transmit power, for use in helping transfer a handicapped person from a bed to a wheelchair. Use either schematic or detailed representation to show the thread in your design sketches.

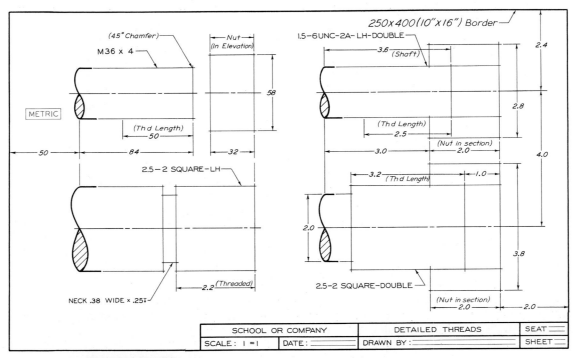

FIGURE 12.47   Draw specified detailed threads arranged as shown. Using Layout B–3 or A3–3. Omit all dimensions and notes given in inclined letters. Letter only the thread notes and the title strip.

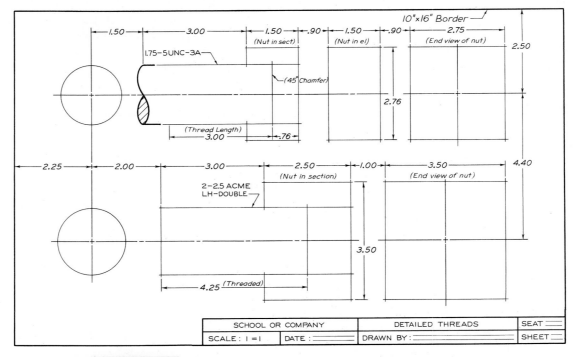

FIGURE 12.48   Draw specified detailed notes given in inclined letters. Letter only the thread notes and the title strip.

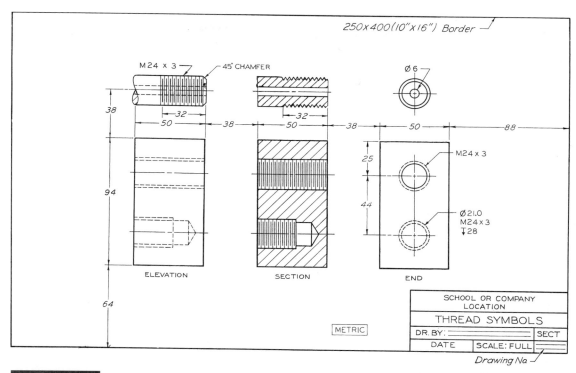

250x400(10"x16") Border

M24 X 3   45° CHAMFER   Ø 6

38   32   M24 x 3
50   38   32   50   38   50   88
94   25   Ø 21.0
M24 x 3
↧ 28
44
64

ELEVATION   SECTION   END

METRIC

SCHOOL OR COMPANY
LOCATION
THREAD SYMBOLS
DR. BY: SECT
DATE   SCALE: FULL
Drawing No

FIGURE 12.49 Draw specified thread symbols, arranged as shown. Draw simplified or schematic symbols, as assigned by instructor. Using Layout B–5 or A3–5. Omit all dimensions and notes given in inclined letters. Letter only the drill and thread notes, the titles of the views, and the title strip.

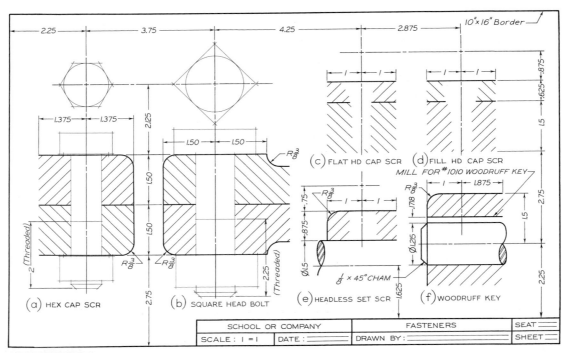

FIGURE 12.50   Draw fasteners, arranged as shown. At (a) draw $\frac{7}{8}$–9 UNC-2A × 4 Hex Cap Screw. At (b) draw $1\frac{1}{8}$–7 UNC-2A × $4\frac{1}{4}$ Sq Hd Bolt. At (c) draw $\frac{3}{8}$–16 UNC-2A × $1\frac{1}{2}$ Flat Hd Cap Screw. At (d) draw $\frac{7}{16}$–14 UNC-2A × 1 Fill Hd Cap Screw. At (e) draw $\frac{1}{2}$ × 1 Headless Slotted Set Screw. At (f) draw front view of No. 1010 Woodruff Key. Draw simplified or schematic thread symbols as assigned. Letter titles under each figure as shown.

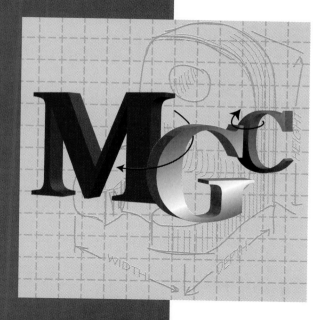

C H A P T E R   1 3

WORKING DRAWINGS

OBJECTIVES

After studying the material in this chapter, you should be able to:

1. Identify the elements of a detail drawing and create a simple detail drawing.
2. List the common elements of a title block and record strip.
3. Create a typical drawing sequence of numbers.
4. Describe the process for revising drawings.
5. List the parts of an assembly drawing.
6. Describe the special requirements of a patent drawing.

OVERVIEW

Working drawings consist of detail drawings, which show all of the necessary information to manufacture the necessary parts, and assembly drawings, which show how multiple parts fit together. They describe the end result of creating individual pieces that must fit together to work. Revising and approving drawings is an important part of the design process. Revisions must be tracked, identified, logged, and saved for future reference. Both paper and electronic storage is an important responsibility for the design team.

■ 13.1 WORKING DRAWINGS

When designing a product or system, a final set of production or working drawings and specifications that provide all of the necessary information must be made, checked, and approved. Working drawings are the specifications for the manufacture of a design and therefore must be correctly made and carefully checked.

Approved production design layouts are used to develop the working drawings. An example of a design layout drawing is shown in Figure 13.1. The necessary views for each part to be made are created, complete with dimensions and notes, so that the drawings describe the parts completely. These drawings of individual parts are known as *detail drawings*. Unaltered standard parts do not require a detail drawing but are shown on the assembly drawing and listed with specifications in the parts list. A detail drawing of one of the parts from the design layout shown in Figure 13.1 is shown in Figure 13.2.

After the parts have been detailed, an *assembly drawing* is made, showing how all the parts go together in the complete product. The assembly may be made directly from the detail drawings by inserting either 2-D views or 3-D models of the parts, or the assembly may be created from the original design layout. When

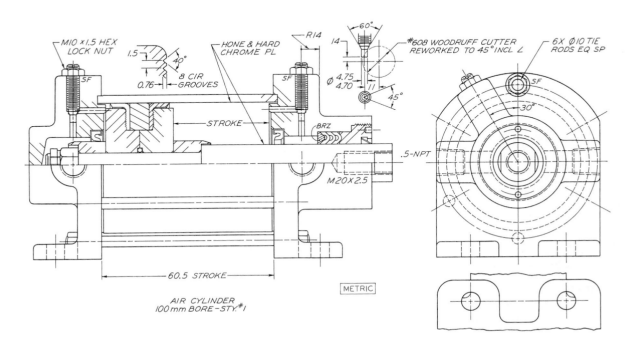

FIGURE 13.1 Design Layout.

2-D or 3-D parts are inserted to create the assembly drawing, it can be a valuable tool for checking fits within the assembly.

Finally, to protect the manufacturer, *patent drawings* are often prepared and filed with the U.S. Patent Office. Patent drawings are usually a type of assembly drawing. They often are shaded pictorial views and must follow rules of the Patent Office.

■ 13.2 DETAIL DRAWINGS

Up to this point in the text you have been learning the skills needed to correctly show individual parts for detail drawings in the design. Detail drawings should show all of the information necessary to manufacture the part. This includes multiple views providing the shape description, dimensions and notes, and the material specification.

■ 13.3 NUMBER OF DETAILS PER SHEET

Most companies show only one detail per sheet, however simple or small. The basic 8-1/2 x 11-inch or

210 mm x 297-mm sheet is most commonly used for details, multiples of these sizes being used for larger details or the assembly. This way if the same part is reused in a different assembly, it is not confusing. If the machine or structure is small or composed of few parts, all the details may be shown on one large sheet, or several details per sheet may be shown.

When several details are drawn on one sheet, you must carefully consider the spacing so that each part is presented clearly. The same scale should be used for all details on a single sheet, if possible. When this is not possible, the scale should be clearly noted under each detail.

■ 13.4 ASSEMBLY DRAWINGS

An assembly drawing shows the assembled machine or structure, with all detail parts in their functional positions. Assembly drawings are of different types: (1) design assemblies, or layouts, (2) general assemblies, (3) working drawing assemblies, (4) outline or installation assemblies, and (5) check assemblies.

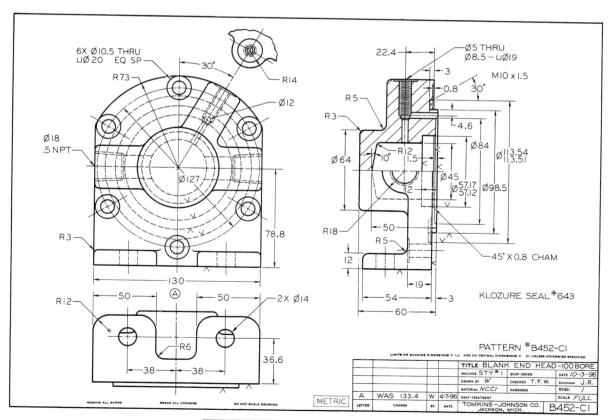

FIGURE 13.2 A Detail Drawing.

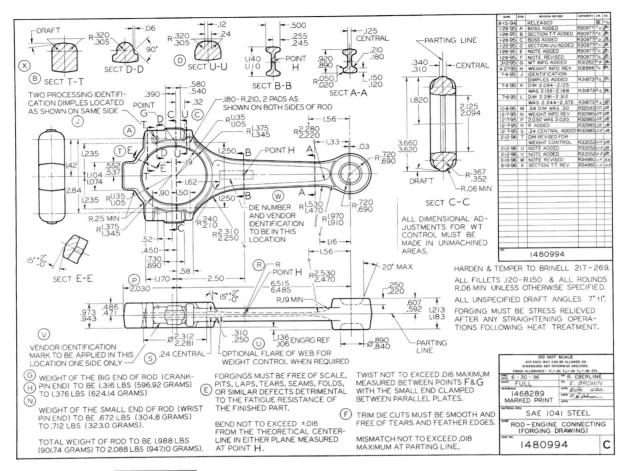

FIGURE 13.3 **Forging Drawing of Connecting Rod.** *Courtesy of Cadillac Motor Car Division.*

■ 13.5 GENERAL ASSEMBLIES

A set of working drawings includes the detail drawings of the individual parts and the assembly drawing of the assembled unit. The detail drawings of an automobile connecting rod are shown in Figures 13.3 and 13.4, and the corresponding assembly drawing is shown in Figure 13.5. Such an assembly, showing only one unit of a larger machine, is often referred to as a *subassembly*.

An example of a complete general assembly appears in Figure 13.6, which shows the assembly of a hand grinder. Another example of a subassembly is shown in Figure 13.7.

VIEWS   In selecting the views for an assembly drawing, keep in mind that the drawing's purpose is to show how the parts fit together and to show the function of the entire unit. Assembly drawings do not need to show details of each individual part. The assembly worker receives the actual finished parts. If information is needed that cannot be obtained from the part itself, the detail drawing must be checked. The assembly drawing purports to show *relationships* of parts, *not shapes*. The view or views selected should be the minimum or partial views that show how the parts fit together. In Figure 13.5, only one view is needed, while in Figure 13.6 two views are necessary.

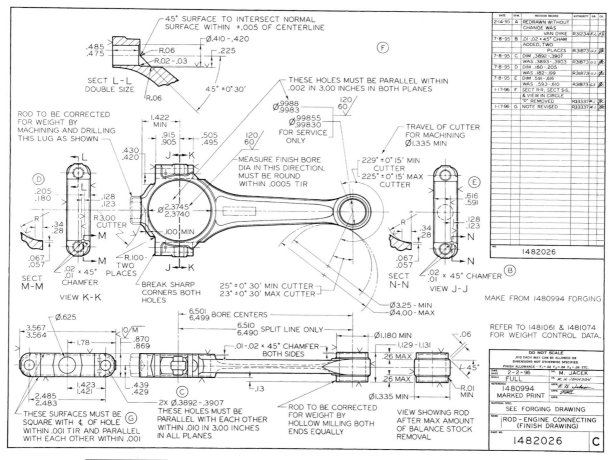

FIGURE 13.4   **Detail Drawing of Connecting Rod.** *Courtesy of Cadillac Motor Car Division.*

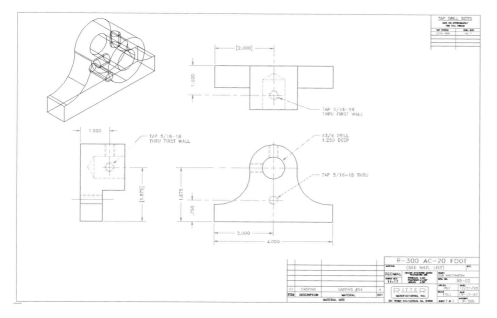

A Detail Drawing Generated from CAD Models. *Courtesy of Ritter Manufacturing.*

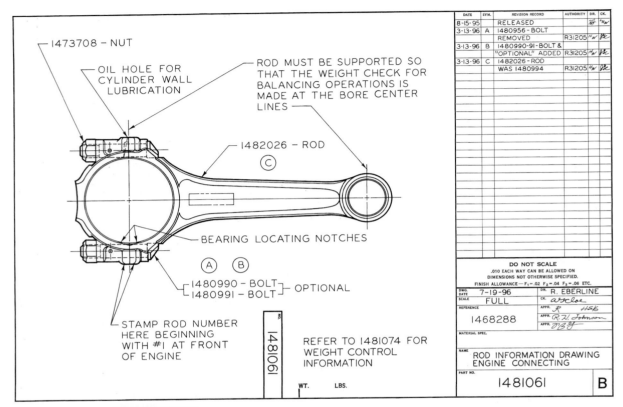

DATE	SYM.	REVISION RECORD	AUTHORITY	DR.	CK.
8-15-95		RELEASED			
3-13-96	A	1480956 - BOLT REMOVED	R31205		
3-13-96	B	1480990-91-BOLT & "OPTIONAL" ADDED	R31205		
3-13-96	C	1482026 - ROD WAS 1480994	R31205		

1473708 – NUT

OIL HOLE FOR CYLINDER WALL LUBRICATION

ROD MUST BE SUPPORTED SO THAT THE WEIGHT CHECK FOR BALANCING OPERATIONS IS MADE AT THE BORE CENTER LINES

1482026 – ROD
C

BEARING LOCATING NOTCHES

A   B

1480990 – BOLT
1480991 – BOLT   OPTIONAL

STAMP ROD NUMBER HERE BEGINNING WITH #1 AT FRONT OF ENGINE

NO. 1481061

REFER TO 1481074 FOR WEIGHT CONTROL INFORMATION

WT.   LBS.

DO NOT SCALE
.010 EACH WAY CAN BE ALLOWED ON DIMENSIONS NOT OTHERWISE SPECIFIED.
FINISH ALLOWANCE — $F_1 = .02$ $F_2 = .04$ $F_3 = .06$ ETC.

DWG. DATE	7-19-96	DR. R. EBERLINE
SCALE	FULL	CK.
REFERENCE	1468288	APPR. R HSB
		APPR. R. H. Johnson
		APPR.

MATERIAL SPEC.

NAME ROD INFORMATION DRAWING ENGINE CONNECTING

PART NO. 1481061   B

FIGURE 13.5 **Assembly Drawing of Connecting Rod.** *Courtesy of Cadillac Motor Car Division.*

SECTIONS   Section views are frequently used because they show interior fits clearly. Any kind of section may be used as needed. A broken-cut section is shown in Figure 13.6, a half section in Figure 13.7.

HIDDEN LINES   Hidden lines are rarely used in assembly drawings because parts fitting into or overlapping others need to be shown clearly. However, they should be used wherever necessary for clarity.

DIMENSIONS   Generally, dimensions are not given on assembly drawings since they are given completely on the detail drawings. Dimensions should be given when they are needed to show some function of the object as a whole, such as the maximum height of a jack or the maximum opening between the jaws of a vise or a distance that must be maintained during assembly. When machining is required in the assembly operation, the necessary dimensions and notes may be given on the assembly drawing.

IDENTIFICATION   Parts in an assembly are identified by circles containing the part numbers and placed adjacent to the parts, with leaders terminated by arrowheads touching the parts, as shown in Figure 13.6. Place these ball tags in neat horizontal or vertical rows as much as possible and not scattered over the sheet. Leaders are never allowed to cross, and adjacent leaders should be parallel or nearly parallel.

The parts list includes the part numbers or symbols, a descriptive title of each part, the number required per machine or unit, the material specified, and frequently other information, such as pattern numbers, stock sizes, and weights.

Another method of identification is to letter the part names, numbers required, and part numbers at the end of leaders, as shown in Figure 13.7. It is more common to give only the part numbers, together with ANSI-approved straight-line leaders.

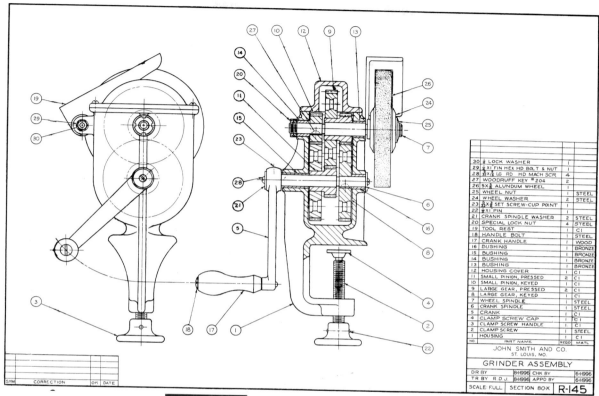

NO.	PART NAME	REQD	MATL
30	¼ LOCK WASHER	1	
29	¼ X1"FIN HEX HD BOLT & NUT	1	
28	¼ X ½ LG RD HD MACH SCR	4	
27	WOODRUFF KEY #204	2	
26	5X½ ALUNDUM WHEEL	1	
25	WHEEL NUT	1	STEEL
24	WHEEL WASHER	2	STEEL
23	⅝X⅜ SET SCREW-CUP POINT	1	
22	¼"X1 PIN	1	
21	CRANK SPINDLE WASHER	2	STEEL
20	SPECIAL LOCK NUT	4	STEEL
19	TOOL REST	1	C1
18	HANDLE BOLT	1	STEEL
17	CRANK HANDLE	1	WOOD
16	BUSHING	1	BRONZE
15	BUSHING	1	BRONZE
14	BUSHING	1	BRONZE
13	BUSHING	1	BRONZE
12	HOUSING COVER	1	C1
11	SMALL PINION, PRESSED	2	C1
10	SMALL PINION, KEYED	1	C1
9	LARGE GEAR, PRESSED	2	C1
8	LARGE GEAR, KEYED	1	C1
7	WHEEL SPINDLE	1	STEEL
6	CRANK SPINDLE	1	STEEL
5	CRANK	1	C1
4	CLAMP SCREW CAP	1	C1
3	CLAMP SCREW HANDLE	1	C1
2	CLAMP SCREW	1	STEEL
1	HOUSING	1	C1

JOHN SMITH AND CO.
ST. LOUIS, MO.

GRINDER ASSEMBLY

DR BY	8H1996	CHK BY		6H1996
TR BY	R D J	8H1996	APPD BY	6H1996
SCALE: FULL	SECTION 80X		R-145	

FIGURE 13.6 Assembly Drawing of Grinder.

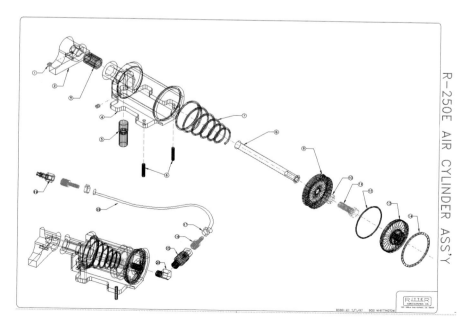

R−250E AIR CYLINDER ASS'Y

RITTER

Assembly Drawing Generated from CAD Models. *Courtesy of Ritter Manufacturing and Autodesk.*

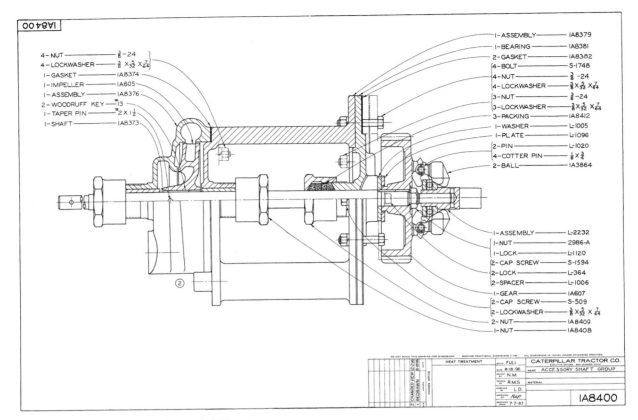

FIGURE 13.7 Subassembly of Accessory Shaft Group.

■ 13.6 PARTS LISTS

A bill of material, or *parts list*, is an itemized list of the various parts of a structure shown on a detail drawing or an assembly drawing (ANSI Y14.34M–1982 (R1988)). This list is often given on a separate sheet, but is frequently lettered directly on the drawing so that it doesn't become separated and lost. The parts list is usually shown in the lower or upper right of the drawing. An example is shown in Figure 13.9.

Parts lists on machine drawings contain the part numbers or symbols, a descriptive title of each part, the number required, the material specified, and frequently other information, such as pattern numbers, stock sizes of materials, and weights of parts.

Parts are listed in general order of size or importance. The main castings or forgings are listed first, parts cut from cold-rolled stock second, and standard parts such as fasteners, bushings, and roller bearings third. If the parts list rests on top of the title box or strip, the order of the items should be from the bottom upward, as shown in Figure 13.9, so that new items can

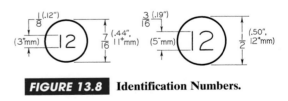

FIGURE 13.8 Identification Numbers.

be added later, if necessary. If the parts list is placed in the upper-right corner, the items should read from the top downward.

Each detail on the drawing may be identified with the parts list by the use of a small circle, or ball tag, containing the part number, placed adjacent to the detail, as shown in Figure 13.9. Figure 13.8 gives two different sizes for drawing ball tags that are commonly used, depending on the size of the drawing.

Standard parts, whether purchased or company produced, do not require detail drawings, but may be

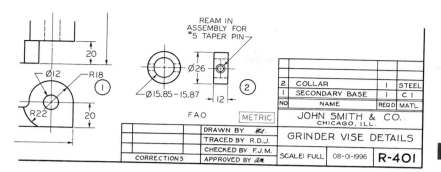

FIGURE 13.9 Identification of Details with Parts List.

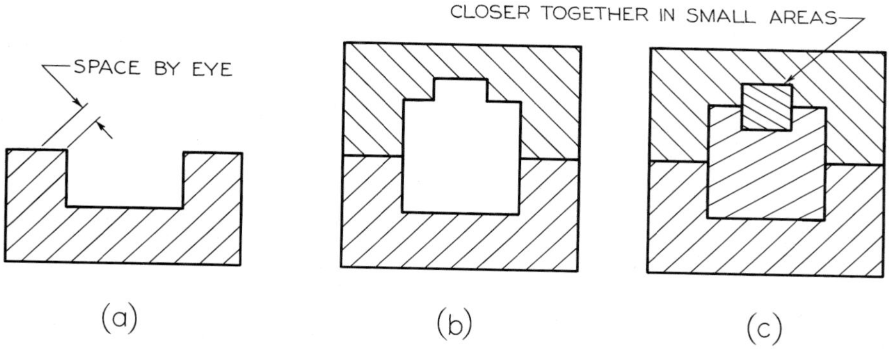

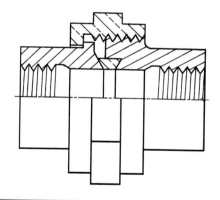

FIGURE 13.10 Section Lining (Full Size).

FIGURE 13.11 Symbolic Section Lining.

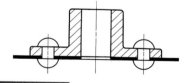

FIGURE 13.12 Sectioning Thin Parts.

shown on the assembly and are included in the parts list. Parts such as bolts, screws, bearings, pins, and keys are identified by the part number from the assembly drawing and specified by name and size or number.

■ 13.7 ASSEMBLY SECTIONING

In assembly sections it is necessary to distinguish between adjacent parts. This is done by drawing the section lines in different directions on different parts, as shown in Figure 13.10. In small areas it is necessary to space the section lines closer together. The section lines in adjacent areas should not meet at the visible lines separating the areas.

For general use, the cast-iron general-purpose section lining is recommended for assemblies. When you want to give a general indication of the materials, use symbolic section lining, as shown in Figure 13.11.

In sectioning relatively thin parts in assembly, such as gaskets and sheet-metal parts, section lining is ineffective, and such parts should be shown in solid black, as in Figure 13.12.

Hands On 13.1
Hatching Assemblies

A full-section assembly drawing is shown without hatching to help identify the different parts. Ball tags identify each part. Use the parts list and look up the material for each part (refer to Chapter 7 for material hatch patterns). Cross-hatch each part using the correct pattern for the material, making the hatching lines a different direction for each different part. Remember that cross-hatching is thin compared to object lines. Use outline cross-hatching when parts are large. Fill small sectioned parts in entirely black when they are too small to cross-hatch effectively. Show bolts and other solid parts in the round.

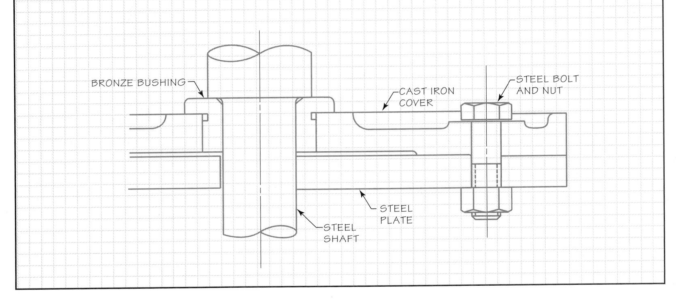

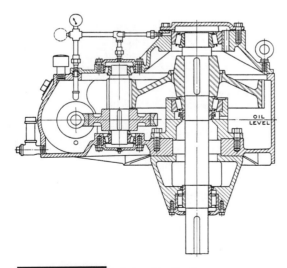

FIGURE 13.13 **Assembly Section.** *Courtesy of Hewitt-Robins, Inc.*

Solid objects and other parts that have interior detail, and therefore do not require sectioning, should be shown unsectioned, or "in the round." These include bolts, nuts, shafts, keys, screws, pins, ball or roller bearings, gear teeth, spokes, and ribs. Several examples are shown in Figure 13.13.

■ 13.8 WORKING DRAWING ASSEMBLY

A working drawing assembly, shown in Figure 13.14, is a combined detail and assembly drawing. When the assembly is simple enough for all its parts to be shown clearly in the single drawing, a combined drawing may be used. In some cases, all but one or two parts can be drawn and dimensioned clearly in the assembly drawing. The few parts that cannot be shown clearly in the assembly are detailed separately on the same sheet. This type of drawing is common in valve drawings, locomotive subassemblies, aircraft subassemblies, and drawings of jigs and fixtures.

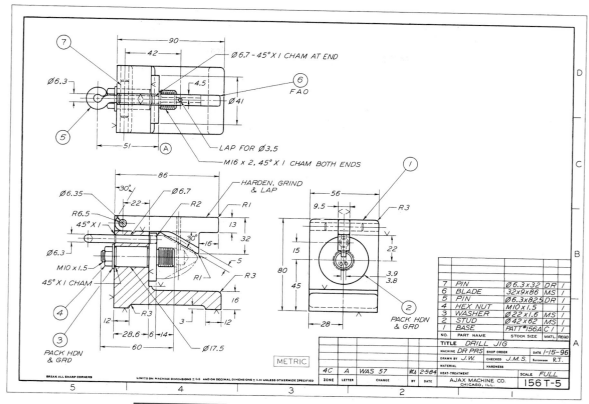

FIGURE 13.14 Working Drawing Assembly of Drill Jig.

13.9 INSTALLATION ASSEMBLIES

A drawing showing how to install or build a machine or structure is an *installation assembly*. This type of drawing is also often called an *outline assembly* because it shows only the outlines and the relationships of exterior surfaces. A typical installation assembly is shown in Figure 13.15. In aircraft design, an installation assembly gives complete information for placing details or subassemblies in their final positions in the airplane.

13.10 CHECK ASSEMBLIES

After all detail drawings of a unit have been made, it may be necessary to make a *check assembly*, especially if a number of changes were made in the details. Check assemblies should be created accurately using CAD in order to graphically check details and their relationship in assembly. After the check assembly has served its purpose, it may be converted into a general assembly drawing.

Practical Tip
Working drawing assemblies

Working drawing assemblies are an excellent way to provide information to manufacturing when parts need to be made quickly to meet a deadline or a design needs to be reworked. Because they include both the detailed dimensions of the parts and the assembly, they show the manufacturing technician the purpose for the tolerances and how the parts will fit together. This understanding of the design helps the manufacturing technician ensure that the parts are made correctly to fit within the assembly.

Detailed part drawings of each part can be included on the same sheet as the assembly for small assemblies or subassemblies. If you do this, make sure to clearly label each part.

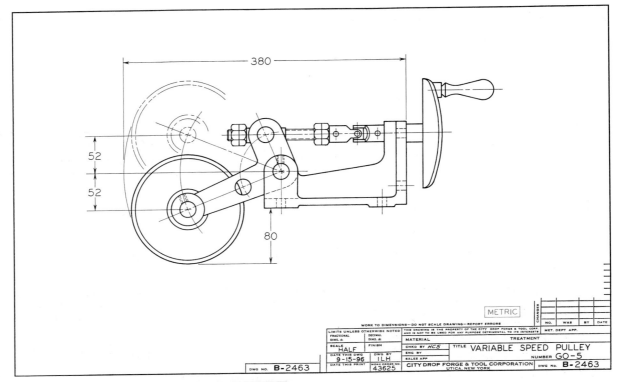

FIGURE 13.15 Installation Assembly.

■ 13.11 TITLE AND RECORD STRIPS

The function of the title and record strip is to show all necessary information not given directly on the drawing with its dimensions and notes. The type of title used depends on the filing system in use, the processes of manufacture, and the requirements of the product. The following information should generally be given in the title form:

1. Descriptive name of the object represented
2. Name and address of the manufacturer
3. Name and address of the purchasing company, if any
4. Signature of the designer who made the drawing and the date of completion
5. Signature of the checker and the date of completion
6. Signature of the chief drafter, chief engineer, or other official, and the date of approval
7. Scale of the drawing
8. Number of the drawing

Other information may be given, such as material, quantity, heat treatment, finish, hardness, pattern number, estimated weight, superseding and superseded drawing numbers, symbol of machine, and many other items, depending on the plant organization and the peculiarities of the product. Some typical commercial titles are shown in Figures 13.16, 13.17, and 13.18. See the inside front cover for traditional title forms and ANSI-approved sheet sizes.

The title form is usually placed along the bottom of the sheet, as shown in Figure 13.16, or in the lower right-hand corner of the sheet, as shown in Figure 13.18, so that the title can be easily found. Sometimes drawings are filed in flat horizontal drawers and the lower right is an easy location to find. However, many different filing systems are used, and the location of the title form is not always standard.

Lettering should be single-stroke vertical or inclined Gothic capitals, whether hand-lettered or lettered using a CAD system. The drawing number should be most noticeable, followed by the name of the object and the name of the company. The date, scale, and drafter's and checker's names are important

REPORT ALL ERRORS TO FOREMAN

		NO. REQUIRED	MATERIAL	HEAT TREATMENT	PART NAME FEED WORM SHAFT	DRAWN BY H.F.	UNIT 3134
		1	SAE 3115	SEE NOTE	DRAWN FOR SIMPLEX & DUPLEX (1200)	TRACED BY E.E.Z.	ALSO USED ON ABOVE MACHINES
		REPLACED BY	REPLACES	OLD PART NO. 563-310	ENGINEERING DEPARTMENT KEARNEY & TRECKER CORPORATION	CHECKED BY C. STB.	FIRST USED ON LOT / LAST USED ON LOT
ALTERATIONS	DATE OF CHG			SCALE FULL SIZE	MILWAUKEE, WISCONSIN, U. S. A.	APPROVED BY / DATE 8-10-1996	17840 B

FIGURE 13.16 Title Strip.

DO NOT SCALE THIS DRAWING FOR DIMENSIONS.   MACHINE FRACTIONAL DIMENSIONS ± 1/64   ALL DIMENSIONS IN INCHES UNLESS OTHERWISE SPECIFIED.

				CHG'D MATL ETC 10-22-83 / WAS #2345 ETC 5-21-83 / CHANGE NOTICE	DATE	HEAT TREATMENT SAE VIII HDN ROCKWELL C-50-56 NOTE 3 TEST LOCATIONS	SCALE FULL DATE 8-10-1996 DRAWN BY S.G. TRACED BY L.R. CHECKED BY n.w. APPROVED BY amB. REDRAWN FROM	CATERPILLAR TRACTOR CO. EXECUTIVE OFFICES — SAN LEANDRO, CALIF. NAME FIRST, FOURTH & THIRD SLIDING PINION MATERIAL C.T. #1E36 STEEL ② ① UPSET FORGING 3.875 ROUND MAX IA4045

FIGURE 13.17 Title Strip.

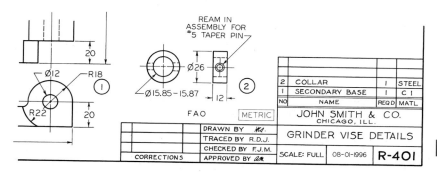

2	COLLAR	1	STEEL
1	SECONDARY BASE	1	C1
NO	NAME	REQ'D	MATL

JOHN SMITH & CO.
CHICAGO, ILL.

	DRAWN BY Hd.	GRINDER VISE DETAILS	
	TRACED BY R.D.J.		
	CHECKED BY F.J.M.		
CORRECTIONS	APPROVED BY am.	SCALE: FULL 08-01-1996	R-401

FIGURE 13.18 Identification of Details with Parts List.

and should be clearly shown but should not overpower the drawing number, title, and company name. Important items are indicated by bold lettering, larger lettering, wider spacing of letters, or by a combination of these methods. See Appendix 16, Letter Sizes for letter heights.

Many companies have adopted their own title forms or those preferred by ANSI and have had them preprinted on standard-size sheets or made into standard CAD symbols to be inserted into the drawing.

Drawings constitute important and valuable information regarding the products of a manufacturer, and carefully designed, well-kept, systematic files are generally maintained for the filing of drawings. Many large companies are using electronic data storage to maintain their drawing records and do not keep paper copies. But electronic drawings must still be systematically stored and backed up, and the revision history must be tracked.

■ 13.12 DRAWING NUMBERS

Every drawing should be numbered. Some companies use serial numbers, such as 60412, or a number with a prefix or suffix letter to indicate the sheet size, as A60412 or 60412-A. The size A sheet would be the standard 8-1/2 x 11 inches or 9 x 12 inches, and the B size is a multiple of the A size sheeet. Many different numbering schemes are in use in which various parts of the drawing number indicate different things, such as model number of the machine and the general nature or use of the part. In general, it is best to use a simple numbering system and not to load the number with too many indications.

GRAPHICS SPOTLIGHT

Technical Document Management Systems

Chris Merritt of CADMAX Consulting in Atlanta, Georgia, specializes in creating usable technical document management systems for CAD-oriented companies. He says, "The goal of any well-designed Technical Document Management (TDM) system is to provide a means for users to create, search for, edit, transfer, store, print and plot documents in a secure, organized, and productive manner."

TOO MANY CAD DRAWINGS?

Often when companies have switched to CAD and have been creating drawings for a number of years they find that they suddenly have too many drawings for anyone to effectively find the drawing they are looking for. It can be difficult to remember file names and the revision history of drawings. Is the drawing that you retrieved the most up-to-date? Even when companies implement a file naming standard, unless there is some kind of search process and automated retrieval, documents cannot be located effectively. Another consideration is that previous revisions of the drawings must be stored so that they can be relocated later, especially if the company has products in the field that used an earlier version and may need technical support or updates. Most engineering documentation is required to be stored for the lifetime of the equipment or permanently.

Chris tells of one manufacturing company he visited who had over 18,000 CAD drawings, 2,000 of which were discarded in the cleanup process as either being exercises or drawings representing removed or discarded equipment. There were over 2,500 duplicate drawings that had to be resolved. The drawings were on floppy disk, individual PCs, and the server with no particular rhyme or reason as to what was where. No one knew where the latest version of many of the drawings were located.

WHY TDM SYSTEMS FAIL

Many times document management systems die in their infancy. Often a lack of understanding about the basic concepts of document management and having unclear goals for the system lead to its early failure. Sophisticated features that may never be used are specified, while important features that are needed every day are overlooked. A frequent problem is that the person responsible for the TDM project lack the necessary authority within the company to gain the acceptance and enforce the changes in habits that are necessary to make the document management system work. Also some companies wait too long in the search phase, while their requirements for the system are ever expanding and the number of unmanaged documents are continuing to grow. Most companies spend about two years making the determination to implement a system.

TDM SOFTWARE

AutoManager-WorkFlow by Cyco International, Inc. is one of the TDM software packages Chris recommends most often. It is easily customizable due to its built in programming language and once it is up and running, it needs very little maintenance. You can design, test, and prototype most of the features on a standalone PC. Most electronically stored information you may have already gathered about your documents can be transferred to its database. And it is relatively quick and easy to design a clear demonstration of it's ability with your own data & documents.

GETTING RESULTS

To implement a TDM system, managers must support the investment of time in a great deal of file system cleanup work. This can be more time consuming than actually getting the software up and running, but is worth the investment. It can be as valuable as the improvements in the ability to manage the documents that are created.

A well-designed technical document management system should have the following components:

- A consistent directory structure for storing all documents
- A document naming convention
- Established procedures for:
  - Creating documents
  - Searching for documents
  - Editing documents
  - Transferring documents to/from a server
  - Printing or plotting of documents

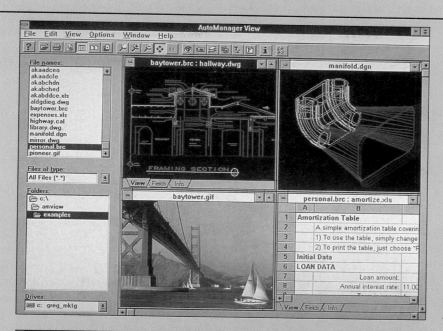

(A) *Courtesy of Cyco International, Inc.*

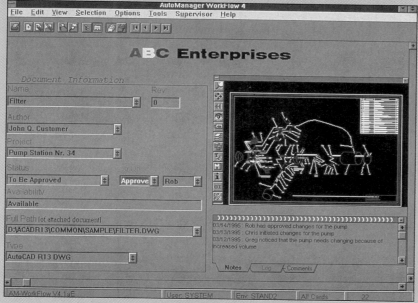

(B) *Courtesy of Cyco International, Inc.*

Part of the benefit of working with CAD is creating a usable database documenting the projects that have been completed. If you are not able to retrieve files down the road, you will not realize many of the benefits of your CAD system. Remember that when you justified the expense of the CAD system, you probably valued the time you would save by not having to recreate drawing geometry, the ability to reuse drawings for different purposes, and the ability to quickly access information. If drawing retrieval and file naming is a mess on your system, a well-designed drawing management system may be the way to realize some of the unachieved gains you expected when you chose CAD in the first place.

The drawing number should be lettered 7 mm (.250 inch) high in the lower right and upper left corners of the sheet.

■ 13.13 ZONING

To make it easy to locate an item on a large or complex drawing, regular ruled intervals are labeled along the margins, often in the right and lower margins only. The intervals on the horizontal margin are labeled from right to left with numerals, and the intervals on the vertical margin are labeled from bottom to top with letters. Zoning on engineering drawings is similar to zoning letters and numbers on maps that help you locate a city or street.

■ 13.14 CHECKING

The importance of accuracy in technical drawing cannot be overestimated. In commercial offices, errors sometimes cause tremendous unnecessary expenditures. The signature on a drawing identifies who is responsible for the accuracy of the work.

In small offices, checking is usually done by the designer. In large offices, experienced engineers or manufacturing technicians, called checkers, devote a major part of their time to checking drawings for release to manufacturing.

The completed drawing is carefully checked by the designer for function, economy, and practicability. The final checker should be able to discover all remaining errors. To be effective, checking should be done in a systematic way. The checker should study the drawing with particular attention to the following points:

1. Soundness of design, with reference to function, strength, materials, economy, manufacturability, serviceability, ease of assembly and repair, and lubrication.

2. Choice of views, partial views, auxiliary views, sections, and lettering.

3. Dimensions, with special reference to repetition, ambiguity, legibility, omissions, errors, and finish marks. Special attention should be given to tolerances.

4. Standard parts. In the interest of economy, as many parts as possible should be standard.

5. Notes, with special reference to clear wording and legibility.

6. Clearances. Moving parts should be checked in all possible positions to assure freedom of movement.

7. Title form information.

■ 13.15 REVISIONS

Changes on drawings are necessitated by changes in design, changes in tools, desires of customers, or errors in design or in production. In order that the sources of all changes on released drawings may be understood, verified, and made noticeable, an accurate record of all changes should be made on the drawings. The record should show the character of the change and by whom, when, and why it was made.

The changes are made directly by changing the original CAD drawing or sketch or sometimes by changing a reproduction of the original. Additions are usually added to the original drawing or CAD file. Removing information by crossing it out is not recommended. If a dimension is changed and the part not redrawn to reflect the correct size, underline the dimension with a heavy line to indicate that it is not to scale. Prints of each issue, microfilms, or a permanent electronic record (such as a file stored on CD-ROM) must be kept to show how the drawing appeared before the revision. New prints are issued to supersede old ones each time a change is made.

If considerable change on a drawing is necessary, a new drawing may be made and the old one then stamped OBSOLETE and placed in the "obsolete" file. In the title block of the old drawing, the words SUPERSEDED BY... or REPLACED BY... are entered followed by the number of the new drawing. On the new drawing, under SUPERSEDES... or REPLACES..., the number of the old drawing is entered.

Various methods are used to reference the area on a drawing where the change is made, with the entry put in the revision block. The most common is to place numbers or letters in small circles or triangles near the

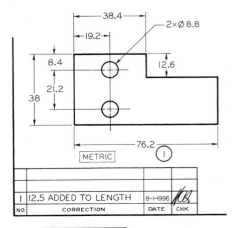

FIGURE 13.19 Revisions.

places where the changes were made and to use the same numbers or letters in the revision block, as shown in Figure 13.19. On zoned drawings, show the zone of the correction in the revision block. In addition, the change should be described briefly, and the date and the initials of the person making the change should be given.

■ 13.16 SIMPLIFIED REPRESENTATION

Time spent generating drawings is a considerable element of the total cost of a product. When possible, simplify the drawing when you can do so without loss of clarity to the user.

The *American National Standard Drafting Manual,* published by the American National Standards Institute, incorporates the best and the most representative practices in this country. These standards advocate simplification in many ways—for example through partial views, half views, thread symbols, piping symbols, and single-line spring drawings. Any line or lettering on a drawing that is not needed for clarity should be omitted. A summary of practices to simplify representation is as follows:

1. Use word description in place of drawing wherever practicable.

2. Never show unnecessary views. Often a view can be eliminated by using abbreviations or symbols such as HEX, SQ, DIA, ∅, □, and ₵.

3. Use partial views instead of full views wherever possible. Show half views of symmetrical parts.

4. Avoid elaborate, pictorial, or repetitive detail as much as possible. Use phantom lines to avoid repeating features.

5. When possible, list rather than show standard parts such as bolts, nuts, keys, and pins.

6. Omit unnecessary hidden lines.

7. Use outline section lining in large sectioned areas when it can be done without loss of clarity.

8. Omit unnecessary duplication of notes and lettering.

9. Wherever possible, use symbolic representation, such as piping symbols and thread symbols.

10. Sketch freehand whenever practicable.

11. Avoid hand-lettering as much as possible. For example, parts lists should be typed.

12. Use laborsaving devices when feasible, such as templates and plastic overlays.

13. Use electronic devices or computer graphics systems wherever feasible for design, drawing, and repetitive work.

Some industries have attempted to simplify their drafting practices even more. Until these practices are accepted generally by industry and in time find their way into ANSI standards, follow the ANSI standards as exemplified throughout this book.

■ 13.17 PATENT DRAWINGS

The patent application for a machine or device must include drawings to illustrate and explain the invention. It is essential that all patent drawings be mechanically correct and constitute complete illustrations of every feature of the invention claimed. The strict requirements of the U.S. Patent Office help that office examine applications and interpret whether to issue a patent. A typical patent drawing is shown in Figure 13.20.

The drawings for patent applications are pictorial and explanatory in nature and therefore are not as detailed as working drawings for production purposes. Centerlines, dimensions, and notes are not included. Views, features, and parts are identified by numbers that refer to the descriptions and explanations given in the specification section of the patent application.

Patent drawings are made with black permanent ink on heavy, smooth, white paper, exactly 10 x 15 inches in size with 1-inch borders on all sides. All lines must

5,511,508
PONTOON RUNNER SYSTEM
John M. Wilson, Sr., and Dean R. Wilson, both of Marrero, La., assignors to Wilco Marsh Buggies & Draglines, Inc., La.
Filed Apr. 21, 1994, Ser. No. 230,618
Int. Cl.⁶ B63B 3/00
U.S. Cl. 114—356   10 Claims

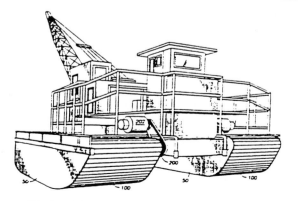

FIGURE 13.20 Pictorial Patent Drawing.

be solid black and suitable for reproduction at a smaller size. Line shading is used whenever it improves readability. A space is left at the top of the drawing of at least 1.25 inches for the heading, title, name, number, and other data to be added by the Patent Office.

The drawings must contain as many figures as necessary to show the invention clearly. There is no restriction on the number of sheets. The figures may be plan, elevation, section, pictorial, and detail views of portions or elements, and they may be drawn to an enlarged scale if necessary. The required signatures must be placed in the lower right-hand corner of the drawing, either inside or outside the border line.

Because of the strict requirements of the Patent Office, if applying, you should find someone with experience to guide you in making your drawing. To help you prepare drawings for submission with patent applications, the *Guide for Patent Draftsmen* can be obtained from the Superintendent of Documents, U.S. Government Printing Office, Washington, DC 20402.

 You can find information on patent laws and rules by looking up KuesterLaw—The Technology Law Resource on the World Wide Web at: http://www.kuesterlaw.com/lawrule/index.html.

■ KEY WORDS

stages of design	assembly drawing	concepts	prototype
models	production drawings	refinement	title blocks
virtual reality	record strips	computer simulation	drawing numbers
detail drawing	checking and proofing	working drawing	patent drawings

■ CHAPTER SUMMARY

- Final drawings created during the design process include assembly drawings, working drawings, design drawings, and patent drawings.

- Assembly drawings do not need to show all necessary views to describe the function of the device. They only need to show enough information so that a worker can assemble the parts correctly.

- Assembly drawings generally do not show dimensions unless they are critical to the assembly of the parts.

- A parts list shows the part number from the ball tag, a description, the material, quantity required, and other information about parts in an assembly.

- Detail drawings are usually shown one per sheet in order to make them easy to reuse. Detail drawings show the necessary views, dimensions, notes, and material specifications needed to manufacture individual parts.

- Stock parts that are easily purchased or supplied are not shown in detail drawings unless they are to be modified for the design.

- There are many revisions to drawings during the design process. You must keep track of each version and what changes were made.

■ REVIEW QUESTIONS

1. What are the special requirements of a patent drawing?

2. What kinds of information are included in an assembly drawing?

3. How is a detail drawing different from an assembly drawing?

4. Why are drawings numbered? Why is this numbering so important?

5. Describe the drawing revision process. Why is it so important to keep track of revisions?

6. How are revised paper drawings stored? How are revised CAD drawings stored?

7. What are the advantages of computer modeling? What are the disadvantages?

■ WORKING DRAWING PROJECTS

The projects in Figures 13.21–81 will give you practice in making regular working drawings of the type used in industry. Many projects, especially those of the assemblies, offer an excellent opportunity for you to redesign or improve on the existing design. Due to the variations in sizes and in scales that may be used, you are required to select the sheet sizes and scales appropriate to the final drawing purpose. Standard sheet layouts are shown inside the front cover of this book.

Use the preferred metric system or the acceptable complete decimal-inch system, as assigned. Either unidirectional or aligned dimensioning may be assigned.

The projects are presented in pictorial form. You should not always follow the placement of dimensions and finish marks as shown in the pictorial drawing. The dimensions given are in most cases those needed to make the parts, but they are not in all cases the dimensions that should be shown on the working drawing. In the pictorial problems, the rough and finished surfaces are shown, but finish marks are usually omitted. You should add all necessary finish marks and place all dimensions in the preferred places in the final drawings.

Each problem should be preceded by a fully dimensioned sketch before continuing on to create CAD drawings or 3-D models. Any of the title blocks shown inside the front cover of this book may adapted, or you may design a title block as assigned by your instructor.

■ DESIGN PROJECT

Design an improvement to our land, sea, or air transportation systems. Vehicles, controls, highways, and airports need further refinement. Make an assembly drawing showing how the parts in your design fit together. Use ball tags to identify the parts and make a parts list.

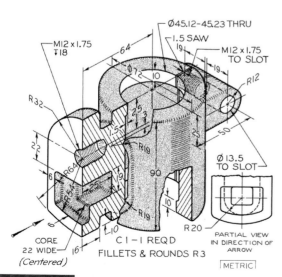

FIGURE 13.21 Table Bracket.
Proj. 13.1: Make detail drawing using size B or A3 sheet.

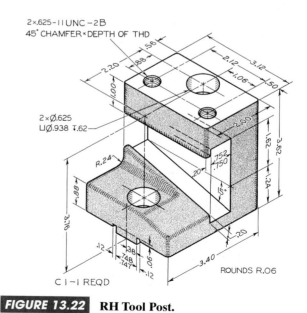

FIGURE 13.22 RH Tool Post.
Proj. 13.2: Make detail drawing using size B or A3 sheet. If assigned, convert dimensions to metric system.

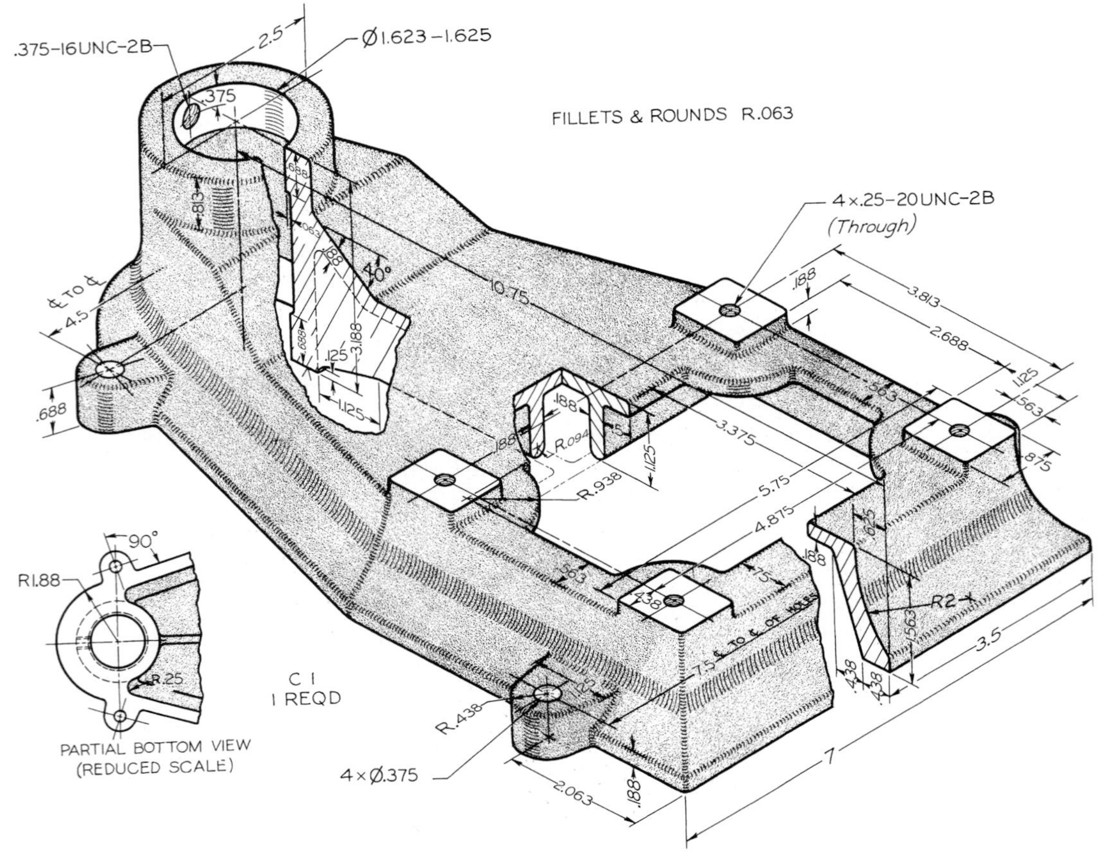

.375–16UNC–2B

2.5

Ø1.623 –1.625

FILLETS & ROUNDS R.063

4×.25–20UNC–2B
(Through)

.375

.813

3.813

2.688

.188

10.75

40°

.125

.188

R.094

3.188

1.125

.188

R.938

3.375

5.75

4.875

.125

.563

.875

R2

3.5

4.5

¢ TO ¢

.688

R1.88

90°

R.25

C 1
1 REQD

PARTIAL BOTTOM VIEW
(REDUCED SCALE)

R.438

4 × Ø.375

2.063

7

.438

.438

.625

FIGURE 13.23 Drill Press Base.
Proj. 13.3: Make detail drawing using size C or A2 sheet. Use unidirectional metric or decimal-inch dimensions.

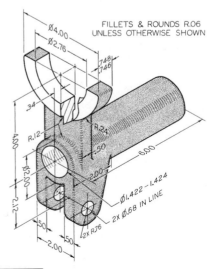

FILLETS & ROUNDS R.06
UNLESS OTHERWISE SHOWN

Ø4.00
Ø2.76

.748
.746

.34

R.24

.50

R.12

6.00

4.00

2.00

Ø2.00

2.12

Ø1.422 –1.424

2× Ø.68 IN LINE

2× R.76

.50

.50

2.00

FIGURE 13.24 Shifter Fork.
Proj. 13. 4: Make detail drawing using size B or A3 sheet. If assigned, convert dimensions to metric system.

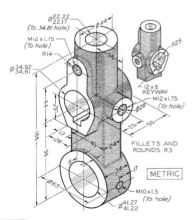

Ø22.22
22.17
(To 34.81 hole)

Ø44

M12 x 1.75
(To hole)

R14

Ø34.92
34.81

R25

12 x 6
KEYWAY

M12 x 1.75
(To hole)

Ø28

FILLETS AND
ROUNDS R3

METRIC

M10 x 1.5
(To hole)

Ø41.27
41.22

Ø65

FIGURE 13.25 Idler Arm.
Proj. 13.5: Make detail drawing using size B or A3 sheet.

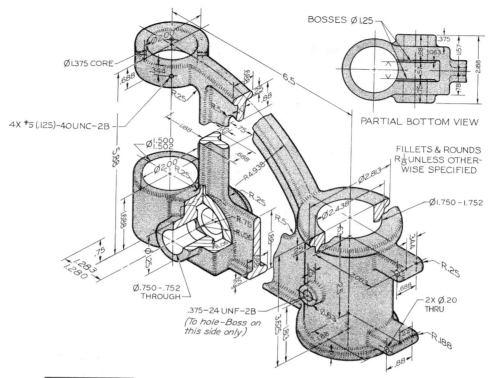

FIGURE 13.26 Drill Press Bracket.
Proj. 13.6: Make detail drawing using size C or A2 sheet. If assigned, convert
dimensions to decimal inches or redesign the part with metric dimensions.

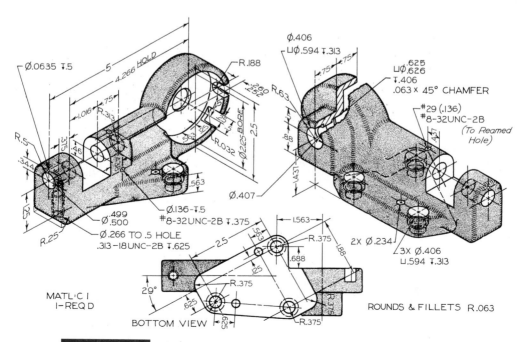

FIGURE 13.27 Dial Holder.
Proj. 13.7: Make detail drawing using size C or A2 sheet. If assigned, convert
dimensions to decimal inches or redesign the part with metric dimensions.

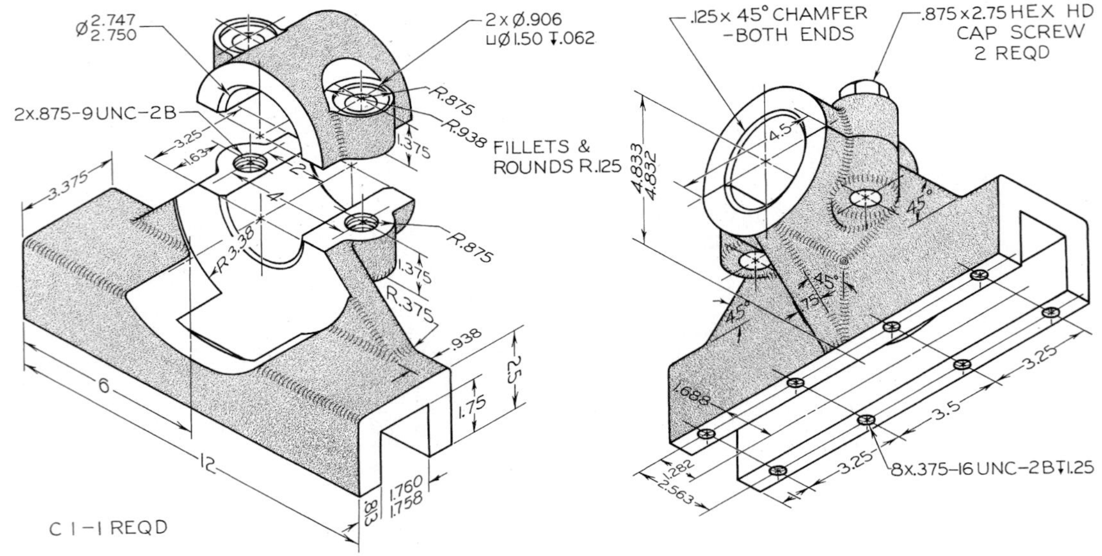

FIGURE 13.28 Rack Slide.

Proj. 13.8: Make detail drawings half size on size B or A3 sheet. If assigned, convert dimensions to decimal inches or redesign the part with metric dimensions.

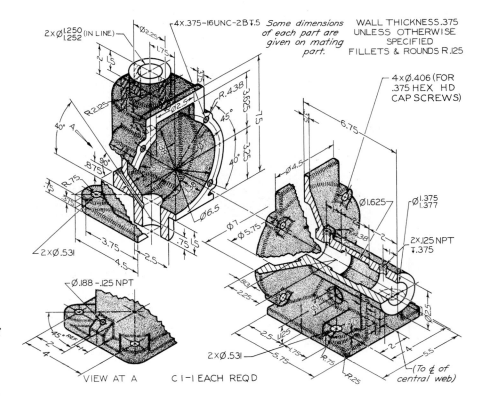

FIGURE 13.29 Conveyor Housing.

Proj. 13.9: Make detail drawings half size on size C or A2 sheets. If assigned, convert dimensions to decimal inches or redesign the parts with metric dimensions.

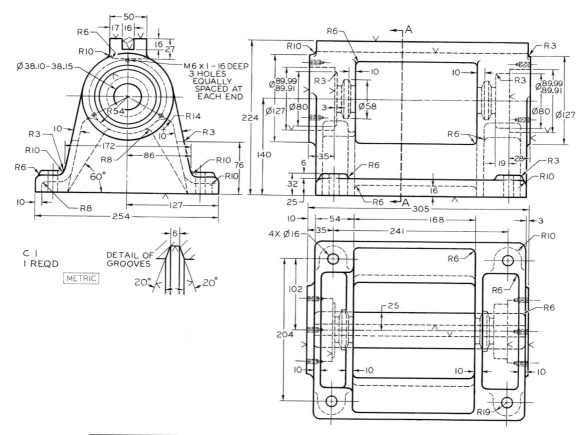

Spindle Housing.
Proj. 13.10: Given: Front, left-side, and bottom views, and partial removed section. Required: Front view in full section, top view, and right-side view in half section on A-A. Draw half size on size C or A2 sheet. If assigned, dimension fully.

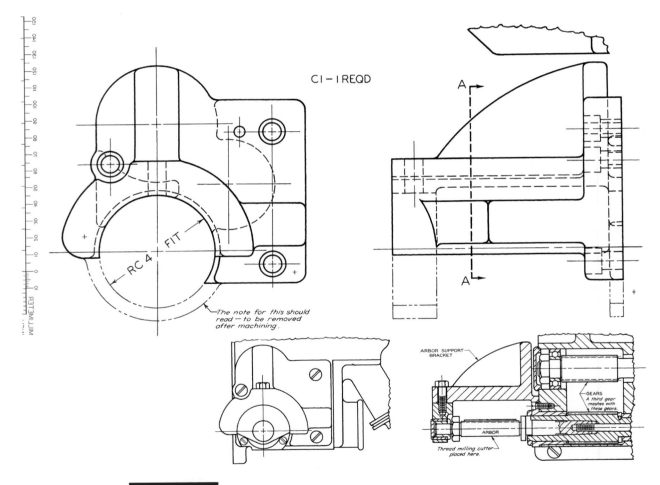

CI — I REQD

RC 4 FIT

—The note for this should
read — to be removed
after machining.

A

A

ARBOR SUPPORT
BRACKET

GEARS
A third gear
meshes with
these gears.

ARBOR

Thread milling cutter
placed here.

FIGURE 13.31 Arbor Support Bracket.
Proj. 13.11: Given: Front and right-side views.
Required: Front, left-side, and bottom views, and a detail section A-A. Use
American National Standard tables for indicated fits and if required convert
to metric values (see Appendixes 5–14). If assigned, dimension in the metric
or decimal-inch system.

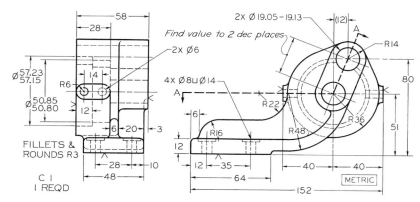

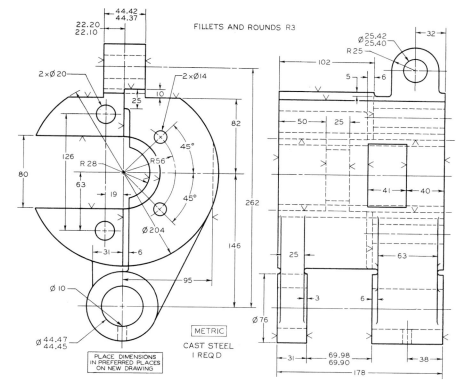

FIGURE 13.32 Pump Bracket for a Thread Milling Machine.
Proj. 13.12: Given: Front and left-side views. Required: Front and right-side views, and top view in section on A-A. Draw full size on size B or A3 sheet. If assigned, dimension fully.

FIGURE 13.33 Fixture Base for 60-Ton Vertical Press.
Proj. 13.13: Given: Front and right-side views. Required: Revolve front view 908 clockwise; then add top and left-side views. Draw half size on size C or A2 sheet. If assigned, complete with dimensions.

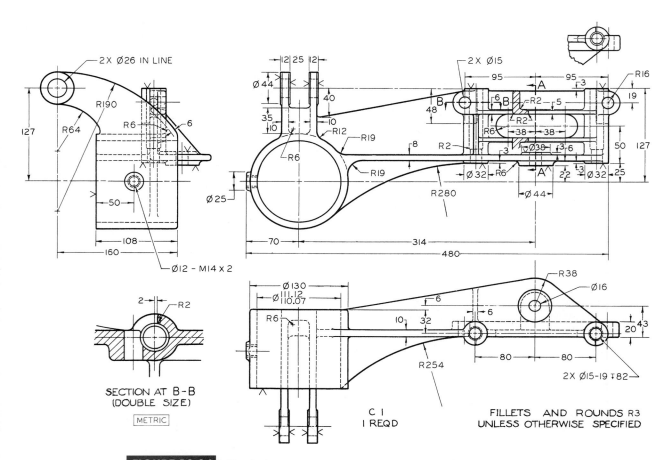

FIGURE 13.34 Bracket.

Proj. 13.14: Given: Front, left-side, and bottom views, and partial removed section. Required: Make detail drawing. Draw front, top, and right-side views, and removed sections A-A and B-B. Draw half size on size C or A2 sheet. Draw section B-B full size. If assigned, complete with dimensions.

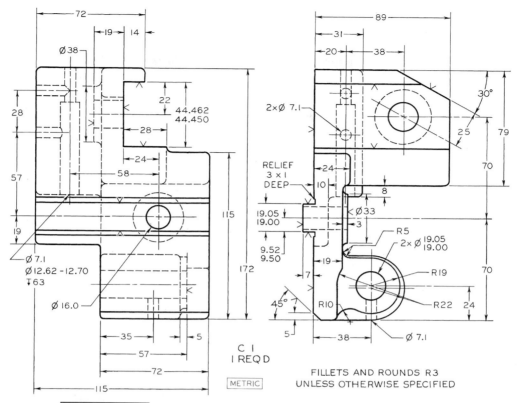

FIGURE 13.35 Roller Rest Bracket for Automatic Screw Machine.
Proj. 13.15: Given: Front and left-side views.
Required: Revolve front view 90° clockwise; then add top and left-side views.
Draw half size on size C or A2 sheet. If assigned, complete with dimensions.

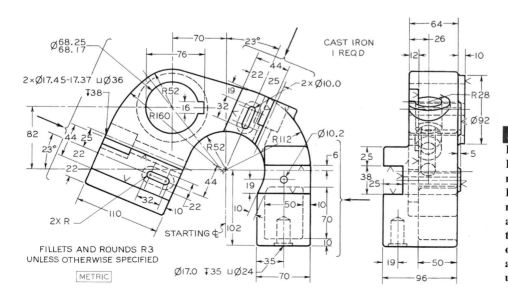

FIGURE 13.36 Guide
Bracket for Gear Shaper.
Proj. 13.16: Given: Front and
right-side views.
Required: Front view, a partial
right-side view, and two partial
auxiliary views taken in direction of arrows. Draw half size
on size C or A2 sheet. If
assigned, complete with
unidirectional dimensions.

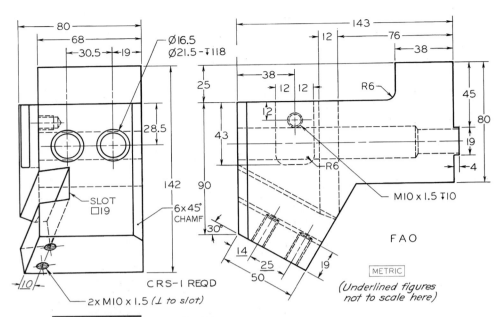

FIGURE 13.37 Rear Tool Post.
Proj. 13.17: Given: Front and left-side views.
Required: Take left-side view as new top view; add front and left-side views, approx. 215 mm apart, a primary auxiliary view, then a secondary view taken so as to show true end view of 19-mm slot. Complete all views, except show only necessary hidden lines in auxiliary views. Draw full size on size C or A2 sheet. If assigned, complete with dimensions.

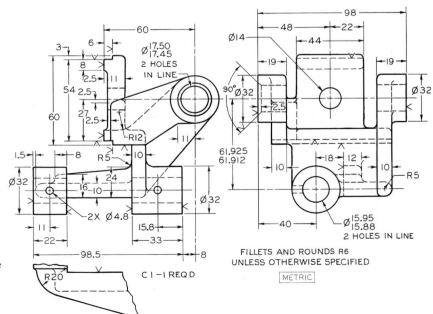

FIGURE 13.38 Bearing for a Worm Gear.
Proj. 13.18: Given: Front and right-side views. Required: Front, top, and left-side views. Draw full size on size C or A2 sheet. If assigned, complete with dimensions.

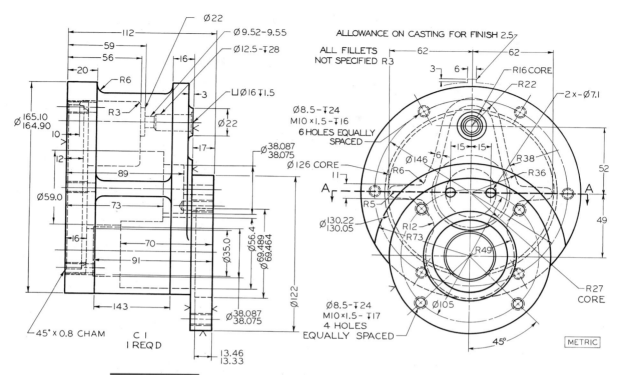

FIGURE 13.39 Generator Drive Housing.
Proj. 13.19: Given: Front and left-side views.
Required: Front view, right-side view in full section, and top view in full section on A-A. Draw full size on size C or A2 sheet. If assigned, complete with dimensions.

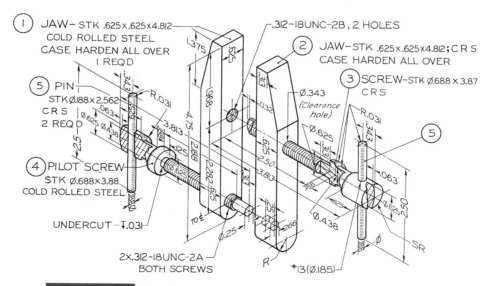

FIGURE 13.40 Machinist's Clamp.
Proj. 13.20: Draw details and assembly. If assigned, use unidirectional two-place decimal-inch dimensions or redesign for metric dimensions.

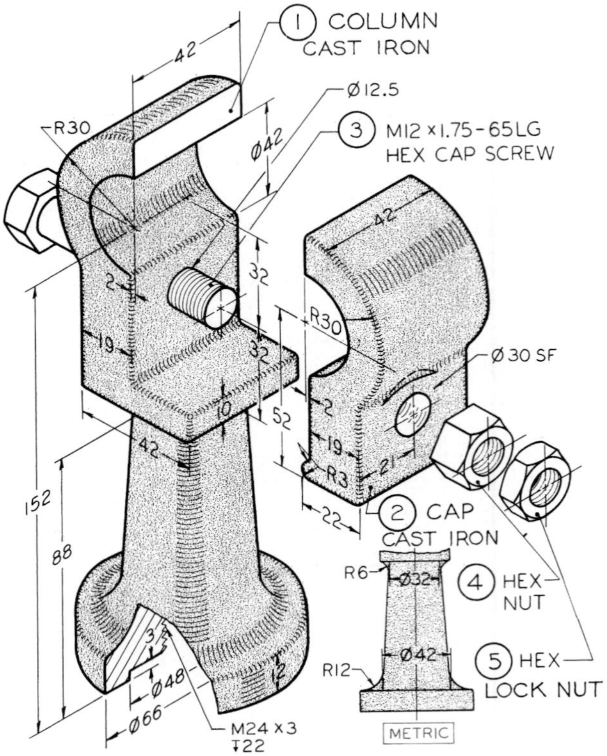

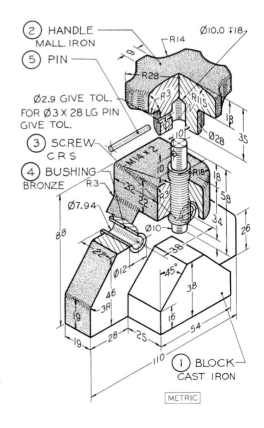

FIGURE 13.41　Hand Rail Column.
Proj. 13.21: (1) Draw details. If assigned, complete
with dimensions. (2) Draw assembly.

FIGURE 13.42　Drill Jig.
Proj. 13.22: (1) Draw details. If assigned, complete with
dimensions. (2) Draw assembly.

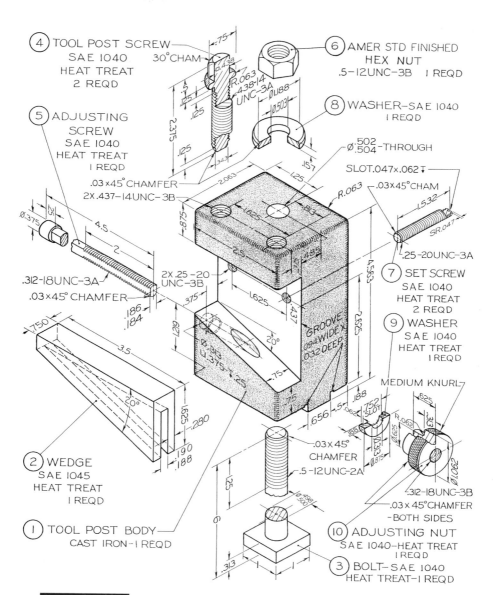

FIGURE 13.43 **Tool Post.**

Proj. 13.23: (1) Draw details. (2) Draw assembly. If assigned, use unidirectional two-place decimals for all fractional dimensions or redesign for all metric dimensions.

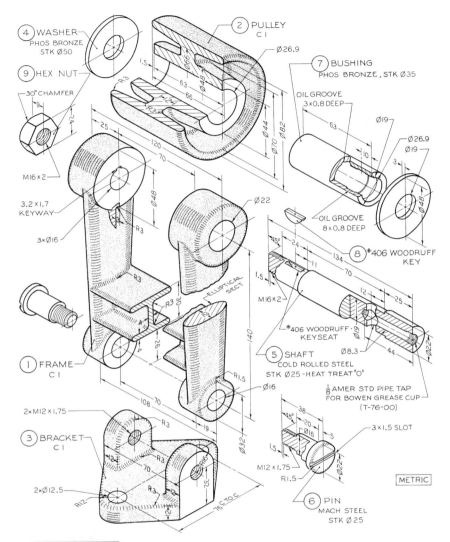

FIGURE 13.44 Belt Tightener.

Proj. 13.24: (1) Draw details. (2) Draw assembly. It is assumed that the parts are to be made in quantity and they are to be dimensioned for interchangeability on the detail drawings. Use tables in Appendixes 11–14 for limit values. Design as follows.

a. **Bushing fit in pulley: Locational interference fit.**

b. **Shaft fit in bushing; Free running fit.**

c. **Shaft fits in frame: Sliding fit.**

d. **Pin fit in frame: Free running fit.**

e. **Pulley hub length plus washers fit in frame: Allowance 0.13 and tolerances 0.10.**

f. **Make bushing 0.25 mm shorter than pulley hub.**

g. **Bracket fit in frame: Same as e above.**

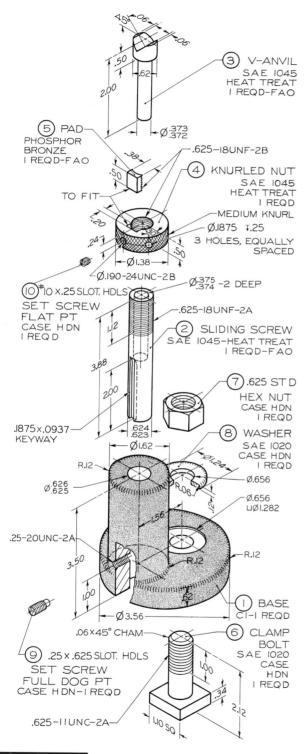

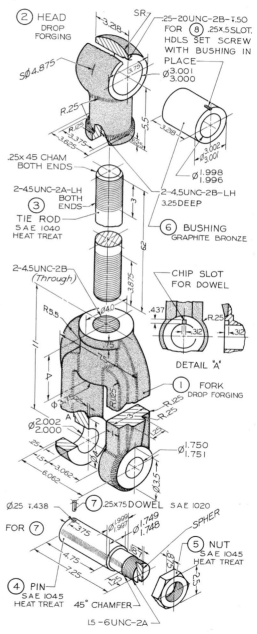

FIGURE 13.45   Milling Jack.
Proj. 13.25: (1) Draw details. (2) Draw assembly. If assigned, convert dimensions to metric or decimal-inch.

FIGURE 13.46   Connecting Bar.
Proj. 13.26: (1) Draw details. (2) Draw assembly. If assigned, convert dimensions to metric or decimal-inch.

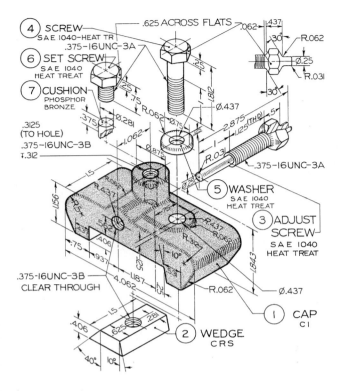

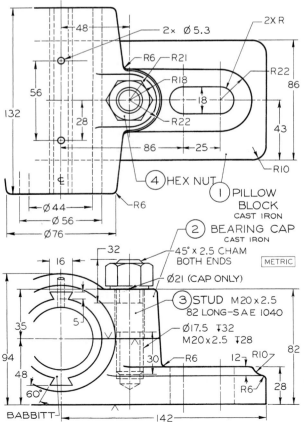

FIGURE 13.47 Clamp Stop.
Proj. 13.27: (1) Draw details. (2) Draw assembly. If assigned, convert dimensions to decimal-inch system or redesign for metric dimensions.

FIGURE 13.48 Pillow Block Bearing.
Proj. 13.28: (1) Draw details. (2) Draw assembly. If assigned, complete with dimensions.

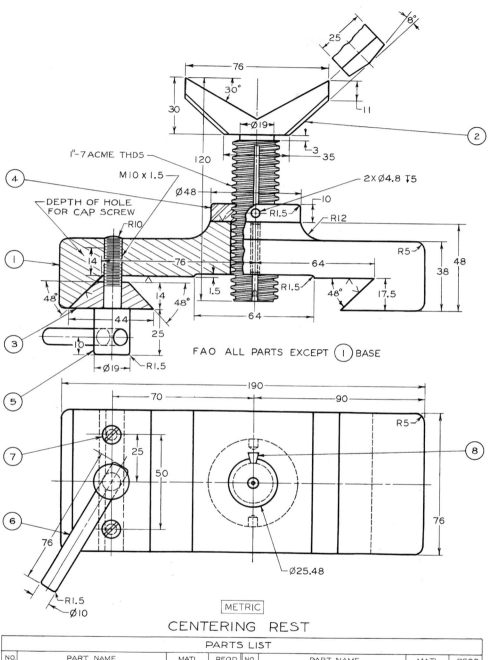

FAO ALL PARTS EXCEPT ① BASE

METRIC

CENTERING REST

NO.	PART NAME	MATL	REQD	NO.	PART NAME	MATL	REQD
					PARTS LIST		
1	BASE	C I	1	5	CLAMP SCREW	SAE 1020	1
2	REST	SAE 1020	1	6	CLAMP HANDLE	SAE 1020	1
3	CLAMP	SAE 1020	1	7	M6X1-25LG FIL HD CAP SCREW		2
4	ADJUSTING NUT	SAE 1020	1	8	5.5 X 5.5 X 3.2 - 25 LG KEY	SAE 1030	1

FIGURE 13.49 Centering Rest.
Proj. 13.29: (1) Draw details. (2) Draw assembly. If assigned, complete with dimensions.

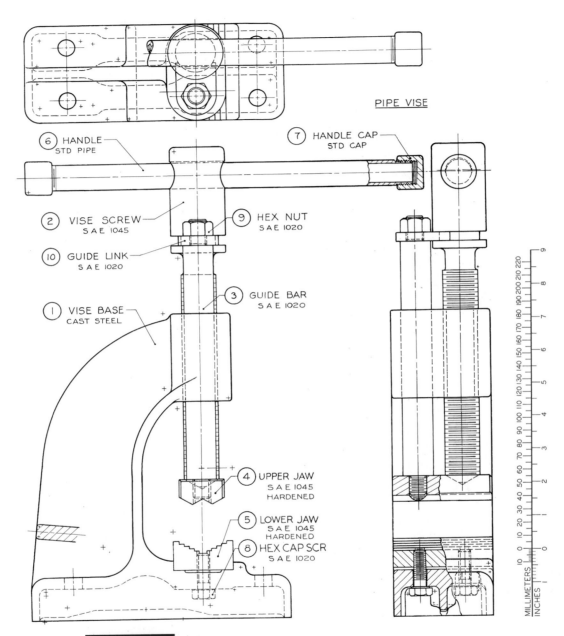

PIPE VISE

⑥ HANDLE
STD PIPE

⑦ HANDLE CAP
STD CAP

② VISE SCREW
S.A.E. 1045

⑨ HEX NUT
S.A.E. 1020

⑩ GUIDE LINK
S.A.E. 1020

③ GUIDE BAR
S.A.E. 1020

① VISE BASE
CAST STEEL

④ UPPER JAW
S.A.E. 1045
HARDENED

⑤ LOWER JAW
S.A.E. 1045
HARDENED

⑧ HEX CAP SCR
S.A.E. 1020

FIGURE 13.50 Pipe Vise.
Proj. 13.30: (1) Draw details. (2) Draw assembly. To obtain dimensions, transfer distances from the figure using a scrap of paper. Use the printed scale provided and read measurements in millimeters or decimal inches as assigned. All threads are general-purpose metric threads (see Appendix 15) or Unified coarse threads except the American National Standard pipe threads on handle and handle caps.

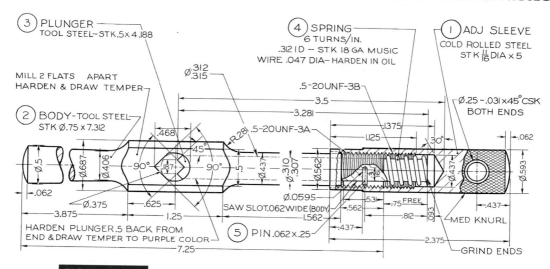

FIGURE 13.51 Tap Wrench.

Proj. 13.31: (1) Draw details. (2) Draw assembly. If assigned, use unidirectional two-place decimals for all fractional dimensions or redesign for metric dimensions.

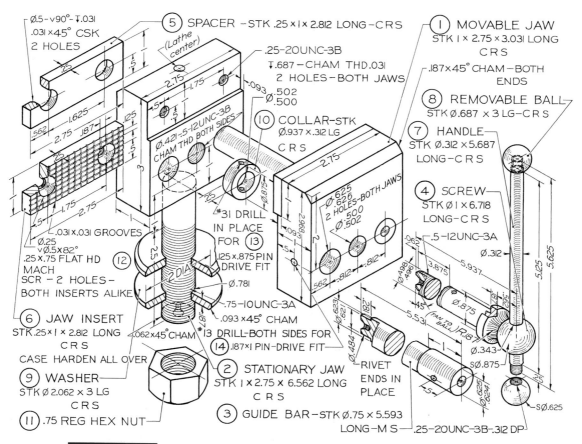

FIGURE 13.52 Machinist's Vise.

Proj. 13.32: (1) Draw details. (2) Draw assembly. If assigned, use unidirectional two-place decimals for all fractional dimensions or redesign for metric dimensions.

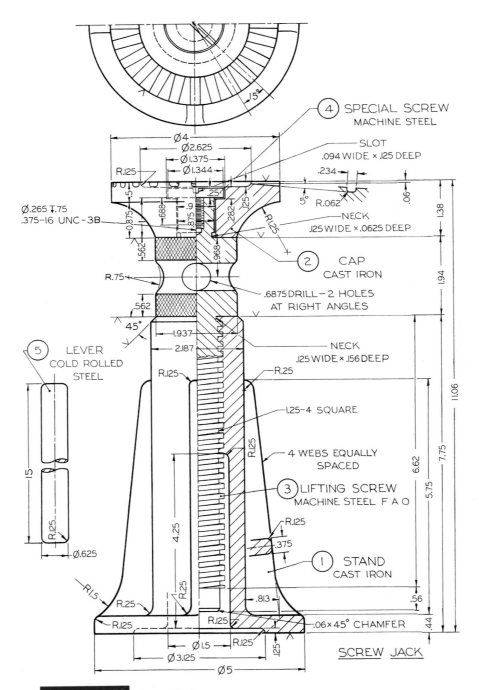

FIGURE 13.53   Screw Jack.

Proj. 13.33: (1) Draw details. See Fig. 13.5, showing "boxed-in" views on sheet layout C–678 or A2–678 (see inside front cover). (2) Draw assembly. If assigned, convert dimensions to decimal inches or redesign for metric dimensions.

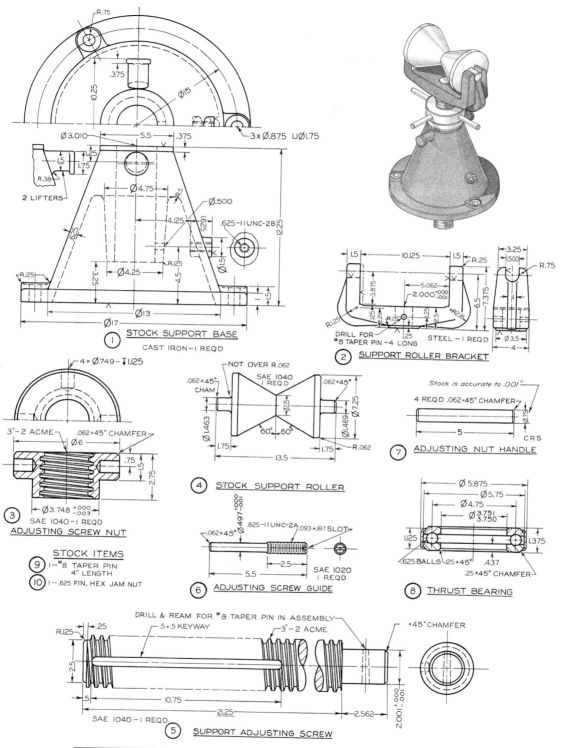

STOCK SUPPORT BASE ①
CAST IRON—1 REQD

SUPPORT ROLLER BRACKET ②
STEEL—1 REQD

ADJUSTING SCREW NUT ③
SAE 1040—1 REQD

STOCK SUPPORT ROLLER ④
SAE 1040 1 REQD

ADJUSTING NUT HANDLE ⑦
4 REQD .062×45° CHAMFER
CRS

ADJUSTING SCREW GUIDE ⑥
SAE 1020 1 REQD

THRUST BEARING ⑧

STOCK ITEMS
⑨ 1—#8 TAPER PIN 4" LENGTH
⑩ 1—.625 FIN. HEX JAM NUT

SUPPORT ADJUSTING SCREW ⑤
SAE 1040—1 REQD

FIGURE 13.54 **Stock Bracket for Cold Saw Machine.**
Proj. 13.34: (1) Draw details. (2) Draw assembly. If assigned, use unidirectional decimal dimensions or redesign for metric dimensions.

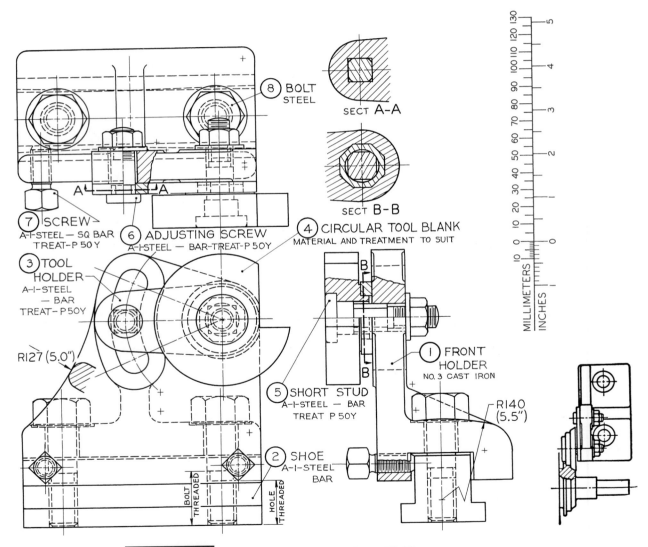

FIGURE 13.55 Front Circular Forming Cutter Holder.
Proj. 13.35: (1) Draw details. (2) Draw assembly. To obtain dimensions, take
distances directly from figure with dividers and set dividers on printed scale.
Use metric or decimal-inch dimensions as assigned.

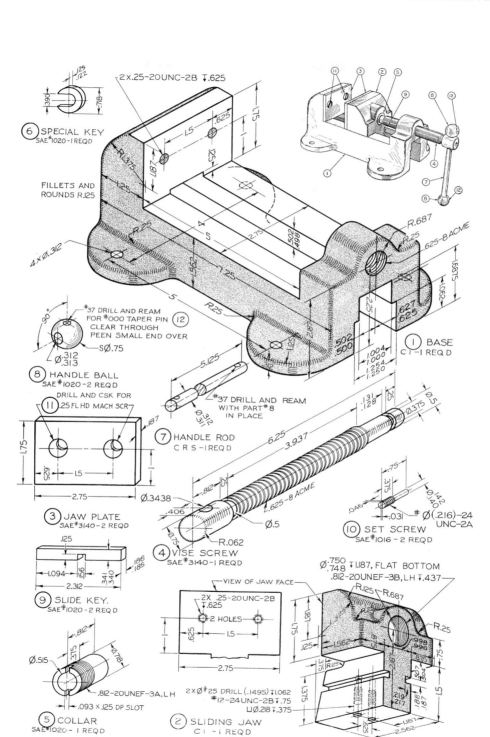

FIGURE 13.56 Machine Vise.
Proj. 13.36: (1) Draw details. (2) Draw assembly. If assigned, convert dimensions to the decimal-inch system or redesign with metric dimensions.

GRINDER WHEEL

STOCK

FIGURE 13.57 Grinder Vise.
Proj. 13.37: See Figs. 13.58 and 13.59.

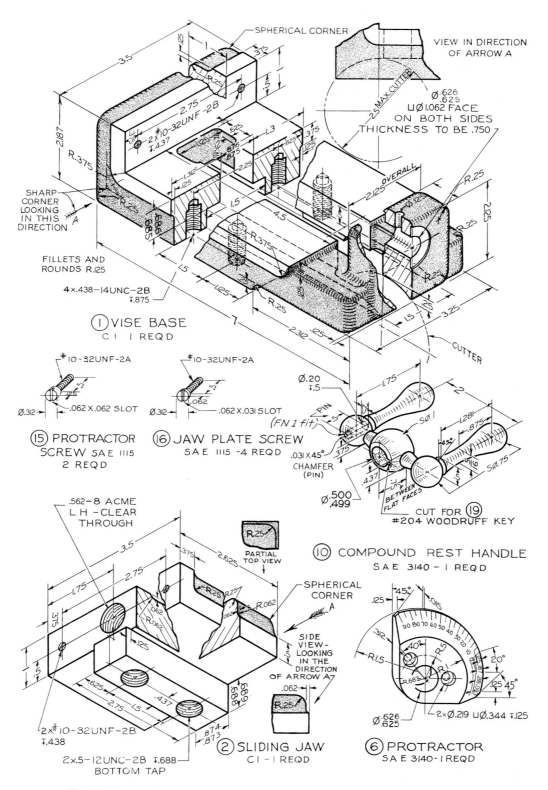

FIGURE 13.58 Grinder Vise.
Proj. 13.37, continued: (1) Draw details. (2) Draw assembly. See Figs. 13.57 and 13.59. If assigned, convert dimensions to decimal inches or redesign with metric dimensions.

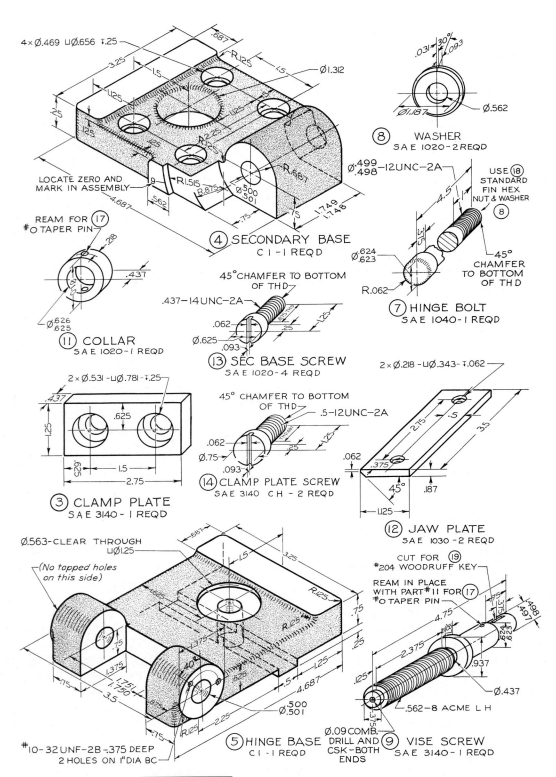

FIGURE 13.59 Grinder Vise.
Proj. 13.37, continued: See Fig. 13.58 for instructions.

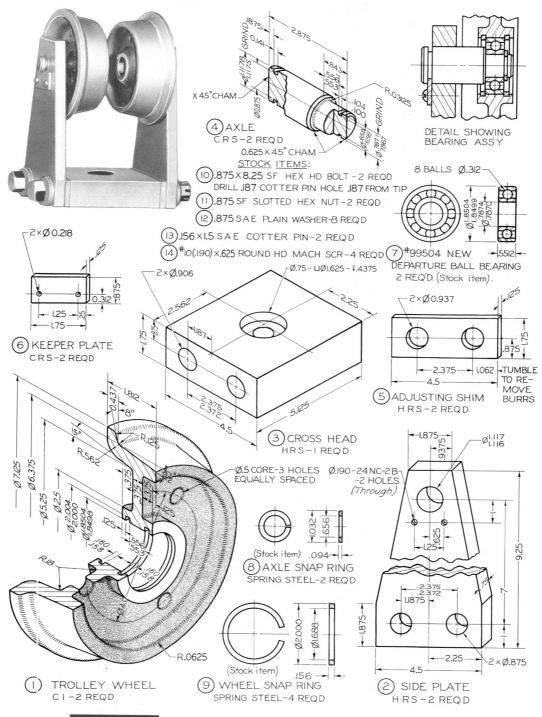

X 45°CHAM

(4) AXLE
C R S – 2 REQD
0.625 × 45° CHAM

0.875 × 8.25 SF HEX HD BOLT - 2 REQD

DETAIL SHOWING
BEARING ASSY

8 BALLS Ø.312

STOCK ITEMS:
(10) .875×8.25 SF HEX HD BOLT - 2 REQD
DRILL .187 COTTER PIN HOLE .187 FROM TIP
(11) .875 SF SLOTTED HEX NUT - 2 REQD
(12) .875 SAE PLAIN WASHER-8 REQD
(13) .156 ×1.5 SAE COTTER PIN-2 REQD
(14) #10(.190)×.625 ROUND HD MACH SCR-4 REQD

(7) #99504 NEW
DEPARTURE BALL BEARING
2 REQ'D. (Stock item).

2 × Ø 0.218

(6) KEEPER PLATE
C R S – 2 REQD

2 × Ø.906

Ø.75 – UØ1.625 – ⊤.4375

(3) CROSS HEAD
H R S – 1 REQD

2 × Ø0.937

(5) ADJUSTING SHIM
H R S – 2 REQD
TUMBLE TO RE-MOVE BURRS

Ø.5 CORE-3 HOLES EQUALLY SPACED   Ø.190-24NC-2B -2 HOLES (Through)

(8) AXLE SNAP RING
SPRING STEEL-2 REQD
(Stock item)

(1) TROLLEY WHEEL
C1–2 REQD

(9) WHEEL SNAP RING
SPRING STEEL-4 REQD
(Stock item)

(2) SIDE PLATE
H R S – 2 REQD

FIGURE 13.60 Trolley.
Proj. 13.38: (1) Draw details, omitting parts 7–14. (2) Draw assembly. If assigned, convert dimensions to decimal inches or redesign for metric dimensions.

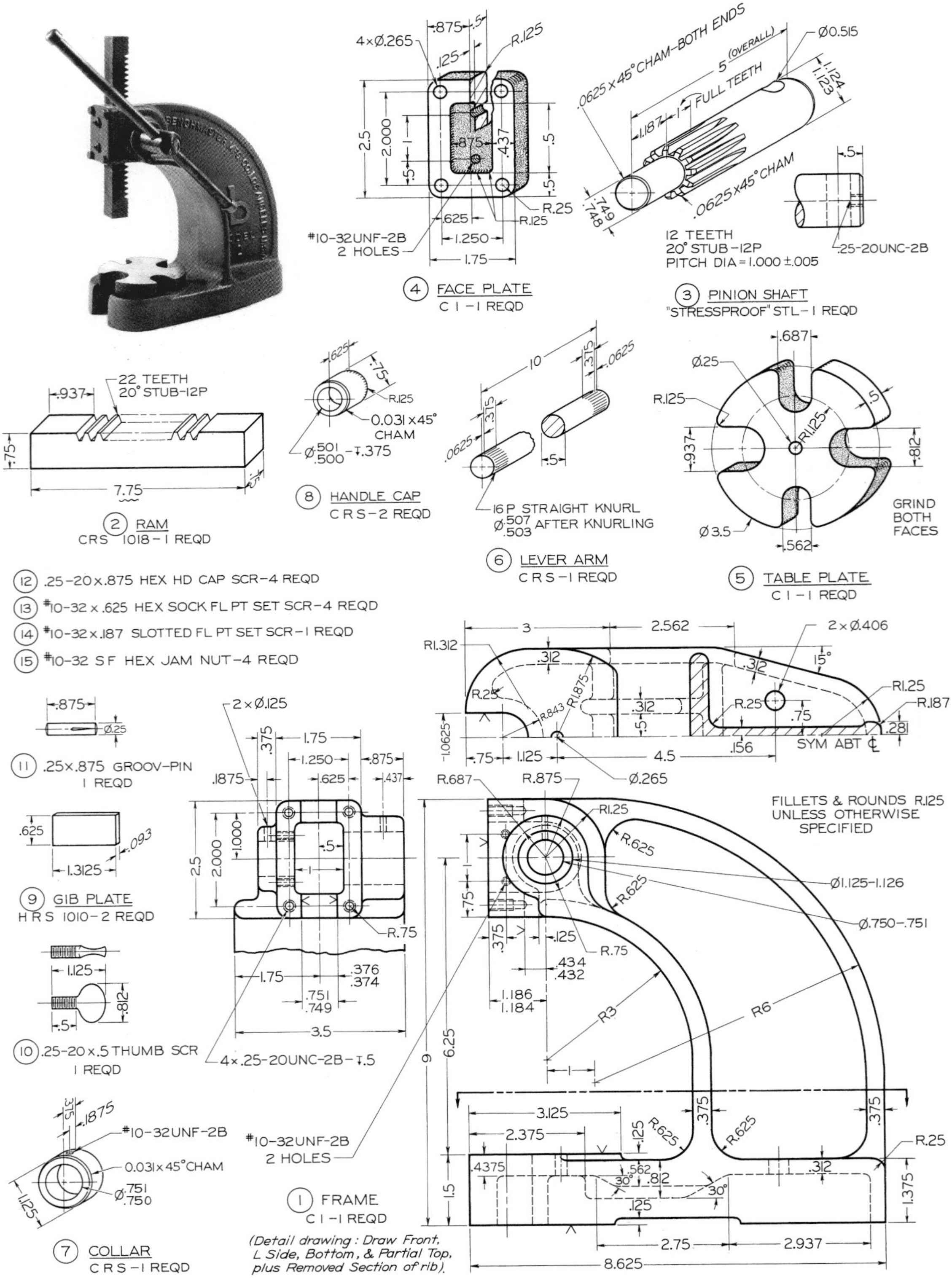

④ FACE PLATE
C I – I REQD

③ PINION SHAFT
"STRESSPROOF" STL – I REQD

② RAM
CRS 1018 – I REQD

⑧ HANDLE CAP
C R S – 2 REQD

⑥ LEVER ARM
C R S – I REQD

⑤ TABLE PLATE
C I – I REQD

⑫ .25–20 x .875 HEX HD CAP SCR–4 REQD

⑬ #10–32 x .625 HEX SOCK FL PT SET SCR–4 REQD

⑭ #10–32 x .187 SLOTTED FL PT SET SCR–I REQD

⑮ #10–32 S F HEX JAM NUT–4 REQD

⑪ .25 x .875 GROOV-PIN
I REQD

⑨ GIB PLATE
H R S 1010 – 2 REQD

⑩ .25–20 x .5 THUMB SCR
I REQD

⑦ COLLAR
C R S – I REQD

① FRAME
C I – I REQD

(Detail drawing : Draw Front,
L Side, Bottom, & Partial Top,
plus Removed Section of rib).

FILLETS & ROUNDS R.125
UNLESS OTHERWISE
SPECIFIED

FIGURE 13.61 Arbor Press.
Proj. 13.39: (1) Draw details. (2) Draw assembly. If assigned, convert dimensions to decimal inches or redesign for metric dimensions.

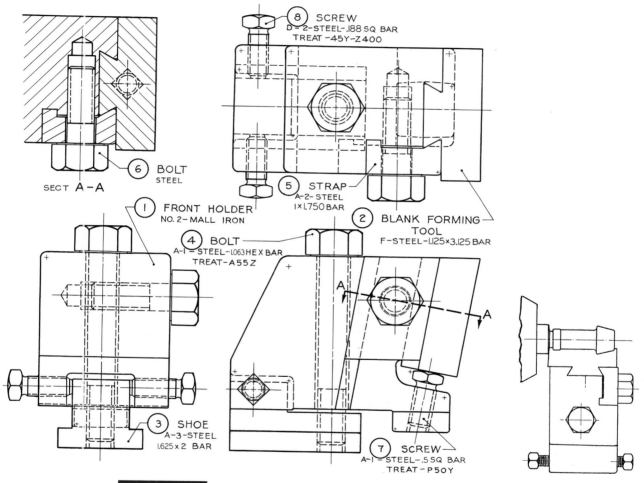

SECT A-A

⑥ BOLT
STEEL

⑧ SCREW
D-2-STEEL-.188 SQ BAR
TREAT -45Y-Z400

⑤ STRAP
A-2-STEEL
1×1.750 BAR

② BLANK FORMING
TOOL
F-STEEL-1.125×3.125 BAR

① FRONT HOLDER
NO. 2- MALL IRON

④ BOLT
A-1 STEEL-1.063 HEX BAR
TREAT-A55Z

A ── A

③ SHOE
A-3-STEEL
1.625×2 BAR

⑦ SCREW
A-1-STEEL-.5 SQ BAR
TREAT -P50Y

FIGURE 13.62 Forming Cutter Holder.
Proj. 13.40: (1) Draw details using decimal or metric dimensions. (2) Draw assembly. Above layout is half size. To obtain dimensions, measure directly from the figure and double your measurements. At left is shown the top view of the forming cutter holder in use on the lathe.

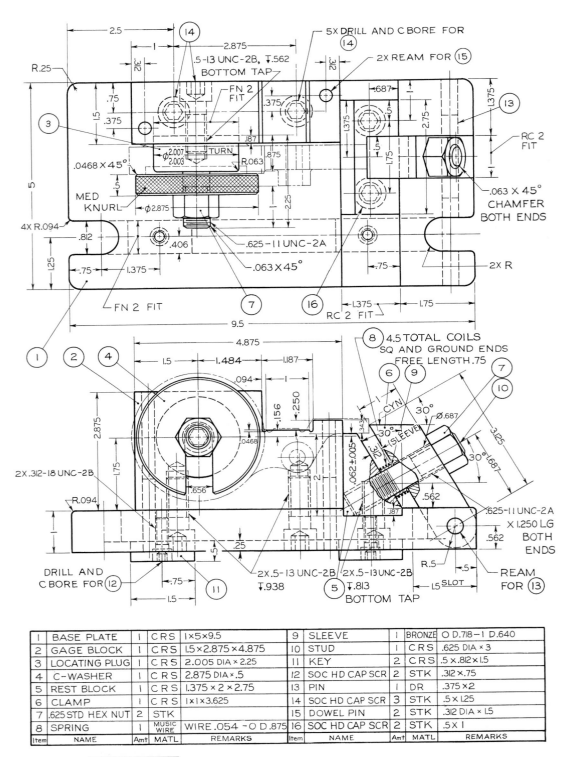

1	BASE PLATE	1	CRS	1×5×9.5	9	SLEEVE	1	BRONZE	O D.718 –I D.640
2	GAGE BLOCK	1	CRS	1.5×2.875×4.875	10	STUD	1	CRS	.625 DIA × 3
3	LOCATING PLUG	1	CRS	2.005 DIA × 2.25	11	KEY	2	CRS	.5 ×.812×1.5
4	C–WASHER	1	CRS	2.875 DIA×.5	12	SOC HD CAP SCR	2	STK	.312 ×.75
5	REST BLOCK	1	CRS	1.375 × 2 ×2.75	13	PIN	1	DR	.375 ×2
6	CLAMP	1	CRS	1×1×3.625	14	SOC HD CAP SCR	3	STK	.5 × 1.25
7	.625 STD HEX NUT	2	STK		15	DOWEL PIN	2	STK	.312 DIA × 1.5
8	SPRING	1	MUSIC WIRE	WIRE .054 – O D.875	16	SOC HD CAP SCR	2	STK	.5×1
Item	NAME	Amt	MATL	REMARKS	Item	NAME	Amt	MATL	REMARKS

FIGURE 13.63 Milling Fixture for Clutch Arm.
Proj. 13.41: (1) Draw details using the decimal-inch system or redesign for metric dimensions, if assigned. (2) Draw assembly.

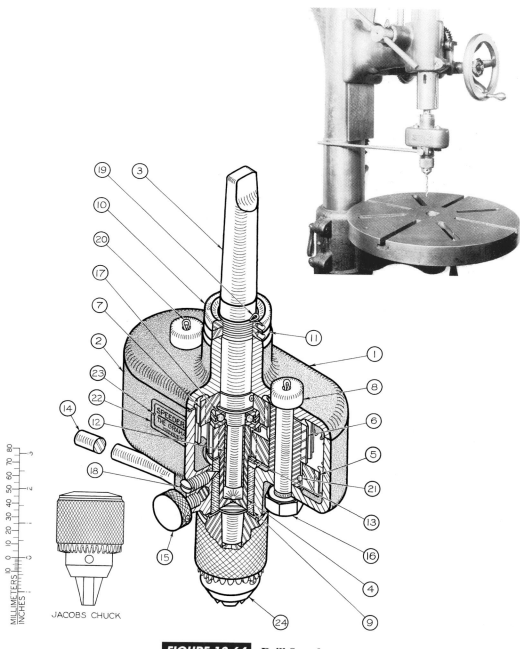

FIGURE 13.64 Drill Speeder.
Proj. 13.42: See Figs. 13.65 and 13.66

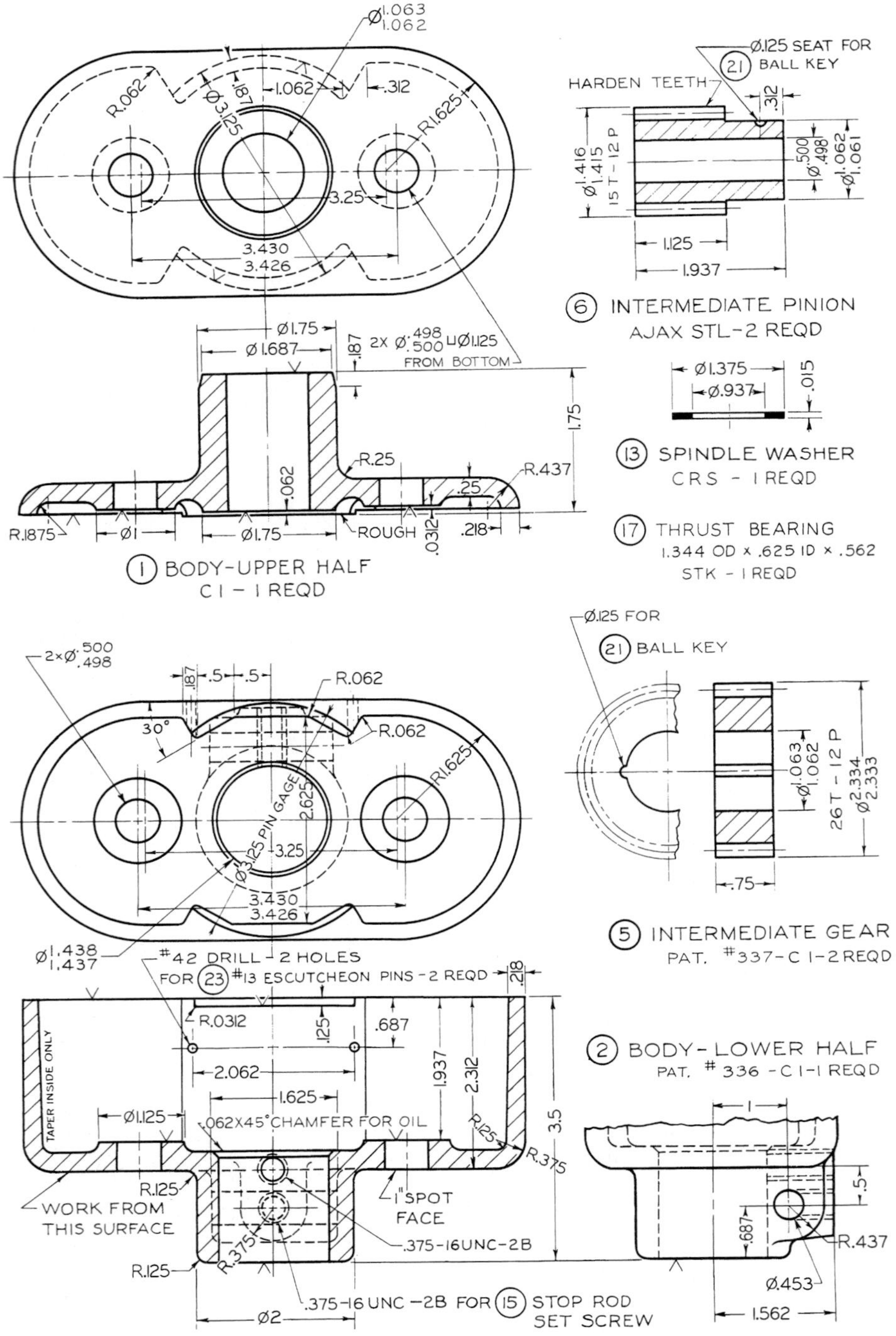

FIGURE 13.65 Drill Speeder.
Proj. 13.42, continued: (1) Draw details. (2) Draw assembly. See Fig.
13.64. If assigned, convert dimensions to decimal inches or redesign with
metric dimensions.

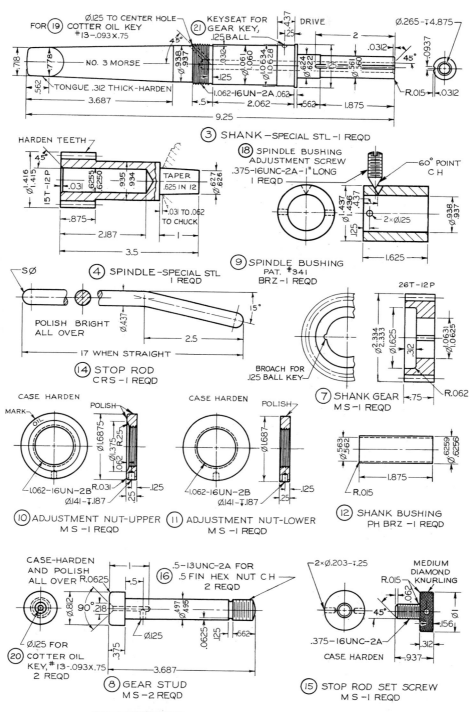

FIGURE 13.66   Drill Speeder.
Proj. 13.42, continued: See Fig. 13.65 for instructions.

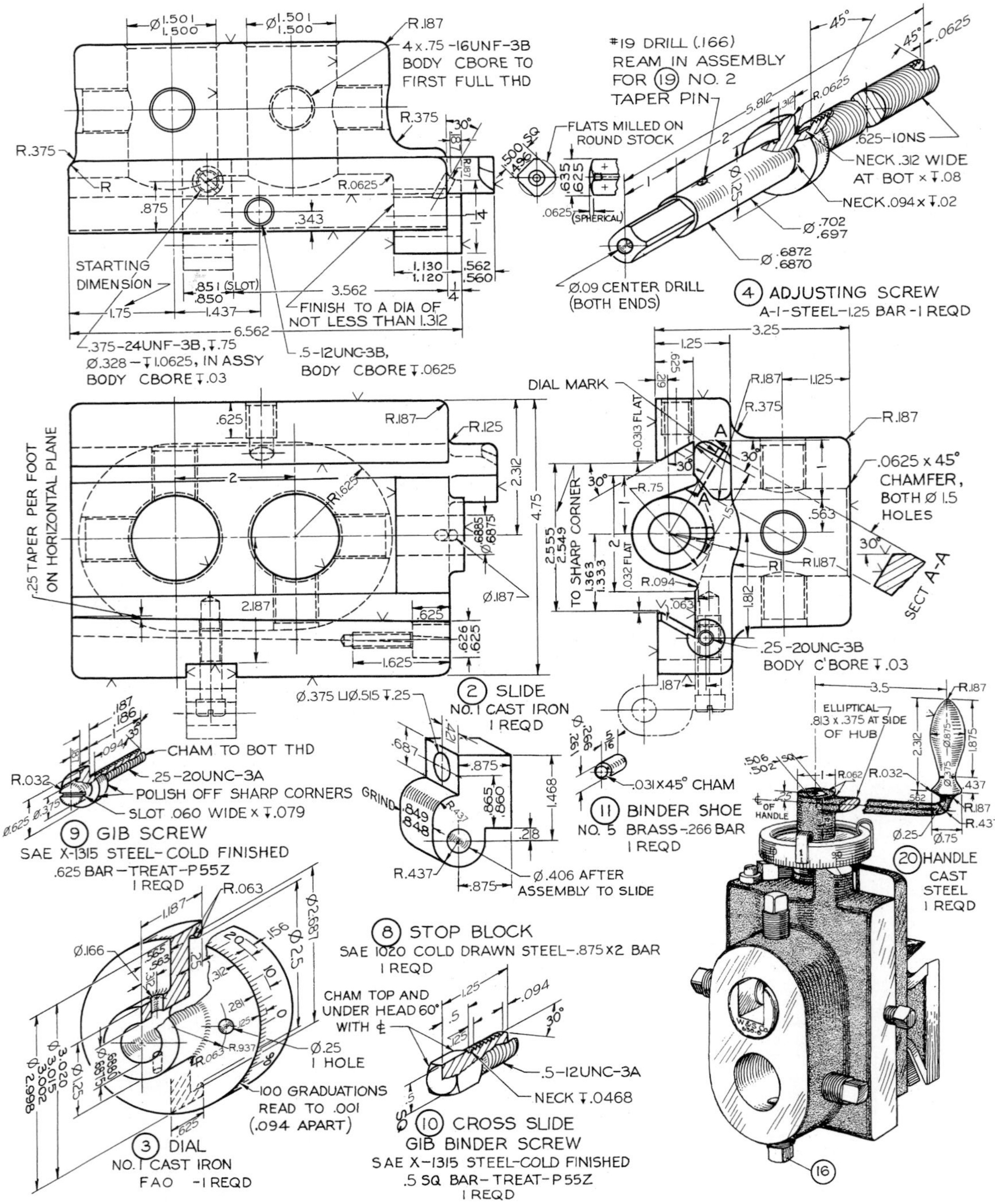

FIGURE 13.67 Vertical Slide Tool.

Proj. 13.43: (1) Draw details. If assigned, convert dimensions to decimal inches or redesign for metric system. (2) Draw assembly. For part 2: Take given top view as front view in the new drawing; then add top and right-side views. See also Fig. 13.68. If assigned, use unidirectional dimensions.

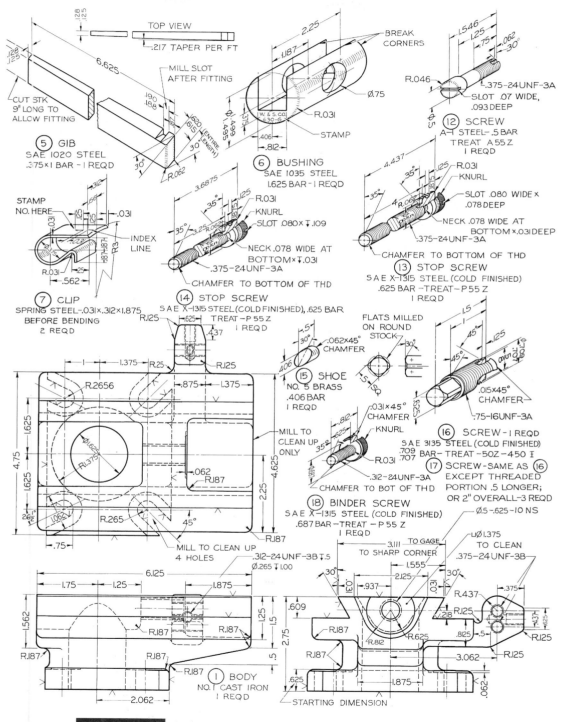

FIGURE 13.68 Vertical Slide Tool.
Proj. 13.43, continued: See Fig. 13.67 for instructions. For part 1: Take top view as front view in the new drawing; then add top and right-side views.

FIGURE 13.69   Slide Tool.
Proj. 13.44: Make assembly drawing. See Figs. 13.67–13.68.

PARTS LIST   NO. OF SHEETS ___2___   SHEET NO. ___1___   MACHINE NO. M-219

NAME ___ NO. 4 SLIDE TOOL (SPECIFY SIZE OF SHANK REQ'D.)   LOT NUMBER

NO. OF PIECES

TOTAL ON MACH.	NO. PCS.	NAME OF PART	PART NO.	CAST FROM PART NO.	TRACING NO.	MATERIAL	ROUGH WEIGHT PER PC.	DIA.	LENGTH	MILL	PART USED ON	NO. REQ. FINISH
	1	Body	219-12		D-17417	A-3-S D F						
	1	Slide	219-6		D-19255	A-3-S D F					219-12	
	1	Nut	219-9		E-19256	#10 BZ					219-6	
	1	Gib	219-1001		C-11129	S A E 1020					219-6	
	1	Slide Screw	219-1002		C-11129	A-3-S					219-12	
	1	Dial Bush.	219-1003		C-11129	A-1-S					219-1002	
	1	Dial Nut	219-1004		C-11129	A-1-S					219-1002	
	1	Handle	219-1011		E-18270	(Buy from Cincinnati Ball Crank Co.)					219-1002	
	1	Stop Screw (Short)	219-1012		E-51950	A-1-S					219-6	
	1	Stop Screw (Long)	219-1013		E-51951	A-1-S					219-6	
	1	Binder Shoe	219-1015		E-51952	#5 Brass					219-6	
	1	Handle Screw	219-1016		E-62322	X-1315 C.F.					219-1011	
	1	Binder Screw	219-1017		E-63927	A-1-S					219-6	
	1	Dial	219-1018		E-39461	A-1-S					219-1002	
	2	Gib Screw	219-1019		E-52777	A-1-S		$\frac{1}{4}$-20	1		219-6	
	1	Binder Screw	280-1010		E-24962	A-1-S					219-1018	
	2	Tool Clamp Screws	683-F-1002		E-19110	D-2-S					219-6	
	1	Fill Hd Cap Scr	1-A			A-1-S		$\frac{3}{8}$	$1\frac{3}{8}$		219-6 219-9	
	1	Key	No. 404 Woodruff								219-1002	

FIGURE 13.70   Proj. 13.45: Slide Tool
Parts List.

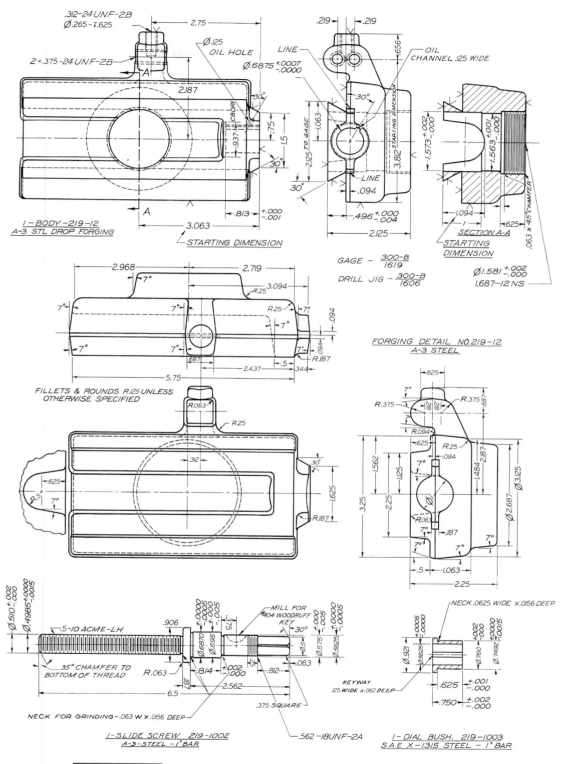

FIGURE 13.71 Slide Tool.

Proj. 13.45, continued: (1) Draw details using decimal-inch dimensions or redesign with metric dimensions, if assigned. (2) Draw assembly. See Fig. 13.70.

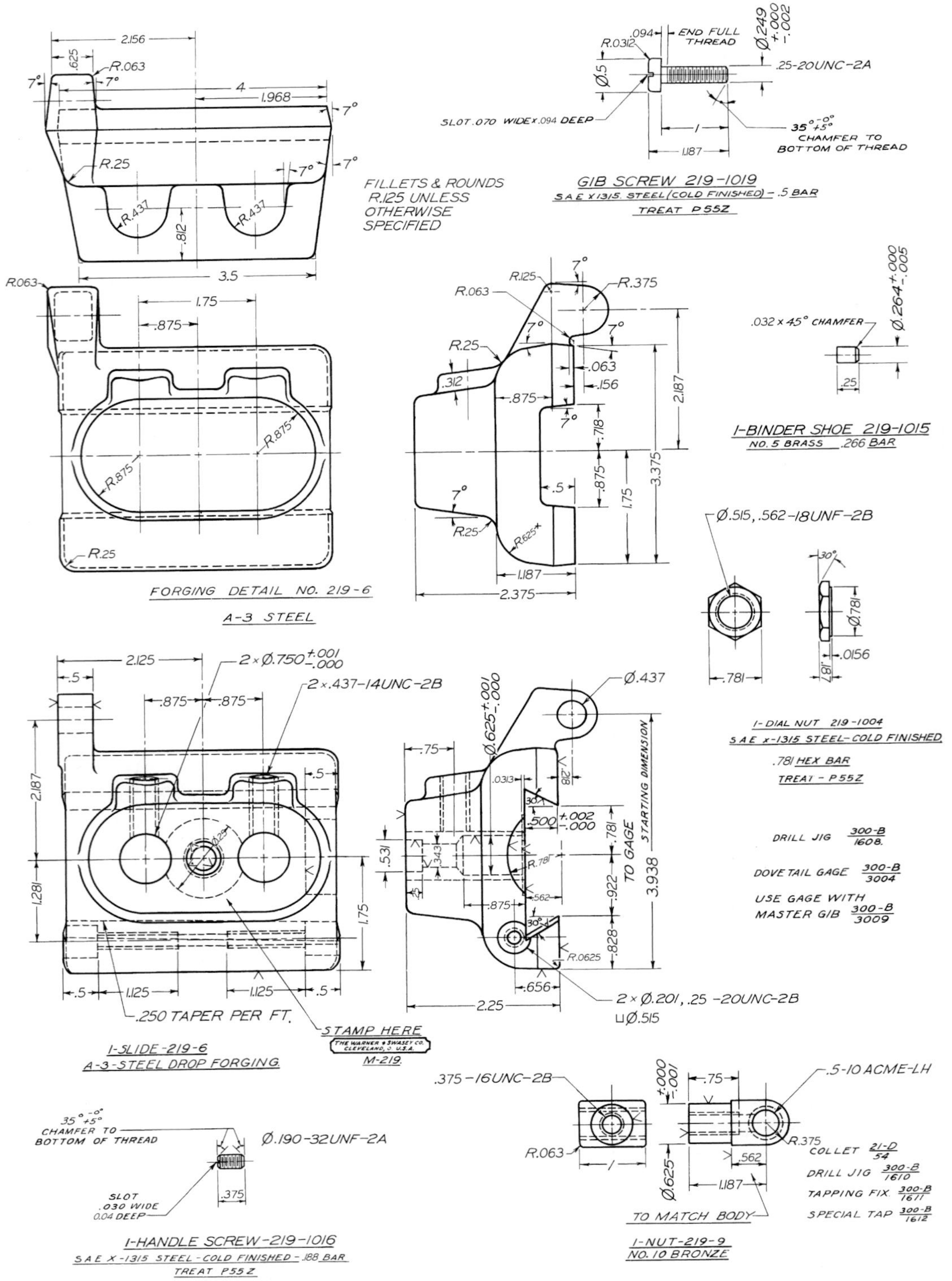
FILLETS & ROUNDS R.125 UNLESS OTHERWISE SPECIFIED
FORGING DETAIL NO. 219-6
A-3 STEEL
GIB SCREW 219-1019
SAE X1315 STEEL (COLD FINISHED) - .5 BAR
TREAT P55Z
SLOT .070 WIDE x .094 DEEP
END FULL THREAD
.25-20UNC-2A
CHAMFER TO BOTTOM OF THREAD
.032 x 45° CHAMFER
1-BINDER SHOE 219-1015
NO. 5 BRASS .266 BAR
Ø.515, .562-18UNF-2B
1-DIAL NUT 219-1004
SAE X-1315 STEEL-COLD FINISHED
.781 HEX BAR
TREAT - P55Z
DRILL JIG 300-B/1608
DOVE TAIL GAGE 300-B/3004
USE GAGE WITH MASTER GIB 300-B/3009
2 x Ø.750 +.001 -.000
2 x .437-14UNC-2B
STARTING DIMENSION TO GAGE
.250 TAPER PER FT.
STAMP HERE
THE WARNER & SWASEY CO. CLEVELAND, O. U.S.A.
M-219
1-SLIDE-219-6
A-3-STEEL DROP FORGING
2 x Ø.201, .25-20UNC-2B
⌴Ø.515
.375-16UNC-2B
.5-10 ACME-LH
COLLET 21-D/54
DRILL JIG 300-B/1610
TAPPING FIX. 300-B/1611
SPECIAL TAP 300-B/1612
TO MATCH BODY
1-NUT-219-9
NO. 10 BRONZE
Ø.190-32UNF-2A
SLOT .030 WIDE 0.04 DEEP
1-HANDLE SCREW-219-1016
SAE X-1315 STEEL-COLD FINISHED-.188 BAR
TREAT P55Z

FIGURE 13.72 **Slide Tool.**
Proj. 13.45, continued: See. Fig. 13.71 for instructions.

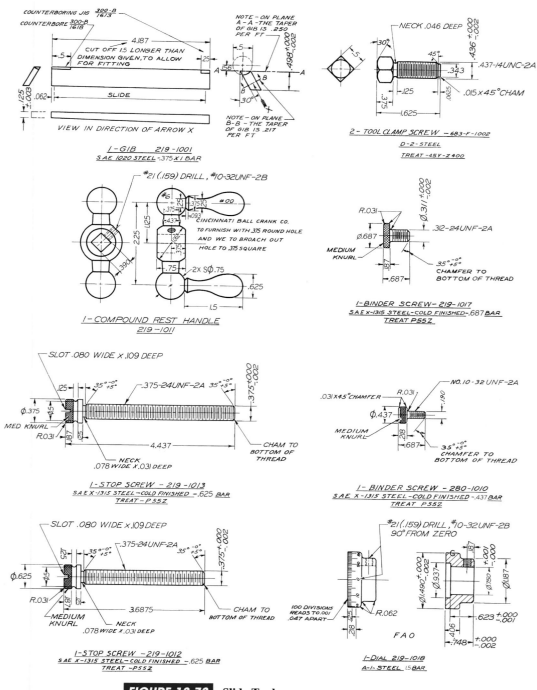

FIGURE 13.73 Slide Tool.
Proj. 13.45, continued: See Fig. 13.71 for instructions.

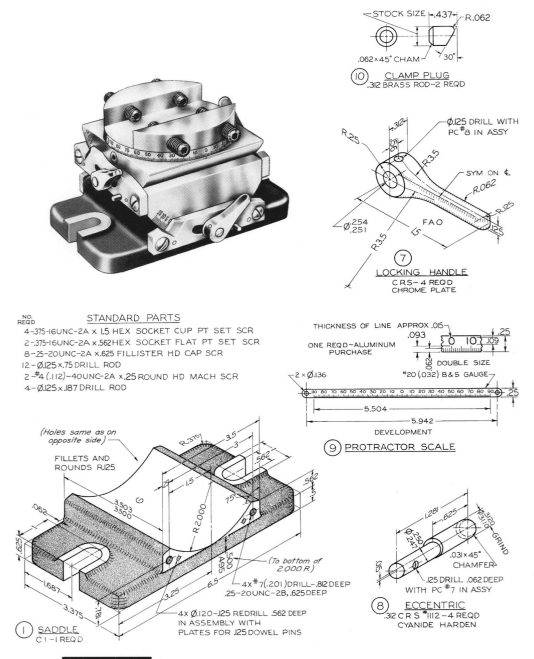

STOCK SIZE .437 R.062

.062×45° CHAM 30°

10 CLAMP PLUG
.312 BRASS ROD–2 REQD

Ø.125 DRILL WITH
PC #8 IN ASSY

R.25 .312

R.3.5

SYM ON ₵

R.062

Ø.254
.251

R.3.5

FAO

R.125

1.5

.125

7 LOCKING HANDLE
C R S– 4 REQD
CHROME PLATE

NO. REQD   STANDARD PARTS
4–.375–16UNC–2A × 1.5 HEX SOCKET CUP PT SET SCR
2–.375–16UNC–2A ×.562 HEX SOCKET FLAT PT SET SCR
8–.25–20UNC–2A ×.625 FILLISTER HD CAP SCR
12–Ø.125×.75 DRILL ROD
2–#4 (.112)–40UNC–2A ×.25 ROUND HD MACH SCR
4–Ø.125×.187 DRILL ROD

THICKNESS OF LINE APPROX .015
ONE REQD–ALUMINUM PURCHASE
.093
0 10
.25
.109
.062
DOUBLE SIZE
#20 (.032) B & S GAUGE
2 × Ø.136
90 80 70 60 50 40 30 20 10 0 10 20 30 40 50 60 70 80 90
.25
5.504
5.942
DEVELOPMENT

9 PROTRACTOR SCALE

(Holes same as on opposite side)

FILLETS AND ROUNDS R.125

R.375   3.5
3
.562
.25
1.5
15°
.562
1.5

.062

.625

3.503
3.500
R.2.000

1.687

.500
.495

1.187

3.375

3.25   6.5

(To bottom of 2.000 R)

4×#7(.201)DRILL–.812 DEEP
.25–20UNC–2B,.625 DEEP

4x Ø.120–.125 REDRILL .562 DEEP
IN ASSEMBLY WITH
PLATES FOR .125 DOWEL PINS

1 SADDLE
C I –1 REQD

1.281   .625

Ø.250
.241

.031×45°
CHAMFER

.031/.120–GRIND

.05

.125 DRILL .062 DEEP
WITH PC #7 IN ASSY

8 ECCENTRIC
.312 C R S #1112 –4 REQD
CYANIDE HARDEN

FIGURE 13.74   "Any Angle" Tool Vise.
**Proj. 13.46: (1) Draw details using decimal-inch dimensions or redesign
with metric dimensions, if assigned. (2) Draw assembly. See also Fig. 13.75.**

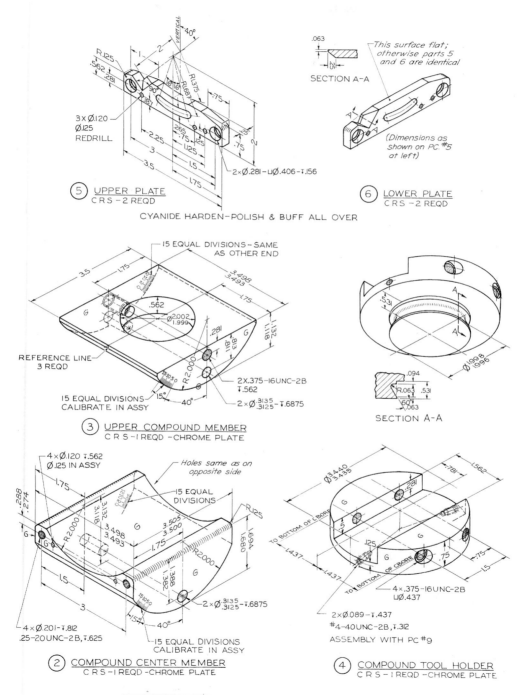

FIGURE 13.75 "Any-Angle" Tool Vise.
Proj. 13.46, continued: See Fig. 13.74 for instructions.

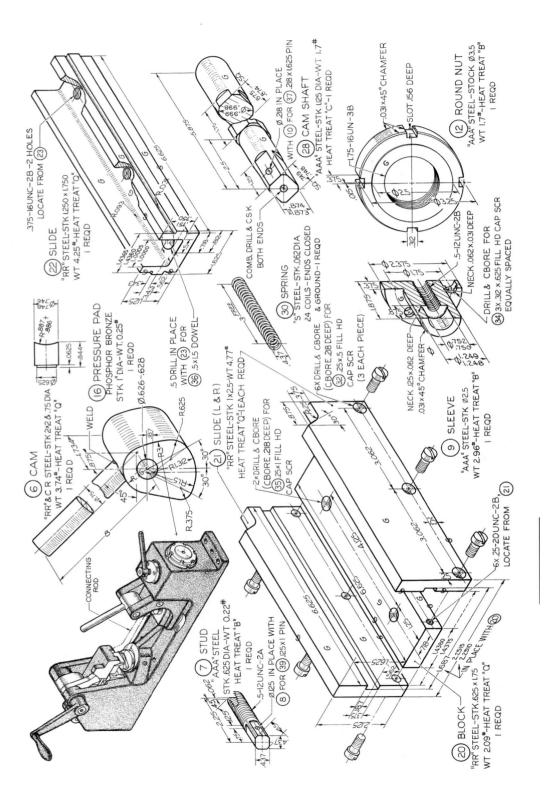

FIGURE 13.76 Fixture for Centering Connecting Rod.
Proj. 13.47: (1) Draw details using decimal-inch dimensions or redesign with metric dimensions, if assigned. (2) Draw assembly. See also Figs. 13.77 and 13.78.

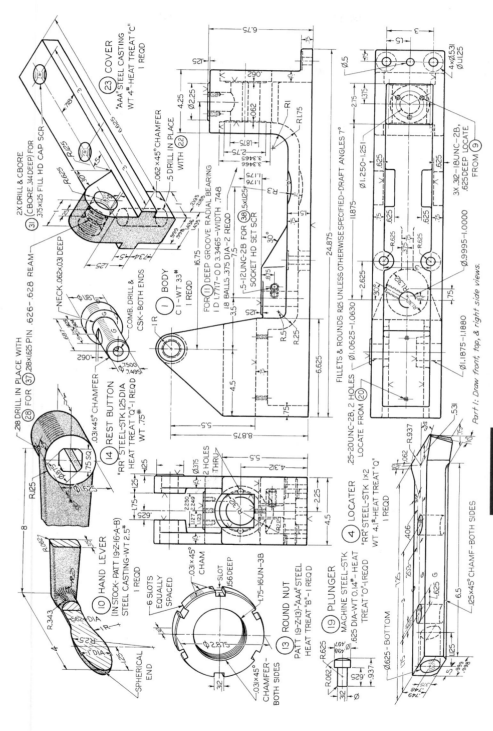

FIGURE 13.77 Fixture for Centering Connecting Rod. Proj. 13.47, continued: See Fig. 13.76 for instructions.

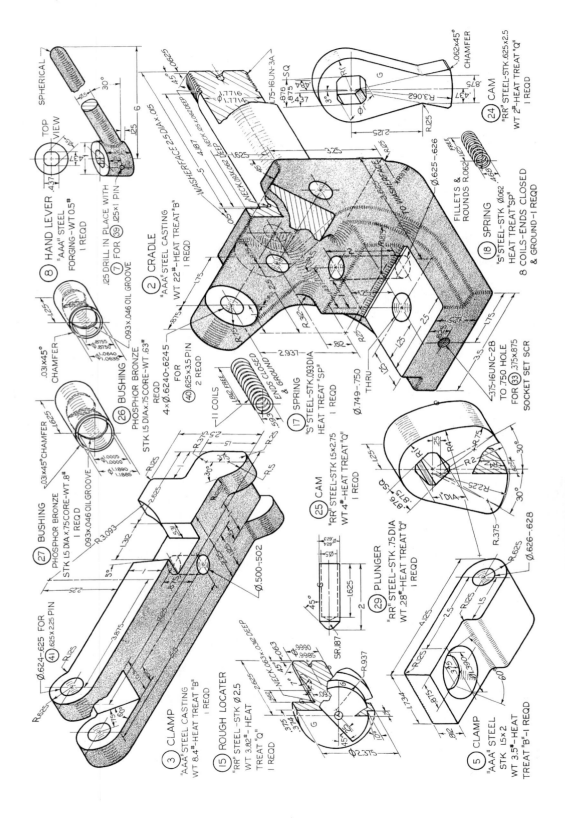

FIGURE 13.78 **Fixture for Centering Connecting Rod.**
Proj. 13.47, continued: See Fig. 13.76 for instructions.

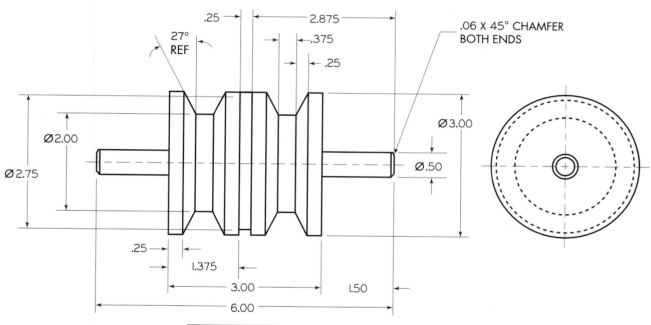

.25

2.875

.375

27°
REF

.25

.06 X 45° CHAMFER
BOTH ENDS

Ø2.00

Ø2.75

Ø3.00

Ø.50

.25

1.375

3.00

1.50

6.00

FIGURE 13.79   Alignment Wheel.
Prob. 14.64. (1) Draw top and left-side views.
(2) Redesign with metric dimensions.

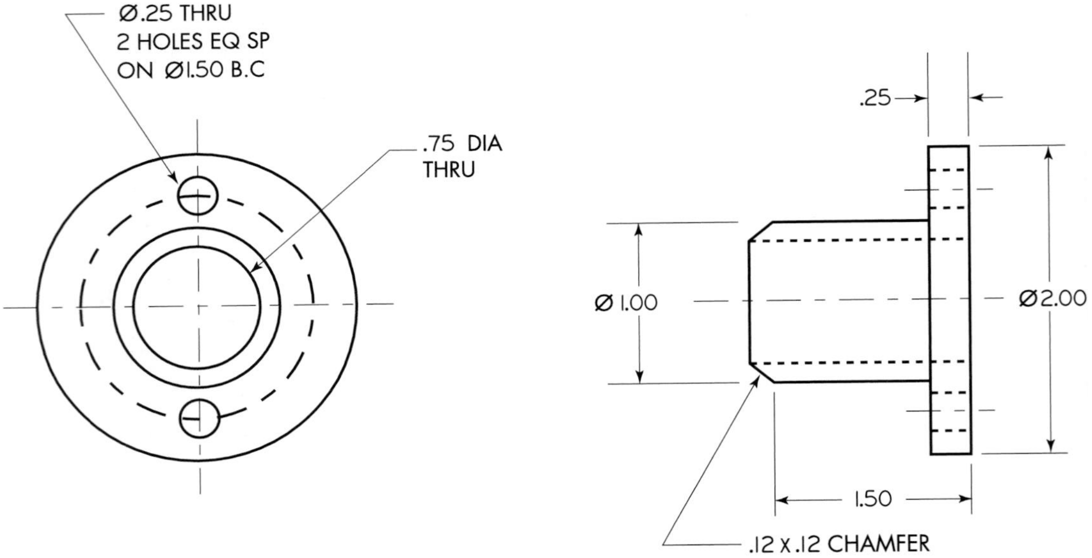

Ø.25 THRU
2 HOLES EQ SP
ON Ø1.50 B.C

.75 DIA
THRU

.25

Ø 1.00

Ø 2.00

1.50

.12 X .12 CHAMFER

FIGURE 13.80   **Rubber Bushing.**
Proj. 13.49. (1) Draw top and left-side views.
(2) Redesign with metric dimensions.

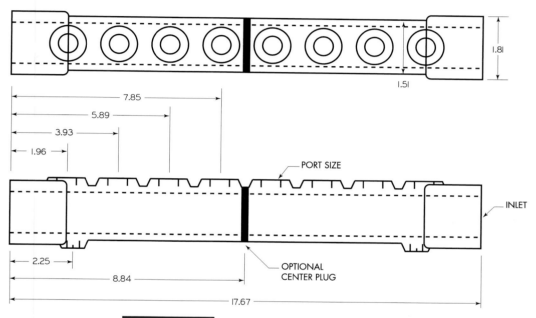

FIGURE 13.81 Proj. 13.50. 8-Port Nylon
Manifold. Draw bottom and left-side view. Redesign
with metric dimensions.

APPENDIX

1 Bibliography of American National Standards

American National Standards Institute, 11 West 42nd St., New York, NY. 10036. For complete listing of standards, see ANSI catalog of American National Standards.

Abbreviations

Abbreviations for Use on Drawings and in Text, ANSI/ASME Y1.1–1989

Bolts, Screws, and Nuts

Bolts, Metric Heavy Hex, ANSI B18.2.3.6M–1979 (R1995)

Bolts, Metric Heavy Hex Structural, ANSI B18.2.3.7M–1979 (R1995)

Bolts, Metric Hex, ANSI B18.2.3.5M–1979 (R1995)

Bolts, Metric Round Head Short Square Neck, ANSI/ASME B18.2.2.1M–1981 (R1995)

Bolts, Metric Round Head Square Neck, ANSI/ASME B18.5.2.2M–1982 (R1993)

Hex Jam Nuts, Metric, ANSI B18.2.4.5M–1979 (R1990)

Hex Nuts, Heavy, Metric, ANSI B18.2.4.6M–1979 (R1990)

Hex Nuts, Slotted, Metric, ANSI/ASME B18.2.4.3M–1979 (R1995)

Hex Nuts, Style 1, Metric, ANSI/ASME B18.2.4.1M–1979 (R1995)

Hex Nuts, Style 2, Metric, ANSI/ASME B18.2.4.2M–1979 (R1995)

Hexagon Socket Flat Countersunk Head Cap Screws (Metric Series), ANSI/ASME B18.3.5M–1986 (R1993)

Mechanical Fasteners, Glossary of Terms, ANSI B18.12–1962 (R1995)

Miniature Screws, ANSI B18.11–1961 (R1992)

Nuts, Metric Hex Flange, ANSI B18.2.4.4M–1982 (R1993)

Plow Bolts, ANSI/ASME B18.9–1958 (R1995)

Round Head Bolts, Metric Round Head Short Square Neck, ANSI/ASME B18.5.2.1M–1981 (R1995)

Screws, Hexagon Socket Button Head Cap, Metric Series, ANSI/ASME B18.3.4M–1986 (R1993)

Screws, Hexagon Socket Head Shoulder, Metric Series, AN-SI/ASME B18.3.3M–1986 (R1993)

Screws, Hexagon Socket Set, Metric Series, ANSI/ASME B18.3.6M–1986 (R1993)

Screws, Metric Formed Hex, ANSI/ASME B18.2.3.2M–1979 (R1995)

Screws, Metric Heavy Hex, ANSI/ASME B18.2.3.3M–1979 (R1995)

Screws, Metric Hex Cap, ANSI/ASME B18.2.3.1M–1979 (R1995)

Screws, Metric Hex Flange, ANSI/ASME B18.2.3.4M–1984 (R1995)

Screws, Metric Hex Lag, ANSI B18.2.3.8M–1981 (R1991)

Screws, Metric Machine, ANSI/ASME B18.6.7M–1985 (R1993)

Screws, Socket Head Cap, Metric Series, ANSI/ASME B18.3.1M–1986 (R1993)

Screws, Tapping and Metallic Drive, Inch Series, Thread Forming and Cutting, ANSI B18.6.4–1981 (R1991)

Slotted and Recessed Head Machine Screws and Machine Screw Nuts, ANSI B18.6.3–1972 (R1991)

Slotted Head Cap Screws, Square Head Set Screws, and Slotted Headless Set Screws, ANSI/ASME B18.6.2–1995

Socket Cap, Shoulder, and Set Screws (Inch Series) ANSI/ASME B18.3–1986 (R1995)

Square and Hex Bolts and Screws, Inch Series, ANSI B18.2.1–1981 (R1992)

Square and Hex Nuts (Inch Series) ANSI/ASME B18.2.2–1987 (R1993)

Track Bolts and Nuts, ANSI/ASME B18.10–1982 (R1992)

Wood Screws, Inch Series, ANSI B18.6.1–1981 (R1991)

Dimensioning and Surface Finish

General Tolerances for Metric Dimensioned Products, ANSI B4.3–1978 (R1994)

Preferred Limits and Fits for Cylindrical Parts, ANSI B4.1–1967 (R1994)

Preferred Metric Limits and Fits, ANSI B4.2–1978 (R1994)

Surface Texture, ANSI/ASME B46.1–1995

Drafting Manual (Y14)

Casting and Forgings, ANSI/ASME Y14.BM–1989

Decimal Inch, Drawing Sheet Size and Format, ANSI/ASME Y14.1–1995

Dimensioning and Tolerancing, ANSI/ASME Y14.5M–1994

Electrical and Electronics Diagrams, ANSI Y14.15–1966 (R1988)

Electrical and Electronics Diagrams—Supplement, ANSI Y14.15a–1971 (R1988)

Electrical and Electronics Diagrams—Supplement, ANSI Y14.15b–1973 (R1988)

Engineering Drawings, Types, and Applications, ANSI/ASME Y14.24M–1989. Revision of Engineering Drawings, ANSI/ASME Y14.35M–1992

Gear and Spline Drawing Standards—Part 2, Bevel and Hypoid Gears, ANSI Y14.7.2–1978 (R1994)

Gear Drawing Standards—Part 1, for Spur, Helical, Double Helical, and Rack, ANSI/ASME Y14.7.1–1971 (R1993)

Line Conventions and Lettering, ANSI/ASME Y14.2M–1992

Mechanical Spring Representation, ANSI/ASME Y14.13M–1981 (R1992)

Metric Drawing Sheet Size and Format, ANSI/ASME Y14.1M–1995

Multiview and Sectional View Drawings, ANSI/ASME Y14.3M–1994

Parts Lists, Data Lists, and Index Lists, ANSI/ASME Y14.34M–1990

Pictorial Drawing, ANSI/ASME Y14.4M–1989 (R1994)

Screw Thread Representation, ANSI/ASME Y14.6–1978 (R1993)

Screw Thread Representation, Metric, ANSI/ASME Y14.6aM–1981 (R1993)

Surface Texture Symbols, ANSI/ASME Y14.36M–1996

Gears

Basic Gear Geometry, ANSI/AGMA 115.01–1989

Gear Nomenclature—Terms, Definitions, Symbols, and Abbreviations, ANSI/AGMA 1012–F90

Nomenclature of Gear-Tooth Failure Modes, ANSI/AGMA 110.04–1980 (R1989)

Design Manual for Bevel Gearing, ANSI/AGMA 2005–B88

Tooth Proportions for Fine-Pitch Spur and Helical Gears, ANSI/AGMA 1003–G93

Graphic Symbols

Public Fire Safety Symbols, ANSI/NFPA 170–1994

Graphic Symbols for Electrical and Electronics Diagrams, ANSI/IEEE 315–1975 (R1994)

Graphic Symbols for Electrical Wiring and Layout Diagrams Used in Architecture and Building Construction, ANSI Y32.9–1972 (R1989)

Graphic Symbols for Fluid Power Diagrams, ANSI/ASME Y32.10–1967 (R1994)

Graphic Symbols for Grid and Mapping Used in Cable Television Systems, ANSI/IEEE 623–1976 (R1989)

Graphic Symbols for Heat-Power Apparatus, ANSI Y32.2.6M–1950 (R1993)

Graphic Symbols for Heating, Ventilating and Air Conditioning, ANSI Y32.2.4–1949 (R1993)

Graphic Symbols for Logic Functions, ANSI/IEEE 91–1984

Graphic Symbols for Pipe Fittings, Valves, and Piping, ANSI/ASME Y32.2.3–1949 (R1994)

Graphic Symbols for Plumbing Fixtures for Diagrams Used in Architecture and Building Construction, ANSI/ASME Y32.4–1977 (R1994)

Graphic Symbols for Process Flow Diagrams in the Petroleum and Chemical Industries, ANSI Y32.11–1961 (R1993)

Graphic Symbols for Railroad Maps and Profiles, ANSI/ASME Y32.7–1972 (R1994)

Instrumentation Symbols and Identification, ANSI/ISA S5.1–1984 (R1992)

Reference Designations for Electrical and Electronics Parts and Equipment, ANSI/IEEE 200–1975 (R1989)

Symbols for Mechanical and Acoustical Elements as Used in Schematic Diagrams, ANSI Y32.18–1972 (R1993)

Symbols for Welding, Brazing, and Nondestructive Examination, ANSI/AWS A2.4–93

Keys and Pins

Clevis Pins and Cotter Pins, ANSI/ASME B18.8.1–1994

Hexagon Keys and Bits (Metric Series), ANSI B18.3.2M–1979 (R1994)

Keys and Keyseats, ANSI B17.1–1967 (R1989)

Pins—Taper Pins, Dowel Pins, Straight Pins, Grooved Pins and Spring Pins (Inch Series), ANSI/ASME B18.8.2–1994

Woodruff Keys and Keyseats, ANSI B17.2–1967 (Rl990)

Piping

Cast Bronze Threaded Fittings, Class 125 and 250, ANSI/ASME B16.15–1985 (R1994)

Cast Copper Alloy Pipe Flanges and Flanged Fittings, ANSI/ASME B16.24–1991

Cast Iron Pipe Flanges and Flanged Fittings, Class 25, 125, 250 and 800, ANSI/ASME B16.1–1989

Gray Iron Threaded Fittings, ANSI/ASME B16.4–1992

Ductile Iron Pipe, Centrifugally Cast, ANSI/AWWA C151/A21.51–91

Factory-Made Wrought Steel Buttwelding Fittings, ANSI/ASME B16.9–1993

Ferrous Pipe Plugs, Bushings, and Locknuts with Pipe Threads, ANSI/ASME B16.14–1991

Flanged Ductile-Iron Pipe with Threaded Flanges, ANSI/AWWA C115/A21.15–94

Malleable-Iron Threaded Fittings, ANSI/ASME B16.3–1992

Pipe Flanges and Flanged Fittings, ANSI/ASME B16.5–1988

Stainless Steel Pipe, ANSI/ASME B36.19M–1985 (R1994)

Welded and Seamless Wrought Steel Pipe, ANSI/ASME B36.10M–1995

Rivets

Large Rivets ($\frac{1}{2}$ Inch Nominal Diameter and Larger), ANSI/ASME B18.1.2–1972 (R1995)

Small Solid Rivets ($\frac{7}{16}$ Inch Nominal Diameter and Smaller), ANSI/ASME B18.1.1–1972 (R1995)

Small Solid Rivets, Metric, ANSI/ASME B18.1.3M–1983 (R1995)

Small Tools and Machine Tool Elements

Jig Bushings, ANSI B94.33–1974 (R1994)

Machine Tapers, ANSI/ASME B5.10–1994

Milling Cutters and End Mills, ANSI/ASME B94.19–1985

Reamers, ANSI/ASME B94.2–1995

T-Slots—Their Bolts, Nuts and Tongues, ANSI/ASME B5.1M–1985 (R1992)

Twist Drills, ANSI/ASME B94.11M–1993

Threads

Acme Screw Threads, ANSI/ASME B1.5–1988 (R1994)

Buttress Inch Screw Threads, ANSI B1.9–1973 (R1992)

Class 5 Interference–Fit Thread, ANSI/ASME B1.12–1987 (R1992)

Dryseal Pipe Threads (Inch), ANSI B1.20.3–1976 (R1991)

Hose Coupling Screw Threads, ANSI/ASME B1.20.7–1991

Metric Screw Threads—M Profile, ANSI/ASME B1.13M–1995

Metric Screw Threads—MJ Profile, ANSI/ASME B1.21M–1978

Nomenclature, Definitions and Letter Symbols for Screw Threads, ANSI/ASME B1.7M–1984 (R1992)

Pipe Threads, General Purpose (Inch), ANSI/ASME B1.20.1–1983 (R1992)

Stub Acme Threads, ANSI/ASME B1.8–1988 (R1994)

Unified Screw Threads (UN and UNR Thread Form), ANSI/ASME B1.1–1989

Unified Miniature Screw Threads, ANSI B1.10–1958 (R1988)

Washers

Lock Washers, Inch, ANSI/ASME B18.21.1–1994

Lock Washers, Metric, ANSI/ASME B18.21.2M–1994

Plain Washers, ANSI B18.22.1–1965 (R1981)

Plain Washers, Metric, ANSI B18.22M–1981

Miscellaneous

Knurling, ANSI/ASME B94.6–1984 (R1995)

Preferred Metric Sizes for Flat Metal Products, ANSI/ASME B32.3M–1984 (R1994)

Preferred Metric Equivalents of Inch Sizes for Tubular Metal Products Other Than Pipe, ANSI/ASME B32.6M–1984 (R1994)

Preferred Metric Sizes for Round, Square, Rectangle and Hexagon Metal Products, ANSI B32.4M–1980 (R1994)

Preferred Metric Sizes for Tubular Metal Products Other Than Pipe, ANSI B32.5–1977 (R1994)

Preferred Thickness for Uncoated Thin Flat Metals (Under 0.250 in.), ANSI B32.1–1952 (R1994)

Surface Texture (Surface Roughness, Waviness and Lay), ANSI/ASME B46.1–1995

Technical Drawings, ISO Handbook, 12–1991

2 Technical Terms

"The beginning of wisdom is to call things by their right names."

—CHINESE PROVERB

n *means a* noun; v *means a* verb

acme (*n*) Screw thread form.

addendum (*n*) Radial distance from pitch circle to top of gear tooth.

allen screw (*n*) Special set screw or cap screw with hexagon socket in head.

allowance (*n*) Minimum clearance between mating parts.

alloy (*n*) Two or more metals in combination, usually a fine metal with a baser metal.

aluminum (*n*) A lightweight but relatively strong metal. Often alloyed with copper to increase hardness and strength.

angle iron (*n*) A structural shape whose section is a right angle.

anneal (*v*) To heat and cool gradually, to reduce brittleness and increase ductility.

arc-weld (*v*) To weld by electric arc. The work is usually the positive terminal.

babbitt (*n*) A soft alloy for bearings, mostly of tin with small amounts of copper and antimony.

bearing (*n*) A supporting member for a rotating shaft.

bevel (*n*) An inclined edge, not at right angle to joining surface.

bolt circle (*n*) A circular center line on a drawing, containing the centers of holes about a common center.

bore (*v*) To enlarge a hole with a boring mill.

boss (*n*) A cylindrical projection on a casting or a forging.

BOSS

brass (*n*) An alloy of copper and zinc.

braze (*v*) To join with hard solder of brass or zinc.

Brinell (*n*) A method of testing hardness of metal.

broach (*n*) A long cutting tool with a series of teeth that gradually increase in size which is forced through a hole or over a surface to produce a desired shape.

bronze (*n*) An alloy of eight or nine parts of copper and one part of tin.

buff (*v*) To finish or polish on a buffing wheel composed of fabric with abrasive powders.

burnish (*v*) To finish or polish by pressure upon a smooth rolling or sliding tool.

burr (*n*) A jagged edge on metal resulting from punching or cutting.

bushing (*n*)   A replaceable lining or sleeve for a bearing.

calipers (*n*)   Instrument (of several types) for measuring diameters.

cam (*n*)   A rotating member for changing circular motion to reciprocating motion.

carburize (*v*)   To heat a low-carbon steel to approximately 2000°F in contact with material which adds carbon to the surface of the steel, and to cool slowly in preparation for heat treatment.

caseharden (*v*)   To harden the outer surface of a carburized steel by heating and then quenching.

castellate (*v*)   To form like a castle, as a castellated shaft or nut.

casting (*n*)   A metal object produced by pouring molten metal into a mold.

cast iron (*n*)   Iron melted and poured into molds.

center drill (*n*)   A special drill to produce bearing holes in the ends of a workpiece to be mounted between centers. Also called a "combined drill and countersink."

COMBINED DRILL
& C SINK

chamfer (*n*)   A narrow inclined surface along the intersection of two surfaces.

CHAMFER

chase (*v*)   To cut threads with an external cutting tool.

cheek (*n*)   The middle portion of a three-piece flask used in molding.

chill (*v*)   To harden the outer surface of cast iron by quick cooling, as in a metal mold.

chip (*v*)   To cut away metal with a cold chisel.

chuck (*n*)   A mechanism for holding a rotating tool or workpiece.

coin (*v*)   To form a part in one stamping operation.

cold-rolled steel (CRS) (*n*)   Open hearth or Bessemer steel containing 0.12–0.20% carbon that has been rolled while cold to produce a smooth, quite accurate stock.

collar (*n*)   A round flange or ring fitted on a shaft to prevent sliding.

COLLAR

colorharden (*v*)   Same as *caseharden*, except that it is done to a shallower depth, usually for appearance only.

cope (*n*)   The upper portion of a flask used in molding.

core (*v*)   To form a hollow portion in a casting by using a dry-sand core or a green-sand core in a mold.

coreprint (*n*)   A projection on a pattern which forms an opening in the sand to hold the end of a core.

cotter pin (*n*)   A split pin used as a fastener, usually to prevent a nut from unscrewing, Appendix 30.

counterbore (*v*)   To enlarge an end of a hole cylindrically with a *counterbore*.

COUNTERBORE

countersink (*v*)   To enlarge an end of a hole conically, usually with a *countersink*.

COUNTERSINK

crown (*n*)   A raised contour, as on the surface of a pulley.

cyanide (*v*)   To surface-harden steel by heating in contact with a cyanide salt, followed by quenching.

dedendum (*n*)   Distance from pitch circle to bottom of tooth space.

development (*n*)   Drawing of the surface of an object unfolded or rolled out on a plane.

diametral pitch (*n*)   Number of gear teeth per inch of pitch diameter.

die (*n*)   (1) Hardened metal piece shaped to cut or form a required shape in a sheet of metal by pressing it against a mating die. (2) Also used for cutting small male threads. In a sense is opposite to a tap.

die casting (*n*)   Process of forcing molten metal under pressure into metal dies or molds, producing a very accurate and smooth casting.

die stamping (*n*)   Process of cutting or forming a piece of sheet metal with a die.

dog (*n*)   A small auxiliary clamp for preventing work from rotating in relation to the face plate of a lathe.

dowel (*n*)   A cylindrical pin, commonly used to prevent sliding between two contacting flat surfaces.

DOWEL

draft (*n*)   The tapered shape of the parts of a pattern to permit it to be easily withdrawn from the sand or, on a forging, to permit it to be easily withdrawn from the dies.

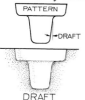

DRAFT

drag (*n*)   Lower portion of a flask used in molding.

draw (*v*)   To stretch or otherwise to deform metal. Also to temper steel.

drill (*v*)   To cut a cylindrical hole with a drill. A *blind hole* does not go through the piece.

drill press (*n*)   A machine for drilling and other hole-forming operations.

drop forge (*v*)   To form a piece while hot between dies in a drop hammer or with great pressure.

face (*v*)   To finish a surface at right angles, or nearly so, to the center line of rotation on a lathe.

FAO Finish all over.

feather key (*n*)   A flat key, which is partly sunk in a shaft and partly in a hub, permitting the hub to slide lengthwise of the shaft.

file (*v*)   To finish or smooth with a file.

fillet (*n*)   An interior rounded intersection between two surfaces.

fin (*n*)   A thin extrusion of metal at the intersection of dies or sand molds.

fit (*n*)   Degree of tightness or looseness between two mating parts, as a *loose fit*, a *snug fit*, or a *tight fit*.

fixture (*n*)   A special device for holding the work in a machine tool, *but not for guiding the cutting tool*.

flange (*n*)   A relatively thin rim around a piece.

FLANGE

flash (*n*)   Same as *fin*.

flask (*n*)   A box made of two or more parts for holding the sand in sand molding.

flute (*n*)   Groove, as on twist drills, reamers, and taps.

forge (*v*)   To force metal while it is hot to take on a desired shape by hammering or pressing.

galvanize (*v*)   To cover a surface with a thin layer of molten alloy, composed mainly of zinc, to prevent rusting.

gasket (*n*)   A thin piece of rubber, metal, or some other material, placed between surfaces to make a tight joint.

gate (*n*)   The opening in a sand mold at the bottom of the *sprue* through which the molten metal passes to enter the cavity or mold.

graduate (*v*)   To set off accurate divisions on a scale or dial.

grind (*v*)   To remove metal by means of an abrasive wheel, often made of carborundum. Use chiefly where accuracy is required.

harden (*v*)   To heat steel above a critical temperature and then quench in water or oil.

heat-treat (*v*)   To change the properties of metals by heating and then cooling.

interchangeable (*adj.*)   Refers to a part made to limit dimensions so that it will fit any mating part similarly manufactured.

jig (*n*)   A device *for guiding a tool* in cutting a piece. Usually it holds the work in position.

journal (*n*)   Portion of a rotating shaft supported by a bearing.

kerf (*n*)   Groove or cut made by a saw.

KERF

key (*n*)   A small piece of metal sunk partly into both shaft and hub to prevent rotation.

keyseat (*n*)   A slot or recess in a shaft to hold a key.

KEYSEAT

keyway (*n*)   A slot in a hub or portion surrounding a shaft to receive a key.

KEYWAY

knurl (*v*)   To impress a pattern of dents in a turned surface with a knurling tool to produce a better hand grip.

lap (*v*)   To produce a very accurate finish by sliding contact with a *lap*, or piece of wood, leather, or soft metal impregnated with abrasive powder.

lathe (*n*)   A machine used to shape metal or other materials by rotating against a tool.

lug (*n*)   An irregular projection of metal, but not round as in the case of a *boss*, usually with a hole in it for a bolt or screw.

malleable casting (*n*)   A casting that has been made less brittle and tougher by annealing.

mill (*v*)   To remove material by means of a rotating cutter on a milling machine.

mold (*n*)   The mass of sand or other material that forms the cavity into which molten metal is poured.

MS (*n*)   Machinery steel, sometimes called *mild steel* with a small percentage of carbon. Cannot be hardened.

neck (*v*)   To cut a groove around a cylindrical piece.

NECK

normalize (*v*) To heat steel above its critical temperature and then to cool it in air.

pack-harden (*v*) To *carburize*, then to *caseharden*.

pad (*n*) A slight projection, usually to provide a bearing surface around one or more holes.

PAD

pattern (*n*) A model, usually of wood, used in forming a mold for a casting. In sheet metal work a pattern is called a *development*.

peen (*v*) To hammer into shape with a ballpeen hammer.

pickle (*v*) To clean forgings or castings in dilute sulphuric acid.

pinion (*n*) The smaller of two mating gears.

pitch circle (*n*) An imaginary circle corresponding to the circumference of the friction gear from which the spur gear was derived.

plane (*v*) To remove material by means of the *planer*.

planish (*v*) To impart a planished surface to sheet metal by hammering with a smooth-surfaced hammer.

plate (*v*) To coat a metal piece with another metal, such as chrome or nickel, by electrochemical methods.

polish (*v*) To produce a highly finished or polished surface by friction, using a very fine abrasive.

profile (*v*) To cut any desired outline by moving a small rotating cutter, usually with a master template as a guide.

punch (*v*) To cut an opening of a desired shape with a rigid tool having the same shape, by pressing the tool through the work.

quench (*v*) To immerse a heated piece of metal in water or oil to harden it.

rack (*n*) A flat bar with gear teeth in a straight line to engage with teeth in a gear.

ream (*v*) To enlarge a finished hole slightly to give it greater accuracy, with a *reamer*.

relief (*n*) An offset of surfaces to provide clearance for machining.

RELIEF

rib (*n*) A relatively thin flat member acting as a brace or support.

RIB

rivet (*v*) To connect with rivets or to clench over the end of a pin by hammering.

round (*n*) An exterior rounded intersection of two surfaces.

SAE Society of Automotive Engineers.

sandblast (*v*) To blow sand at high velocity with compressed air against castings or forgings to clean them.

scleroscope (*n*) An instrument for measuring hardness of metals.

scrape (*v*) To remove metal by scraping with a hand scraper, usually to fit a bearing.

shape (*v*) To remove metal from a piece with a *shaper*.

shear (*v*) To cut metal by means of shearing with two blades in sliding contact.

sherardize (*v*) To galvanize a piece with a coating of zinc by heating it in a drum with zinc powder, to a temperature of 575–850°F.

shim (*n*) A thin piece of metal or other material used as a spacer in adjusting two parts.

solder (*v*) To join with solder, usually composed of lead and tin.

spin (*v*) To form a rotating piece of sheet metal into a desired shape by pressing it with a smooth tool against a rotating form.

spline (*n*) A keyway, usually one of a series cut around a shaft or hole.

SPLINED HOLE

spotface (*v*) To produce a round *spot* or bearing surface around a hole, usually with a *spotfacer*. The spotface may be on top of a boss or it may be sunk into the surface.

SPOTFACE

sprue (*n*) A hole in the sand leading to the *gate* which leads to the mold, through which the metal enters.

steel casting (*n*) Like cast-iron casting except that in the furnace scrap steel has been added to the casting.

swage (*v*)   To hammer metal into shape while it is held over a *swage*, or die, which fits in a hole in the *swage block*, or anvil.

sweat (*v*)   To fasten metal together by the use of solder between the pieces and by the application of heat and pressure.

tap (*v*)   To cut relatively small internal threads with a *tap*.

tape (*n*)   Conical form given to a shaft or a hole. Also refers to the slope of a plane surface.

taper pin (*n*)   A small tapered pin for fastening, usually to prevent a collar or hub from rotating on a shaft.

TAPER PIN

taper reamer (*n*)   A tapered reamer for producing accurate tapered holes, as for a taper pin.

temper (*v*)   To reheat hardened steel to bring it to a desired degree of hardness.

template or *templet* (*n*)   A guide or pattern used to mark out the work, guide the tool in cutting it, or check the finished product.

tin (*n*)   A silvery metal used in alloys and for coating other metals, such as tin plate.

tolerance (*n*)   Total amount of variation permitted in limit dimension of a part.

trepan (*v*)   To cut a circular groove in the flat surface at one end of a hole.

tumble (*v*)   To clean rough castings or forgings in a revolving drum filled with scrap metal.

turn (*v*)   To produce, on a lathe, a cylindrical surface parallel to the center line.

twist drill (*n*)   A drill for use in a drill press.

undercut (*n*)   A recessed cut or a cut with inwardly sloping sides.

UNDERCUT

upset (*v*)   To form a head or enlarged end on a bar or rod by pressure or by hammering between dies.

web (*n*)   A thin flat part joining larger parts. Also known as a *rib*.

weld (*v*)   Uniting metal pieces by pressure or fusion welding processes.

Woodruff key (*n*)   A semicircular flat key.

WOODRUFF KEYS

wrought iron (*n*)   Iron of low carbon content useful because of its toughness, ductility, and malleability.

3   CAD/CAM Glossary *

access time (or disk access time)   One measure of system response. The time interval between the instant that data is called for from storage and the instant that delivery is completed—i.e., read time. See also *response time*.

alphanumeric (or alphameric)   A term that encompasses letters, digits, and special characters that are machine-processable.

alphanumeric display (or alphameric display)   A work-station device consisting of a CRT on which text can be viewed. An alphanumeric display is capable of showing a fixed set of letters, digits, and special characters. It allows the designer to observe entered commands and to receive messages from the system.

alphanumeric keyboard (or alphameric keyboard)   A work-station device consisting of a typewriter-like keyboard that allows the designer to communicate with the system using an English-like command language.

American Standard Code for Information Interchange (ASCII)   An industry-standard character code widely used for information interchange among data processing systems, communications systems, and associated equipment.

analog   Applied to an electrical or computer system, this denotes the capability to represent data in continuously varying physical quantities.

annotation   Process of inserting text or a special note or identification (such as a flag) on a drawing, map, or diagram constructed on a CAD/CAM system. The text can be generated and positioned on the drawing using the system.

application program (or package)   A computer program or collection of programs to perform a task or tasks specific to a particular user's need or class of needs.

archival storage   Refers to memory (on magnetic tape,

---

* Extracted from *The CAD/CAM Glossary*, 1983 edition, published by the Computervision Corporation, Bedford, MA 01730; reproduced with permission of the publisher.

disks, printouts, or drums) used to store data on completed designs or elements outside of main memory.

array (*v*) To create automatically on a CAD system an arrangement of identical elements or components. The designer defines the element once, then indicates the starting location and spacing for automatic generation of the array. (*n*) An arrangement created in the above manner. A series of elements or sets of elements arranged in a pattern—i.e., matrix.

ASCII See *American National Standard Code for Information Exchange.*

assembler A computer program that converts (i.e., translates) programmer-written symbolic instructions, usually in mnemonic form, into machine-executable (computer or binary-coded) instructions. This conversion is typically one-to-one (one symbolic instruction converts to one machine-executable instruction). A software programming aid.

associative dimensioning A CAD capability that links dimension entities to geometric entities being dimensioned. This allows the value of a dimension to be automatically updated as the geometry changes.

attribute A nongraphic characteristic of a part, component, or entity under design on a CAD system. Examples include: dimension entities associated with geometry, text with text nodes, and nodal lines with connect nodes. Changing one entity in an association can produce automatic changes by the system in the associated entity; e.g., moving one entity can cause moving or stretching of the other entity.

automatic dimensioning A CAD capability that computes the dimensions in a displayed design, or in a designated section, and automatically places dimensions, dimensional lines, and arrowheads where required. In the case of mapping, this capability labels the linear feature with length and azimuth.

auxiliary storage Storage that supplements main memory devices such as disk or drum storage. Contrast with *archival storage.*

benchmark The program(s) used to test, compare, and evaluate in real time the performance of various CAD/CAM systems prior to selection and purchase. A *synthetic* benchmark has preestablished parameters designed to exercise a set of system features and resources. A *live* benchmark is drawn from the prospective user's workload as a model of the entire workload.

bit The smallest unit of information that can be stored and processed by a digital computer. A bit may assume only one of two values: 0 or 1 (i.e., ON/ OFF or YES/NO). Bits are organized into larger units called *words* for access by computer instructions.

Computers are often categorized by word size in bits, i.e., the maximum word size that can be processed as a unit during an instruction cycle (e.g., 16-bit computers or 32-bit computers). The number of bits in a word is an indication of the processing power of the system, especially for calculations or for high-precision data.

bit rate The speed at which bits are transmitted, usually expressed in bits per second.

bits per inch (bpi) The number of bits that can be stored per inch of a magnetic tape. A measure of the data storage capacity of a magnetic tape.

blank A CAD command that causes a predefined entity to go temporarily blank on the CRT. The reversing command is *unblank.*

blinking A CAD design aid that makes a predefined graphic entity blink on the CRT to attract the attention of the designer.

boot up Start up (a system).

B-spline A sequence of parametric polynomial curves (typically quadratic or cubic polynomials) forming a smooth fit between a sequence of points in 3D space. The piece-wise defined curve maintains a level of mathematical continuity dependent upon the polynomial degree chosen. It is used extensively in mechanical design applications in the automotive and aerospace industries.

bug A flaw in the design or implementation of a software program or hardware design that causes erroneous results or malfunctions.

bulk memory A memory device for storing a large amount of data, e.g., disk, drum, or magnetic tape. It is not randomly accessible as main memory is.

byte A sequence of adjacent bits, usually eight, representing a character that is operated on as a unit. Usually shorter than a word. A measure of the memory capacity of a system, or of an individual storage unit (as a 300-million-byte disk).

CAD See *computer-aided design.*

CAD/CAM See *computer-aided design/computer-aided manufacturing.*

CADDS® Computervision Corporation's registered trademark for its prerecorded software programs.

CAE See *computer-aided engineering.*

CAM See *computer-aided manufacturing.*

CAMACS™ (CAM Asynchronous Communications Software) Computervision's communications link, which enables users to exchange machine control data with other automation systems and devices, or to interact directly with local or remote manufacturing systems and machines. CAMACS tailors CAD/CAM data automatically for a wide range of machine tools, robots, coordinate measurement systems, and off-line storage devices.

cathode ray tube (CRT) The principal component in a CAD display device. A CRT displays graphic representations of geometric entities and designs and can be of various types: storage tube, raster scan, or refresh. These tubes create images by means of a controllable beam of electrons striking a screen. The term *CRT* is often used to denote the entire display device.

central processing unit (CPU) The computer brain of a CAD/CAM system that controls the retrieval, decoding, and processing of information, as well as the interpretation and execution of operating instructions—the building blocks of application and other computer programs. A CPU comprises arithmetic, control, and logic elements.

character An alphabetical, numerical, or special graphic symbol used as part of the organization, control, or representation of CAD/CAM data.

characters per second (cps) A measure of the speed with which an alphanumeric terminal can process data.

chip See *integrated circuit*.

code A set of specific symbols and rules for representing data (usually instructions) so that the data can be understood and executed by a computer. A code can be in binary (machine) language, assembly language, or a high-level language. Frequently refers to an industry-standard code such as ANSI, ASCII, IPC, or Standard Code for Information Exchange. Many application codes for CAD/CAM are written in FORTRAN.

color display A CAD/CAM display device. Color raster-scan displays offer a variety of user-selectable, contrasting colors to make it easier to discriminate among various groups of design elements on different layers of a large, complex design. Color speeds up the recognition of specific areas and subassemblies, helps the designer interpret complex surfaces, and highlights interference problems. Color displays can be of the penetration type, in which various phosphor layers give off different colors (refresh display) or the TV-type with red, blue, and green electron guns (raster-scan display).

command A control signal or instruction to a CPU or graphics processor, commonly initiated by means of a menu/tablet and electronic pen or by an alphanumeric keyboard.

command language A language for communicating with a CAD/CAM system in order to perform specific functions or tasks.

communication link The physical means, such as a telephone line, for connecting one system module or peripheral to another in a different location in order to transmit and receive data. See also *data link*.

compatibility The ability of a particular hardware module or software program, code, or language to be used in a CAD/CAM system without prior modification or special interfaces. *Upward compatible* denotes the ability of a system to interface with new hardware or software modules or enhancements (i.e., the system vendor provides with each new module a reasonable means of transferring data,

programs, and operator skills from the user's present system to the new enhancements).

compiler A computer program that converts or translates a high-level, user-written language (e.g., PASCAL, COBOL, VARPRO, or FORTRAN) or source, into a language that a computer can understand. The conversion is typically one to many (i.e., one user instruction to many machine-executable instructions). A software programming aid, the compiler allows the designer to write programs in an English-like language with relatively few statements, thus saving program development time.

component A physical entity, or a symbol used in CAD to denote such an entity. Depending on the application, a component might refer to an IC or part of a wiring circuit (e.g., a resistor), or a valve, elbow, or vee in a plant layout, or a substation or cable in a utility map. Also applies to a subassembly or part that goes into higher level assemblies.

computer-aided design (CAD) A process that uses a computer system to assist in the creation, modification, and display of a design.

computer-aided design/computer-aided manufacturing (CAD/CAM) Refers to the integration of computers into the entire design-to-fabrication cycle of a product or plant.

computer-aided engineering (CAE) Analysis of a design for basic error checking, or to optimize manufacturability, performance, and economy (for example, by comparing various possible materials or designs). Information drawn from the CAD/CAM design data base is used to analyze the functional characteristics of a part, product, or system under design and to simulate its performance under various conditions. In electronic design, CAE enables users of the Computervision Designer system to detect and correct potentially costly design flaws. CAE permits the execution of complex circuit loading analyses and simulation during the circuit definition stage. CAE can be used to determine section properties, moments of inertia, shear and bending moments, weight, volume, surface area, and center of gravity. CAE can precisely determine loads, vibration, noise, and service life early in the design cycle so that components can be optimized to meet those criteria. Perhaps the most powerful CAE technique is finite element modeling. See also *kinematics*.

computer-aided manufacturing (CAM) The use of computer and digital technology to generate manufacturing-oriented data. Data drawn from a CAD/ CAM data base can assist in or control a portion or all of a manufacturing process, including numerically controlled machines, computer-assisted parts programming, computer-assisted process planning, robotics, and programmable logic controllers, CAM can involve production programming, manufacturing engineering, industrial engineering, facilities engineering, and reliability engineering (quality control).

CAM techniques can be used to produce process plans for fabricating a complete assembly, to program robots, and to coordinate plant operation.

computer graphics A general term encompassing any discipline or activity that uses computers to generate, process, and display graphic images. The essential technology of CAD/CAM systems. See also *computer-aided design*.

computer network An interconnected complex (arrangement or configuration) of two or more systems. See also *network*.

computer program A specific set of software commands in a form acceptable to a computer and used to achieve a desired result. Often called a *software program* or *package*.

configuration A particular combination of a computer, software and hardware modules, and peripherals at a single installation and interconnected in such a way as to support certain application(s).

connector A termination point for a signal entering or leaving a PC board or a cabling system.

convention Standardized methodology or accepted procedure for executing a computer program. In CAD, the term denotes a standard rule or mode of execution undertaken to provide consistency. For example, a drafting convention might require all dimensions to be in metric units.

core (core memory) A largely obsolete term for *main storage*.

CPU See *central processing unit*.

CRT See *cathode ray tube*.

cursor A visual tracking symbol, usually an underline or cross hairs, for indicating a location or entity selection on the CRT display. A text cursor indicates the alphanumeric input; a graphics cursor indicates the next geometric input. A cursor is guided by an electronic or light pen, joystick, keyboard, etc., and follows every movement of the input device.

cycle A preset sequence of events (hardware or software) initiated by a single command.

data base A comprehensive collection of interrelated information stored on some kind of mass data storage device, usually a disk. Generally consists of information organized into a number of fixed-format record types with logical links between associated records. Typically includes operating systems instructions, standard parts libraries, completed designs and documentation, source code, graphic and application programs, as well as current user tasks in progress.

data communication The transmission of data (usually digital) from one point (such as a CAD/CAM workstation or CPU) to another point via communication channels such as telephone lines.

data link The communication line(s), related controls, and interface(s) for the transmission of data between two or more computer systems. Can include modems, telephone lines, or dedicated transmission media such as cable or optical fiber.

data tablet A CAD/CAM input device that allows the designer to communicate with the system by placing an electronic pen or stylus on the tablet surface. There is a direct correspondence between positions on the tablet and addressable points on the display surface of the CRT. Typically used for indicating positions on the CRT, for digitizing input of drawings, or for menu selection. See also *graphic tablet*.

debug To detect, locate, and correct any bugs in a system's software or hardware.

dedicated Designed or intended for a single function or use. For example, a dedicated work station might be used exclusively for engineering calculations or plotting.

default The predetermined value of a parameter required in a CAD/CAM task or operation. It is automatically supplied by the system whenever that value (e.g., text, height, or grid size) is not specified.

density (1) A measure of the complexity of an electronic design. For example, IC density can be measured by the number of gates or transistors per unit area or by the number of square inches per component. (2) Magnetic tape storage capacity. High capacity might be 1600 bits/inch; low, 800 bits/inch.

device A system hardware module external to the CPU and designed to perform a specific function—i.e., a CRT, plotter, printer, hard-copy unit, etc. See also *peripheral*.

diagnostics Computer programs designed to test the status of a system or its key components and to detect and isolate malfunctions.

dial up To initiate station-to-station communication with a computer via a dial telephone, usually from a workstation to a computer.

digital Applied to an electrical or computer system, this denotes the capability to represent data in the form of digits.

digitize (1) General description: to convert a drawing into digital form (i.e., coordinate locations) so that it can be entered into the data base for later processing. A digitizer, available with many CAD systems, implements the conversion process. This is one of the primary ways of entering existing drawings, crude graphics, lines, and shapes into the system. (2) Computervision usage: to specify a coordinate location or entity using an electronic pen or other device; or a single coordinate value or entity pointer generated by a digitizing operation.

digitizer A CAD input device consisting of a data tablet on which is mounted the drawing or design to be digitized into the system. The designer moves a puck or electronic pen to selected points on the drawing and enters coordinate data for lines and shapes by simply pressing down the digitize button with the puck or pen.

dimensioning automatic   A CAD capability that will automatically compute and insert the dimensions of a design or drawing, or a designated section of it.

direct access (linkage)   Retrieval or storage of data in the system by reference to its location on a tape, disk, or cartridge, without the need for processing on a CPU.

direct-view storage tube (DVST)   One of the most widely used graphics display devices, DVST generates a long-lasting, flicker-tree image with high resolution and no refreshing. It handles an almost unlimited amount of data. However, display dynamics are limited since DVSTs do not permit selective erase. The image is not as bright as with refresh or raster. Also called *storage tube*.

directory   A named space on the disk or other mass storage device in which are stored the names of files and some summary information about them.

discrete components   Components with a single functional capability per package—for example, transistors and diodes.

disk (storage)   A device on which large amounts of information can be stored in the data base. Synonymous with *magnetic disk storage* or *magnetic disk memory*.

display   A CAD/CAM work station device for rapidly presenting a graphic image so that the designer can react to it, making changes interactively in real time. Usually refers to a CRT.

dot-matrix plotter   A CAD peripheral device for generating graphic plots. Consists of a combination of wire nibs (styli) spaced 100 to 200 styli per inch, which place dots where needed to generate a drawing. Because of its high speed, it is typically used in electronic design applications. Accuracy and resolution are not as great as with pen plotters. Also known as *electrostatic plotter*.

drum plotter   An electromechanical pen plotter that draws an image on paper or film mounted on a rotatable drum. In this CAD peripheral device a combination of plotting-head movement and drum rotation provides the motion.

dynamic (motion)   Simulation of movement using CAD software, so that the designer can see on the CRT screen 3D representations of the parts in a piece of machinery as they interact dynamically. Thus, any collision or interference problems are revealed at a glance.

dynamic menuing   This feature of Computervision's Instaview terminal allows a particular function or command to be initiated by touching an electronic pen to the appropriate key word displayed in the status text area on the screen.

dynamics   The capability of a CAD system to zoom, scroll, and rotate.

edit   To modify, refine, or update an emerging design or text on a CAD system. This can be done on-line interactively.

electrostatic plotter   See *dot-matrix plotter*.

element   The basic design entity in computer-aided design whose logical, positional, electrical, or mechanical function is identifiable.

enhancements   Software or hardware improvements, additions, or updates to a CAD/CAM system.

entity   A geometric primitive—the fundamental building block used in constructing a design or drawing, such as an arc, circle, line, text, point, spline, figure, or nodal line. Or a group of primitives processed as an identifiable unit. Thus, a square may be defined as a discrete entity consisting of four primitives (vectors), although each side of the square could be defined as an entity in its own right. See also *primitive*.

feedback   (1) The ability of a system to respond to an operator command in real time either visually or with a message on the alphanumeric display or CRT. This message registers the command, indicates any possible errors, and simultaneously displays the updated design on the CRT. (2) The signal or data fed back to a commanding unit from a controlled machine or process to denote its response to a command. (3) The signal representing the difference between actual response and desired response and used by the commanding unit to improve performance of the controlled machine or process. See also *prompt*.

figure   A symbol or a part that may contain primitive entities, other figures, nongraphic properties, and associations. A figure can be incorporated into other parts or figures.

file   A collection of related information in the system that may be accessed by a unique name. May be stored on a disk, tape, or other mass storage media.

file protection   A technique for preventing access to or accidental erasure of data within a file on the system.

firmware   Computer programs, instructions, or functions implemented in user-modifiable hardware, i.e., a microprocessor with read-only memory. Such programs or instructions, stored permanently in programmable read-only memories, constitute a fundamental part of system hardware. The advantage is that a frequently used program or routine can be invoked by a single command instead of multiple commands as in a software program.

flatbed plotter   A CAD/CAM peripheral device that draws an image on paper, glass, or film mounted on a flat table. The plotting head provides all the motion.

flat-pattern generation   A CAD/CAM capability for automatically unfolding a 3D design of a sheet metal part into its corresponding flat-pattern design. Calculations for material bending and stretching are performed automatically for any specified material. The reverse flat-pattern generation package automatically folds a flat-pattern design into its 3D version. Flat-pattern generation eliminates major bottlenecks for sheet metal fabricators.

flicker An undesired visual effect on a CRT when the refresh rate is low.

font, line Repetitive pattern used in CAD to give a displayed line appearance characteristics that make it more easily distinguishable, such as a solid, dashed, or dotted line. A line font can be applied to graphic images in order to provide meaning, either graphic (e.g., hidden lines) or functional (roads, tracks, wires, pipes, etc.). It can help a designer to identify and define specific graphic representations of entities that are view-dependent. For example, a line may be solid when drawn in the top view of an object but, when a line font is used, becomes dotted in the side view where it is not normally visible.

font, text Sets of type faces of various styles and sizes. In CAD, fonts are used to create text for drawings, special characters such as Greek letters, and mathematical symbols.

FORTRAN FORmula TRANslation, a high-level programming language used primarily for scientific or engineering applications.

fracturing The division of IC graphics by CAD into simple trapezoidal or rectangular areas for pattern-generation purposes.

function key A specific square on a data tablet, or a key on a function key box, used by the designer to enter a particular command or other input. See also *data tablet*.

function keyboard An input device located at a CAD/CAM workstation and containing a number of function keys.

gap The gap between two entities on a computer-aided design is the length of the shortest line segment that can be drawn from the boundary of one entity to the other without intersecting the boundary of the other. CAD/CAM design-rules checking programs can automatically perform gap checks.

graphic tablet A CAD/CAM input device that enables graphic and location instruments to be entered into the system using an electronic pen on the tablet. See also *data tablet*.

gray scales In CAD systems with a monochromatic display, variations in brightness level (gray scale) are employed to enhance the contrast among various design elements. This feature is very useful in helping the designer discriminate among complex entities on different layers displayed concurrently on the CRT.

grid A network of uniformly spaced points or crosshatch optionally displayed on the CRT and used for exactly locating and digitizing a position, inputting components to assist in the creation of a design layout, or constructing precise angles. For example, the coordinate data supplied by digitizers is automatically calculated by the CPU from the closest grid point. The grid determines the minimum accuracy with which design entities are described or connected. In the mapping environment, a grid is used to describe the distribution network of utility resources.

hard copy A copy on paper of an image displayed on the CRT—for example, a drawing, printed report, plot, listing, or summary. Most CAD/CAM systems can automatically generate hard copy through an on-line printer or plotter.

hardware The physical components, modules, and peripherals comprising a system—computer disk, magnetic tape, CRT terminal(s), plotter(s), etc.

hard-wired link A technique of physically connecting two systems by fixed circuit interconnections using digital signals.

high-level language A problem-oriented programming language using words, symbols, and command statements that closely resemble English-language statements. Each statement typically represents a series of computer instructions. Relatively easy to learn and use, a high-level language permits the execution of a number of subroutines through a simple command. Examples are BASIC, FORTRAN, PL/I, PASCAL, and COBOL.

A high-level language must be translated or compiled into machine language before it can be understood and processed by a computer. See also *assembler: low-level language*.

host computer The primary or controlling computer in a multicomputer network. Large-scale host computers typically are equipped with mass memory and a variety of peripheral devices, including magnetic tape, line printers, card readers, and possibly hard-copy devices. Host computers may be used to support, with their own memory and processing capabilities, not only graphics programs running on a CAD/CAM system but also related engineering analysis.

host-satellite system A CAD/CAM system configuration characterized by a graphic workstation with its own computer (typically holding the display file) that is connected to another, usually larger, computer for more extensive computation or data manipulation. The computer local to the display is a satellite to the larger host computer, and the two comprise a host-satellite system.

IC See *integrated circuit*.

IGES See *Initial Graphics Exchange Specification*.

inches per second (ips) Measure of the speed of a device (i.e., the number of inches of magnetic tape that can be processed per second, or the speed of a pen plotter).

Initial Graphics Exchange Specification (IGES) An interim CAD/CAM data base specification until the American National Standards Institute develops its own specification. IGES attempts to standardize communication of drawing and geometric product information between computer systems.

initialize To set counters, switches, and addresses on a

computer to zero or to other starting values at the beginning of, or at predetermined stages in, a program or routine.

input (data)   (1) The data supplied to a computer program for processing by the system. (2) The process of entering such data into the system.

input devices   A variety of devices (such as data tablets or keyboard devices) that allow the user to communicate with the CAD/CAM system, for example, to pick a function from many presented, to enter text and/or numerical data, to modify the picture shown on the CRT, or to construct the desired design.

input/output (I/O)   A term used to describe a CAD/CAM communications device as well as the process by which communications take place in a CAD/CAM system. An I/O device is one that makes possible communications between a device and a workstation operator or between devices on the system (such as workstations or controllers). By extension, input/output also denotes the process by which communications takes place. Input refers to the data transmitted to the processor for manipulation, and output refers to the data transmitted from the processor to the workstation operator or to another device (i.e., the results). Contrast with the other major parts of a CAD/CAM system: the CPU or central processing unit, which performs arithmetic and logical operations, and data storage devices (such as memories, disks, or tapes).

insert   To create and place entities, figures, or information on a CRT or into an emerging design on the display.

instruction set   (1) All the commands to which a CAD/CAM computer will respond. (2) The repertoire of functions the computer can perform.

integrated circuit (IC)   A tiny complex of electronic components and interconnections comprising a circuit that may vary in functional complexity from a simple logic gate to a microprocessor. An IC is usually packaged in a single substrate such as a slice of silicon. The complexity of most IC designs and the many repetitive elements have made computer-aided design an economic necessity. Also called a *chip*.

integrated system   A CAD/CAM system that integrates the entire product development cycle—analysis, design, and fabrication—so that all processes flow smoothly from concept to production.

intelligent work station/terminal   A workstation in a system that can perform certain data processing functions in a stand-alone mode, independent of another computer. Contains a built-in computer, usually a microprocessor or minicomputer, and dedicated memory.

interactive   Denotes two-way communications between a CAD/CAM system or workstation and its operators. An operator can modify or terminate a program and receive feedback from the system for guidance and verification. See also *feedback*.

interactive graphics system (IGS) or interactive computer graphics (ICG)   A CAD/CAM system in which the workstations are used interactively for computer-aided design and/or drafting, as well as for CAM, all under full operator control, and possibly also for text-processing, generation of charts and graphs, or computer-aided engineering. The designer (operator) can intervene to enter data and direct the course of any program, receiving immediate visual feedback via the CRT. Bilateral communication is provided between the system and the designer(s). Often used synonymously with *CAD*.

interface (*n*)   (1) A hardware and/or software link that enables two systems, or a system and its peripherals, to operate as a single, integrated system. (2) The input devices and visual feedback capabilities that allow bilateral communication between the designer and the system. The interface to a large computer can be a communications link (hardware) or a combination of software and hard-wired connections. An interface might be a portion of storage accessed by two or more programs or a link between two subroutines in a program.

I/O   See *input/output*.

ips   See *inches per second*.

jaggies   A CAD jargon term used to refer to straight or curved lines that appear to be jagged or sawtoothed on the CRT screen.

joystick   A CAD data-entering device employing a hand-controlled lever to manually enter the coordinates of various points on a design being digitized into the system.

key file   A disk file that provides user-defined definitions for a tablet menu. See *menu*.

kinematics   A computer-aided engineering (CAE) process for plotting or animating the motion of parts in a machine or a structure under design on the system. CAE simulation programs allow the motion of mechanisms to be studied for interference, acceleration, and force determinations while still in the design stage.

layering   A method of logically organizing data in a CAD/CAM data base. Functionally different classes of data (e.g., various graphic/geometric entities) are segregated on separate layers, each of which can be displayed individually or in any desired combination. Layering helps the designer distinguish among different kinds of data in creating a complex product such as a multilayered PC board or IC.

layers   User-defined logical subdivisions of data in a CAD/CAM data base that may be viewed on the CRT individually or overlaid and viewed in groups.

learning curve   A concept that projects the expected improvement in operator productivity over a period of time.

Usually applied in the first 1 to $1\frac{1}{2}$ years of a new CAD/CAM facility as part of a cost-justification study, or when new operators are introduced. An accepted tool of management for predicting manpower requirements and evaluating training programs.

library, graphics (or parts library)   A collection of standard, often-used symbols, components, shapes, or parts stored in the CAD data base as templates or building blocks to speed up future design work on the system. Generally an organization of files under a common library name.

light pen   A hand-held photosensitive CAD input device used on a refreshed CRT screen for identifying display elements, or for designating a location on the screen where an action is to take place.

line font   See *font, line*.

line printer   A CAD/CAM peripheral device used for rapid printing of data.

line smoothing   An automated mapping capability for the interpolation and insertion of additional points along a linear entity yielding a series of shorter linear segments to generate a smooth curved appearance to the original linear component. The additional points or segments are created only for display purposes and are interpolated from a relatively small set of stored representative points. Thus, data storage space is minimized.

low-level language   A programming language in which statements translate on a one-for-one basis. See also *machine language*.

machine   A computer, CPU, or other processor.

machine instruction   An instruction that a machine (computer) can recognize and execute.

machine language   The complete set of command instructions understandable to and used directly by a computer when it performs operations.

macro   (1) A sequence of computer instructions executable as a single command. A frequently used, multistep operation can be organized into a macro, given a new name, and remain in the system for easy use, thus shortening program development time. (2) In Computer-vision's IC design system, macro refers to macroexpansion of a cell. This system capability enables the designer to replicate the contents of a cell as primitives without the original cell grouping.

magnetic disk   A flat circular plate with a magnetic surface on which information can be stored by selective magnetization of portions of the flat surface. Commonly used for temporary working storage during computer-aided design. See also *disk*.

magnetic tape   A tape with a magnetic surface on which information can be stored by selective polarization of portions of the surface. Commonly used in CAD/CAM for off-line storage of completed design files and other archival material.

mainframe (computer)   A large central computer facility.

main memory/storage   The computer's general-purpose storage from which instructions may be executed and data loaded directly into operating registers.

mass storage   Auxiliary large-capacity memory for storing large amounts of data readily accessible by the computer. Commonly a disk or magnetic tape.

matrix   A 2D or 3D rectangular array (arrangement) of identical geometric or symbolic entities. A matrix can be generated automatically on a CAD system by specifying the building block entity and the desired locations. This process is used extensively in computer-aided electrical/electronic design.

memory   Any form of data storage where information can be read and written. Standard memories include RAM, ROM, and PROM. See also *programmable read-only memory; random access memory; read-only memory; storage*.

menu   A common CAD/CAM input device consisting of a checkerboard pattern of squares printed on a sheet of paper or plastic placed over a data tablet. These squares have been preprogrammed to represent a part of a command, a command, or a series of commands. Each square, when touched by an electronic pen, initiates the particular function or command indicated on that square. See also *data tablet, dynamic menuing*.

merge   To combine two or more sets of related data into one, usually in a specified sequence. This can be done automatically on a CAD/CAM system to generate lists and reports.

microcomputer   A smaller, lower-cost equivalent of a full-scale minicomputer. Includes a microprocessor (CPU), memory, and necessary interface circuits. Consists of one or more ICs (chips) comprising a chip set.

microprocessor   The central control element of a microcomputer, implemented in a single integrated circuit. It performs instruction sequencing and processing, as well as all required computations. It requires additional circuits to function as a microcomputer. See *microcomputer*.

minicomputer   A general-purpose, single processor computer of limited flexibility and memory performance.

mirroring   A CAD design aid that automatically creates a mirror image of a graphic entity on the CRT by flipping the entity or drawing on its x or y axis.

mnemonic symbol   An easily remembered symbol that assists the designer in communicating with the system (e.g., an abbreviation such as MPY for *multiply*).

model, geometric   A complete, geometrically accurate 3D or 2D representation of a shape, a part, a geographic area, a plant or any part of it, designed on a CAD system and stored in the data base. A mathematical or analytic model of a physical system used to determine the response of that system to a stimulus or load. See *modeling, geometric.*

modeling, geometric Constructing a mathematical or analytic model of a physical object or system for the purpose of determining the response of that object or system to a stimulus or load. First, the designer describes the shape under design using a geometric model constructed on the system. The computer then converts this pictorial representation on the CRT into a mathematical model later used for other CAD functions such as design optimization.

modeling, solid A type of 3D modeling in which the solid characteristics of an object under design are built into the data base so that complex internal structures and external shapes can be realistically represented. This makes computer-aided design and analysis of solid objects easier, clearer, and more accurate than with wire-frame graphics.

modem *MOdulator-DEMulator*, a device that converts digital signals to analog signals, and vice versa, for long-distance transmission over communications circuits such as telephone lines, dedicated wires, optical fiber, or microwave.

module A separate and distinct unit of hardware or software that is part of a system.

mouse A hand-held data-entering device used to position a cursor on a data tablet. See *cursor.*

multiprocessor A computer whose architecture consists of more than one processing unit. See *central processing unit: microcomputer.*

network An arrangement of two or more interconnected computer systems to facilitate the exchange of information in order to perform a specific function. For example, a CAD/CAM system might be connected to a mainframe computer to off-load heavy analytic tasks. Also refers to a piping network in computer-aided plant design.

numerical control (NC) A technique of operating machine tools or similar equipment in which motion is developed in response to numerically coded commands. These commands may be generated by a CAD/CAM system on punched tapes or other communications media. Also, the processes involved in generating the data or tapes necessary to guide a machine tool in the manufacture of a part.

off-line Refers to peripheral devices not currently connected to and under the direct control of the system's computer.

on-line Refers to peripheral devices connected to and under the direct control of the system's computer, so that operator-system interaction, feedback, and output are all in real time.

operating system A structured set of software programs that control the operation of the computer and associated peripheral devices in a CAD/CAM system, as well as the execution of computer programs and data flow to and from peripheral devices. May provide support for activities and programs such as scheduling, debugging, input/output control, accounting, editing, assembly, compilation, storage assignment, data management, and diagnostics. An operating system may assign task priority levels, support a file system, provide drives for I/O devices, support standard system commands or utilities for on-line programming, process commands, and support both networking and diagnostics.

output The end result of a particular CAD/CAM process or series of processes. The output of a CAD cycle can be artwork and hard-copy lists and reports. The output of a total design-to-manufacturing CAD/CAM system can also include numerical control tapes for manufacturing.

overlay A segment of code or data to be brought into the memory of a computer to replace existing code or data.

paint To fill in a bounded graphic figure on a raster display using a combination of repetitive patterns or line fonts to add meaning or clarity. See *font, line.*

paper-tape punch/reader A peripheral device that can read as well as punch a perforated paper tape generated by a CAD/CAM system. These tapes are the principal means of supplying data to an NC machine.

parallel processing Executing more than one element of a single process concurrently on multiple processors in a computer system.

password protection A security feature of certain CAD/CAM systems that prevents access to the system or to files within the system without first entering a password, i.e., a special sequence of characters.

PC board See *printed circuit board.*

pen plotter An electromechanical CAD output device that generates hard copy of displayed graphic data by means of a ballpoint pen or liquid ink. Used when a very accurate final drawing is required. Provides exceptional uniformity and density of lines, precise positional accuracy, as well as various user-selectable colors.

peripheral (device) Any device, distinct from the basic system modules, that provides input to and/or output from the CPU. May include printers, keyboards, plotters, graphics display terminals, paper-tape reader/punches, analog-to-digital converters, disks, and tape drives.

permanent storage A method or device for storing the results of a completed program outside the CPU—usually in the form of magnetic tape or punched cards.

photo plotter A CAD output device that generates high-precision artwork masters photographically for PC board design and IC masks.

pixel The smallest portion of a CRT screen that can be individually referenced. An individual dot on a display image. Typically, pixels are evenly spaced, horizontally and vertically, on the display.

plotter A CAD peripheral device used to output for external use the image stored in the data base. Generally makes large, accurate drawings substantially better than what is displayed. Plotter types include pen, drum, electrostatic, and flatbed.

postprocessor A software program or procedure that formats graphic or other data processed on the system for some other purpose. For example, a postprocessor might format cutter centerline data into a form that a machine controller can interpret.

precision The degree of accuracy. Generally refers to the number of significant digits of information to the right of the decimal point for data represented within a computer system. Thus, the term denotes the degree of discrimination with which a design or design element can be described in the data base.

preplaced line (or bus) A run (or line) between a set of points on a PC board layout that has been predefined by the designer and must be avoided by a CAD automatic routing program.

preprocessor A computer program that takes a specific set of instructions from an external source and translates it into the format required by the system.

primitive A design element at the lowest stage of complexity. A fundamental graphic entity. It can be a vector, a point, or a text string. The smallest definable object in a display processor's instruction set.

printed circuit (PC) board A baseboard made of insulating materials and an etched copper-foil circuit pattern on which are mounted ICs and other components required to implement one or more electronic functions. PC boards plug into a rack or subassembly of electronic equipment to provide the brains or logic to control the operation of a computer, or a communications system, instrumentation, or other electronic systems. The name derives from the fact that the circuitry is connected not by wires but by copper-foil lines, paths, or traces actually etched onto the board surface. CAD/CAM is used extensively in PC board design, testing, and manufacture.

process simulation A program utilizing a mathematical model created on the system to try out numerous process design iterations with real-time visual and numerical feedback. Designers can see on the CRT what is taking place at every stage in the manufacturing process. They can therefore optimize a process and correct problems that could affect the actual manufacturing process downstream.

processor In CAD/CAM system hardware, any device that performs a specific function. Most often used to refer to the CPU. In software, it refers to a complex set of instructions to perform a general function. See also *central processing unit*.

productivity ratio A widely accepted means of measuring CAD/CAM productivity (throughput per hour) by comparing the productivity of a design/engineering group before and after installation of the system or relative to some standard norm or potential maximum. The most common way of recording productivity is Actual Manual Hours/Actual CAD Hours, expressed as 4 : 1, 6 : 1, etc.

program (*n*) A precise sequential set of instructions that direct a computer to perform a particular task or action or to solve a problem. A complete program includes plans for the transcription of data, coding for the computer, and plans for the absorption of the results into the system. (*v*) To develop a program. See also *computer program*.

Programmable Read-Only Memory (PROM) A memory that, once programmed with permanent data or instructions, becomes a ROM. See *read-only memory*.

PROM See *programmable read-only memory*.

prompt A message or symbol generated automatically by the system, and appearing on the CRT, to inform the user of (a) a procedural error or incorrect input to the program being executed or (b) the next expected action, option(s), or input. See also *tutorial*.

puck A hand-held, manually controlled input device that allows coordinate data to be digitized into the system from a drawing placed on the data tablet or digitizer surface. A puck has a transparent window containing cross hairs.

RAM See *random access memory*.

random access memory (RAM) A main memory read/write storage unit that provides the CAD/CAM operator direct access to the stored information. The time required to access any word stored in the memory is the same as for any other word.

raster display A CAD workstation display in which the entire CRT surface is scanned at a constant refresh rate. The bright, flicker-free image can be selectively written and erased. Also called a digital TV display.

raster scan (video) Currently, the dominant technology in CAD graphic displays. Similar to conventional television, it involves a line-by-line sweep across the entire CRT surface to generate the image. Raster-scan features include good brightness, accuracy, selective erase, dynamic motion capabilities, and the opportunity for unlimited color. The device can display a large amount of information without flicker, although resolution is not as good as with storage-tube displays.

read-only memory (ROM) A memory that cannot be modified or reprogrammed. Typically used for control and execute programs. See also *programmable read-only memory*.

real time   Refers to tasks or functions executed so rapidly by a CAD/CAM system that the feedback at various stages in the process can be used to guide the designer in completing the task. Immediate visual feedback through the CRT makes possible real time, interactive operation of a CAD/CAM system.

rectangular array   Insertion of the same entity at multiple locations on a CRT using the system's ability to copy design elements and place them at user-specified intervals to create a rectangular arrangement or matrix. A feature of PC and IC design systems.

refresh (or vector refresh)   A CAD display technology that involves frequent redrawing of an image displayed on the CRT to keep it bright, crisp, and clear. Refresh permits a high degree of movement in the displayed image as well as high resolution. Selective erase or editing is possible at any time without erasing and repainting the entire image. Although substantial amounts of high-speed memory are required, large, complex images may flicker.

refresh rate   The rate at which the graphic image on a CRT is redrawn in a refresh display, i.e., the time needed for one refresh of the displayed image.

registration   The degree of accuracy in the positioning of one layer or overlay in a CAD display or artwork, relative to another layer, as reflected by the clarity and sharpness of the resulting image.

repaint   A CAD feature that automatically redraws a design displayed on the CRT.

resolution   The smallest spacing between two display elements that will allow the elements to be distinguished visually on the CRT. The ability to define very minute detail. For example, the resolution of Computervision's IC design system is one part in 33.5 million. As applied to an electrostatic plotter, resolution means the number of dots per square inch.

response time   The elapsed time from initiation of an operation at a workstation to the receipt of the results at that workstation. Includes transmission of data to the CPU, processing, file access, and transmission of results back to the initiating workstation.

restart   To resume a computer program interrupted by operator intervention.

restore   To bring back to its original state a design currently being worked on in a CAD/CAM system after editing or modification that the designer now wants to cancel or rescind.

resume   A feature of some application programs that allows the designer to suspend the data-processing operation at some logical break point and restart it later from that point.

reticle   The photographic plate used to create an IC mask. See also *photo plotter*.

rotate   To turn a displayed 2D or 3D construction about an axis through a predefined angle relative to the original position.

robotics   The use of computer-controlled manipulators or arms to automate a variety of manufacturing processes such as welding, material handling, painting and assembly.

ROM   See *read-only memory*.

routine   A computer program, or a subroutine in the main program. The smallest separately compilable source code unit. See *computer program: source*.

rubber banding   A CAD capability that allows a component to be tracked (dragged) across the CRT screen, by means of an electronic pen, to a desired location, while simultaneously stretching all related interconnections to maintain signal continuity. During tracking the interconnections associated with the component stretch and bend, providing an excellent visual guide for optimizing the location of a component to best fit into the flow of the PC board, or other entity, minimizing total interconnect length and avoiding areas of congestion.

satellite   A remote system connected to another, usually larger, host system. A satellite differs from a remote intelligent work station in that it contains a full set of processors, memory, and mass storage resources to operate independently of the host. See *host-satellite system*.

scale   (*v*) To enlarge or diminish the size of a displayed entity without changing its shape, i.e., to bring it into a user-specified ratio to its original dimensions. Scaling can be done automatically by a CAD system. (*n*) Denotes the coordinate system for representing an object.

scissoring   The automatic erasing of all portions of a design on the CRT that lie outside user-specified boundaries.

scroll   To automatically roll up, as on a spool, a design or text message on a CRT to permit the sequential viewing of a message or drawing too large to be displayed all at once on the screen. New data appear on the CRT at one edge as other data disappear at the opposite edge. Graphics can be scrolled up, down, left, or right.

selective erase   A CAD feature for deleting portions of a display without affecting the remainder or having to repaint the entire CRT display.

shape fill   The automatic painting-in of an area, defined by user-specified boundaries, on an IC or PC board layout, for example, the area to be filled by copper when the PC board is manufactured. Can be done on-line by CAD.

smoothing   Fitting together curves and surfaces so that a smooth, continuous geometry results.

software   The collection of executable computer programs including application programs, operating systems, and languages.

source   A text file written in a high-level language and containing a computer program. It is easily read and understood by people but must be compiled or assembled to generate machine-recognizable instructions. Also known as *source code*. See also *high-level language*.

source language   A symbolic language comprised of statements and formulas used in computer processing. It is translated into object language (object code) by an assembler or compiler for execution by a computer.

spline   A subset of a B-spline where in a sequence of curves is restricted to a plane. An interpolation routine executed on a CAD/CAM system automatically adjusts a curve by design iteration until the curvature is continuous over the length of the curve. See also *B-spline*.

storage   The physical repository of all information relating to products designed on a CAD/CAM system. It is typically in the form of a magnetic tape or disk. Also called *memory*.

storage tube   A common type of CRT that retains an image continuously for a considerable period of time without redrawing (refreshing). The image will not flicker regardless of the amount of information displayed. However, the display tends to be slow relative to raster scan, the image is rather dim, and no single element by itself can be modified or deleted without redrawing. See also *direct view storage tube*.

stretch   A CAD design/editing aid that enables the designer to automatically expand a displayed entity beyond its original dimensions.

string   A linear sequence of entities, such as characters or physical elements, in a computer-aided design.

stylus   A hand-held pen used in conjunction with a data table to enter commands and coordinate input into the system. Also called an *electronic pen*.

subfigure   A part or a design element that may be extracted from a CAD library and inserted intact into another part displayed on the CRT.

surface machining   Automatic generation of NC tool paths to cut 3D shapes. Both the tool paths and the shapes may be constructed using the mechanical design capabilities of a CAD/CAM system.

symbol   Any recognizable sign, mark, shape or pattern used as a building block for designing meaningful structures. A set of primitive graphic entities (line, point, arc, circle, text, etc.) that form a construction that can be expressed as one unit and assigned a meaning. Symbols may be combined or nested to form larger symbols and/or drawings. They can be as complex as an entire PC board or as simple as a single element, such as a pad. Symbols are commonly used to represent physical things. For example, a particular graphic shape may be used to represent a complete device or a certain kind of electrical component in a schematic. To simplify the preparation of drawings of piping systems and flow diagrams, standard symbols are used to represent various types of fittings and components in common use.

Symbols are also basic units in a language. The recognizable sequence of characters END may inform a compiler that the routine it is compiling is completed. In computer-aided mapping, a symbol can be a diagram, design, letter, character, or abbreviation placed on maps and charts, that, by convention or reference to a legend, is understood to stand for or represent a specific characteristic or feature. In a CAD environment, symbol libraries contribute to the quick maintenance, placement, and interpretation of symbols.

syntax   (1) A set of rules describing the structure of statements allowed in a computer language. To make grammatical sense, commands and routines must be written in conformity to these rules. (2) The structure of a computer command language, i.e., the English-sentence structure of a CAD/CAM command language, e.g., verb, noun, modifiers.

system   An arrangement of CAD/CAM data processing, memory, display, and plotting modules—coupled with appropriate software—to achieve specific objectives. The term CAD/CAM system implies both hardware and software. See also *operating system* (a purely software term).

tablet   An input device on which a designer can digitize coordinate data or enter commands into a CAD/CAM system by means of an electronic pen. See also *data tablet*.

task   (1) A specific project that can be executed by a CAD/CAM software program. (2) A specific portion of memory assigned to the user for executing that project.

template   The pattern of a standard, commonly used component or part that serves as a design aid. Once created, it can be subsequently traced instead of redrawn whenever needed. The CAD equivalent of a designer's template might be a standard part in the data-base library that can be retrieved and inserted intact into an emerging drawing on the CRT.

temporary storage   Memory locations for storing immediate and partial results obtained during the execution of a program on the system.

terminal   See *workstation*.

text editor   An operating system program used to create and modify text files on the system.

text file   A file stored in the system in text format that can be printed and edited on-line as required.

throughput   The number of units of work performed by a CAD/CAM system or a work station during a given period of time. A quantitative measure of system productivity.

time-sharing   The use of a common CPU memory and processing capabilities by two or more CAD/CAM terminals to execute different tasks simultaneously.

tool path   Centerline of the tip of an NC cutting tool as it moves over a part produced on a CAD/CAM system. Tool paths can be created and displayed interactively or automatically by a CAD/CAM system, and reformatted into NC tapes, by means of postprocessor, to guide or control machining equipment. See also *surface machining*.

track ball A CAD graphics input device consisting of a ball recessed into a surface. The designer can rotate it in any direction to control the position of the cursor used for entering coordinate data into the system.

tracking Moving a predefined (tracking) symbol across the surface of the CRT with a light pen or an electronic pen.

transform To change an image displayed on the CRT by, for example, scaling, rotating, translating, or mirroring.

transformation The process of transforming a CAD display image. Also the matrix representation of a geometric space.

translate (1) To convert CAD/CAM output from one language to another, for example, by means of a postprocessor such as Computervision's IPC-to-Numerics Translator program. (2) Also, by an editing command, to move a CAD display entity a specified distance in a specified direction.

trap The area that is searched around each digitize to find a hit on a graphics entity to be edited. See also *digitize*.

turnaround time The elapsed time between the moment a task or project is input into the CAD/CAM system and the moment the required output is obtained.

turnkey A CAD/CAM system for which the supplier/vendor assumes total responsibility for building, installing, and testing both hardware and software, and the training of user personnel. Also, loosely, a system that comes equipped with all the hardware and software required for a specific application or applications. Usually implies a commitment by the vendor to make the system work and to provide preventive and remedial maintenance of both hardware and software. Sometimes used interchangeably with stand-alone, although stand-alone applies more to system architecture than to terms of purchase.

tutorial A characteristic of CAD/CAM systems. If the user is not sure how to execute a task, the system will show how. A message is displayed to provide information and guidance.

utilities Another term for system capabilities and/or features that enable the user to perform certain processes.

vector A quantity that has magnitude and direction and that, in CAD, is commonly represented by a directed line segment.

verification (1) A system-generated message to a work station acknowledging that a valid instruction or input has been received. (2) The process of checking the accuracy, viability, and/or manufacturability of an emerging design on the system.

view port A user-selected, rectangular view of a part, as-

sembly, etc., that presents the contents of a window on the CRT. See also *window*.

window A temporary, usually rectangular, bounded area on the CRT that is user-specified to include particular entities for modification, editing, or deletion.

wire-frame graphics A computer-aided design technique for displaying a 3D object on the CRT screen as a series of lines outlining its surface.

wiring diagram (1) Graphic representation of all circuits and device elements of an electrical system and its associated apparatus or any clearly defined functional portion of that system. A wiring diagram may contain not only wiring system components and wires but also nongraphic information such as wire number, wire size, color, function, component label, and pin number. (2) Illustration of device elements and their interconnectivity as distinguished from their physical arrangement. (3) Drawing that shows how to hook things up.

Wiring diagrams can be constructed, annotated, and documented on a CAD system.

word A set of bits (typically 16 to 32) that occupies a single storage location and is treated by the computer as a unit. See also *bit*.

working storage That part of the system's internal storage reserved for intermediate results (i.e., while a computer program is still in progress). Also called *temporary storage*.

workstation The work area and equipment used for CAD/CAM operations. It is where the designer interacts (communicates) with the computer. Frequently consists of a CRT display and an input device as well as, possibly, a digitizer and a hard-copy device. In a distributed processing system, a work station would have local processing and mass storage capabilities. Also called a *terminal* or *design terminal*.

write To transfer information from CPU main memory to a peripheral device, such as a mass storage device.

write-protect A security feature in a CAD/CAM data storage device that prevents new data from being written over existing data.

zero The origin of all coordinate dimensions defined in an absolute system as the intersection of the baselines of the x, y, and z axes.

zero offset On an NC unit, this features allows the zero point on an axis to be relocated anywhere within a specified range, thus temporarily redefining the coordinate frame of reference.

zoom A CAD capability that proportionately enlarges or reduces a figure displayed on a CRT screen.

4 Abbreviations for Use on Drawings and in Text— American National Standard

(Selected from ANSI/ASME Y1.1–1989)

A

absolute	ABS
accelerate	ACCEL
accessory	ACCESS.
account	ACCT
accumulate	ACCUM
actual	ACT.
adapter	ADPT
addendum	ADD.
addition	ADD.
adjust	ADJ
advance	ADV
after	AFT.
aggregate	AGGR
air condition	AIR COND
airplane	APL
allowance	ALLOW
alloy	ALY
alteration	ALT
alternate	ALT
alternating current	AC
altitude	ALT
aluminum	AL
American National Standard	AMER NATL STD
American wire gage	AWG
amount	AMT
ampere	AMP
amplifier	AMPL
anneal	ANL
antenna	ANT.
apartment	APT.
apparatus	APP
appendix	APPX
approved	APPD
approximate	APPROX
arc weld	ARC/W
area	A
armature	ARM.
armor plate	ARM-PL
army navy	AN
arrange	ARR.
artificial	ART.
asbestos	ASB
asphalt	ASPH
assemble	ASSEM
assembly	ASSY

assistant	ASST
associate	ASSOC
association	ASSN
atomic	AT
audible	AUD
audio frequency	AF
authorized	AUTH
automatic	AUTO
auto-transformer	AUTO TR
auxiliary	AUX
avenue	AVE
average	AVG
aviation	AVI
azimuth	AZ

B

Babbitt	BAB
back feed	BF
back pressure	BP
back to back	B to B
backface	BF
balance	BAL
ball bearing	BB
barometer	BAR
base line	BL
base plate	BP
bearing	BRG
bench mark	BM
bending moment	M
bent	BT
bessemer	BESS
between	BET.
between centers	BC
between perpendiculars	BP
bevel	BEV
bill of material	B/M
Birmingham wire gage	BWG
blank	BLK
block	BLK
blueprint	BP
board	BD
boiler	BLR
boiler feed	BF
boiler horsepower	BHP
boiling point	BP
bolt circle	BC
both faces	BF

both sides	BS
both ways	BW
bottom	BOT
bottom chord	BC
bottom face	BF
bracket	BRKT
brake	BK
brake horsepower	BHP
brass	BRS
brazing	BRZG
break	BRK
Brinell hardness	BH
British Standard	BR STD
British thermal unit	BTU
broach	BRO
bronze	BRZ
Brown & Sharpe (wire gage, same as AWG)	B&S
building	BLDG
bulkhead	BHD
burnish	BNH
bushing	BUSH.
button	BUT.

C

cabinet	CAB.
calculate	CALC
calibrate	CAL
cap screw	CAP SCR
capacity	CAP
carburetor	CARB
carburize	CARB
carriage	CRG
case harden	CH
cast iron	CI
cast steel	CS
casting	CSTG
castle nut	CAS NUT
catalogue	CAT.
cement	CEM
center	CTR
center line	CL
center of gravity	CG
center of pressure	CP
center to center	C to C
centering	CTR
chamfer	CHAM

change	CHG
channel	CHAN
check	CHK
check valve	CV
chord	CHD
circle	CIR
circular	CIR
circular pitch	CP
circumference	CIRC
clear	CLR
clearance	CL
clockwise	CW
coated	CTD
cold drawn	CD
cold-drawn steel	CDS
cold finish	CF
cold punched	CP
cold rolled	CR
cold-rolled steel	CRS
combination	COMB.
combustion	COMB
commercial	COML
company	CO
complete	COMPL
compress	COMP
concentric	CONC
concrete	CONC
condition	COND
connect	CONN
constant	CONST
construction	CONST
contact	CONT
continue	CONT
copper	COP.
corner	COR
corporation	CORP
correct	CORR
corrugate	CORR
cotter	COT
counter	CTR
counterbore	CBORE
counter clockwise	CCW
counterdrill	CDRILL
counterpunch	CPUNCH
countersink	CSK
coupling	CPLG
cover	COV
cross section	XSECT
cubic	CU
cubic foot	CU FT
cubic inch	CU IN.
current	CUR
customer	CUST
cyanide	CYN

D

decimal	DEC
dedendum	DED
deflect	DEFL
degree	(°) DEG
density	D
department	DEPT
design	DSGN
detail	DET
develop	DEV
diagonal	DIAG
diagram	DIAG
diameter	DIA
diametral pitch	DP
dimension	DIM.
discharge	DISCH
distance	DIST
division	DIV
double	DBL
dovetail	DVTL
dowel	DWL
down	DN
dozen	DOZ
drafting	DFTG
drawing	DWG
drill or drill rod	DR
drive	DR
drive fit	DF
drop	D
drop forge	DF
duplicate	DUP

E

each	EA
east	E
eccentric	ECC
effective	EFF
elbow	ELL
electric	ELEC
elementary	ELEM
elevate	ELEV
elevation	EL
engine	ENG
engineer	ENGR
engineering	ENGRG
entrance	ENT
equal	EQ
equation	EQ
equipment	EQUIP
equivalent	EQUIV
estimate	EST
exchange	EXCH
exhaust	EXH

existing	EXIST.
exterior	EXT
extra heavy	X HVY
extra strong	X STR
extrude	EXTR

F

fabricate	FAB
face to face	F to F
Fahrenheit	F
far side	FS
federal	FED.
feed	FD
feet	(′) FT
figure	FIG.
fillet	FIL
fillister	FIL
finish	FIN.
finish all over	FAO
flange	FLG
flat	F
flat head	FH
floor	FL
fluid	FL
focus	FOC
foot	(′) FT
force	F
forged steel	FST
forging	FORG
forward	FWD
foundry	FDRY
frequency	FREQ
front	FR
furnish	FURN

G

gage or gauge	GA
gallon	GAL
galvanize	GALV
galvanized iron	GI
galvanized steel	GS
gasket	GSKT
general	GEN
glass	GL
government	GOVT
governor	GOV
grade	GR
graduation	GRAD
graphite	GPH
grind	GRD
groove	GRV
ground	GRD

H

half-round	$\frac{1}{2}$RD
handle	HDL
hanger	HGR
hard	H
harden	HDN
hardware	HDW
head	HD
headless	HDLS
heat	HT
heat-treat	HT TR
heavy	HVY
hexagon	HEX
high-pressure	HP
high-speed	HS
horizontal	HOR
horsepower	HP
hot rolled	HR
hot-rolled steel	HRS
hour	HR
housing	HSG
hydraulic	HYD

I

illustrate	ILLUS
inboard	INBD
inch	(″) IN.
inches per second	IPS
inclosure	INCL
include	INCL
inside diameter	ID
instrument	INST
interior	INT
internal	INT
intersect	INT
iron	I
irregular	IREG

J

joint	JT
joint army-navy	JAN
journal	JNL
junction	JCT

K

key	K
keyseat	KST
Keyway	KWY

L

laboratory	LAB
laminate	LAM
lateral	LAT
left	L
left hand	LH
length	LG
length over all	LOA
letter	LTR
light	LT
line	L
locate	LOC
logarithm	LOG.
long	LG
lubricate	LUB
lumber	LBR

M

machine	MACH
machine steel	MS
maintenance	MAINT
malleable	MALL
malleable iron	MI
manual	MAN.
manufacture	MFR
manufactured	MFD
manufacturing	MFG
material	MATL
maximum	MAX
mechanical	MECH
mechanism	MECH
median	MED
metal	MET.
meter	M
miles	MI
miles per hour	MPH
millimeter	MM
minimum	MIN
minute	(′) MIN
miscellaneous	MISC
month	MO
Morse taper	MOR T
motor	MOT
mounted	MTD
mounting	MTG
multiple	MULT
music wire gage	MWG

N

national	NATL
natural	NAT
near face	NF
near side	NS
negative	NEG
neutral	NEUT
nominal	NOM
normal	NOR
north	N
not to scale	NTS
number	NO.

O

obsolete	OBS
octagon	OCT
office	OFF.
on center	OC
opposite	OPP
optical	OPT
original	ORIG
outlet	OUT.
outside diameter	OD
outside face	OF
outside radius	OR
overall	OA

P

pack	PK
packing	PKG
page	P
paragraph	PAR.
part	PT
patent	PAT.
pattern	PATT
permanent	PERM
perpendicular	PERP
piece	PC
piece mark	PC MK
pint	PT
pitch	P
pitch circle	PC
pitch diameter	PD
plastic	PLSTC
plate	PL
plumbing	PLMB
point	PT
point of curve	PC
point of intersection	PI
point of tangent	PT
polish	POL
position	POS
potential	POT.
pound	LB
pounds per square inch	PSI
power	PWR
prefabricated	PREFAB
preferred	PFD
prepare	PREP
pressure	PRESS.

Term	Abbr.	Term	Abbr.	Term	Abbr.
process	PROC	screw	SCR	thread	THD
production	PROD	second	SEC	threads per inch	TPI
profile	PF	section	SECT	through	THRU
propeller	PROP	semi-steel	SS	time	T
publication	PUB	separate	SEP	tolerance	TOL
push button	PB	set screw	SS	tongue & groove	T & G
		shaft	SFT	tool steel	TS
Q		sheet	SH	tooth	T
quadrant	QUAD	shoulder	SHLD	total	TOT
quality	QUAL	side	S	transfer	TRANS
quarter	QTR	single	S	typical	TYP
		sketch	SK		
R		sleeve	SLV	**U**	
radial	RAD	slide	SL	ultimate	ULT
radius	R	slotted	SLOT.	unit	U
railroad	RR	small	SM	universal	UNIV
ream	RM	socket	SOC		
received	RECD	space	SP	**V**	
record	REC	special	SPL	vacuum	VAC
rectangle	RECT	specific	SP	valve	V
reduce	RED.	spot faced	SF	variable	VAR
reference line	REF L	spring	SPG	versus	VS
reinforce	REINF	square	SQ	vertical	VERT
release	REL	standard	STD	volt	V
relief	REL	station	STA	volume	VOL
remove	REM	stationary	STA		
require	REQ	steel	STL	**W**	
required	REQD	stock	STK	wall	W
return	RET.	straight	STR	washer	WASH.
reverse	REV	street	ST	watt	W
revolution	REV	structural	STR	week	WK
revolutions per minute	RPM	substitute	SUB	weight	WT
right	R	summary	SUM.	west	W
right hand	RH	support	SUP	width	W
rivet	RIV	surface	SUR	wood	WD
Rockwell hardness	RH	symbol	SYM	Woodruff	WDF
roller bearing	RB	system	SYS	working point	WP
room	RM			working pressure	WP
root diameter	RD	**T**		wrought	WRT
root mean square	RMS	tangent	TAN.	wrought iron	WI
rough	RGH	taper	TPR		
round	RD	technical	TECH		
		template	TEMP		
S		tension	TENS.	**Y**	
schedule	SCH	terminal	TERM.	yard	YD
schematic	SCHEM	thick	THK	year	YR
scleroscope hardness	SH	thousand	M		

5   Running and Sliding Fits[a]—American National Standard

RC 1   *Close sliding fits* are intended for the accurate location of parts which must assemble without perceptible play.

RC 2   *Sliding fits* are intended for accurate location, but with greater maximum clearance than class RC 1. Parts made to this fit move and turn easily but are not intended to run freely, and in the larger sizes may seize with small temperature changes.

RC 3   *Precision running fits* are about the closest fits which can be expected to run freely and are intended for precision work at slow speeds and light journal pressures, but they are not suitable where appreciable temperature differences are likely to be encountered.

RC 4   *Close running fits* are intended chiefly for running fits on accurate machinery with moderate surface speeds and journal pressures, where accurate location and minimum play are desired.

Basic hole system. Limits are in thousandths of an inch.
Limits for hole and shaft are applied algebraically to the basic size to obtain the limits of size for the parts.
Data in **boldface** are in accordance with ABC agreements.
Symbols H5, g5, etc., are hole and shaft designations used in ABC System.

Nominal Size Range, inches Over To	Class RC 1 Limits of Clearance	Class RC 1 Standard Limits Hole H5	Class RC 1 Standard Limits Shaft g4	Class RC 2 Limits of Clearance	Class RC 2 Standard Limits Hole H6	Class RC 2 Standard Limits Shaft g5	Class RC 3 Limits of Clearance	Class RC 3 Standard Limits Hole H7	Class RC 3 Standard Limits Shaft f6	Class RC 4 Limits of Clearance	Class RC 4 Standard Limits Hole H8	Class RC 4 Standard Limits Shaft f7
0–0.12	0.1 0.45	+0.2 −0	−0.1 −0.25	0.1 0.55	+0.25 −0	−0.1 −0.3	0.3 0.95	+0.4 −0	−0.3 −0.55	0.3 1.3	+0.6 −0	−0.3 −0.7
0.12–0.24	0.15 0.5	+0.2 −0	−0.15 −0.3	0.15 0.65	+0.3 −0	−0.15 −0.35	0.4 1.12	+0.5 −0	−0.4 −0.7	0.4 1.6	+0.7 −0	−0.4 −0.9
0.24–0.40	0.2 0.6	+0.25 −0	−0.2 −0.35	0.2 0.85	+0.4 −0	−0.2 −0.45	0.5 1.5	+0.6 −0	−0.5 −0.9	0.5 2.0	+0.9 −0	−0.5 −1.1
0.40–0.71	0.25 0.75	+0.3 −0	−0.25 −0.45	0.25 0.95	+0.4 −0	−0.25 −0.55	0.6 1.7	+0.7 −0	−0.6 −1.0	0.6 2.3	+1.0 −0	−0.6 −1.3
0.71–1.19	0.3 0.95	+0.4 −0	−0.3. −0.55	0.3 1.2	+0.5 −0	−0.3 −0.7	0.8 2.1	+0.8 −0	−0.8 −1.3	0.8 2.8	+1.2 −0	−0.8 −1.6
1.19–1.97	0.4 1.1	+0.4 −0	−0.4 −0.7	0.4 1.4	+0.6 −0	−0.4 −0.8	1.0 2.6	+1.0 −0	−1.0 −1.6	1.0 3.6	+1.6 −0	−1.0 −2.0
1.97–3.15	0.4 1.2	+0.5 −0	−0.4 −0.7	0.4 1.6	+0.7 −0	−0.4 −0.9	1.2 3.1	+1.2 −0	−1.2 −1.9	1.2 4.2	+1.8 −0	−1.2 −2.4
3.15–4.73	0.5 1.5	+0.6 −0	−0.5 −0.9	0.5 2.0	+0.9 −0	−0.5 −1.1	1.4 3.7	+1.4 −0	−1.4 −2.3	1.4 5.0	+2.2 −0	−1.4 −2.8
4.73–7.09	0.6 1.8	+0.7 −0	−0.6 −1.1	0.6 2.3	+1.0 −0	−0.6 −1.3	1.6 4.2	+1.6 −0	−1.6 −2.6	1.6 5.7	+2.5 −0	−1.6 −3.2
7.09–9.85	0.6 2.0	+0.8 −0	−0.6 −1.2	0.6 2.6	+1.2 −0	−0.6 −1.4	2.0 5.0	+1.8 −0	−2.0 −3.2	2.0 6.6	+2.8 −0	−2.0 −3.8
9.85–12.41	0.8 2.3	+0.9 −0	−0.8 −1.4	0.8 2.9	+1.2 −0	−0.8 −1.7	2.5 5.7	+2.0 −0	−2.5 −3.7	2.5 7.5	+3.0 −0	−2.5 −4.5
12.41–15.75	1.0 2.7	+1.0 −0	−1.0 −1.7	1.0 3.4	+1.4 −0	−1.0 −2.0	3.0 6.6	+2.2 −0	−3.0 −4.4	3.0 8.7	+3.5 −0	−3.0 −5.2

[a] From ANSI B4.1–1967 (R1994). For larger diameters, see the standard.

5 Running and Sliding Fits[a]—American National Standard (continued)

RC 5
RC 6 *Medium running fits* are intended for higher running speeds, or heavy journal pressures, or both.

RC 7 *Free running fits* are intended for use where accuracy is not essential, or where large temperature variations are likely to be encountered, or under both these conditions.

RC 8
RC 9 *Loose running fits* are intended for use where wide commercial tolerances may be necessary, together with an allowance, on the external member.

Nominal Size Range, inches Over To	Class RC 5 Limits of Clearance	Class RC 5 Standard Limits Hole H8	Class RC 5 Standard Limits Shaft e7	Class RC 6 Limits of Clearance	Class RC 6 Standard Limits Hole H9	Class RC 6 Standard Limits Shaft e8	Class RC 7 Limits of Clearance	Class RC 7 Standard Limits Hole H9	Class RC 7 Standard Limits Shaft d8	Class RC 8 Limits of Clearance	Class RC 8 Standard Limits Hole H10	Class RC 8 Standard Limits Shaft c9	Class RC 9 Limits of Clearance	Class RC 9 Standard Limits Hole H11	Class RC 9 Standard Limits Shaft
0–0.12	0.6	+0.6	−0.6	0.6	+1.0	−0.6	1.0	+1.0	−1.0	2.5	+1.6	−2.5	4.0	+2.5	−4.0
	1.6	−0	−1.0	2.2	−0	−1.2	2.6	−0	−1.6	5.1	−0	−3.5	8.1	−0	−5.6
0.12–0.24	0.8	+0.7	−0.8	0.8	+1.2	−0.8	1.2	+1.2	−1.2	2.8	+1.8	−2.8	4.5	+3.0	−4.5
	2.0	−0	−1.3	2.7	−0	−1.5	3.1	−0	−1.9	5.8	−0	−4.0	9.0	−0	−6.0
0.24–0.40	1.0	+0.9	−1.0	1.0	+1.4	−1.0	1.6	+1.4	−1.6	3.0	+2.2	−3.0	5.0	+3.5	−5.0
	2.5	−0	−1.6	3.3	−0	−1.9	3.9	−0	−2.5	6.6	−0	−4.4	10.7	−0	−7.2
0.40–0.71	1.2	+1.0	−1.2	1.2	+1.6	−1.2	2.0	+1.6	−2.0	3.5	+2.8	−3.5	6.0	+4.0	−6.0
	2.9	−0	−1.9	3.8	−0	−2.2	4.6	−0	−3.0	7.9	−0	−5.1	12.8	−0	−8.8
0.71–1.19	1.6	+1.2	−1.6	1.6	+2.0	−1.6	2.5	+2.0	−2.5	4.5	+3.5	−4.5	7.0	+5.0	−7.0
	3.6	−0	−2.4	4.8	−0	−2.8	5.7	−0	−3.7	10.0	−0	−6.5	15.5	−0	−10.5
1.19–1.97	2.0	+1.6	−2.0	2.0	+2.5	−2.0	3.0	+2.5	−3.0	5.0	+4.0	−5.0	8.0	+6.0	−8.0
	4.6	−0	−3.0	6.1	−0	−3.6	7.1	−0	−4.6	11.5	−0	−7.5	18.0	−0	−12.0
1.97–3.15	2.5	+1.8	−2.5	2.5	+3.0	−2.5	4.0	+3.0	−4.0	6.0	+4.5	−6.0	9.0	+7.0	−9.0
	5.5	−0	−3.7	7.3	−0	−4.3	8.8	−0	−5.8	13.5	−0	−9.0	20.5	−0	−13.5
3.15–4.73	3.0	+2.2	−3.0	3.0	+3.5	−3.0	5.0	+3.5	−5.0	7.0	+5.0	−7.0	10.0	+9.0	−10.0
	6.6	−0	−4.4	8.7	−0	−5.2	10.7	−0	−7.2	15.5	−0	−10.5	24.0	−0	−15.0
4.73–7.09	3.5	+2.5	−3.5	3.5	+4.0	−3.5	6.0	+4.0	−6.0	8.0	+6.0	−8.0	12.0	+10.0	−12.0
	7.6	−0	−5.1	10.0	−0	−6.0	12.5	−0	−8.5	18.0	−0	−12.0	28.0	−0	−18.0
7.09–9.85	4.0	+2.8	−4.0	4.0	+4.5	−4.0	7.0	+4.5	−7.0	10.0	+7.0	−10.0	15.0	+12.0	−15.0
	8.6	−0	−5.8	11.3	−0	−6.8	14.3	−0	−9.8	21.5	−0	−14.5	34.0	−0	−22.0
9.85–12.41	5.0	+3.0	−5.0	5.0	+5.0	−5.0	8.0	+5.0	−8.0	12.0	+8.0	−12.0	18.0	+12.0	−18.0
	10.0	−0	−7.0	13.0	−0	−8.0	16.0	−0	−11.0	25.0	−0	−17.0	38.0	−0	−26.0
12.41–15.75	6.0	+3.5	−6.0	6.0	+6.0	−6.0	10.0	+6.0	−10.0	14.0	+9.0	−14.0	22.0	+14.0	−22.0
	11.7	−0	−8.2	15.5	−0	−9.5	19.5	−0	13.5	29.0	−0	−20.0	45.0	−0	−31.0

[a] From ANSI B4.1–1967 (R1994). For larger diameters, see the standard.

6 Clearance Locational Fits[a]—American National Standard

LC   *Locational clearance fits* are intended for parts which are normally stationary but which can be freely assembled or disassembled They run from snug fits for parts requiring accuracy of location, through the medium clearance fits for parts such as spigots, to the looser fastener fits, where freedom of assembly is of prime importance.

Basic hole system. Limits are in thousandths of an inch.
Limits for hole and shaft are applied algebraically to the basic size to obtain the limits of size for the parts.
Data in **boldface** are in accordance with ABC agreements.
Symbols H6, H5, etc., are hole and shaft designations used in ABC System.

Nominal Size Range, inches		Class LC 1			Class LC 2			Class LC 3			Class LC 4			Class LC 5		
	Limits of Clearance	Standard Limits		Limits of Clearance	Standard Limits		Limits of Clearance	Standard Limits		Limits of Clearance	Standard Limits		Limits of Clearance	Standard Limits	
Over To		Hole H6	Shaft h5		Hole H7	Shaft h6		Hole H8	Shaft h7		Hole H10	Shaft h9		Hole H7	Shaft g6
0–0.12	0 0.45	+0.25 −0	+0 −0.2	0 0.65	+0.4 −0	+0 −0.25	0 1	+0.6 −0	+0 −0.4	0 2.6	+1.6 −0	+0 −1.0	0.1 0.75	+0.4 −0	−0.1 −0.35
0.12–0.24	0 0.5	+0.3 −0	+0 −0.2	0 0.8	+0.5 −0	+0 −0.3	0 1.2	+0.7 −0	+0 −0.5	0 3.0	+1.8 −0	+0 −1.2	0.15 0.95	+0.5 −0	−0.15 −0.45
0.24–0.40	0 0.65	+0.4 −0	+0 −0.25	0 1.0	+0.6 −0	+0 −0.4	0 1.5	+0.9 −0	+0 −0.6	0 3.6	+2.2 −0	+0 −1.4	0.2 1.2	+0.6 −0	−0.2 −0.6
0.40–0.71	0 0.7	+0.4 −0	+0 −0.3	0 1.1	+0.7 −0	+0 −0.4	0 1.7	+1.0 −0	+0 −0.7	0 4.4	+2.8 −0	+0 −1.6	0.25 1.35	+0.7 −0	−0.25 −0.65
0.71–1.19	0 0.9	+0.5 −0	+0 −0.4	0 1.3	+0.8 −0	+0 −0.5	0 2	+1.2 −0	+0 −0.8	0 5.5	+3.5 −0	+0 −2.0	0.3 1.6	+0.8 −0	−0.3 −0.8
1.19–1.97	0 1.0	+0.6 −0	+0 −0.4	0 1.6	+1.0 −0	+0 −0.6	0 2.6	+1.6 −0	+0 −1	0 6.5	+4.0 −0	+0 −2.5	0.4 2.0	+1.0 −0	−0.4 −1.0
1.97–3.15	0 1.2	+0.7 −0	+0 −0.5	0 1.9	+1.2 −0	+0 −0.7	0 3	+1.8 −0	+0 −1.2	0 7.5	+4.5 −0	+0 −3	0.4 2.3	+1.2 −0	−0.4 −1.1
3.15–4.73	0 1.5	+0.9 −0	+0 −0.6	0 2.3	+1.4 −0	+0 −0.9	0 3.6	+2.2 −0	+0 −1.4	0 8.5	+5.0 −0	+0 −3.5	0.5 2.8	+1.4 −0	−0.5 −1.4
4.73–7.09	0 1.7	+1.0 −0	+0 −0.7	0 2.6	+1.6 −0	+0 −1.0	0 4.1	+2.5 −0	+0 −1.6	0 10	+6.0 −0	+0 −4	0.6 3.2	+1.6 −0	−0.6 −1.6
7.09–9.85	0 2.0	+1.2 −0	+0 −0.8	0 3.0	+1.8 −0	+0 −1.2	0 4.6	+2.8 −0	+0 −1.8	0 11.5	+7.0 −0	+0 −4.5	0.6 3.6	+1.8 −0	−0.6 −1.8
9.85–12.41	0 2.1	+1.2 −0	+0 −0.9	0 3.2	+2.0 −0	+0 −1.2	0 5	+3.0 −0	+0 −2.0	0 13	+8.0 −0	+0 −5	0.7 3.9	+2.0 −0	−0.7 −1.9
12.41–15.75	0 2.4	+1.4 −0	+0 −1.0	0 3.6	+2.2 −0	+0 −1.4	0 5.7	+3.5 −0	+0 −2.2	0 15	+9.0 −0	+0 −6	0.7 4.3	+2.2 −0	−0.7 −2.1

[a] From ANSI B4.1–1967 (R1994). For larger diameters, see the standard.

Nominal Size Range, inches Over To	Class LC 6 Limits of Clearance	Class LC 6 Standard Limits Hole H9	Class LC 6 Standard Limits Shaft f8	Class LC 7 Limits of Clearance	Class LC 7 Standard Limits Hole H10	Class LC 7 Standard Limits Shaft e9	Class LC 8 Limits of Clearance	Class LC 8 Standard Limits Hole H10	Class LC 8 Standard Limits Shaft d9	Class LC 9 Limits of Clearance	Class LC 9 Standard Limits Hole H11	Class LC 9 Standard Limits Shaft c10	Class LC 10 Limits of Clearance	Class LC 10 Standard Limits Hole H12	Class LC 10 Standard Limits Shaft	Class LC 11 Limits of Clearance	Class LC 11 Standards Limits Hole H13	Class LC 11 Standards Limits Shaft
0–0.12	0.3 / 1.9	+1.0 / −0	−0.3 / −0.9	0.6 / 3.2	+1.6 / −0	−0.6 / −1.6	1.0 / 3.6	+1.6 / −0	−1.0 / −2.0	2.5 / 6.6	+2.5 / −0	−2.5 / −4.1	4 / 12	+4 / −0	−4 / −8	5 / 17	+6 / −0	−5 / −11
0.12–0.24	0.4 / 2.3	+1.2 / −0	−0.4 / −1.1	0.8 / 3.8	+1.8 / −0	−0.8 / −2.0	1.2 / 4.2	+1.8 / −0	−1.2 / −2.4	2.8 / 7.6	+3.0 / −0	−2.8 / −4.6	4.5 / 14.5	+5 / −0	−4.5 / −9.5	6 / 20	+7 / −0	−6 / −13
0.24–0.40	0.5 / 2.8	+1.4 / −0	−0.5 / −1.4	1.0 / 4.6	+2.2 / −0	−1.0 / −2.4	1.6 / 5.2	+2.2 / −0	−1.6 / −3.0	3.0 / 8.7	+3.5 / −0	−3.0 / −5.2	5 / 17	+6 / −0	−5 / −11	7 / 25	+9 / −0	−7 / −16
0.40–0.71	0.6 / 3.2	+1.6 / −0	−0.6 / −1.6	1.2 / 5.6	+2.8 / −0	−1.2 / −2.8	2.0 / 6.4	+2.8 / −0	−2.0 / −3.6	3.5 / 10.3	+4.0 / −0	−3.5 / −6.3	6 / 20	+7 / −0	−6 / −13	8 / 28	+10 / −0	−8 / −18
071–1.19	0.8 / 4.0	+2.0 / −0	−0.8 / −2.0	1.6 / 7.1	+3.5 / −0	−1.6 / −3.6	2.5 / 8.0	+3.5 / −0	−2.5 / −4.5	4.5 / 13.0	+5.0 / −0	−4.5 / −8.0	7 / 23	+8 / −0	−7 / −15	10 / 34	+12 / −0	−10 / −22
1.19–1.97	1.0 / 5.1	+2.5 / −0	−1.0 / −2.6	2.0 / 8.5	+4.0 / −0	−2.0 / −4.5	3.0 / 9.5	+4.0 / −0	−3.0 / −5.5	5 / 15	+6 / −0	−5 / −9	8 / 28	+10 / −0	−8 / −18	12 / 44	+16 / −0	−12 / −28
1.97–3.15	1.2 / 6.0	+3.0 / −0	−1.2 / −3.0	2.5 / 10.0	+4.5 / −0	−2.5 / −5.5	4.0 / 11.5	+4.5 / −0	−4.0 / −7.0	6 / 17.5	+7 / −0	−6 / −10.5	10 / 34	+12 / −0	−10 / −22	14 / 50	+18 / −0	−14 / −32
3.15–4.73	1.4 / 7.1	+3.5 / −0	−1.4 / −3.6	3.0 / 11.5	+5.0 / −0	−3.0 / −6.5	5.0 / 13.5	+5.0 / −0	−5.0 / −8.5	7 / 21	+9 / −0	−7 / −12	11 / 39	+14 / −0	−11 / −25	16 / 60	+22 / −0	−16 / −38
4.73–7.09	1.6 / 8.1	+4.0 / −0	−1.6 / −4.1	3.5 / 13.5	+6.0 / −0	−3.5 / −7.5	6 / 16	+6 / −0	−6 / −10	8 / 24	+10 / −0	−8 / −14	12 / 44	+16 / −0	−12 / −28	18 / 68	+25 / −0	−18 / −43
7.09–9.85	2.0 / 9.3	+4.5 / −0	−2.0 / −4.8	4.0 / 15.5	+7.0 / −0	−4.0 / −8.5	7 / 18.5	+7 / −0	−7 / −11.5	10 / 29	+12 / −0	−10 / −17	16 / 52	+18 / −0	−16 / −34	22 / 78	+28 / −0	−22 / −50
9.85–12.41	2.2 / 10.2	+5.0 / −0	−2.2 / −5.2	4.5 / 17.5	+8.0 / −0	−4.5 / −9.5	7 / 20	+8 / −0	−7 / −12	12 / 32	+12 / −0	−12 / −20	20 / 60	+20 / −0	−20 / −40	28 / 88	+30 / −0	−28 / −58
12.41–15.75	2.5 / 12.0	+6.0 / −0	−2.5 / −6.0	5.0 / 20.0	+9.0 / −0	−5 / −11	8 / 23	+9 / −0	−8 / −14	14 / 37	+14 / −0	−14 / −23	22 / 66	+22 / −0	−22 / −44	30 / 100	+35 / −0	−30 / −65

[a] From ANSI B4.1–1967 (R1994). For larger diameters, see the standard.

7 Transition Locational Fits[a]—American National Standard

LT Transition fits are a compromise between clearance and interference fits, for application where accuracy of location is important, but either a small amount of clearance or interference is permissible.

Basic hole system. Limits are in thousandths of an inch.
Limits for hole and shaft are applied algebraically to the basic size to obtain the limits of size for the mating parts.
Data in **boldface** are in accordance with ABC agreements.
"Fit" represents the maximum interference (minus values) and the maximum clearance (plus values).
Symbols H7, js6, etc., are hole and shaft designations used in ABC System.

Nominal Size Range, inches (Over To)	LT 1 Fit	LT 1 Hole H7	LT 1 Shaft js6	LT 2 Fit	LT 2 Hole H8	LT 2 Shaft js7	LT 3 Fit	LT 3 Hole H7	LT 3 Shaft k6	LT 4 Fit	LT 4 Hole H8	LT 4 Shaft k7	LT 5 Fit	LT 5 Hole H7	LT 5 Shaft n6	LT 6 Fit	LT 6 Hole H7	LT 6 Shaft n7
0–0.12	−0.10 / +0.50	+0.4 / −0	+0.10 / −0.10	−0.2 / +0.8	+0.6 / −0	+0.2 / −0.2							−0.5 / +0.15	+0.4 / −0	+0.5 / +0.25	−0.65 / +0.15	+0.4 / −0	+0.65 / +0.25
0.12–0.24	−0.15 / +0.65	+0.5 / −0	+0.15 / −0.15	−0.25 / +0.95	+0.7 / −0	+0.25 / −0.25							−0.6 / +0.2	+0.5 / −0	+0.6 / +0.3	−0.8 / +0.2	+0.5 / −0	+0.8 / +0.3
0.24–0.40	−0.2 / +0.8	+0.6 / −0	+0.2 / −0.2	−0.3 / +1.2	+0.9 / −0	+0.3 / −0.3	−0.5 / +0.5	+0.6 / −0	+0.5 / +0.1	−0.7 / +0.8	+0.9 / −0	+0.7 / +0.1	−0.8 / +0.2	+0.6 / −0	+0.8 / +0.4	−1.0 / +0.2	+0.6 / −0	+1.0 / +0.4
0.40–0.71	−0.2 / +0.9	+0.7 / −0	+0.2 / −0.2	−0.35 / +1.35	+1.0 / −0	+0.35 / −0.35	−0.5 / +0.6	+0.7 / −0	+0.5 / +0.1	−0.8 / +0.9	+1.0 / −0	+0.8 / +0.1	−0.9 / +0.2	+0.7 / −0	+0.9 / +0.5	−1.2 / +0.2	+0.7 / −0	+1.2 / +0.5
0.71–1.19	−0.25 / +1.05	+0.8 / −0	+0.25 / −0.25	−0.4 / +1.6	+1.2 / −0	+0.4 / −0.4	−0.6 / +0.7	+0.8 / −0	+0.6 / +0.1	−0.9 / +1.1	+1.2 / −0	+0.9 / +0.1	−1.1 / +0.2	+0.8 / −0	+1.1 / +0.6	−1.4 / +0.2	+0.8 / −0	+1.4 / +0.6
1.19–1.97	−0.3 / +1.3	+1.0 / −0	+0.3 / −0.3	−0.5 / +2.1	+1.6 / −0	+0.5 / −0.5	−0.7 / +0.9	+1.0 / −0	+0.7 / +0.1	−1.1 / +1.5	+1.6 / −0	+1.1 / +0.1	−1.3 / +0.3	+1.0 / −0	+1.3 / +0.7	−1.7 / +0.3	+1.0 / −0	+1.7 / +0.7
1.97–3.15	−0.3 / +1.5	+1.2 / −0	+0.3 / −0.3	−0.6 / +2.4	+1.8 / −0	+0.6 / −0.6	−0.8 / +1.1	+1.2 / −0	+0.8 / +0.1	−1.3 / +1.7	+1.8 / −0	+1.3 / +0.1	−1.5 / +0.4	+1.2 / −0	+1.5 / +0.8	−2.0 / +0.4	+1.2 / −0	+2.0 / +0.8
3.15–4.73	−0.4 / +1.8	+1.4 / −0	+0.4 / −0.4	−0.7 / +2.9	+2.2 / −0	+0.7 / −0.7	−1.0 / +1.3	+1.4 / −0	+1.0 / +0.1	−1.5 / +2.1	+2.2 / −0	+1.5 / +0.1	−1.9 / +0.4	+1.4 / −0	+1.9 / +1.0	−2.4 / +0.4	+1.4 / −0	+2.4 / +1.0
4.73–7.09	−0.5 / +2.1	+1.6 / −0	+0.5 / −0.5	−0.8 / +3.3	+2.5 / −0	+0.8 / −0.8	−1.1 / +1.5	+1.6 / −0	+1.1 / +0.1	−1.7 / +2.4	+2.5 / −0	+1.7 / +0.1	−2.2 / +0.4	+1.6 / −0	+2.2 / +1.2	−2.8 / +0.4	+1.6 / −0	+2.8 / +1.2
7.09–9.85	−0.6 / +2.4	+1.8 / −0	+0.6 / −0.6	−0.9 / +3.7	+2.8 / −0	+0.9 / −0.9	−1.4 / +1.6	+1.8 / −0	+1.4 / +0.2	−2.0 / +2.6	+2.8 / −0	+2.0 / +0.2	−2.6 / +0.4	+1.8 / −0	+2.6 / +1.4	−3.2 / +0.4	+1.8 / −0	+3.2 / +1.4
9.85–12.41	−0.6 / +2.6	+2.0 / −0	+0.6 / −0.6	−1.0 / +4.0	+3.0 / −0	+1.0 / −1.0	−1.4 / +1.8	+2.0 / −0	+1.4 / +0.2	−2.2 / +2.8	+3.0 / −0	+2.2 / +0.2	−2.6 / +0.6	+2.0 / −0	+2.6 / +1.4	−3.4 / +0.6	+2.0 / −0	+3.4 / +1.4
12.41–15.75	−0.7 / +2.9	+2.2 / −0	+0.7 / −0.7	−1.0 / +4.5	+3.5 / −0	+1.0 / −1.0	−1.6 / +2.0	+2.2 / −0	+1.6 / +0.2	−2.4 / +3.3	+3.5 / −0	+2.4 / +0.2	−3.0 / +0.6	+2.2 / −0	+3.0 / +1.6	−3.8 / +0.6	+2.2 / −0	+3.8 / +1.6

[a] From ANSI B4.1-1967 (R1994). For larger diameters, see the standard.

8 Interference Locational Fits[a]—American National Standard

LN *Locational interference fits* are used where accuracy of location is of prime importance and for parts requiring rigidity and alignment with no special requirements for bore pressure. Such fits are not intended for parts designed to transmit frictional loads from one part to another by virtue of the tightness of fit, as these conditions are covered by force fits.

Basic hole system. Limits are in thousandths of an inch.
Limits for hole and shaft are applied algebraically to the basic size to obtain the limits of size for the parts.
Data in **boldface** are in accordance with ABC agreements.
Symbols H7, p6, etc., are hole and shaft designations used in ABC System.

Nominal Size Range, inches Over To	Class LN 1 Limits of Interference	Class LN 1 Standard Limits Hole H6	Class LN 1 Standard Limits Shaft n5	Class LN 2 Limits of Interference	Class LN 2 Standard Limits Hole H7	Class LN 2 Standard Limits Shaft p6	Class LN 3 Limits of Interference	Class LN 3 Standard Limits Hole H7	Class LN 3 Standard Limits Shaft r6
0–0.12	0 0.45	+0.25 −0	+0.45 +0.25	0 0.65	+0.4 −0	+0.65 +0.4	0.1 0.75	+0.4 −0	+0.75 +0.5
0.12–0.24	0 0.5	+0.3 −0	+0.5 +0.3	0 0.8	+0.5 −0	+0.8 +0.5	0.1 0.9	+0.5 0	+0.9 +0.6
0.24–0.40	0 0.65	+0.4 −0	+0.65 +0.4	0 1.0	+0.6 −0	+1.0 +0.6	0.2 1.2	+0.6 −0	+1.2 +0.8
0.40–0.71	0 0.8	+0.4 −0	+0.8 +0.4	0 1.1	+0.7 −0	+1.1 +0.7	0.3 1.4	+0.7 −0	+1.4 +1.0
0.71–1.19	0 1.0	+0.5 −0	+1.0 +0.5	0 1.3	+0.8 −0	+1.3 +0.8	0.4 1.7	+0.8 −0	+1.7 +1.2
1.19–1.97	0 1.1	+0.6 −0	+1.1 +0.6	0 1.6	+1.0 −0	+1.6 +1.0	0.4 2.0	+1.0 −0	+2.0 +1.4
1.97–3.15	0.1 1.3	+0.7 −0	+1.3 +0.7	0.2 2.1	+1.2 −0	+2.1 +1.4	0.4 2.3	+1.2 −0	+2.3 +1.6
3.15–4.73	0.1 1.6	+0.9 −0	+1.6 +1.0	0.2 2.5	+1.4 −0	+2.5 +1.6	0.6 2.9	+1.4 −0	+2.9 +2.0
4.73–7.09	0.2 1.9	+1.0 −0	+1.9 +1.2	0.2 2.8	+1.6 −0	+2.8 +1.8	0.9 3.5	+1.6 −0	+3.5 +2.5
7.09–9.85	0.2 2.2	+1.2 −0	+2.2 +1.4	0.2 3.2	+1.8 −0	+3.2 +2.0	1.2 4.2	+1.8 −0	+4.2 +3.0
9.85–12.41	0.2 2.3	+1.2 −0	+2.3 +1.4	0.2 3.4	+2.0 −0	+3.4 +2.2	1.5 4.7	+2.0 −0	+4.7 +3.5

[a] From ANSI B4.1–1967 (R1994). For larger diameters, see the standard.

FN 1   *Light drive fits* are those requiring light assembly pressures, and produce more or less permanent assemblies. They are suitable for thin sections or long fits, or in cast-iron external members.

FN 2   *Medium drive fits* are suitable for ordinary steel parts, or for shrink fits on light sections. They are about the tightest fits that can be used with high-grade cast-iron external members.

FN 3   *Heavy drive fits* are suitable for heavier steel parts or for shrink fits in medium sections.

FN 4 }
FN 5 }   *Force fits* are suitable for parts which can be highly stressed, or for shrink fits where the heavy pressing forces required are impractical.

Basic hole system. Limits are in thousandths of an inch.
Limits for hole and shaft are applied algebraically to the basic size to obtain the limits of size for the parts.
Data in **boldface** are in accordance with ABC agreements.
Symbols H7, s6, etc., are hole and shaft designations used in ABC System.

Nominal Size Range, inches Over — To	Class FN 1 Limits of Interference	Class FN 1 Standard Limits Hole H6	Class FN 1 Standard Limits Shaft	Class FN 2 Limits of Interference	Class FN 2 Standard Limits Hole H7	Class FN 2 Standard Limits Shaft s6	Class FN 3 Limits of Interference	Class FN 3 Standard Limits Hole H7	Class FN 3 Standard Limits Shaft t6	Class FN 4 Limits of Interference	Class FN 4 Standard Limits Hole H7	Class FN 4 Standard Limits Shaft u6	Class FN 5 Limits of Interference	Class FN 5 Standard Limits Hole H8	Class FN 5 Standard Limits Shaft x7
0–0.12	0.05 / 0.5	+0.25 / −0	+0.5 / +0.3	**0.2** / **0.85**	**+0.4** / **−0**	**+0.85** / **+0.6**				**0.3** / **0.95**	**+0.4** / **−0**	**+0.95** / **+0.7**	**0.3** / **1.3**	**+0.6** / **−0**	**+1.3** / **+0.9**
0.12–0.24	0.1 / 0.6	+0.3 / −0	+0.6 / +0.4	**0.2** / **1.0**	**+0.5** / **−0**	**+1.0** / **+0.7**				**0.4** / **1.2**	**+0.5** / **−0**	**+1.2** / **+0.9**	**0.5** / **1.7**	**+0.7** / **−0**	**+1.7** / **+1.2**
0.24–0.40	0.1 / 0.75	+0.4 / −0	+0.75 / +0.5	**0.4** / **1.4**	**+0.6** / **−0**	**+1.4** / **+1.0**				**0.6** / **1.6**	**+0.6** / **−0**	**+1.6** / **+1.2**	**0.5** / **2.0**	**+0.9** / **−0**	**+2.0** / **+1.4**
0.40–0.56	0.1 / 0.8	+0.4 / −0	+0.8 / +0.5	**0.5** / **1.6**	**+0.7** / **−0**	**+1.6** / **+1.2**				**0.7** / **1.8**	**+0.7** / **−0**	**+1.8** / **+1.4**	**0.6** / **2.3**	**+1.0** / **−0**	**+2.3** / **+1.6**
0.56–0.71	0.2 / 0.9	+0.4 / −0	+0.9 / +0.6	**0.5** / **1.6**	**+0.7** / **−0**	**+1.6** / **+1.2**				**0.7** / **1.8**	**+0.7** / **−0**	**+1.8** / **+1.4**	**0.8** / **2.5**	**+1.0** / **−0**	**+2.5** / **+1.8**
0.71–0.95	0.2 / 1.1	+0.5 / −0	+1.1 / +0.7	**0.6** / **1.9**	**+0.8** / **−0**	**+1.9** / **+1.4**				**0.8** / **2.1**	**+0.8** / **−0**	**+2.1** / **+1.6**	**1.0** / **3.0**	**+1.2** / **−0**	**+3.0** / **+2.2**
0.95–1.19	0.3 / 1.2	+0.5 / −0	+1.2 / +0.8	**0.6** / **1.9**	**+0.8** / **−0**	**+1.9** / **+1.4**	**0.8** / **2.1**	**+0.8** / **−0**	**+2.1** / **+1.6**	**1.0** / **2.3**	**+0.8** / **−0**	**+2.3** / **+1.8**	**1.3** / **3.3**	**+1.2** / **−0**	**+3.3** / **+2.5**
1.19–1.58	0.3 / 1.3	+0.6 / −0	+1.3 / +0.9	**0.8** / **2.4**	**+1.0** / **−0**	**+2.4** / **+1.8**	**1.0** / **2.6**	**+1.0** / **−0**	**+2.6** / **+2.0**	**1.5** / **3.1**	**+1.0** / **−0**	**+3.1** / **+2.5**	**1.4** / **4.0**	**+1.6** / **−0**	**+4.0** / **+3.0**

[a] ANSI B4.1-1967 (R1994).

Values in thousandths of an inch. Each cell shows upper / lower limit.

Nominal Size Range, inches (Over–To)	Class FN 1 Limits of Interference	Class FN 1 Standard Limits Hole H6	Class FN 1 Standard Limits Shaft	Class FN 2 Limits of Interference	Class FN 2 Standard Limits Hole H7	Class FN 2 Standard Limits Shaft s6	Class FN 3 Limits of Interference	Class FN 3 Standard Limits Hole H7	Class FN 3 Standard Limits Shaft t6	Class FN 4 Limits of Interference	Class FN 4 Standard Limits Hole H7	Class FN 4 Standard Limits Shaft u6	Class FN 5 Limits of Interference	Class FN 5 Standard Limits Hole H8	Class FN 5 Standard Limits Shaft x7
1.58–1.97	0.4 / 1.4	+0.6 / −0	+1.4 / −1.0	0.8 / 2.4	+1.0 / −0	+2.4 / +1.8	1.2 / 2.8	+1.0 / −0	+2.8 / +2.2	1.8 / 3.4	+1.0 / −0	+3.4 / +2.8	2.4 / 5.0	+1.6 / −0	+5.0 / +4.0
1.97–2.56	0.6 / 1.8	+0.7 / −0	+1.8 / +1.3	0.8 / 2.7	+1.2 / −0	+2.7 / +2.0	1.3 / 3.2	+1.2 / −0	+3.2 / +2.5	2.3 / 4.2	+1.2 / −0	+4.2 / +3.5	3.2 / 6.2	+1.8 / −0	+6.2 / +5.0
2.56–3.15	0.7 / 1.9	+0.7 / −0	+1.9 / +1.4	1.0 / 2.9	+1.2 / −0	+2.9 / +2.2	1.8 / 3.7	+1.2 / −0	+3.7 / +3.0	2.8 / 4.7	+1.2 / −0	+4.7 / +4.0	4.2 / 7.2	+1.8 / −0	+7.2 / +6.0
3.15–3.94	0.9 / 24	+0.9 / −0	+2.4 / +1.8	1.4 / 3.7	+1.4 / −0	+3.7 / +2.8	2.1 / 4.4	+1.4 / −0	+4.4 / +3.5	3.6 / 5.9	+1.4 / −0	+5.9 / +5.0	4.8 / 8.4	+2.2 / −0	+8.4 / +7.0
3.94–4.73	1.1 / 2.6	+0.9 / −0	+2.6 / +2.0	1.6 / 3.9	+1.4 / −0	+3.9 / +3.0	2.6 / 4.9	+1.4 / −0	+4.9 / +4.0	4.6 / 6.9	+1.4 / −0	+6.9 / +6.0	5.8 / 9.4	+2.2 / −0	+9.4 / +8.0
4.73–5.52	1.2 / 2.9	+1.0 / −0	+2.9 / +2.2	1.9 / 4.5	+1.6 / −0	+4.5 / +3.5	3.4 / 6.0	+1.6 / −0	+6.0 / +5.0	5.4 / 8.0	+1.6 / −0	+8.0 / +7.0	7.5 / 11.6	+2.5 / −0	+11.6 / +10.0
5.52–6.30	1.5 / 3.2	+1.0 / −0	+3.2 / +2.5	2.4 / 5.0	+1.6 / −0	+5.0 / +4.0	3.4 / 6.0	+1.6 / −0	+6.0 / +5.0	5.4 / 8.0	+1.6 / −0	+8.0 / +7.0	9.5 / 13.6	+2.5 / −0	+13.6 / +12.0
6.30–7.09	1.8 / 3.5	+1.0 / −0	+3.5 / +2.8	2.9 / 5.5	+1.6 / −0	+5.5 / +4.5	4.4 / 7.0	+1.6 / −0	+7.0 / +6.0	6.4 / 9.0	+1.6 / −0	+9.0 / +8.0	9.5 / 13.6	+2.5 / −0	+13.6 / +12.0
7.09–7.88	1.8 / 3.8	+1.2 / −0	+3.8 / +3.0	3.2 / 6.2	+1.8 / −0	+6.2 / +5.0	5.2 / 8.2	+1.8 / −0	+8.2 / +7.0	7.2 / 10.2	+1.8 / −0	+10.2 / +9.0	11.2 / 15.8	+2.8 / −0	+15.8 / +14.0
7.88–8.86	2.3 / 4.3	+1.2 / −0	+4.3 / +3.5	3.2 / 6.2	+1.8 / −0	+6.2 / +5.0	5.2 / 8.2	+1.8 / −0	+8.2 / +7.0	8.2 / 11.2	+1.8 / −0	+11.2 / +10.0	13.2 / 17.8	+2.8 / −0	+17.8 / +16.0
8.86–9.85	2.3 / 4.3	+1.2 / −0	+4.3 / +3.5	4.2 / 7.2	+1.8 / −0	+7.2 / +6.0	6.2 / 9.2	+1.8 / −0	+9.2 / +8.0	10.2 / 13.2	+1.8 / −0	+13.2 / +12.0	13.2 / 17.8	+2.8 / −0	+17.8 / +16.0
9.85–11.03	2.8 / 4.9	+1.2 / −0	+4.9 / +4.0	4.0 / 7.2	+2.0 / −0	+7.2 / +6.0	7.0 / 10.2	+2.0 / −0	+10.2 / +9.0	10.0 / 13.2	+2.0 / −0	+13.2 / +12.0	15.0 / 20.0	+3.0 / −0	+20.0 / +18.0
11.03–12.41	2.8 / 4.9	+1.2 / −0	+4.9 / +4.0	5.0 / 8.2	+2.0 / −0	+8.2 / +7.0	7.0 / 10.2	+2.0 / −0	+10.2 / +9.0	12.0 / 15.2	+2.0 / −0	+15.2 / +14.0	17.0 / 22.0	+3.0 / −0	+22.0 / +20.0
12.41–13.98	3.1 / 5.5	+1.4 / −0	+5.5 / +4.5	5.8 / 9.4	+2.2 / −0	+9.4 / +8.0	7.8 / 11.4	+2.2 / −0	+11.4 / +10.0	13.8 / 17.4	+2.2 / −0	+17.4 / +16.0	18.5 / 24.2	+3.5 / +0	+24.2 / +22.0

[a] From ANSI B4.1-1967 (R1994). For larger diameters, see the standard.

Dimensions are in millimeters.

Basic sizes		Tolerance grades[b]																		
Over	Up to and Including	IT01	IT0	IT1	IT2	IT3	IT4	IT5	IT6	IT7	IT8	IT9	IT10	IT11	IT12	IT13	IT14	IT15	IT16
0	3	0.0003	0.0005	0.0008	0.0012	0.002	0.003	0.004	0.006	0.010	0.014	0.025	0.040	0.060	0.100	0.140	0.250	0.400	0.600
3	6	0.0004	0.0006	0.001	0.0015	0.0025	0.004	0.005	0.008	0.012	0.018	0.030	0.048	0.075	0.120	0.180	0.300	0.480	0.750
6	10	0.0004	0.0006	0.001	0.0015	0.0025	0.004	0.006	0.009	0.015	0.022	0.036	0.058	0.090	0.150	0.220	0.360	0.580	0.900
10	18	0.0005	0.0008	0.0012	0.002	0.003	0.005	0.008	0.011	0.018	0.027	0.043	0.070	0.110	0.180	0.270	0.430	0.700	1.100
18	30	0.0006	0.001	0.0015	0.0025	0.004	0.006	0.009	0.013	0.021	0.033	0.052	0.084	0.130	0.210	0.330	0.520	0.840	1.300
30	50	0.0006	0.001	0.0015	0.0025	0.004	0.007	0.011	0.016	0.025	0.039	0.062	0.100	0.160	0.250	0.390	0.620	1.000	1.600
50	80	0.0008	0.0012	0.002	0.003	0.005	0.008	0.013	0.019	0.030	0.046	0.074	0.120	0.190	0.300	0.460	0.740	1.200	1.900
80	120	0.001	0.0015	0.0025	0.004	0.006	0.010	0.015	0.022	0.035	0.054	0.087	0.140	0.220	0.350	0.540	0.870	1.400	2.200
120	180	0.0012	0.002	0.0035	0.005	0.008	0.012	0.018	0.025	0.040	0.063	0.100	0.160	0.250	0.400	0.630	1.000	1.600	2.500
180	250	0.002	0.003	0.0045	0.007	0.010	0.014	0.020	0.029	0.046	0.072	0.115	0.185	0.290	0.460	0.720	1.150	1.850	2.900
250	315	0.0025	0.004	0.006	0.008	0.012	0.016	0.023	0.032	0.052	0.081	0.130	0.210	0.320	0.520	0.810	1.300	2.100	3.200
315	400	0.003	0.005	0.007	0.009	0.013	0.018	0.025	0.036	0.057	0.089	0.140	0.230	0.360	0.570	0.890	1.400	2.300	3.600
400	500	0.004	0.006	0.008	0.010	0.015	0.020	0.027	0.040	0.063	0.097	0.155	0.250	0.400	0.630	0.970	1.550	2.500	4.000
500	630	0.0045	0.006	0.009	0.011	0.016	0.022	0.030	0.044	0.070	0.110	0.175	0.280	0.440	0.700	1.100	1.750	2.800	4.400
630	800	0.005	0.007	0.010	0.013	0.018	0.025	0.035	0.050	0.080	0.125	0.200	0.320	0.500	0.800	1.250	2.000	3.200	5.000
800	1000	0.0055	0.008	0.011	0.015	0.021	0.029	0.040	0.056	0.090	0.140	0.230	0.360	0.560	0.900	1.400	2.300	3.600	5.600
1000	1250	0.0065	0.009	0.013	0.018	0.024	0.034	0.046	0.066	0.105	0.165	0.260	0.420	0.660	1.050	1.650	2.600	4.200	6.600
1250	1600	0.008	0.011	0.015	0.021	0.029	0.040	0.054	0.078	0.125	0.195	0.310	0.500	0.780	1.250	1.950	3.100	5.000	7.800
1600	2000	0.009	0.013	0.018	0.025	0.035	0.048	0.065	0.092	0.150	0.230	0.370	0.600	0.920	1.500	2.300	3.700	6.000	9.200
2000	2500	0.011	0.015	0.022	0.030	0.041	0.057	0.077	0.110	0.175	0.280	0.440	0.700	1.100	1.750	2.800	4.400	7.000	11.000
2500	3150	0.013	0.018	0.026	0.036	0.050	0.069	0.093	0.135	0.210	0.330	0.540	0.860	1.350	2.100	3.300	5.400	8.600	13.500

[a] From ANSI B4.2–1978 (R1994).
[b] IT Values for tolerance grades larger than IT16 can be calculated by using the formulas: IT17 = IT × 10, IT18 = IT13 × 10, etc.

Dimensions are in millimeters.

Basic Size		Loose Running Hole H11	Shaft c11	Fit	Free Running Hole H9	Shaft d9	Fit	Close Running Hole H8	f7	Fit	Sliding Hole H7	Shaft g6	Fit	Locational Clearance Hole H7	Shaft h6	Fit
1	Max	1.060	0.940	0.180	1.025	0.980	0.070	1.014	0.994	0.030	1.010	0.998	0.018	1.010	1.000	0.016
	Min	1.060	0.880	0.060	1.000	0.955	0.020	1.000	0.984	0.006	1.000	0.992	0.002	1.000	0.994	0.000
1.2	Max	1.260	1.140	0.180	1.225	1.180	0.070	1.214	1.194	0.030	1.210	1.198	0.018	1.210	1.200	0.016
	Min	1.200	1.080	0.060	1.200	1.155	0.020	1.200	1.184	0.006	1.200	1.192	0.002	1.200	1.194	0.000
1.6	Max	1.660	1.540	0.180	1.625	1.580	0.070	1.614	1.594	0.030	1.610	1.598	0.018	1.610	1.600	0.016
	Min	1.600	1.480	0.060	1.600	1.555	0.020	1.600	1.584	0.006	1.600	1.592	0.002	1.600	1.594	0.000
2	Max	2.060	1.940	0.180	2.025	1.980	0.070	2.014	1.994	0.030	2.010	1.998	0.018	2.010	2.000	0.016
	Min	2.000	1.880	0.060	2.000	1.955	0.020	2.000	1.984	0.006	2.000	1.992	0.002	2.000	1.994	0.000
2.5	Max	2.560	2.440	0.180	2.525	2.480	0.070	2.514	2.494	0.030	2.510	2.498	0.018	2.510	2.500	0.016
	Min	2.500	2.380	0.060	2.500	2.455	0.020	2.500	2.484	0.006	2.500	2.492	0.002	2.500	2.494	0.000
3	Max	3.060	2.940	0.180	3.025	2.980	0.070	3.014	2.994	0.030	3.010	2.998	0.018	3.010	3.000	0.016
	Min	3.000	2.880	0.060	3.000	2.955	0.020	3.000	2.984	0.006	3.000	2.992	0.002	3.000	2.994	0.000
4	Max	4.075	3.930	0.220	4.030	3.970	0.090	4.018	3.990	0.040	4.012	3.996	0.024	4.012	4.000	0.020
	Min	4.000	3.855	0.070	4.000	3.940	0.030	4.000	3.978	0.010	4.000	3.988	0.004	4.000	3.992	0.000
5	Max	5.075	4.930	0.220	5.030	4.970	0.090	5.018	4.990	0.040	5.012	4.996	0.024	5.012	5.000	0.020
	Min	5.000	4.855	0.070	5.000	4.940	0.030	5.000	4.978	0.010	5.000	4.988	0.004	5.000	4.992	0.000
6	Max	6.075	5.930	0.220	6.030	5.970	0.090	6.018	5.990	0.040	6.012	5.996	0.024	6.012	6.000	0.020
	Min	6.000	5.855	0.070	6.000	5.940	0.030	6.000	5.978	0.010	6.000	5.988	0.004	6.000	5.992	0.000
8	Max	8.090	7.920	0.260	8.036	7.960	0.112	8.022	7.987	0.050	8.015	7.995	0.029	8.015	8.000	0.024
	Min	8.000	7.830	0.080	8.000	7.924	0.040	8.000	7.972	0.013	8.000	7.986	0.005	8.000	7.991	0.000
10	Max	10.090	9.920	0.260	10.036	9.960	0.112	10.022	9.987	0.050	10.015	9.995	0.029	10.015	10.000	0.024
	Min	10.000	9.830	0.080	10.000	9.924	0.040	10.000	9.972	0.013	10.000	9.986	0.005	10.000	9.991	0.000
12	Max	12.110	11.905	0.315	12.043	11.950	0.136	12.027	11.984	0.061	12.018	11.994	0.035	12.018	12.000	0.029
	Min	12.000	11.795	0.095	12.000	11.907	0.050	12.000	11.966	0.016	12.000	11.983	0.006	12.000	11.989	0.000
16	Max	16.110	15.905	0.315	16.043	15.950	0.136	16.027	15.984	0.061	16.018	15.994	0.035	16.018	16.000	0.029
	Min	16.000	15.795	0.095	16.000	15.907	0.050	16.000	15.966	0.016	16.000	15.983	0.006	16.000	15.989	0.000
20	Max	20.130	19.890	0.370	20.052	19.935	0.169	20.033	19.980	0.074	20.021	19.993	0.041	20.021	20.000	0.034
	Min	20.000	19.760	0.110	20.000	19.883	0.065	20.000	19.959	0.020	20.000	19.980	0.007	20.000	19.987	0.000
25	Max	25.130	24.890	0.370	25.052	24.935	0.169	25.033	24.980	0.074	25.021	24.993	0.041	25.021	25.000	0.034
	Min	25.000	24.760	0.110	25.000	24.883	0.065	25.000	24.959	0.020	25.000	24.980	0.007	25.000	24.987	0.000
30	Max	30.130	29.890	0.370	30.052	29.935	0.169	30.033	29.980	0.074	30.021	29.993	0.041	30.021	30.000	0.034
	Min	30.000	29.760	0.110	30.000	29.883	0.065	30.000	29.959	0.020	30.000	29.980	0.007	30.000	29.987	0.000

[a] From ANSI B4.2-1978 (R1994).

Dimensions are in millimeters.

Basic Size		Loose Running			Free Running			Close Running			Sliding			Locational Clearance		
		Hole H11	Shaft c11	Fit	Hole H9	Shaft d9	Fit	Hole H8	Shaft f7	Fit	Hole H7	Shaft g6	Fit	Hole H7	Shaft h6	Fit
40	Max	40.160	39.880	0.440	40.062	39.920	0.204	40.039	39.975	0.089	40.025	39.991	0.050	40.025	40.000	0.041
	Min	40.000	39.720	0.120	40.000	39.858	0.080	40.000	39.950	0.025	40.000	39.975	0.009	40.000	39.984	0.000
50	Max	50.160	49.870	0.450	50.062	49.920	0.204	50.039	49.975	0.089	50.025	49.991	0.050	50.025	50.000	0.041
	Min	50.000	49.710	0.130	50.000	49.858	0.080	50.000	49.950	0.025	50.000	49.975	0.009	50.000	49.984	0.000
60	Max	60.190	59.860	0.520	60.074	59.900	0.248	60.046	59.970	0.106	60.030	59.990	0.059	60.030	60.000	0.049
	Min	60.000	59.670	0.140	60.000	59.826	0.100	60.000	59.940	0.030	60.000	59.971	0.010	60.000	59.981	0.000
80	Max	80.190	79.950	0.530	80.074	79.900	0.248	80.046	79.970	0.106	80.030	79.990	0.059	80.030	80.000	0.049
	Min	80.000	79.660	0.150	80.000	79.826	0.100	80.000	79.940	0.030	80.000	79.971	0.010	80.000	79.981	0.000
100	Max	100.220	99.830	0.610	100.087	99.880	0.294	100.054	99.964	0.125	100.035	99.988	0.069	100.035	100.000	0.057
	Min	100.000	99.610	0.170	100.000	99.793	0.120	100.000	99.929	0.036	100.000	99.966	0.012	100.000	99.978	0.000
120	Max	120.220	119.820	0.620	120.087	119.880	0.294	120.054	119.964	0.125	120.035	119.988	0.069	120.035	120.000	0.057
	Min	120.000	119.600	0.180	120.000	119.793	0.120	120.000	119.929	0.036	120.000	119.966	0.012	120.000	119.978	0.000
160	Max	160.250	159.790	0.710	160.100	159.855	0.345	160.063	159.957	0.146	160.040	159.986	0.079	160.040	160.000	0.065
	Min	160.000	159.540	0.210	160.000	159.755	0.145	160.000	159.917	0.043	160.000	159.961	0.014	160.000	159.975	0.000
200	Max	200.290	199.760	0.820	200.115	199.830	0.400	200.072	199.950	0.168	200.046	199.985	0.090	200.046	200.000	0.075
	Min	200.000	199.470	0.240	200.000	199.715	0.170	200.000	199.904	0.050	200.000	199.956	0.015	200.000	199.971	0.000
250	Max	250.290	249.720	0.860	250.115	249.830	0.400	250.072	249.950	0.168	250.046	249.985	0.090	250.046	250.000	0.075
	Min	250.000	249.430	0.280	250.000	249.715	0.170	250.000	249.904	0.050	250.000	249.956	0.015	250.000	249.971	0.000
300	Max	300.320	299.670	0.970	300.130	299.810	0.450	300.081	299.944	0.189	300.052	299.983	0.101	300.052	300.000	0.084
	Min	300.000	299.350	0.330	300.000	299.680	0.190	300.000	299.892	0.056	300.000	299.951	0.017	300.000	299.968	0.000
400	Max	400.360	399.600	1.120	400.140	399.790	0.490	400.089	399.938	0.208	400.057	399.982	0.111	400.057	400.000	0.093
	Min	400.000	399.240	0.400	400.000	399.650	0.210	400.000	399.881	0.062	400.000	399.946	0.018	400.000	399.964	0.000
500	Max	500.400	499.520	1.280	500.155	499.770	0.540	500.097	499.932	0.228	500.063	499.980	0.123	500.063	500.000	0.103
	Min	500.000	499.120	0.480	500.000	499.615	0.230	500.000	499.869	0.068	500.000	499.940	0.020	500.000	499.960	0.000

[a] From ANSI B4.2–1978 (R1994).

12 Preferred Metric Hole Basis Transition and Interference Fits[a]— American National Standard

Dimensions are in millimeters.

Basic Size		Locational Transn.			Locational Transn.			Locational Interf.			Medium Drive			Force		
		Hole H7	Shaft k6	Fit	Hole H7	Shaft n6	Fit	Hole H7	Shaft p6	Fit	Hole H7	Shaft s6	Fit	Hole H7	Shaft u6	Fit
1	Max	1.010	1.006	0.010	1.010	1.010	0.006	1.010	1.012	0.004	1.010	1.020	−0.004	1.010	1.024	−0.008
	Min	1.000	1.000	−0.006	1.000	1.004	−0.010	1.000	1.006	−0.012	1.000	1.014	−0.020	1.000	1.018	−0.024
1.2	Max	1.210	1.206	0.010	1.210	1.210	0.006	1.210	1.212	0.004	1.210	1.220	−0.004	1.210	1.224	−0.008
	Min	1.200	1.200	−0.006	1.200	1.204	−0.010	1.200	1.206	−0.012	1.200	1.214	−0.020	1.200	1.218	−0.024
1.6	Max	1.610	1.606	0.010	1.610	1.610	0.006	1.610	1.612	0.004	1.610	1.620	−0.004	1.610	1.624	−0.008
	Min	1.600	1.600	−0.006	1.600	1.604	−0.010	1.600	1.606	−0.012	1.600	1.614	−0.020	1.600	1.618	−0.024
2	Max	2.010	2.006	0.010	2.010	2.010	0.006	2.010	2.012	0.004	2.010	2.020	−0.004	2.010	2.024	−0.008
	Min	2.000	2.000	−0.006	2.000	2.004	−0.010	2.000	2.006	−0.012	2.000	2.014	−0.020	2.000	2.018	−0.024
2.5	Max	2.510	2.506	0.010	2.510	2.510	0.006	2.510	2.512	0.004	2.510	2.520	−0.004	2.510	2.524	−0.008
	Min	2.500	2.500	−0.006	2.500	2.504	−0.010	2.500	2.506	−0.012	2.500	2.514	−0.020	2.500	2.518	−0.024
3	Max	3.010	3.006	0.010	3.010	3.010	0.006	3.010	3.012	0.004	3.010	3.020	−0.004	3.010	3.024	−0.008
	Min	3.000	3.000	−0.006	3.000	3.004	−0.010	3.000	3.006	−0.012	3.000	3.014	−0.020	3.000	3.018	−0.024
4	Max	4.012	4.009	0.011	4.012	4.016	0.004	4.012	4.020	0.000	4.012	4.027	−0.007	4.012	4.031	−0.011
	Min	4.000	4.001	−0.009	4.000	4.008	−0.016	4.000	4.012	−0.020	4.000	4.019	−0.027	4.000	4.023	−0.031
5	Max	5.012	5.009	0.011	5.012	5.016	0.004	5.012	5.020	0.000	5.012	5.027	−0.007	5.012	5.031	−0.011
	Min	5.000	5.001	−0.009	5.000	5.008	−0.016	5.000	5.012	−0.020	5.000	5.019	−0.027	5.000	5.023	−0.031
6	Max	6.012	6.009	0.011	6.012	6.016	0.004	6.012	6.020	0.000	6.012	6.027	−0.007	6.012	6.031	−0.011
	Min	6.000	6.001	−0.009	6.000	6.008	−0.016	6.000	6.012	−0.020	6.000	6.019	−0.027	6.000	6.023	−0.031
8	Max	8.015	8.010	0.014	8.015	8.019	0.005	8.015	8.024	0.000	8.015	8.032	−0.008	8.015	8.037	−0.013
	Min	8.000	8.001	−0.010	8.000	8.010	−0.019	8.000	8.015	−0.024	8.000	8.023	−0.032	8.000	8.028	−0.037
10	Max	10.015	10.010	0.014	10.015	10.019	0.005	10.015	10.024	0.000	10.015	10.032	−0.008	10.015	10.037	−0.013
	Min	10.000	10.001	−0.010	10.000	10.010	−0.019	10.000	10.015	−0.024	10.000	10.023	−0.032	10.000	10.028	−0.037
12	Max	12.018	12.012	0.017	12.018	12.023	0.006	12.018	12.029	0.000	12.018	12.039	−0.010	12.018	12.044	−0.015
	Min	12.000	12.001	−0.012	12.000	12.012	−0.023	12.000	12.018	−0.029	12.000	12.028	−0.039	12.000	12.033	−0.044
16	Max	16.018	16.012	0.017	16.018	16.023	0.006	16.018	16.029	0.000	16.018	16.039	−0.010	16.018	16.044	−0.015
	Min	16.000	16.001	−0.012	16.000	16.012	−0.023	16.000	16.018	−0.029	16.000	16.028	−0.039	16.000	16.033	−0.044
20	Max	20.021	20.015	0.019	20.021	20.028	0.006	20.021	20.035	−0.001	20.021	20.048	−0.014	20.021	20.054	−0.020
	Min	20.000	20.002	−0.015	20.000	20.015	−0.028	20.000	20.022	−0.035	20.000	20.035	−0.048	20.000	20.041	−0.054
25	Max	25.021	25.015	0.019	25.021	25.028	0.006	25.021	25.035	−0.001	25.021	25.048	−0.014	25.021	25.061	−0.027
	Min	25.000	25.002	−0.015	25.000	25.015	−0.028	25.000	25.022	−0.035	25.000	25.035	−0.048	25.000	25.048	−0.061
30	Max	30.021	30.015	0.019	30.021	30.028	0.006	30.021	30.035	−0.001	30.021	30.048	−0.014	30.021	30.061	−0.027
	Min	30.000	30.002	−0.015	30.000	30.015	−0.028	30.000	30.022	−0.035	30.000	30.035	−0.048	30.000	30.048	−0.061

[a] From ANSI B4.2-1978 (R1994).

Dimensions are in millimeters.

Basic Size		Locational Transn.			Locational Transn.			Locational Interf.			Medium Drive			Force		
		Hole H7	Shaft k6	Fit	Hole H7	Shaft n6	Fit	Hole H7	Shaft p6	Fit	Hole H7	Shaft s6	Fit	Hole H7	Shaft u6	Fit
40	Max	40.025	40.018	0.023	40.025	40.033	0.008	40.025	40.042	−0.001	40.025	40.059	−0.018	40.025	40.076	−0.035
	Min	40.000	40.002	−0.018	40.000	40.017	−0.033	40.000	40.026	−0.042	40.000	40.043	−0.059	40.000	40.060	−0.076
50	Max	50.025	50.018	0.023	50.025	50.033	0.008	50.025	50.042	−0.001	50.025	50.059	−0.018	50.025	50.086	−0.045
	Min	50.000	50.002	−0.018	50.000	50.017	−0.033	50.000	50.026	−0.042	50.000	50.043	−0.059	50.000	50.070	−0.086
60	Max	60.030	60.021	0.028	60.030	60.039	0.010	60.030	60.051	−0.002	60.030	60.072	−0.023	60.030	60.106	−0.057
	Min	60.000	60.002	−0.021	60.000	60.020	−0.039	60.000	60.032	−0.051	60.000	60.053	−0.072	60.000	60.087	−0.106
80	Max	80.030	80.021	0.028	80.030	80.039	0.010	80.030	80.051	−0.002	80.030	80.078	−0.029	80.030	80.121	−0.072
	Min	80.000	80.002	−0.021	80.000	80.020	−0.039	80.000	80.032	−0.051	80.000	80.059	−0.078	80.000	80.102	−0.121
100	Max	100.035	100.025	0.032	100.035	100.045	0.012	100.035	100.059	−0.002	100.035	100.093	−0.036	100.035	100.146	−0.089
	Min	100.000	100.003	−0.025	100.000	100.023	−0.045	100.000	100.037	−0.059	100.000	100.071	−0.093	100.000	100.124	−0.146
120	Max	120.035	120.025	0.032	120.035	120.045	0.012	120.035	120.059	−0.002	120.035	120.101	−0.044	120.035	120.166	−0.109
	Min	120.000	120.003	−0.025	120.000	120.023	−0.045	120.000	120.037	−0.059	120.000	120.079	−0.101	120.000	120.144	−0.166
160	Max	160.040	160.028	0.037	160.040	160.052	0.013	160.040	160.068	−0.003	160.040	160.125	−0.060	160.040	160.215	−0.150
	Min	160.000	160.003	−0.028	160.000	160.027	−0.052	160.000	160.043	−0.068	160.000	160.100	−0.125	160.000	160.190	−0.215
200	Max	200.046	200.033	0.042	200.046	200.060	0.015	200.046	200.079	−0.004	200.046	200.151	−0.076	200.046	200.265	−0.190
	Min	200.000	200.004	−0.033	200.000	200.031	−0.060	200.000	200.050	−0.079	200.000	200.122	−0.151	200.000	200.236	−0.265
250	Max	250.046	250.033	0.042	250.046	250.060	0.015	250.046	250.079	−0.004	250.046	250.169	−0.094	250.046	250.313	−0.238
	Min	250.000	250.004	−0.033	250.000	250.031	−0.060	250.000	250.050	−0.079	250.000	250.140	−0.169	250.000	250.284	−0.313
300	Max	300.052	300.036	0.048	300.052	300.066	0.018	300.052	300.088	−0.004	300.052	300.202	−0.118	300.052	300.382	−0.298
	Min	300.000	300.004	−0.036	300.000	300.034	−0.066	300.000	300.056	−0.088	300.000	300.170	−0.202	300.000	300.350	−0.382
400	Max	400.057	400.040	0.053	400.057	400.073	0.020	400.057	400.098	−0.005	400.057	400.244	−0.151	400.057	400.471	−0.378
	Min	400.000	400.004	−0.040	400.000	400.037	−0.073	400.000	400.062	−0.098	400.000	400.208	−0.244	400.000	400.435	−0.471
500	Max	500.063	500.045	0.058	500.063	500.080	0.023	500.063	500.108	−0.005	500.063	500.292	−0.189	500.063	500.580	−0.477
	Min	500.000	500.005	−0.045	500.000	500.040	−0.080	500.000	500.068	−0.108	500.000	500.252	−0.292	500.000	500.540	−0.580

a From ANSI B4.2–1978 (R1994).

Dimensions are in millimeters.

Basic Size		Loose Running			Free Running			Close Running			Sliding			Locational Clearance		
		Hole C11	Shaft h11	Fit	Hole D9	Shaft h9	Fit	Hole F8	Shaft h7	Fit	Hole G7	Shaft h6	Fit	Hole H7	Shaft h6	Fit
1	Max	1.120	1.000	0.180	1.045	1.000	0.070	1.020	1.000	0.030	1.012	1.000	0.018	1.010	1.000	0.016
	Min	1.060	0.940	0.060	1.020	0.975	0.020	1.006	0.990	0.006	1.002	0.994	0.002	1.000	0.994	0.000
1.2	Max	1.320	1.200	0.180	1.245	1.200	0.070	1.220	1.200	0.030	1.212	1.200	0.018	1.210	1.200	0.016
	Min	1.260	1.140	0.060	1.220	1.175	0.020	1.206	1.190	0.006	1.202	1.194	0.002	1.200	1.194	0.000
1.6	Max	1.720	1.600	0.180	1.645	1.600	0.070	1.620	1.600	0.030	1.612	1.600	0.018	1.610	1.600	0.016
	Min	1.660	1.540	0.060	1.620	1.575	0.020	1.606	1.590	0.006	1.602	1.594	0.002	1.600	1.594	0.000
2	Max	2.120	2.000	0.180	2.045	2.000	0.070	2.020	2.000	0.030	2.012	2.000	0.018	2.010	2.000	0.016
	Min	2.060	1.940	0.060	2.020	1.975	0.020	2.006	1.990	0.006	2.002	1.994	0.002	2.000	1.994	0.000
2.5	Max	2.620	2.500	0.180	2.545	2.500	0.070	2.520	2.500	0.030	2.512	2.500	0.018	2.510	2.500	0.016
	Min	2.560	2.440	0.060	2.520	2.475	0.020	2.506	2.490	0.006	2.502	2.494	0.002	2.500	2.494	0.000
3	Max	3.120	3.000	0.180	3.045	3.000	0.070	3.020	3.000	0.030	3.012	3.000	0.018	3.010	3.000	0.016
	Min	3.060	2.940	0.060	3.020	2.975	0.020	3.006	2.990	0.006	3.002	2.994	0.002	3.000	2.994	0.000
4	Max	4.145	4.000	0.220	4.060	4.000	0.090	4.028	4.000	0.040	4.016	4.000	0.024	4.012	4.000	0.020
	Min	4.070	3.925	0.070	4.030	3.970	0.030	4.010	3.988	0.010	4.004	3.992	0.004	4.000	3.992	0.000
5	Max	5.145	5.000	0.220	5.060	5.000	0.090	5.028	5.000	0.040	5.016	5.000	0.024	5.012	5.000	0.020
	Min	5.070	4.925	0.070	5.030	4.970	0.030	5.010	4.988	0.010	5.004	4.992	0.004	5.000	4.992	0.000
6	Max	6.145	6.000	0.220	6.060	6.000	0.090	6.028	6.000	0.040	6.016	6.000	0.024	6.012	6.000	0.020
	Min	6.070	5.925	0.070	6.030	5.970	0.030	6.010	5.988	0.010	6.004	5.992	0.004	6.000	5.992	0.000
8	Max	8.170	8.000	0.260	8.076	8.000	0.112	8.035	8.000	0.050	8.020	8.000	0.029	8.015	8.000	0.024
	Min	8.080	7.910	0.080	8.040	7.964	0.040	8.013	7.985	0.013	8.005	7.991	0.005	8.000	7.991	0.000
10	Max	10.170	10.000	0.260	10.076	10.000	0.112	10.035	10.000	0.050	10.020	10.000	0.029	10.015	10.000	0.024
	Min	10.080	9.910	0.080	10.040	9.964	0.040	10.013	9.985	0.013	10.005	9.991	0.005	10.000	9.991	0.000
12	Max	12.205	12.000	0.315	12.093	12.000	0.136	12.043	12.000	0.061	12.024	12.000	0.035	12.018	12.000	0.029
	Min	12.095	11.890	0.095	12.050	11.957	0.050	12.016	11.982	0.016	12.006	11.989	0.006	12.000	11.989	0.000
16	Max	16.205	16.000	0.315	16.093	16.000	0.136	16.043	16.000	0.061	16.024	16.000	0.035	16.018	16.000	0.029
	Min	16.095	15.890	0.095	16.050	15.957	0.050	16.016	15.982	0.016	16.006	15.989	0.006	16.000	15.989	0.000
20	Max	20.240	20.000	0.370	20.117	20.000	0.169	20.053	20.000	0.074	20.028	20.000	0.041	20.021	20.000	0.034
	Min	20.110	19.870	0.110	20.065	19.948	0.065	20.020	19.979	0.020	20.007	19.987	0.007	20.000	19.987	0.000
25	Max	25.240	25.000	0.370	25.117	25.000	0.169	25.053	25.000	0.074	25.028	25.000	0.041	25.021	25.000	0.034
	Min	25.110	24.870	0.110	25.065	24.948	0.065	25.020	24.979	0.020	25.007	24.987	0.007	25.000	24.987	0.000
30	Max	30.240	30.000	0.370	30.117	30.000	0.169	30.053	30.000	0.074	30.028	30.000	0.041	30.021	30.000	0.034
	Min	30.110	29.870	0.110	30.065	29.948	0.065	30.020	29.979	0.020	30.007	29.987	0.007	30.000	29.987	0.000

[a] From ANSI B4.2–1978 (R1994).

Dimensions are in millimeters.

Basic Size		Loose Running			Free Running			Close Running			Sliding			Locational Clearance		
		Hole C11	Shaft h11	Fit	Hole D9	Shaft h9	Fit	Hole F8	Shaft h7	Fit	Hole G7	Shaft h6	Fit	Hole H7	Shaft h6	Fit
40	Max	40.280	40.000	0.440	40.142	40.000	0.204	40.064	40.000	0.089	40.034	40.000	0.050	40.025	40.000	0.041
	Min	40.120	39.840	0.120	40.080	39.938	0.080	40.025	39.975	0.025	40.009	39.984	0.009	40.000	39.984	0.000
50	Max	50.290	50.000	0.450	50.142	50.000	0.204	50.064	50.000	0.089	50.034	50.000	0.050	50.025	50.000	0.041
	Min	50.130	49.840	0.130	50.080	49.938	0.080	50.025	49.975	0.025	50.009	49.984	0.009	50.000	49.984	0.000
60	Max	60.330	60.000	0.520	60.174	60.000	0.248	60.076	60.000	0.106	60.040	60.000	0.059	60.030	60.000	0.049
	Min	60.140	59.810	0.140	60.100	59.926	0.100	60.030	59.970	0.030	60.010	59.981	0.010	60.000	59.981	0.000
80	Max	80.340	80.000	0.530	80.174	80.000	0.248	80.076	80.000	0.106	80.040	80.000	0.059	80.030	80.000	0.049
	Min	80.150	79.810	0.150	80.100	79.926	0.100	80.030	79.970	0.030	80.010	79.981	0.010	80.000	79.981	0.000
100	Max	100.390	100.000	0.610	100.207	100.000	0.294	100.090	100.000	0.125	100.047	100.000	0.069	100.035	100.000	0.057
	Min	100.170	99.780	0.170	100.120	99.913	0.120	100.036	99.965	0.036	100.012	99.978	0.012	100.000	99.978	0.000
120	Max	120.400	120.000	0.620	120.207	120.000	0.294	120.090	120.000	0.125	120.047	120.000	0.069	120.035	120.000	0.057
	Min	120.180	119.780	0.180	120.120	119.913	0.120	120.036	119.965	0.036	120.012	119.978	0.012	120.000	119.978	0.000
160	Max	160.460	160.000	0.710	160.245	160.000	0.345	160.106	160.000	0.146	160.054	160.000	0.079	160.040	160.000	0.065
	Min	160.210	159.750	0.210	160.145	159.900	0.145	160.043	159.960	0.043	160.014	159.975	0.014	160.000	159.975	0.000
200	Max	200.530	200.000	0.820	200.285	200.000	0.400	200.122	200.000	0.168	200.061	200.000	0.090	200.046	200.000	0.075
	Min	200.240	199.710	0.240	200.170	199.885	0.170	200.050	199.954	0.050	200.015	199.971	0.015	200.000	199.971	0.000
250	Max	250.570	250.000	0.860	250.285	250.000	0.400	250.122	250.000	0.168	250.061	250.000	0.090	250.046	250.000	0.075
	Min	250.280	249.710	0.280	250.170	249.885	0.170	250.050	249.954	0.050	250.015	249.971	0.015	250.000	249.971	0.000
300	Max	300.650	300.000	0.970	300.320	300.000	0.450	300.137	300.000	0.189	300.069	300.000	0.101	300.052	300.000	0.084
	Min	300.330	299.680	0.330	300.190	299.870	0.190	300.056	299.948	0.056	300.017	299.968	0.017	300.000	299.968	0.000
400	Max	400.760	400.000	1.120	400.350	400.000	0.490	400.151	400.000	0.208	400.075	400.000	0.111	400.057	400.000	0.093
	Min	400.400	399.640	0.400	400.210	399.860	0.210	400.062	399.943	0.062	400.018	399.964	0.018	400.000	399.964	0.000
500	Max	500.880	500.000	1.280	500.385	500.000	0.540	500.165	500.000	0.228	500.083	500.000	0.123	500.063	500.000	0.103
	Min	500.480	499.600	0.480	500.230	499.845	0.230	500.068	499.937	0.068	500.020	499.960	5.020	500.000	499.960	0.000

[a] From ANSI B4.2–1978 (R1994).

a39

14 Preferred Metric Shaft Basis Transition and Interference Fits[a]— American National Standard

Dimensions are in millimeters.

Basic Size		Locational Transn. Hole K7	Shaft h6	Fit	Locational Transn. Hole N7	Shaft h6	Fit	Locational Interf. Hole P7	Shaft h6	Fit	Medium Drive Hole S7	Shaft h6	Fit	Force Hole U7	Shaft h6	Fit
1	Max	1.000	1.000	0.006	0.996	1.000	0.002	0.994	1.000	0.000	0.986	1.000	−0.008	0.982	1.000	−0.012
	Min	0.990	0.994	−0.010	0.986	0.994	−0.014	0.984	0.994	−0.016	0.976	0.994	−0.024	0.972	0.994	−0.028
1.2	Max	1.200	1.200	0.006	1.196	1.200	0.002	1.194	1.200	0.000	1.186	1.200	−0.008	1.182	1.200	−0.012
	Min	1.190	1.194	−0.010	1.186	1.194	−0.014	1.184	1.194	−0.016	1.176	1.194	−0.024	1.172	1.194	−0.028
1.6	Max	1.600	1.600	0.006	1.596	1.600	0.002	1.594	1.600	0.000	1.586	1.600	−0.008	1.582	1.600	−0.012
	Min	1.590	1.594	−0.010	1.586	1.594	−0.014	1.584	1.594	−0.016	1.576	1.594	−0.024	1.572	1.594	−0.028
2	Max	2.000	2.000	0.006	1.996	2.000	0.002	1.994	2.000	0.000	1.986	2.000	−0.008	1.982	2.000	−0.012
	Min	1.990	1.994	−0.010	1.986	1.994	−0.014	1.984	1.994	−0.016	1.976	1.994	−0.024	1.972	1.994	−0.028
2.5	Max	2.500	2.500	0.006	2.496	2.500	0.002	2.494	2.500	0.000	2.486	2.500	−0.008	2.482	2.500	−0.012
	Min	2.490	2.494	−0.010	2.486	2.494	−0.014	2.484	2.494	−0.016	2.476	2.494	−0.024	2.472	2.494	−0.028
3	Max	3.000	3.000	0.006	2.996	3.000	0.002	2.994	3.000	0.000	2.986	3.000	−0.008	2.982	3.000	−0.012
	Min	2.990	2.994	−0.010	2.986	2.994	−0.014	2.984	2.994	−0.016	2.976	2.994	−0.024	2.972	2.994	−0.028
4	Max	4.003	4.000	0.011	3.996	4.000	0.004	3.992	4.000	0.000	3.985	4.000	−0.007	3.981	4.000	−0.011
	Min	3.991	3.992	−0.009	3.984	3.992	−0.016	3.980	3.992	−0.020	3.973	3.992	−0.027	3.969	3.992	−0.031
5	Max	5.003	5.000	0.011	4.996	5.000	0.004	4.992	5.000	0.000	4.985	5.000	−0.007	4.981	5.000	−0.011
	Min	4.991	4.992	−0.009	4.984	4.992	−0.016	4.980	4.992	−0.020	4.973	4.992	−0.027	4.969	4.992	−0.031
6	Max	6.003	6.000	0.011	5.996	6.000	0.004	5.992	6.000	0.000	5.985	6.000	−0.007	5.981	6.000	−0.011
	Min	5.991	5.992	−0.009	5.984	5.992	−0.016	5.980	5.992	−0.020	5.973	5.992	−0.027	5.969	5.992	−0.031
8	Max	8.005	8.000	0.014	7.996	8.000	0.005	7.991	8.000	0.000	7.983	8.000	−0.008	7.978	8.000	−0.013
	Min	7.990	7.991	−0.010	7.981	7.991	−0.019	7.976	7.991	−0.024	7.968	7.991	−0.032	7.963	7.991	−0.037
10	Max	10.005	10.000	0.014	9.996	10.000	0.005	9.991	10.000	0.000	9.983	10.000	−0.008	9.978	10.000	−0.013
	Min	9.990	9.991	−0.010	9.981	9.991	−0.019	9.976	9.991	−0.024	9.968	9.991	−0.032	9.963	9.991	−0.037
12	Max	12.006	12.000	0.017	11.995	12.000	0.006	11.989	12.000	0.000	11.979	12.000	−0.010	11.974	12.000	−0.015
	Min	11.988	11.989	−0.012	11.977	11.989	−0.023	11.971	11.989	−0.029	11.961	11.989	−0.039	11.956	11.989	−0.044
16	Max	16.006	16.000	0.017	15.995	16.000	0.006	15.989	16.000	0.000	15.979	16.000	−0.010	15.974	16.000	−0.015
	Min	15.988	15.989	−0.012	15.977	15.989	−0.023	15.971	15.989	−0.029	15.961	15.989	−0.039	15.956	15.989	−0.044
20	Max	20.006	20.000	0.019	19.993	20.000	0.006	19.986	20.000	−0.001	19.973	20.000	−0.014	19.967	20.000	−0.020
	Min	19.985	19.987	−0.015	19.972	19.987	−0.028	19.965	19.987	−0.035	19.952	19.987	−0.048	19.946	19.987	−0.054
25	Max	25.006	25.000	0.019	24.993	25.000	0.006	24.986	25.000	−0.001	24.973	25.000	−0.014	24.960	25.000	−0.027
	Min	24.985	24.987	−0.015	24.972	24.987	−0.028	24.965	24.987	−0.035	24.952	24.987	−0.048	24.939	24.987	−0.061
30	Max	30.006	30.000	0.019	29.993	30.000	0.006	29.986	30.000	−0.001	29.973	30.000	−0.014	29.960	30.000	−0.027
	Min	29.985	29.987	−0.015	29.972	29.987	−0.028	29.965	29.987	−0.035	29.952	29.987	−0.048	29.939	29.987	−0.061

[a] From ANSI B4.2-1978 (R1994).

Dimensions are in millimeters.

Basic Size		Locational Transn.			Locational Transn.			Locational Interf.			Medium Drive			Force		
		Hole K7	Shaft h6	Fit	Hole N7	Shaft h6	Fit	Hole P7	Shaft h6	Fit	Hole S7	Shaft h6	Fit	Hole U7	Shaft h6	Fit
40	Max	40.007	40.000	0.023	39.992	40.000	0.008	39.983	40.000	−0.001	39.966	40.000	−0.018	39.949	40.000	−0.035
	Min	39.982	39.984	−0.018	39.967	39.984	−0.033	39.958	39.984	−0.042	39.941	39.984	−0.059	39.924	39.984	−0.076
50	Max	50.007	50.000	0.023	49.992	50.000	0.008	49.983	50.000	−0.001	49.966	50.000	−0.018	49.939	50.000	−0.045
	Min	49.982	49.984	−0.018	49.967	49.984	−0.033	49.958	49.984	−0.042	49.941	49.984	−0.059	49.914	49.984	−0.086
60	Max	60.009	60.000	0.028	59.991	60.000	0.010	59.979	60.000	−0.002	59.958	60.000	−0.023	59.924	60.000	−0.057
	Min	59.979	59.981	−0.021	59.961	59.981	−0.039	59.949	59.981	−0.051	59.928	59.981	−0.072	59.894	59.981	−0.106
80	Max	80.009	80.000	0.028	79.991	80.000	0.010	79.979	80.000	−0.002	79.952	80.000	−0.029	79.909	80.000	−0.072
	Min	79.979	79.981	−0.021	79.961	79.981	−0.039	79.949	79.981	−0.051	79.922	79.981	−0.078	79.879	79.981	−0.121
100	Max	100.010	100.000	0.032	99.990	100.000	0.012	99.976	100.000	−0.002	99.942	100.000	−0.036	99.889	100.000	−0.089
	Min	99.975	99.978	−0.025	99.955	99.978	−0.045	99.941	99.978	−0.059	99.907	99.978	−0.093	99.854	99.978	−0.146
120	Max	120.010	120.000	0.032	119.990	120.000	0.012	119.976	120.000	−0.002	119.934	120.000	−0.044	119.869	120.000	−0.109
	Min	119.975	119.978	−0.025	119.955	119.978	−0.045	119.941	119.978	−0.059	119.899	119.978	−0.101	119.834	119.978	−0.166
160	Max	160.012	160.000	0.037	159.988	160.000	0.013	159.972	160.000	−0.003	159.915	160.000	−0.060	159.825	160.000	−0.150
	Min	159.972	159.975	−0.028	159.948	159.975	−0.052	159.932	159.975	−0.068	159.875	159.975	−0.125	159.785	159.975	−0.215
200	Max	200.013	200.000	0.042	199.986	200.000	0.015	199.967	200.000	−0.004	199.895	200.000	−0.076	199.781	200.000	−0.190
	Min	199.967	199.971	−0.033	199.940	199.971	−0.060	199.921	199.971	−0.079	199.849	199.971	−0.151	199.735	199.971	−0.265
250	Max	250.013	250.000	0.042	249.986	250.000	0.015	249.967	250.000	−0.004	249.877	250.000	−0.094	249.733	250.000	−0.238
	Min	249.967	249.971	−0.033	249.940	249.971	−0.060	249.921	249.971	−0.079	249.831	249.971	−0.169	249.687	249.971	−0.313
300	Max	300.016	300.000	0.048	299.986	300.000	0.018	299.964	300.000	−0.004	299.850	300.000	−0.118	299.670	300.000	−0.298
	Min	299.964	299.968	−0.036	299.934	299.968	−0.066	299.912	299.968	−0.088	299.798	299.968	−0.202	299.618	299.968	−0.382
400	Max	400.017	400.000	0.053	399.984	400.000	0.020	399.959	400.000	−0.005	399.813	400.000	−0.151	399.586	400.000	−0.378
	Min	399.960	399.964	−0.040	399.927	399.964	−0.073	399.902	399.964	−0.098	399.756	399.964	−0.244	399.529	399.964	−0.471
500	Max	500.018	500.000	0.058	499.983	500.000	0.023	499.955	500.000	−0.005	499.771	500.000	−0.189	499.483	500.000	−0.477
	Min	499.955	499.960	−0.045	499.920	499.960	−0.080	499.892	499.960	−0.108	499.708	499.960	−0.292	499.420	499.960	−0.580

[a] From ANSI B4.2-1978 (R1994).

15 Screw Threads, American National, Unified, and Metric

AMERICAN NATIONAL STANDARD UNIFIED AND AMERICAN NATIONAL SCREW THREADS[a]

Nominal Diameter	Coarse[b] NC UNC Thds. per Inch	Coarse[b] NC UNC Tap Drill[d]	Fine[b] NF UNF Thds. per Inch	Fine[b] NF UNF Tap Drill[d]	Extra Fine[c] NEF UNEF Thds. per Inch	Extra Fine[c] NEF UNEF Tap Drill[d]
0 (.060)			80	$\frac{3}{64}$		
1 (.073)	64	No. 53	72	No. 53	...	...
2 (.086)	56	No. 50	64	No. 50	...	...
3 (.099)	48	No. 47	56	No. 45	...	...
4 (.112)	40	No. 43	48	No. 42	...	...
5 (.125)	40	No. 38	44	No. 37	...	...
6 (.138)	32	No. 36	40	No. 33	...	...
8 (.164)	32	No. 29	36	No. 29	...	...
10 (.190)	24	No. 25	32	No. 21	...	...
12 (.216)	24	No. 16	28	No. 14	32	No. 13
$\frac{1}{4}$	20	No. 7	28	No. 3	32	$\frac{7}{32}$
$\frac{5}{16}$	18	F	24	I	32	$\frac{9}{32}$
$\frac{3}{8}$	16	$\frac{5}{16}$	24	Q	32	$\frac{11}{32}$
$\frac{7}{16}$	14	U	20	$\frac{25}{64}$	28	$\frac{13}{32}$
$\frac{1}{2}$	13	$\frac{27}{64}$	20	$\frac{29}{64}$	28	$\frac{15}{32}$
$\frac{9}{16}$	12	$\frac{31}{64}$	18	$\frac{33}{64}$	24	$\frac{33}{64}$
$\frac{5}{8}$	11	$\frac{17}{32}$	18	$\frac{37}{64}$	24	$\frac{37}{64}$
$\frac{11}{16}$	...	...	...	...	24	$\frac{41}{64}$
$\frac{3}{4}$	10	$\frac{21}{32}$	16	$\frac{11}{16}$	20	$\frac{45}{64}$
$\frac{13}{16}$	...	...	...	...	20	$\frac{49}{64}$
$\frac{7}{8}$	9	$\frac{49}{64}$	14	$\frac{13}{16}$	20	$\frac{53}{64}$
$\frac{15}{16}$	...	...	...	...	20	$\frac{57}{64}$

Nominal Diameter	Coarse[b] NC UNC Thds. per Inch	Coarse[b] NC UNC Tap Drill[d]	Fine[b] NF UNF Thds. per Inch	Fine[b] NF UNF Tap Drill[d]	Extra Fine[c] NEF UNEF Thds. per Inch	Extra Fine[c] NEF UNEF Tap Drill[d]
1	8	$\frac{7}{8}$	12	$\frac{59}{64}$	20	$\frac{61}{64}$
$1\frac{1}{16}$	...	...	...	...	18	1
$1\frac{1}{8}$	7	$\frac{63}{64}$	12	$1\frac{3}{64}$	18	$1\frac{5}{64}$
$1\frac{3}{16}$	...	...	...	...	18	$1\frac{9}{64}$
$1\frac{1}{4}$	7	$1\frac{7}{64}$	12	$1\frac{11}{64}$	18	$1\frac{3}{16}$
$1\frac{5}{16}$	...	...	...	...	18	$1\frac{17}{64}$
$1\frac{3}{8}$	6	$1\frac{7}{32}$	12	$1\frac{19}{64}$	18	$1\frac{5}{16}$
$1\frac{7}{16}$	...	...	...	...	18	$1\frac{3}{8}$
$1\frac{1}{2}$	6	$1\frac{11}{32}$	12	$1\frac{27}{64}$	18	$1\frac{7}{16}$
$1\frac{9}{16}$	...	...	...	...	18	$1\frac{1}{2}$
$1\frac{5}{8}$	...	...	...	...	18	$1\frac{9}{16}$
$1\frac{11}{16}$	...	...	...	...	18	$1\frac{5}{8}$
$1\frac{3}{4}$	5	$1\frac{9}{16}$	...	...	...	...
2	$4\frac{1}{2}$	$1\frac{25}{32}$	...	...	...	...
$2\frac{1}{4}$	$4\frac{1}{2}$	$2\frac{1}{32}$	...	...	...	...
$2\frac{1}{2}$	4	$2\frac{1}{4}$	...	...	...	...
$2\frac{3}{4}$	4	$2\frac{1}{2}$	...	...	...	...
3	4	$2\frac{3}{4}$	...	...	...	...
$3\frac{1}{4}$	4	...	...	...	...	...
$3\frac{1}{2}$	4	...	...	...	...	...
$3\frac{3}{4}$	4	...	...	...	...	...
4	4	...	...	...	...	...

[a] ANSI/ASME B1.1–1989. For 8-, 12-, and 16-pitch thread series, see next page.
[b] Classes 1A, 2A, 3A, 1B, 2B, 3B, 2, and 3.
[c] Classes 2A, 2B, 2, and 3.
[d] For approximate 75% full depth of thread. For decimal sizes of numbered and lettered drills, see Appendix 16.

15 Screw Threads, American National, Unified, and Metric (continued)

AMERICAN NATIONAL STANDARD UNIFIED AND AMERICAN NATIONAL SCREW THREADS[a] (continued)

Nominal Diameter	8-Pitch[b] Series 8N and 8UN		12-Pitch[b] Series 12N and 12UN		16-Pitch[b] Series 16N and 16UN	
	Thds. per Inch	Tap Drill[c]	Thds. per Inch	Tap Drill[c]	Thds. per Inch	Tap Drill[c]
$\frac{1}{2}$	...	...	12	$\frac{27}{64}$	...	...
$\frac{9}{16}$	...	...	12[e]	$\frac{31}{64}$	...	...
$\frac{5}{8}$	...	...	12	$\frac{35}{64}$	...	...
$\frac{11}{16}$	...	...	12	$\frac{39}{64}$	...	...
$\frac{3}{4}$	...	...	12	$\frac{43}{64}$	16[e]	$\frac{11}{16}$
$\frac{13}{16}$	...	...	12	$\frac{47}{64}$	16	$\frac{3}{4}$
$\frac{7}{8}$	...	...	12	$\frac{51}{64}$	16	$\frac{13}{16}$
$\frac{15}{16}$	...	...	12	$\frac{55}{64}$	16	$\frac{7}{8}$
1	8[e]	$\frac{7}{8}$	12	$\frac{59}{64}$	16	$\frac{15}{16}$
$1\frac{1}{16}$	...	...	12	$\frac{63}{64}$	16	1
$1\frac{1}{8}$	8	1	12[e]	$1\frac{3}{64}$	16	$1\frac{1}{16}$
$1\frac{3}{16}$	...	...	12	$1\frac{7}{64}$	16	$1\frac{1}{8}$
$1\frac{1}{4}$	8	$1\frac{1}{8}$	12	$1\frac{11}{64}$	16	$1\frac{3}{16}$
$1\frac{5}{16}$	...	...	12	$1\frac{15}{64}$	16	$1\frac{1}{4}$
$1\frac{3}{8}$	8	$1\frac{1}{4}$	12[e]	$1\frac{19}{64}$	16	$1\frac{5}{16}$
$1\frac{7}{16}$	...	...	12	$1\frac{23}{64}$	16	$1\frac{3}{8}$
$1\frac{1}{2}$	8	$1\frac{3}{8}$	12[e]	$1\frac{27}{64}$	16	$1\frac{7}{16}$
$1\frac{9}{16}$	...	...	...	...	16	$1\frac{1}{2}$
$1\frac{5}{8}$	8	$1\frac{1}{2}$	12	$1\frac{35}{64}$	16	$1\frac{9}{16}$
$1\frac{11}{16}$	...	...	...	...	16	$1\frac{5}{8}$
$1\frac{3}{4}$	8	$1\frac{5}{8}$	12	$1\frac{43}{64}$	16[e]	$1\frac{11}{16}$
$1\frac{13}{16}$	...	...	...	...	16	$1\frac{3}{4}$
$1\frac{7}{8}$	8	$1\frac{3}{4}$	12	$1\frac{51}{64}$	16	$1\frac{13}{16}$
$1\frac{15}{16}$	...	...	...	...	16	$1\frac{7}{8}$
2	8	$1\frac{7}{8}$	12	$1\frac{59}{64}$	16[e]	$1\frac{15}{16}$

Nominal Diameter	8-Pitch[b] Series 8N and 8UN		12-Pitch[b] Series 12N and 12UN		16-Pitch[c] Series 16N and 16UN	
	Thds. per Inch	Tap Drill[c]	Thds. per Inch	Tap Drill[c]	Thds. per Inch	Tap Drill[c]
$2\frac{1}{16}$	...	...	...	...	**16**	2
$2\frac{1}{8}$	...	...	12	$2\frac{3}{64}$	16	$2\frac{1}{16}$
$2\frac{3}{16}$	...	...	...	...	**16**	$2\frac{1}{8}$
$2\frac{1}{4}$	8	$2\frac{1}{8}$	12	$2\frac{17}{64}$	16	$2\frac{3}{16}$
$2\frac{5}{16}$	...	...	...	...	**16**	$2\frac{1}{4}$
$2\frac{3}{8}$	...	...	12	$2\frac{19}{64}$	16	$2\frac{5}{16}$
$2\frac{7}{16}$	...	...	...	...	**16**	$2\frac{3}{8}$
$2\frac{1}{2}$	8	$2\frac{3}{8}$	12	$2\frac{27}{64}$	16	$2\frac{7}{16}$
$2\frac{5}{8}$	...	...	12	$2\frac{35}{64}$	16	$2\frac{9}{16}$
$2\frac{3}{4}$	8	$2\frac{5}{8}$	12	$2\frac{43}{64}$	16	$2\frac{11}{16}$
$2\frac{7}{8}$	...	...	12	...	16	...
3	8	$2\frac{7}{8}$	12	...	16	...
$3\frac{1}{8}$	...	...	12	...	16	...
$3\frac{1}{4}$	8	...	12	...	16	...
$3\frac{3}{8}$	...	...	12	...	16	...
$3\frac{1}{2}$	8	...	12	...	16	...
$3\frac{5}{8}$	...	...	12	...	16	...
$3\frac{3}{4}$	8	...	12	...	16	...
$3\frac{7}{8}$	...	...	12	...	16	...
4	8	...	12	...	16	...
$4\frac{1}{4}$	8	...	12	...	16	...
$4\frac{1}{2}$	8	...	12	...	16	...
$4\frac{3}{4}$	8	...	12	...	16	...
5	8	...	12	...	16	...
$5\frac{1}{4}$	8	...	12	...	16	...

[a] ANSI/ASME B1.1–1989.
[b] Classes 2A, 3A, 2B, 3B, 2, and 3.
[c] For approximate 75% full depth of thread.
[d] Boldface type indicates Amrican National Threads only.
[e] This is a standard size of the Unified or American National threads of the coarse, fine, or extra fine series. See preceding page.

15 Screw Threads, American National, Unified, and Metric (continued)

METRIC SCREW THREADS[a]

Preferred sizes for commercial threads and fasteners are shown in **boldface** type.

Coarse (general purpose)		Fine	
Nominal Size & Thd Pitch	Tap Drill Diameter, mm	Nominal Size & Thd Pitch	Tap Drill Diameter, mm
M1.6 × 0.35	1.25	—	—
M1.8 × 0.35	1.45	—	—
M2 × 0.4	1.6	—	—
M2.2 × 0.45	1.75	—	—
M2.5 × 0.45	2.05	—	—
M3 × 0.5	2.5	—	—
M3.5 × 0.6	2.9	—	—
M4 × 0.7	3.3	—	—
M4.5 × 0.75	3.75	—	—
M5 × 0.8	4.2	—	—
M6 × 1	5.0	—	—
M7 × 1	6.0	—	—
M8 × 1.25	6.8	**M8 × 1**	7.0
M9 × 1.25	7.75	—	—
M10 × 1.5	8.5	**M10 × 1.25**	8.75
M11 × 1.5	9.50	—	—
M12 × 1.75	10.30	**M12 × 1.25**	10.5
M14 × 2	12.00	**M14 × 1.5**	12.5
M16 × 2	14.00	**M16 × 1.5**	14.5
M18 × 2.5	15.50	**M18 × 1.5**	16.5
M20 × 2.5	17.5	**M20 × 1.5**	18.5
M22 × 25[b]	19.5	**M22 × 1.5**	20.5
M24 × 3	21.0	**M24 × 2**	22.0
M27 × 3[b]	24.0	**M27 × 2**	25.0
M30 × 3.5	26.5	**M30 × 2**	28.0
M33 × 3.5	29.5	**M30 × 2**	31.0
M36 × 4	32.0	**M36 × 2**	33.0
M39 × 4	35.0	M39 × 2	36.0
M42 × 4.5	37.5	**M42 × 2**	39.0
M45 × 4.5	40.5	M45 × 1.5	42.0
M48 × 5	43.0	**M48 × 2**	45.0
M52 × 5	47.0	M52 × 2	49.0
M56 × 5.5	50.5	**M56 × 2**	52.0
M60 × 5.5	54.5	M60 × 1.5	56.0
M64 × 6	58.0	**M64 × 2**	60.0
M68 × 6	62.0	M68 × 2	64.0
M72 × 6	66.0	**M72 × 2**	68.0
M80 × 6	74.0	**M80 × 2**	76.0
M90 × 6	84.0	**M90 × 2**	86.0
M100 × 6	94.0	**M100 × 2**	96.0

[a] ANSI/ASME B1.13M–1995.
[b] Only for high strength structural steel fasteners.

16 Twist Drill Sizes—American National Standard and Metric

AMERICAN NATIONAL STANDARD DRILL SIZES[a]

All dimensions are in inches.

Drills designated in common fractions are available in diameters $\frac{1}{64}''$ to $1\frac{3}{4}''$ in $\frac{1}{64}''$ increments, $1\frac{3}{4}''$ to $2\frac{1}{4}''$ in $\frac{1}{32}''$ increments. $2\frac{1}{4}''$ to $3''$ in $\frac{1}{16}''$ increments and $3''$ to $3\frac{1}{2}''$ in $\frac{1}{8}''$ increments. Drills larger than $3\frac{1}{2}''$ are seldom used, and are regarded as special drills.

Size	Drill Diameter	Size	Drill Diameter	Size	Drill Diameter	Size	Drill Diameter	Size	Drill Diameter	Size	Drill Diameter
1	.2280	17	.1730	33	.1130	49	.0730	65	.0350	81	.0130
2	.2210	18	.1695	34	.1110	50	.0700	66	.0330	82	.0125
3	.2130	19	.1660	35	.1100	51	.0670	67	.0320	83	.0120
4	.2090	20	.1610	36	.1065	52	.0635	68	.0310	84	.0115
5	.2055	21	.1590	37	.1040	53	.0595	69	.0292	85	.0110
6	.2040	22	.1570	38	.1015	54	.0550	70	.0280	86	.0105
7	.2010	23	.1540	39	.0995	55	.0520	71	.0260	87	.0100
8	.1990	24	.1520	40	.0980	56	.0465	72	.0250	88	.0095
9	.1960	25	.1495	41	.0960	57	.0430	73	.0240	89	.0091
10	.1935	26	.1470	42	.0935	58	.0420	74	.0225	90	.0087
11	.1910	27	.1440	43	.0890	59	.0410	75	.0210	91	.0083
12	.1890	28	.1405	44	.0860	60	.0400	76	.0200	92	.0079
13	.1850	29	.1360	45	.0820	61	.0390	77	.0180	93	.0075
14	.1820	30	.1285	46	.0810	62	.0380	78	.0160	94	.0071
15	.1800	31	.1200	47	.0785	63	.0370	79	.0145	95	.0067
16	.1770	32	.1160	48	.0760	64	.0360	80	.0135	96	.0063
										97	.0059

LETTER SIZES

A	.234	G	.261	L	.290	Q	.332	V	.377
B	.238	H	.266	M	.295	R	.339	W	.386
C	.242	I	.272	N	.302	S	.348	X	.397
D	.246	J	.277	O	.316	T	.358	Y	.404
E	.250	K	.281	P	.323	U	.368	Z	.413
F	.257								

[a] ANSI/ASME B94.11M–1993.

METRIC DRILL SIZES

Decimal-inch equivalents are for reference only.

Drill Diameter		Drill Diameter		Drill Diameter		Drill Diameter		Drill Diameter		Drill Diameter	
mm	in.	mm	in.	mm	in.	mm	in.	mm	in.	mm	in.
0.40	.0157	1.95	.0768	4.70	.1850	8.00	.3150	13.20	.5197	25.50	1.0039
0.42	.0165	2.00	.0787	4.80	.1890	8.10	.3189	13.50	.5315	26.00	1.0236
0.45	.0177	2.05	.0807	4.90	.1929	8.20	.3228	13.80	.5433	26.50	1.0433
0.48	.0189	2.10	.0827	5.00	.1969	8.30	.3268	14.00	.5512	27.00	1.0630
0.50	.0197	2.15	.0846	5.10	.2008	8.40	.3307	14.25	.5610	27.50	1.0827
0.55	.0217	2.20	.0866	5.20	.2047	8.50	.3346	14.50	.5709	28.00	1.1024
0.60	.0236	2.25	.0886	5.30	.2087	8.60	.3386	14.75	.5807	28.50	1.1220
0.65	.0256	2.30	.0906	5.40	.2126	8.70	.3425	15.00	.5906	29.00	1.1417
0.70	.0276	2.35	.0925	5.50	.2165	8.80	.3465	15.25	.6004	29.50	1.1614
0.75	.0295	2.40	.0945	5.60	.2205	8.90	.3504	15.50	.6102	30.00	1.1811
0.80	.0315	2.45	.0965	5.70	.2244	9.00	.3543	15.75	.6201	30.50	1.2008
0.85	.0335	2.50	.0984	5.80	.2283	9.10	.3583	16.00	.6299	31.00	1.2205
0.90	.0354	2.60	.1024	5.90	.2323	9.20	.3622	16.25	.6398	31.50	1.2402
0.95	.0374	2.70	.1063	6.00	.2362	9.30	.3661	16.50	.6496	32.00	1.2598
1.00	.0394	2.80	.1102	6.10	.2402	9.40	.3701	16.75	.6594	32.50	1.2795
1.05	.0413	2.90	.1142	6.20	.2441	9.50	.3740	17.00	.6693	33.00	1.2992
1.10	.0433	3.00	.1181	6.30	.2480	9.60	.3780	17.25	.6791	33.50	1.3189
1.15	.0453	3.10	.1220	6.40	.2520	9.70	.3819	17.50	.6890	34.00	1.3386
1.20	.0472	3.20	.1260	6.50	.2559	9.80	.3858	18.00	.7087	34.50	1.3583
1.25	.0492	3.30	.1299	6.60	.2598	9.90	.3898	18.50	.7283	35.00	1.3780
1.30	.0512	3.40	.1339	6.70	.2638	10.00	.3937	19.00	.7480	35.50	1.3976
1.35	.0531	3.50	.1378	6.80	.2677	10.20	.4016	19.50	.7677	36.00	1.4173
1.40	.0551	3.60	.1417	6.90	.2717	10.50	.4134	20.00	.7874	36.50	1.4370
1.45	.0571	3.70	.1457	7.00	.2756	10.80	.4252	20.50	.8071	37.00	1.4567
1.50	.0591	3.80	.1496	7.10	.2795	11.00	.4331	21.00	.8268	37.50	1.4764
1.55	.0610	3.90	.1535	7.20	.2835	11.20	.4409	21.50	.8465	38.00	1.4961
1.60	.0630	4.00	.1575	7.30	.2874	11.50	.4528	22.00	.8661	40.00	1.5748
1.65	.0650	4.10	.1614	7.40	.2913	11.80	.4646	22.50	.8858	42.00	1.6535
1.70	.0669	4.20	.1654	7.50	.2953	12.00	.4724	23.00	.9055	44.00	1.7323
1.75	.0689	4.30	.1693	7.60	.2992	12.20	.4803	23.50	.9252	46.00	1.8110
1.80	.0709	4.40	.1732	7.70	.3031	12.50	.4921	24.00	.9449	48.00	1.8898
1.85	.0728	4.50	.1772	7.80	.3071	12.50	.5039	24.50	.9646	50.00	1.9685
1.90	.0748	4.60	.1811	7.90	.3110	13.00	.5118	25.00	.9843		

17 Acme Threads, General Purpose[a]

Size	Threads per Inch	Size	Threads per Inch	Size	Threads per Inch	Size	Threads per Inch
$\frac{1}{4}$	16	$\frac{3}{4}$	6	$1\frac{1}{2}$	4	3	2
$\frac{5}{16}$	14	$\frac{7}{8}$	6	$1\frac{3}{4}$	4	$3\frac{1}{2}$	2
$\frac{3}{8}$	12	1	5	2	4	4	2
$\frac{7}{16}$	12	$1\frac{1}{8}$	5	$2\frac{1}{4}$	3	$4\frac{1}{2}$	2
$\frac{1}{2}$	10	$1\frac{1}{4}$	5	$2\frac{1}{2}$	3	5	2
$\frac{5}{8}$	8	$1\frac{3}{8}$	4	$2\frac{3}{4}$	3	...	...

[a] ANSI/ASME B1.5–1988 (R1994).

18 Bolts, Nuts, and Cap Screws—Square and Hexagon— American National Standard and Metric

AMERICAN NATIONAL STANDARD SQUARE AND HEXAGON BOLTS[a] AND NUTS[b] AND HEXAGON CAP SCREWS[c]

Boldface type indicates product features unified dimensionally with British and Canadian standards. All dimensions are in inches.

Nominal Size D Body Diameter of Bolt	Regular Bolts					Heavy Bolts		
	Width Across Flats W		Height H			Width Across Flats W	Height H	
	Sq.	Hex.	Sq. (Unfin.)	Hex (Unfin.)	Hex Cap Scr.[c] (Fin.)		Hex. (Unfin.)	Hex Screw (Fin.)
$\frac{1}{4}$ 0.2500	$\frac{3}{8}$	$\frac{7}{16}$	$\frac{11}{64}$	$\frac{11}{64}$	$\frac{5}{32}$	...	...	...
$\frac{5}{16}$ 0.3125	$\frac{1}{2}$	$\frac{1}{2}$	$\frac{13}{64}$	$\frac{7}{32}$	$\frac{13}{64}$	...	...	...
$\frac{3}{8}$ 0.3750	$\frac{9}{16}$	$\frac{9}{16}$	$\frac{1}{4}$	$\frac{1}{4}$	$\frac{15}{64}$	...	...	...
$\frac{7}{16}$ 0.4375	$\frac{5}{8}$	$\frac{5}{8}$	$\frac{19}{64}$	$\frac{19}{64}$	$\frac{9}{32}$	...	...	...
$\frac{1}{2}$ 0.5000	$\frac{3}{4}$	$\frac{3}{4}$	$\frac{21}{64}$	$\frac{11}{32}$	$\frac{5}{16}$	$\frac{7}{8}$	$\frac{11}{32}$	$\frac{5}{16}$
$\frac{9}{16}$ 0.5625	...	$\frac{13}{16}$	...	...	$\frac{23}{64}$	...	...	...
$\frac{5}{8}$ 0.6250	$\frac{15}{16}$	$\frac{15}{16}$	$\frac{27}{64}$	$\frac{27}{64}$	$\frac{25}{64}$	$1\frac{1}{16}$	$\frac{27}{64}$	$\frac{25}{64}$
$\frac{3}{4}$ 0.7500	$1\frac{1}{8}$	$1\frac{1}{8}$	$\frac{1}{2}$	$\frac{1}{2}$	$\frac{15}{32}$	$1\frac{1}{4}$	$\frac{1}{2}$	$\frac{15}{32}$
$\frac{7}{8}$ 0.8750	$1\frac{5}{16}$	$1\frac{5}{16}$	$\frac{19}{32}$	$\frac{37}{64}$	$\frac{35}{64}$	$1\frac{7}{16}$	$\frac{37}{64}$	$\frac{35}{64}$
1 **1.000**	$1\frac{1}{2}$	$1\frac{1}{2}$	$\frac{21}{32}$	$\frac{43}{64}$	$\frac{39}{64}$	$1\frac{5}{8}$	$\frac{43}{64}$	$\frac{39}{64}$
$1\frac{1}{8}$ 1.1250	$1\frac{11}{16}$	$1\frac{11}{16}$	$\frac{3}{4}$	$\frac{3}{4}$	$\frac{11}{16}$	$1\frac{13}{16}$	$\frac{3}{4}$	$\frac{11}{16}$
$1\frac{1}{4}$ 1.2500	$1\frac{7}{8}$	$1\frac{7}{8}$	$\frac{27}{32}$	$\frac{27}{32}$	$\frac{25}{32}$	2	$\frac{27}{32}$	$\frac{25}{32}$
$1\frac{3}{8}$ 1.3750	$2\frac{1}{16}$	$2\frac{1}{16}$	$\frac{29}{32}$	$\frac{29}{32}$	$\frac{27}{32}$	$2\frac{3}{16}$	$\frac{29}{32}$	$\frac{27}{32}$
$1\frac{1}{2}$ 1.5000	$2\frac{1}{4}$	$2\frac{1}{4}$	**1**	**1**	$\frac{15}{16}$	$2\frac{3}{8}$	1	$\frac{15}{16}$
$1\frac{3}{4}$ 1.7500	...	$2\frac{5}{8}$	...	$1\frac{5}{32}$	$1\frac{3}{32}$	$2\frac{3}{4}$	$1\frac{5}{32}$	$1\frac{3}{32}$
2 **2.0000**	...	**3**	...	$1\frac{11}{32}$	$1\frac{7}{32}$	$3\frac{1}{8}$	$1\frac{11}{32}$	$1\frac{7}{32}$
$2\frac{1}{4}$ 2.2500	...	$3\frac{3}{8}$	...	$1\frac{1}{2}$	$1\frac{3}{8}$	$3\frac{1}{2}$	$1\frac{1}{2}$	$1\frac{3}{8}$
$2\frac{1}{2}$ 2.5000	...	$3\frac{3}{4}$	...	$1\frac{21}{32}$	$1\frac{17}{32}$	$3\frac{7}{8}$	$1\frac{21}{32}$	$1\frac{17}{32}$
$2\frac{3}{4}$ 2.7500	...	$4\frac{1}{8}$	...	$1\frac{13}{16}$	$1\frac{11}{16}$	$4\frac{1}{4}$	$1\frac{13}{16}$	$1\frac{11}{16}$
3 3.0000	...	$4\frac{1}{2}$	...	2	$1\frac{7}{8}$	$4\frac{5}{8}$	2	$1\frac{7}{8}$
$3\frac{1}{4}$ 3.2500	...	$4\frac{7}{8}$	...	$2\frac{3}{16}$	...	...	...	...
$3\frac{1}{2}$ 3.5000	...	$5\frac{1}{4}$	...	$2\frac{5}{16}$	...	...	...	...
$3\frac{3}{4}$ 3.7500	...	$5\frac{5}{8}$	...	$2\frac{1}{2}$	...	...	...	...
4 4.0000	...	6	...	$2\frac{11}{16}$	...	...	...	...

[a] ANSI B18.2.1–1981 (R1992).
[b] ANSI/ASME B18.2.2.–1987 (R1993).
[c] Hexagon cap screws and finished hexagon bolts are combined as a single product.

18 Bolts, Nuts, and Cap Screws—Square and Hexagon— American National Standard and Metric (continued)

AMERICAN NATIONAL STANDARD SQUARE AND HEXAGON BOLTS AND NUTS AND HEXAGON CAP SCREWS (continued)

See ANSI B18.2.2 for jam nuts, slotted nuts, thick nuts, thick slotted nuts, and castle nuts.

Nominal Size D	Body Diameter of Bolt	Regular Bolts Width Across Flats W Sq.	Hex.	Thickness T Sq. (Unfin.)	Hex. Flat (Unfin.)	Hex. (Fin.)	Heavy Nuts Width Across Flats W	Thickness T Sq. (Unfin.)	Hex. Flat (Unfin.)	Hex. (Fin.)
1/4	0.2500	7/16	7/16	7/32	7/32	7/32	1/2	1/4	15/64	15/64
5/16	0.3125	9/16	1/2	17/64	17/64	17/64	9/16	5/16	19/64	19/64
3/8	0.3750	5/8	9/16	21/64	...	21/64	11/16	3/8	23/64	23/64
7/16	0.4375	3/4	11/16	3/8	3/8	3/8	3/4 a	7/16	27/64	27/64
1/2	0.5000	13/16	3/4	7/16	7/16	7/16	7/8 a	1/2	31/64	31/64
9/16	0.5625	...	7/8	...	31/64	31/64	15/16	...	35/64	35/64
5/8	0.6250	1	15/16	35/64	35/64	35/64	1 1/16 a	5/8	39/64	39/64
3/4	0.7500	1 1/8	1 1/8	21/32	41/64	41/64	1 1/4 a	3/4	47/64	47/64
7/8	0.8750	1 5/16	1 5/16	49/64	3/4	3/4	1 7/16 a	7/8	55/64	55/64
1	1.0000	1 1/2	1 1/2	7/8	55/64	55/64	1 5/8 a	1	63/64	63/64
1 1/8	1.1250	1 11/16	1 11/16	1	1	31/32	1 13/16 a	1 1/8	1 1/8	1 7/64
1 1/4	1.2500	1 7/8	1 7/8	1 3/32	1 3/32	1 1/16	2 a	1 1/4	1 1/4	1 7/32
1 3/8	1.3750	2 1/16	2 1/16	1 13/64	1 13/64	1 11/64	2 3/16 a	1 3/8	1 3/8	1 11/32
1 1/2	1.5000	2 1/4	2 1/4	1 5/16	1 5/16	1 9/32	2 3/8 a	1 1/2	1 1/2	1 15/32
1 5/8	1.6250	...	...	...	...	...	2 9/16	...	...	1 19/32
1 3/4	1.7500	...	...	...	...	...	2 3/4	...	1 3/4	1 23/32
1 7/8	1.8750	...	...	...	...	...	2 15/16	...	...	1 27/32
2	2.0000	...	...	...	...	...	3 1/8	...	2	1 31/32
2 1/4	2.2500	...	...	...	...	...	3 1/2	...	2 1/4	2 13/64
2 1/2	2.5000	...	...	...	...	...	3 7/8	...	2 1/2	2 29/64
2 3/4	2.7500	...	...	...	...	...	4 1/4	...	2 3/4	2 45/64
3	3.0000	...	...	...	...	...	4 5/8	...	3	2 61/64
3 1/4	3.2500	...	...	...	...	...	5	...	3 1/4	3 3/16
3 1/2	3.5000	...	...	...	...	...	5 3/8	...	3 1/2	3 7/16
3 3/4	3.7500	...	...	...	...	...	5 3/4	...	3 3/4	3 11/16
4	4.0000	...	...	...	...	...	6 1/8	...	4	3 15/16

a Product feature not unified for heavy square nut.

18 Bolts, Nuts, and Cap Screws—Square and Hexagon—American National Standard and Metric (continued)

METRIC HEXAGON BOLTS, HEXAGON CAP SCREWS,
HEXAGON STRUCTURAL BOLTS, AND HEXAGON NUTS

Nominal Size D, mm	Width Across Flats W (max)		Thickness T (max)			
Body Dia and Thd Pitch	Bolts,[a] Cap Screws,[b] and Nuts[c]	Heavy Hex & Hex Structural Bolts[a] & Nuts[c]	Bolts (Unfin.)	Cap Screw (Fin.)	Nut (Fin. or Unfin.)	
					Style 1	Style 2
M5 × 0.8	8.0		3.88	3.65	4.7	5.1
M6 × 1	10.0		4.38	4.47	5.2	5.7
M8 × 1.25	13.0		5.68	5.50	6.8	7.5
M10 × 1.5	16.0		6.85	6.63	8.4	9.3
M12 × 1.75	18.0	21.0	7.95	7.76	10.8	12.0
M14 × 2	21.0	24.0	9.25	9.09	12.8	14.1
M16 × 2	24.0	27.0	10.75	10.32	14.8	16.4
M20 × 2.5	30.0	34.0	13.40	12.88	18.0	20.3
M24 × 3	36.0	41.0	15.90	15.44	21.5	23.9
M30 × 3.5	46.0	50.0	19.75	19.48	25.6	28.6
M36 × 4	55.0	60.0	23.55	23.38	31.0	34.7
M42 × 4.5	65.0		27.05	26.97	...	...
M48 × 5	75.0		31.07	31.07	...	...
M56 × 5.5	85.0		36.20	36.20	...	...
M64 × 6	95.0		41.32	41.32	...	...
M72 × 6	105.0		46.45	46.45	...	...
M80 × 6	115.0		51.58	51.58	...	...
M90 × 6	130.0		57.74	57.74	...	...
M100 × 6	145.0		63.90	63.90	...	...

HIGH STRENGTH STRUCTURAL HEXAGON BOLTS[a] (FIN.) AND HEXAGON NUTS[c]

Nominal Size	W Bolts/Nuts	Heavy Hex	Bolts (Unfin.)	Cap Screw (Fin.)	Style 1	Style 2
M16 × 2	27.0	...	10.75	...	...	17.1
M20 × 2.5	34.0	...	13.40	...	...	20.7
M22 × 2.5	36.0	...	14.9	...	...	23.6
M24 × 3	41.0	...	15.9	...	...	24.2
M27 × 3	46.0	...	17.9	...	...	27.6
M30 × 3.5	50.0	...	19.75	...	...	31.7
M36 × 4	60.0	...	23.55	...	...	36.6

[a] ANSI/ASME B18.2.3.5M–1979 (R1995), B18.2.3.6M–1979 (R1995), B18.2.3.7M–1979 (R1995).
[b] ANSI/ASME B18.2.3.1M–1979 (R1995).
[c] ANSI/ASME B18.2.4.1M–1979 (R1995), B18.2.4.2M–1979 (R1995).

19 Cap Screws, Slotted[a] and Socket Head[b]— American National Standard and Metric

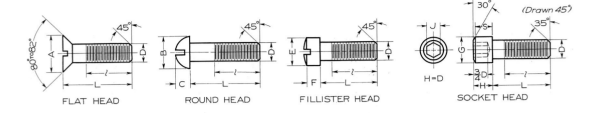

Nominal Size D	Flat Head[a]	Round Head[a]		Fillister Head[a]		Socket Head[b]		
	A	B	C	E	F	G	J	S
0 (.060)	...	...	...	...	...	.096	.05	.054
1 (.073)	...	...	...	...	...	.118	$\frac{1}{16}$	.066
2 (.086)	...	...	...	...	...	.140	$\frac{5}{64}$	.077
3 (.099)	...	...	...	...	...	.161	$\frac{5}{64}$	.089
4 (.112)	...	...	...	...	...	.183	$\frac{3}{32}$	.101
5 (.125)	...	...	...	...	...	.205	$\frac{3}{32}$	.112
6 (.138)	...	...	...	...	...	.226	$\frac{7}{64}$	.124
8 (.164)	...	...	...	...	...	.270	$\frac{9}{64}$	.148
10 (.190)	...	...	...	...	...	.312	$\frac{5}{32}$	.171
$\frac{1}{4}$	.500	.437	.191	.375	.172	.375	$\frac{3}{16}$	.225
$\frac{5}{16}$	.625	.562	.245	.437	.203	.469	$\frac{1}{4}$	.281
$\frac{3}{8}$	.750	.675	.273	.562	.250	.562	$\frac{5}{16}$	.337
$\frac{7}{16}$	.812	.750	.328	.625	.297	.656	$\frac{3}{8}$	.394
$\frac{1}{2}$	.875	.812	.354	.750	.328	.750	$\frac{3}{8}$	.450
$\frac{9}{16}$	1.000	.937	.409	.812	.375	...	...	...
$\frac{5}{8}$	1.125	1.000	.437	.875	.422	.938	$\frac{1}{2}$	.562
$\frac{3}{4}$	1.375	1.250	.546	1.000	.500	1.125	$\frac{5}{8}$	.675
$\frac{7}{8}$	1.625	...	...	1.125	.594	1.312	$\frac{3}{4}$	.787
1	1.875	...	...	1.312	.656	1.500	$\frac{3}{4}$	.900
$1\frac{1}{8}$	2.062	...	...	...	...	1.688	$\frac{7}{8}$	1.012
$1\frac{1}{4}$	2.312	...	...	...	...	1.875	$\frac{7}{8}$	1.125
$1\frac{3}{8}$	2.562	...	...	...	...	2.062	1	1.237
$1\frac{1}{2}$	2.812	...	...	...	...	2.250	1	1.350

[a] ANSI/ASME B18.6.2–1995.
[b] ANSI/ASME B18.3–1986 (R1995). For hexagon-head screws, see Appendix 18.

19 Cap Screws, Slotted[a] and Socket Head[b]—
American National Standard and Metric (continued)

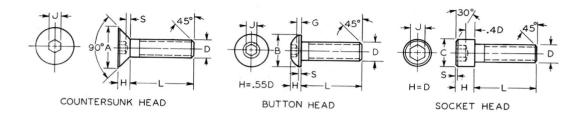

COUNTERSUNK HEAD   BUTTON HEAD   SOCKET HEAD

	Metric Socket Head Cap Screws								
Nominal Size D	Countersunk Head[a]			Button Head[a]			Socket Head[b]		Hex Socket Size
	A (max)	H	S	B	S	G	C	S	J
M1.6 × 0.35	...	...	...	...	...	...	3.0	0.16	1.5
M2 × 0.4	...	...	...	...	...	...	3.8	0.2	1.5
M2.5 × 0.45	...	...	...	...	...	...	4.5	0.25	2.0
M3 × 0.5	6.72	1.86	0.25	5.70	0.38	0.2	5.5	0.3	2.5
M4 × 0.7	8.96	2.48	0.45	7.6	0.38	0.3	7.0	0.4	3.0
M5 × 0.8	11.2	3.1	0.66	9.5	0.5	0.38	8.5	0.5	4.0
M6 × 1	13.44	3.72	0.7	10.5	0.8	0.74	10.0	0.6	5.0
M8 × 1.25	17.92	4.96	1.16	14.0	0.8	1.05	13.0	0.8	6.0
M10 × 1.5	22.4	6.2	1.62	17.5	0.8	1.45	16.0	1.0	8.0
M12 × 1.75	26.88	7.44	1.8	21.0	0.8	1.63	18.0	1.2	10.0
M14 × 2	30.24	8.12	2.0	...	...	...	21.0	1.4	12.0
M16 × 2	33.6	8.8	2.2	28.0	1.5	2.25	24.0	1.6	14.0
M20 × 2.5	40.32	10.16	2.2	...	...	...	30.0	2.0	17.0
M24 × 3	...	...	...	...	...	...	36.0	2.4	19.0
M30 × 3.5	...	...	...	...	...	...	45.0	3.0	22.0
M36 × 4	...	...	...	...	...	...	54.0	3.6	27.0
M42 × 4.5	...	...	...	...	...	...	63.0	4.2	32.0
M48 × 5	...	...	...	...	...	...	72.0	4.8	36.0

[a] ANSI/ASME B18.3.4M–1986 (R1993).
[b] ANSI/ASME B18.3.1M–1986 (R1993).

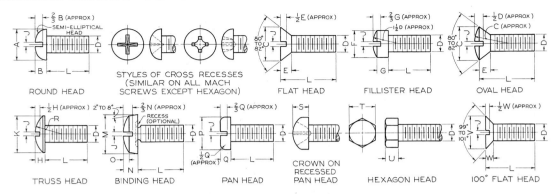

ROUND HEAD   STYLES OF CROSS RECESSES (SIMILAR ON ALL MACH SCREWS EXCEPT HEXAGON)   FLAT HEAD   FILLISTER HEAD   OVAL HEAD

TRUSS HEAD   BINDING HEAD   PAN HEAD   CROWN ON RECESSED PAN HEAD   HEXAGON HEAD   100° FLAT HEAD

AMERICAN NATIONAL STANDARD MACHINE SCREWS[a]

Length of Thread: On screws 2″ long and shorter, the threads extend to within two threads of the head and closer if practicable; longer screws have minimum thread length of $1\frac{3}{4}″$.

Points: Machine screws are regularly made with plain sheared ends, not chamfered.

Threads: Either Coarse or Fine Thread Series, Class 2 fit.

Recessed Heads: Two styles of cross recesses are available on all screws except hexagon head.

Nominal Size	Max Diameter D	Round Head		Flat Heads & Oval Head		Fillister Head		Truss Head			Slot Width
		A	B	C	E	F	G	K	H	R	J
0	0.060	0.113	0.053	0.119	0.035	0.096	0.045	0.131	0.037	0.087	0.023
1	0.073	0.138	0.061	0.146	0.043	0.118	0.053	0.164	0.045	0.107	0.026
2	0.086	0.162	0.069	0.172	0.051	0.140	0.062	0.194	0.053	0.129	0.031
3	0.099	0.187	0.078	0.199	0.059	0.161	0.070	0.226	0.061	0.151	0.035
4	0.112	0.211	0.086	0.225	0.067	0.183	0.079	0.257	0.069	0.169	0.039
5	0.125	0.236	0.095	0.252	0.075	0.205	0.088	0.289	0.078	0.191	0.043
6	0.138	0.260	0.103	0.279	0.083	0.226	0.096	0.321	0.086	0.211	0.048
8	0.164	0.309	0.120	0.332	0.100	0.270	0.113	0.384	0.102	0.254	0.054
10	0.190	0.359	0.137	0.385	0.116	0.313	0.130	0.448	0.118	0.283	0.060
12	0.216	0.408	0.153	0.438	0.132	0.357	0.148	0.511	0.134	0.336	0.067
$\frac{1}{4}$	0.250	0.472	0.175	0.507	0.153	0.414	0.170	0.573	0.150	0.375	0.075
$\frac{5}{16}$	0.3125	0.590	0.216	0.635	0.191	0.518	0.211	0.698	0.183	0.457	0.084
$\frac{3}{8}$	0.375	0.708	0.256	0.762	0.230	0.622	0.253	0.823	0.215	0.538	0.094
$\frac{7}{16}$	0.4375	0.750	0.328	0.812	0.223	0.625	0.265	0.948	0.248	0.619	0.094
$\frac{1}{2}$	0.500	0.813	0.355	0.875	0.223	0.750	0.297	1.073	0.280	0.701	0.106
$\frac{9}{16}$	0.5625	0.938	0.410	1.000	0.260	0.812	0.336	1.198	0.312	0.783	0.118
$\frac{5}{8}$	0.625	1.000	0.438	1.125	0.298	0.875	0.375	1.323	0.345	0.863	0.133
$\frac{3}{4}$	0.750	1.250	0.547	1.375	0.372	1.000	0.441	1.573	0.410	1.024	0.149

Nominal Size	Max Diameter D	Binding Head			Pan Head			Hexagon Head		100° Flat Head		Slot Width
		M	N	O	P	Q	S	T	U	V	W	J
2	0.086	0.181	0.050	0.018	0.167	0.053	0.062	0.125	0.050	...	...	0.031
3	0.099	0.208	0.059	0.022	0.193	0.060	0.071	0.187	0.055	...	...	0.035
4	0.112	0.235	0.068	0.025	0.219	0.068	0.080	0.187	0.060	0.225	0.049	0.039
5	0.125	0.263	0.078	0.029	0.245	0.075	0.089	0.187	0.070	...	...	0.043
6	0.138	0.290	0.087	0.032	0.270	0.082	0.097	0.250	0.080	0.279	0.060	0.048
8	0.164	0.344	0.105	0.039	0.322	0.096	0.115	0.250	0.110	0.332	0.072	0.054
10	0.190	0.399	0.123	0.045	0.373	0.110	0.133	0.312	0.120	0.385	0.083	0.060
12	0.216	0.454	0.141	0.052	0.425	0.125	0.151	0.312	0.155	...	...	0.067
$\frac{1}{4}$	0.250	0.513	0.165	0.061	0.492	0.144	0.175	0.375	0.190	0.507	0.110	0.075
$\frac{5}{16}$	0.3125	0.641	0.209	0.077	0.615	0.178	0.218	0.500	0.230	0.635	0.138	0.084
$\frac{3}{8}$	0.375	0.769	0.253	0.094	0.740	0.212	0.261	0.562	0.295	0.762	0.165	0.094
$\frac{7}{16}$	.4375	...	...	...	.865	.247	.305	...	...	...	...	.094
$\frac{1}{2}$	.500	...	...	...	.987	.281	.348	...	...	...	...	.106
$\frac{9}{16}$	.5625	...	...	...	1.041	.315	.391	...	...	...	...	.118
$\frac{5}{8}$	.625	...	...	...	1.172	.350	.434	...	...	...	...	.133
$\frac{3}{4}$	.750	...	...	...	1.435	.419	.521	...	...	...	...	.149

20 Machine Screws—American National Standard and Metric (continued)

METRIC MACHINE SCREWS

Length of Thread: On screws 36 mm long or shorter, the threads extend to within one thread of the head: on longer screws the thread extends to within two threads of the head.

Points: Machine screws are regularly made with sheared ends, not chamfered.

Threads: Coarse (genera purpose) threads series are given.

Recessed Heads: Two styles of cross-recesses are available on all screws except hexagon head.

Nominal Size & Thd Pitch	Max. Dia. D. mm	Flat Heads & Oval Head		Pan Heads			Hex Head		Slot Width
		C	E	P	Q	S	T	U	J
M2 × M	2.0	3.5	1.2	4.0	1.3	1.6	3.2	1.6	0.7
M2.5 × 0.45	2.5	4.4	1.5	5.0	1.5	2.1	4.0	2.1	0.8
M3 × 0.5	3.0	5.2	1.7	5.6	1.8	2.4	5.0	2.3	1.0
M3.5 × 0.6	3.5	6.9	2.3	7.0	2.1	2.6	5.5	2.6	1.2
M4 × 0.7	4.0	8.0	2.7	8.0	2.4	3.1	7.0	3.0	1.5
M5 × 0.8	5.0	8.9	2.7	9.5	3.0	3.7	8.0	3.8	1.5
M6 × 1	6.0	10.9	3.3	12.0	3.6	4.6	10.0	4.7	1.9
M8 × 1.25	8.0	15.14	4.6	16.0	4.8	6.0	13.0	6.0	2.3
M10 × 1.5	10.0	17.8	5.0	20.0	6.0	7.5	15.0	7.5	2.8
M12 × 1.75	12.0	…	…	…	…	…	18.0	9.0	…

Nominal Size	Metric Machine Screw Lengths—L																					
	2.5	3	4	5	6	8	10	13	16	20	25	30	35	40	45	50	55	60	65	70	80	90
M2 × 0.4	PH	A	A	A	A	A	A	A	A	A												
M2.5 × 0.45		PH	A	A	A	A	A	A	A	A	A											
M3 × 0.5			PH	A	A	A	A	A	A	A	A	A										
M3.5 × 0.6				PH	A	A	A	A	A	A	A	A	A									
M4 × 0.7				PH	A	A	A	A	A	A	A	A	A	A								
M5 × 0.8					PH	A	A	A	A	A	A	A	A	A	A							
M6 × 1						A	A	A	A	A	A	A	A	A	A	A	A					
M8 × 1.25						A	A	A	A	A	A	A	A	A	A	A	A	A	A	A		
M10 × 1.5							A	A	A	A	A	A	A	A	A	A	A	A	A	A	A	
M12 × 1.75								A	A	A	A	A	A	A	A	A	A	A	A	A	A	A

Min. Thd Length—28 mm

Min. Thd Length—38 mm

[a]PH = recommended lengths for only pan and hex head metric screws.

A = recommended lengths for all metric screw head-styles.

21 Keys—Square, Flat, Plain Taper,[a] and Gib Head

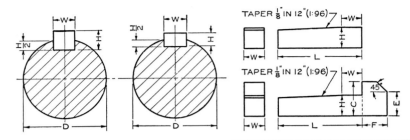

Shaft Diameters	Square Stock Key	Flat Stock Key	Gib Head Taper Stock Key					
			Square			Flat		
			Height	Length	Height to Chamfer	Height	Length	Height to Chamfer
D	$W = H$	$W \times H$	C	F	E	C	F	E
$\frac{1}{2}$ to $\frac{9}{16}$	$\frac{1}{8}$	$\frac{1}{8} \times \frac{3}{32}$	$\frac{1}{4}$	$\frac{7}{32}$	$\frac{5}{32}$	$\frac{3}{16}$	$\frac{1}{8}$	$\frac{1}{8}$
$\frac{5}{8}$ to $\frac{7}{8}$	$\frac{3}{16}$	$\frac{3}{16} \times \frac{1}{8}$	$\frac{5}{16}$	$\frac{9}{32}$	$\frac{7}{32}$	$\frac{1}{4}$	$\frac{3}{16}$	$\frac{5}{32}$
$\frac{15}{16}$ to $1\frac{1}{4}$	$\frac{1}{4}$	$\frac{1}{4} \times \frac{3}{16}$	$\frac{7}{16}$	$\frac{11}{32}$	$\frac{11}{32}$	$\frac{5}{16}$	$\frac{1}{4}$	$\frac{3}{16}$
$1\frac{5}{16}$ to $1\frac{3}{8}$	$\frac{5}{16}$	$\frac{5}{16} \times \frac{1}{4}$	$\frac{9}{16}$	$\frac{13}{32}$	$\frac{13}{32}$	$\frac{3}{8}$	$\frac{5}{16}$	$\frac{1}{4}$
$1\frac{7}{16}$ to $1\frac{3}{4}$	$\frac{3}{8}$	$\frac{3}{8} \times \frac{1}{4}$	$\frac{11}{16}$	$\frac{15}{32}$	$\frac{15}{32}$	$\frac{7}{16}$	$\frac{3}{8}$	$\frac{5}{16}$
$1\frac{13}{16}$ to $2\frac{1}{4}$	$\frac{1}{2}$	$\frac{1}{2} \times \frac{3}{8}$	$\frac{7}{8}$	$\frac{19}{32}$	$\frac{5}{8}$	$\frac{5}{8}$	$\frac{1}{2}$	$\frac{7}{16}$
$2\frac{5}{16}$ to $2\frac{3}{4}$	$\frac{5}{8}$	$\frac{5}{8} \times \frac{7}{16}$	$1\frac{1}{16}$	$\frac{23}{32}$	$\frac{3}{4}$	$\frac{3}{4}$	$\frac{5}{8}$	$\frac{1}{2}$
$2\frac{7}{8}$ to $3\frac{1}{4}$	$\frac{3}{4}$	$\frac{3}{4} \times \frac{1}{2}$	$1\frac{1}{4}$	$\frac{7}{8}$	$\frac{7}{8}$	$\frac{7}{8}$	$\frac{3}{4}$	$\frac{5}{8}$
$3\frac{3}{8}$ to $3\frac{3}{4}$	$\frac{7}{8}$	$\frac{7}{8} \times \frac{5}{8}$	$1\frac{1}{2}$	1	1	$1\frac{1}{16}$	$\frac{7}{8}$	$\frac{3}{4}$
$3\frac{7}{8}$ to $4\frac{1}{2}$	1	$1 \times \frac{3}{4}$	$1\frac{3}{4}$	$1\frac{3}{16}$	$1\frac{3}{16}$	$1\frac{1}{4}$	1	$\frac{13}{16}$
$4\frac{3}{4}$ to $5\frac{1}{2}$	$1\frac{1}{4}$	$1\frac{1}{4} \times \frac{7}{8}$	2	$1\frac{7}{16}$	$1\frac{7}{16}$	$1\frac{1}{2}$	$1\frac{1}{4}$	1
$5\frac{3}{4}$ to 6	$1\frac{1}{2}$	$1\frac{1}{2} \times 1$	$2\frac{1}{2}$	$1\frac{3}{4}$	$1\frac{3}{4}$	$1\frac{3}{4}$	$1\frac{1}{2}$	1

[a] Plain taper square and flat keys have the same dimensions as the plain parallel stock keys, with the addition of the taper on top. Gib head taper square and flat keys have the same dimensions as the plain taper keys, with the addition of the gib head.

Stock lengths for plain taper and gib head taper keys: The minimum stock length equals 4W, and the maximum equals 16W. The increments of increase of length equal 2W.

22 Screw Threads,[a] Square and Acme

Size	Threads per Inch	Size	Threads per Inch	Size	Threads per Inch	Size	Threads per Inch
$\frac{3}{8}$	12	$\frac{7}{8}$	5	2	$2\frac{1}{2}$	$3\frac{1}{2}$	$1\frac{1}{3}$
$\frac{7}{16}$	10	1	5	$2\frac{1}{4}$	2	$3\frac{3}{4}$	$1\frac{1}{3}$
$\frac{1}{2}$	10	$1\frac{1}{8}$	4	$2\frac{1}{2}$	2	4	$1\frac{1}{3}$
$\frac{9}{16}$	8	$1\frac{1}{4}$	4	$2\frac{3}{4}$	2	$4\frac{1}{4}$	$1\frac{1}{3}$
$\frac{5}{8}$	8	$1\frac{1}{2}$	3	3	$1\frac{1}{2}$	$4\frac{1}{2}$	1
$\frac{3}{4}$	6	$1\frac{3}{4}$	$2\frac{1}{2}$	$3\frac{1}{4}$	$1\frac{1}{2}$	over $4\frac{1}{2}$	1

[a] See Appendix 17 for General-Purpose Acme Threads.

23 Woodruff Keys[a]—American National Standard

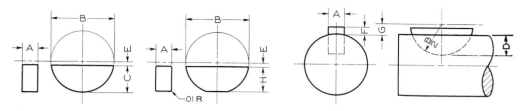

Key No.[b]	Nominal Sizes A × B	E	F	G	Maximum Sizes H	D	C
204	$\frac{1}{16} \times \frac{1}{2}$	$\frac{3}{64}$	$\frac{1}{32}$	$\frac{5}{64}$	.194	.1718	.203
304	$\frac{3}{32} \times \frac{1}{2}$	$\frac{3}{64}$	$\frac{3}{64}$	$\frac{3}{32}$	.194	.1561	.203
305	$\frac{3}{32} \times \frac{5}{8}$	$\frac{1}{16}$	$\frac{3}{64}$	$\frac{7}{64}$	.240	.2031	.250
404	$\frac{1}{8} \times \frac{1}{2}$	$\frac{3}{64}$	$\frac{1}{16}$	$\frac{7}{64}$	.194	.1405	.203
405	$\frac{1}{8} \times \frac{5}{8}$	$\frac{1}{16}$	$\frac{1}{16}$	$\frac{1}{8}$	.240	.1875	.250
406	$\frac{1}{8} \times \frac{3}{4}$	$\frac{1}{16}$	$\frac{1}{16}$	$\frac{1}{8}$	.303	.2505	.313
505	$\frac{5}{32} \times \frac{5}{8}$	$\frac{1}{16}$	$\frac{5}{64}$	$\frac{9}{64}$	.240	.1719	.250
506	$\frac{5}{32} \times \frac{3}{4}$	$\frac{1}{16}$	$\frac{5}{64}$	$\frac{9}{64}$	.303	.2349	.313
507	$\frac{5}{32} \times \frac{7}{8}$	$\frac{1}{16}$	$\frac{5}{64}$	$\frac{9}{64}$	.365	.2969	.375
606	$\frac{3}{16} \times \frac{3}{4}$	$\frac{1}{16}$	$\frac{3}{32}$	$\frac{5}{32}$	.303	.2193	.313
607	$\frac{3}{16} \times \frac{7}{8}$	$\frac{1}{16}$	$\frac{3}{32}$	$\frac{5}{32}$	.365	.2813	.375
608	$\frac{3}{16} \times 1$	$\frac{1}{16}$	$\frac{3}{32}$	$\frac{5}{32}$	.428	.3443	.438
609	$\frac{3}{16} \times 1\frac{1}{8}$	$\frac{5}{64}$	$\frac{3}{32}$	$\frac{11}{64}$	.475	.3903	.484
807	$\frac{1}{4} \times \frac{7}{8}$	$\frac{1}{16}$	$\frac{1}{8}$	$\frac{3}{16}$	.365	.2500	.375
808	$\frac{1}{4} \times 1$	$\frac{1}{16}$	$\frac{1}{8}$	$\frac{3}{16}$	.428	.3130	.438
809	$\frac{1}{4} \times 1\frac{1}{8}$	$\frac{5}{64}$	$\frac{1}{8}$	$\frac{13}{64}$	.475	.3590	.484
810	$\frac{1}{4} \times 1\frac{1}{4}$	$\frac{5}{64}$	$\frac{1}{8}$	$\frac{13}{64}$	.537	.4220	.547
811	$\frac{1}{4} \times 1\frac{3}{8}$	$\frac{3}{32}$	$\frac{1}{8}$	$\frac{7}{32}$	.584	.4690	.594
812	$\frac{1}{4} \times 1\frac{1}{2}$	$\frac{7}{64}$	$\frac{1}{8}$	$\frac{15}{64}$	.631	.5160	.641
1008	$\frac{5}{16} \times 1$	$\frac{1}{16}$	$\frac{5}{32}$	$\frac{7}{32}$	.428	.2818	.438
1009	$\frac{5}{16} \times 1\frac{1}{8}$	$\frac{5}{64}$	$\frac{5}{32}$	$\frac{15}{64}$	.475	.3278	.484
1010	$\frac{5}{16} \times 1\frac{1}{4}$	$\frac{5}{64}$	$\frac{5}{32}$	$\frac{15}{64}$	.537	.3908	.547
1011	$\frac{5}{16} \times 1\frac{3}{8}$	$\frac{3}{32}$	$\frac{5}{32}$	$\frac{8}{32}$	.584	.4378	.594
1012	$\frac{5}{16} \times 1\frac{1}{2}$	$\frac{7}{64}$	$\frac{5}{32}$	$\frac{17}{64}$	.631	.4848	.641
1210	$\frac{3}{8} \times 1\frac{1}{4}$	$\frac{5}{64}$	$\frac{3}{16}$	$\frac{17}{64}$	.537	.3595	.547
1211	$\frac{3}{8} \times 1\frac{3}{8}$	$\frac{3}{32}$	$\frac{3}{16}$	$\frac{9}{32}$	.584	.4065	.594
1212	$\frac{3}{8} \times 1\frac{1}{2}$	$\frac{7}{64}$	$\frac{3}{16}$	$\frac{19}{64}$	.631	.4535	.641

[a] ANSI B17.2–1967 (R1990).
[b] Key numbers indicate nominal key dimensions. The last two digits give the nominal diameter B in eighths of an inch, and the digits before the last two give the nominal width A in thirty-seconds of an inch.

24 Woodruff Key Sizes for Different Shaft Diameters[a]

Shaft Diameter	$\frac{5}{16}$ to $\frac{3}{8}$	$\frac{7}{16}$ to $\frac{1}{2}$	$\frac{9}{16}$ to $\frac{3}{4}$	$\frac{13}{16}$ to $\frac{15}{16}$	1 to $1\frac{3}{16}$	$1\frac{1}{4}$ to $1\frac{7}{16}$	$1\frac{1}{2}$ to $1\frac{3}{4}$	$1\frac{13}{16}$ to $2\frac{1}{8}$	$2\frac{3}{16}$ to $2\frac{1}{2}$
Key Numbers	204	304 305	404 405 406	505 506 507	606 607 608 609	807 808 809	810 811 812	1011 1012	1211 1212

[a] Suggested sizes; not standard.

25 Pratt and Whitney Round-End Keys

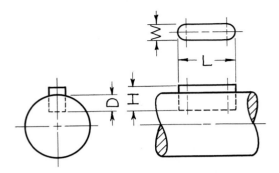

KEYS MADE WITH ROUND ENDS AND KEYWAYS CUT IN SPLINE MILLER

Maximum length of slot is 4″ + W. Note that key is sunk two-thirds into shaft in all cases.

Key No.	L^a	W or D	H	Key No.	L^a	W or D	H
1	$\frac{1}{2}$	$\frac{1}{16}$	$\frac{3}{32}$	22	$1\frac{3}{8}$	$\frac{1}{4}$	$\frac{3}{8}$
2	$\frac{1}{2}$	$\frac{3}{32}$	$\frac{9}{64}$	23	$1\frac{1}{38}$	$\frac{5}{16}$	$\frac{15}{32}$
3	$\frac{1}{2}$	$\frac{1}{8}$	$\frac{3}{16}$	F	$1\frac{3}{8}$	$\frac{3}{8}$	$\frac{9}{16}$
4	$\frac{5}{8}$	$\frac{3}{32}$	$\frac{9}{64}$	24	$1\frac{1}{2}$	$\frac{1}{4}$	$\frac{3}{8}$
5	$\frac{5}{8}$	$\frac{1}{8}$	$\frac{3}{16}$	25	$1\frac{1}{2}$	$\frac{5}{16}$	$\frac{15}{32}$
6	$\frac{5}{8}$	$\frac{5}{32}$	$\frac{15}{64}$	G	$1\frac{1}{2}$	$\frac{3}{8}$	$\frac{9}{16}$
7	$\frac{3}{4}$	$\frac{1}{8}$	$\frac{3}{16}$	51	$1\frac{3}{4}$	$\frac{1}{4}$	$\frac{3}{8}$
8	$\frac{3}{4}$	$\frac{5}{32}$	$\frac{15}{64}$	52	$1\frac{3}{4}$	$\frac{5}{16}$	$\frac{15}{32}$
9	$\frac{3}{4}$	$\frac{3}{16}$	$\frac{9}{32}$	53	$1\frac{3}{4}$	$\frac{3}{8}$	$\frac{9}{16}$
10	$\frac{7}{8}$	$\frac{5}{32}$	$\frac{15}{64}$	26	2	$\frac{3}{16}$	$\frac{9}{32}$
11	$\frac{7}{8}$	$\frac{3}{16}$	$\frac{9}{32}$	27	2	$\frac{1}{4}$	$\frac{3}{8}$
12	$\frac{7}{8}$	$\frac{7}{32}$	$\frac{21}{64}$	28	2	$\frac{5}{16}$	$\frac{15}{32}$
A	$\frac{7}{8}$	$\frac{1}{4}$	$\frac{3}{8}$	29	2	$\frac{3}{8}$	$\frac{9}{16}$
13	1	$\frac{3}{16}$	$\frac{9}{32}$	54	$2\frac{1}{4}$	$\frac{1}{4}$	$\frac{3}{8}$
14	1	$\frac{7}{32}$	$\frac{21}{64}$	55	$2\frac{1}{4}$	$\frac{5}{16}$	$\frac{15}{32}$
15	1	$\frac{1}{4}$	$\frac{3}{8}$	56	$2\frac{1}{4}$	$\frac{3}{8}$	$\frac{9}{16}$
B	1	$\frac{5}{16}$	$\frac{15}{32}$	57	$2\frac{1}{4}$	$\frac{7}{16}$	$\frac{21}{32}$
16	$1\frac{1}{8}$	$\frac{3}{16}$	$\frac{9}{32}$	58	$2\frac{1}{2}$	$\frac{5}{16}$	$\frac{15}{32}$
17	$1\frac{1}{8}$	$\frac{7}{32}$	$\frac{21}{64}$	59	$2\frac{1}{2}$	$\frac{3}{8}$	$\frac{9}{16}$
18	$1\frac{1}{8}$	$\frac{1}{4}$	$\frac{3}{8}$	60	$2\frac{1}{2}$	$\frac{7}{16}$	$\frac{21}{32}$
C	$1\frac{1}{8}$	$\frac{5}{16}$	$\frac{15}{32}$	61	$2\frac{1}{2}$	$\frac{1}{2}$	$\frac{3}{4}$
19	$1\frac{1}{4}$	$\frac{3}{16}$	$\frac{9}{32}$	30	3	$\frac{3}{8}$	$\frac{9}{16}$
20	$1\frac{1}{4}$	$\frac{7}{32}$	$\frac{21}{64}$	31	3	$\frac{7}{16}$	$\frac{21}{32}$
21	$1\frac{1}{4}$	$\frac{1}{4}$	$\frac{3}{8}$	32	3	$\frac{1}{2}$	$\frac{3}{4}$
D	$1\frac{1}{4}$	$\frac{5}{16}$	$\frac{15}{32}$	33	3	$\frac{9}{16}$	$\frac{27}{32}$
E	$1\frac{1}{4}$	$\frac{3}{8}$	$\frac{9}{16}$	34	3	$\frac{5}{8}$	$\frac{15}{16}$

a The length L may vary from the table, but equals at least 2W.

For parts lists, etc., give inside diameter, outside diameter, and the thickness; for example, .344 × .688 × .065 TYPE A PLAIN WASHER.

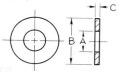

PREFERRED SIZES OF TYPE A PLAIN WASHERS[b]

Nominal Washer Size[c]			Inside Diameter A	Outside Diameter B	Nominal Thickness C
...	...		0.078	0.188	0.020
...	...		0.094	0.250	0.020
...	...		0.125	0.312	0.032
No. 6	0.138		0.156	0.375	0.049
No. 8	0.164		0.188	0.438	0.049
No. 10	0.190		0.219	0.500	0.049
$\frac{3}{16}$	0.188		0.250	0.562	0.049
No. 12	0.216		0.250	0.562	0.065
$\frac{1}{4}$	0.250	N	0.281	0.625	0.065
$\frac{1}{4}$	0.250	W	0.312	0.734	0.065
$\frac{5}{16}$	0.312	N	0.344	0.688	0.065
$\frac{5}{16}$	0.312	W	0.375	0.875	0.083
$\frac{3}{8}$	0.375	N	0.406	0.812	0.065
$\frac{3}{8}$	0.375	W	0.438	1.000	0.083
$\frac{7}{16}$	0.438	N	0.469	0.922	0.065
$\frac{7}{16}$	0.438	W	0.500	1.250	0.083
$\frac{1}{2}$	0.500	N	0.531	1.062	0.095
$\frac{1}{2}$	0.500	W	0.562	1.375	0.109
$\frac{9}{16}$	0.562	N	0.594	1.156	0.095
$\frac{9}{16}$	0.562	W	0.625	1.469	0.109
$\frac{5}{8}$	0.625	N	0.656	1.312	0.095
$\frac{5}{8}$	0.625	W	0.688	1.750	0.134
$\frac{3}{4}$	0.750	N	0.812	1.469	0.134
$\frac{3}{4}$	0.750	W	0.812	2.000	0.148
$\frac{7}{8}$	0.875	N	0.938	1.750	0.134
$\frac{7}{8}$	0.875	W	0.938	2.250	0.165
1	1.000	N	1.062	2.000	0.134
1	1.000	W	1.062	2.500	0.165
$1\frac{1}{8}$	1.125	N	1.250	2.250	0.134
$1\frac{1}{8}$	1.125	W	1.250	2.750	0.165
$1\frac{1}{4}$	1.250	N	1.375	2.500	0.165
$1\frac{1}{4}$	1.250	W	1.375	3.000	0.165
$1\frac{3}{8}$	1.375	N	1.500	2.750	0.165
$1\frac{3}{8}$	1.375	W	1.500	3.250	0.180
$1\frac{1}{2}$	1.500	N	1.625	3.000	0.165
$1\frac{1}{2}$	1.500	W	1.625	3.500	0.180
$1\frac{5}{8}$	1.625		1.750	3.750	0.180
$1\frac{3}{4}$	1.750		1.875	4.000	0.180
$1\frac{7}{8}$	1.875		2.000	4.250	0.180
2	2.000		2.125	4.500	0.180
$2\frac{1}{4}$	2.250		2.375	4.750	0.220
$2\frac{1}{2}$	2.500		2.625	5.000	0.238
$2\frac{3}{4}$	2.750		2.875	5.250	0.259
3	3000		3.125	5.500	0.284

[a] From ANSI B18.22.1–1965 (R1981). For complete listings, see the standard.
[b] Preferred sizes are for the most part from series previously designated "Standard Plate" and "SAE." Where common sizes existed in the two series, the SAE size is designated "N" (narrow) and the Standard Plate "W" (wide).
[c] Nominal washer sizes are intended for use with comparable nominal screw or bolt sizes.

27 Washers,ᵃ Lock—American National Standard

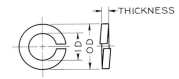

For parts lists, etc., give nominal size and series; for example, $\frac{1}{4}$ REGULAR LOCK WASHER

PREFERRED SERIES

Nominal Washer Sizeᵇ		Inside Diameter, Min.	Regular		Extra Duty		Hi-Collar	
			Outside Diameter, Max.	Thick-ness, Min.	Outside Diameter, Max.	Thick-ness, Min.	Outside Diameter, Max.	Thick-ness, Min.
No. 2	0.086	0.088	0.172	0.020	0.208	0.027	. . .	. . .
No. 3	0.099	0.101	0.195	0.025	0.239	0.034	. . .	. . .
No. 4	0.112	0.115	0.209	0.025	0.253	0.034	0.173	0.022
No. 5	0.125	0.128	0.236	0.031	0.300	0.045	0.202	0.030
No. 6	0.138	0.141	0.250	0.031	0.314	0.045	0.216	0.030
No. 8	0.164	0.168	0.293	0.040	0.375	0.057	0.267	0.047
No. 10	0.190	0.194	0.334	0.047	0.434	0.068	0.294	0.047
No. 12	0.216	0.221	0.377	0.056	0.497	0.080	. . .	. . .
$\frac{1}{4}$	0.250	0.255	0.489	0.062	0.535	0.084	0.365	0.078
$\frac{5}{16}$	0.312	0.318	0.586	0.078	0.622	0.108	0.460	0.093
$\frac{3}{8}$	0.375	0.382	0.683	0.094	0.741	0.123	0.553	0.125
$\frac{7}{16}$	0.438	0.446	0.779	0.109	0.839	0.143	0.647	0.140
$\frac{1}{2}$	0.500	0.509	0.873	0.125	0.939	0.162	0.737	0.172
$\frac{9}{16}$	0.562	0.572	0.971	0.141	1.041	0.182	. . .	. . .
$\frac{5}{8}$	0.625	0.636	1.079	0.156	1.157	0.202	0.923	0.203
$\frac{11}{16}$	0.688	0.700	1.176	0.172	1.258	0.221	. . .	. . .
$\frac{3}{4}$	0.750	0.763	1.271	0.188	1.361	0.241	1.111	0.218
$\frac{13}{16}$	0.812	0.826	1.367	0.203	1.463	0.261	. . .	. . .
$\frac{7}{8}$	0.875	0.890	1.464	0.219	1.576	0.285	1.296	0.234
$\frac{15}{16}$	0.938	0.954	1.560	0.234	1.688	0.308	. . .	. . .
1	1.000	1.017	1.661	0.250	1.799	0.330	1.483	0.250
$1\frac{1}{16}$	1.062	1.080	1.756	0.266	1.910	0.352	. . .	. . .
$1\frac{1}{8}$	1.125	1.144	1.853	0.281	2.019	0.375	1.669	0.313
$1\frac{3}{16}$	1.188	1.208	1.950	0.297	2.124	0.396	. . .	. . .
$1\frac{1}{4}$	1.250	1.271	2.045	0.312	2.231	0.417	1.799	0.313
$1\frac{5}{16}$	1.312	1.334	2.141	0.328	2.335	0.438	. . .	. . .
$1\frac{3}{8}$	1.375	1.398	2.239	0.344	2.439	0.458	2.041	0.375
$1\frac{7}{16}$	1.438	1.462	2.334	0.359	2.540	0.478	. . .	. . .
$1\frac{1}{2}$	1.500	1.525	2.430	0.375	2.638	0.496	2.170	0.375

ᵃ From ANSI/ASME B18.21.1–1994. For complete listing, see the standard.
ᵇ Nominal washer sizes are intended for use with comparable nominal screw or bolt sizes.

28 Wire Gage Standards[a]

Dimensions of sizes in decimal parts of an inch.[b]

No. of Wire	American or Brown & Sharpe for Non-ferrous Metals	Birmingham, or Stubs' Iron Wire[c]	American S. & W. Co.'s (Washburn & Moen) Std. Steel Wire	American S. & W. Co.'s Music Wire	Imperial Wire	Stubs' Steel Wire[c]	Steel Manufacturers' Sheet Gage[b]	No. of Wire
7–0's	.651354	...	.4900	...	.500	...	...	7–0's
6–0's	.580049	...	.4615	.004	.464	...	...	6–0's
5–0's	.516549	.500	.4305	.005	.432	...	...	5–0's
4–0's	.460	.454	.3938	.006	.400	...	...	4–0's
000	.40964	.425	.3625	.007	.372	...	...	000
00	.3648	.380	.3310	.008	.348	...	...	00
0	.32486	.340	.3065	.009	.324	...	...	0
1	.2893	.300	.2830	.010	.300	.227	...	1
2	.25763	.284	.2625	.011	.276	.219	...	2
3	.22942	.259	.2437	.012	.252	.212	.2391	3
4	.20431	.238	.2253	.013	.232	.207	.2242	4
6	.16202	.203	.1920	.016	.192	.201	.1943	6
7	.14428	.180	.1770	.018	.176	.199	.1793	7
8	.12849	.165	.1620	.020	.160	.197	.1644	8
9	.11443	.148	.1483	.022	.144	.194	.1495	9
10	.10189	.134	.1350	.024	.128	.191	.1345	10
11	.090742	.120	.1205	.026	.116	.188	.1196	11
12	.080808	.109	.1055	.029	.104	.185	.1046	12
13	.071961	.095	.0915	.031	.092	.182	.0897	13
14	.064084	.083	.0800	.033	.080	.180	.0747	14
15	.057068	.072	.0720	.035	.072	.178	.0763	15
16	.05082	.065	.0625	.037	.064	.175	.0598	16
17	.045257	.058	.0540	.039	.056	.172	.0538	17
18	.040303	.049	.0475	.041	.048	.168	.0478	18
19	.03589	.042	.0410	.043	.040	.164	.0418	19
20	.031961	.035	.0348	.045	.036	.161	.0359	20
21	.028462	.032	.0317	.047	.032	.157	.0329	21
22	.025347	.028	.0286	.049	.028	.155	.0299	22
23	.022571	.025	.0258	.051	.024	.153	.0269	23
24	.0201	.022	.0230	.055	.022	.151	.0239	24
25	.0179	.020	.0204	.059	.020	.148	.0209	25
26	.01594	.018	.0181	.063	.018	.146	.0179	26
27	.014195	.016	.0173	.067	.0164	.143	.0164	27
28	.012641	.014	.0162	.071	.0149	.139	.0149	28
29	.011257	.013	.0150	.075	.0136	.134	.0135	29
30	.010025	.012	.0140	.080	.0124	.127	.0120	30
31	.008928	.010	.0132	.085	.0116	.120	.0105	31
32	.00795	.009	.0128	.090	.0108	.115	.0097	32
33	.00708	.008	.0118	.095	.0100	.112	.0090	33
34	.006304	.007	.0104	...	.0092	.110	.0082	34
35	.005614	.005	.0095	...	.0084	.108	.0075	35
36	.005	.004	.0090	...	.0076	.106	.0067	36
37	.004453	...	.0085	...	.0068	.103	.0064	37
38	.003965	...	.0080	...	.0060	.101	.0060	38
39	.003531	...	.0075	...	.0052	.099	...	39
40	.003144	...	.0070	...	.0048	.097	...	40

[a] Courtesy Brown & Sharpe Mfg. Co.
[b] Now used by steel manufacturers in place of old U.S. Standard Gage.
[c] The difference between the Stubs' Iron Wire Gage and the Stubs' Steel Wire Gage should be noted, the first being commonly known as the English Standard Wire, or Birmingham Gage, which designates the Stubs' soft wire sizes and the second being used in measuring drawn steel wire or drill rods of Stubs' make.

29 Taper Pins[a]—American National Standard

TAPER .25 PER FT · L (MAX) · D

To find small diameter of pin, multiply the length by .02083 and subtract the result from the larger diameter.
All dimensions are given in inches.
Standard reamers are available for pins given above the heavy line.

Number	8	7	6	5	4	3	2	1	0	2/0	3/0	4/0	5/0	6/0	7/0
Size (Large End)	.4920	.4090	.3410	.2890	.2500	.2190	.1930	.1720	.1560	.1410	.1250	.1090	.0940	.0780	.0625
Shaft Diameter (Approx)[b]	$1\frac{1}{2}$	$1\frac{1}{4}$	1	$\frac{7}{8}$	$\frac{13}{16}$	$\frac{3}{4}$	$\frac{5}{8}$	$\frac{9}{16}$	$\frac{1}{2}$	$\frac{7}{16}$	$\frac{3}{8}$	$\frac{5}{16}$	$\frac{1}{4}$	$\frac{7}{32}$	
Drill Size (Before Reamer)[b]	.3125	.2344	.2188	.1562	.1562	.1250	.1250	.1094	.0938	.0938	.0781	.0625	.0625	.0312	.0312
Length L															
.250											×	×	×	×	×
.375										×	×	×	×	×	×
.500									×	×	×	×	×	×	×
.625									×	×	×	×	×	×	×
.750					×	×		×	×	×	×	×	×	×	×
.875				×	×	×	×	×	×	×	×	×	×	×	×
1.000				×	×	×	×	×	×	×	×	×	×	×	×
1.250			×	×	×	×	×	×	×	×	×	×	×	×	⋮
1.500		×	×	×	×	×	×	×	×	×	×	×	×	×	⋮
1.750	×	×	×	×	×	×	×	×	×	×	×	×	⋮	⋮	⋮
2.000	×	×	×	×	×	×	×	×	×	×	×	⋮	⋮	⋮	⋮
2.250	×	×	×	×	×	×	×	×	×	×	⋮	⋮	⋮	⋮	⋮
2.500	×	×	×	×	×	×	×	×	×	⋮	⋮	⋮	⋮	⋮	⋮
2.750	×	×	×	×	×	×	×	×	⋮	⋮	⋮	⋮	⋮	⋮	⋮
3.000	×	×	×	×	×	×	×	⋮	⋮	⋮	⋮	⋮	⋮	⋮	⋮
3.250	×	×	×	×	×	×	⋮	⋮	⋮	⋮	⋮	⋮	⋮	⋮	⋮
3.500	×	×	×	×	×	⋮	⋮	⋮	⋮	⋮	⋮	⋮	⋮	⋮	⋮
3.750	×	×	×	×	×	⋮	⋮	⋮	⋮	⋮	⋮	⋮	⋮	⋮	⋮
4.000	×	×	×	×	⋮	⋮	⋮	⋮	⋮	⋮	⋮	⋮	⋮	⋮	⋮
4.250	×	×	×	×	⋮	⋮	⋮	⋮	⋮	⋮	⋮	⋮	⋮	⋮	⋮
4.500	×	×	×	×	⋮	⋮	⋮	⋮	⋮	⋮	⋮	⋮	⋮	⋮	⋮

[a] ANSI/ASME B18.8.2–1994. For Nos. 9 and 10, see the standard. Pins Nos 11 (size .8600), 12 (size 1.032), 13 (size 1.241), and 14 (size 1.523) are special sizes; hence their lengths are special.
[b] Suggested sizes; not American National Standard.

30 Cotter Pins[a]—American National Standard

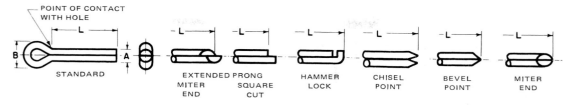

POINT OF CONTACT WITH HOLE

STANDARD

EXTENDED MITER END | PRONG SQUARE CUT | HAMMER LOCK | CHISEL POINT | BEVEL POINT | MITER END

PREFERRED POINT TYPES

All dimensions are given in inches.

Nominal Size or Pin Diameter		Diameter A		Outside Eye Diameter B Min.	Extended Prong Length Min.	Hole Sizes Recommended
		Max.	Min.			
$\frac{1}{32}$	.031	.032	.028	.06	.01	.047
$\frac{3}{64}$	.047	.048	.044	.09	.02	.062
$\frac{1}{16}$	.062	.060	.056	.12	.03	.078
$\frac{5}{64}$	.078	.076	.072	.16	.04	.094
$\frac{3}{32}$	.094	.090	.086	.19	.04	.109
$\frac{7}{64}$	.109	.104	.100	.22	.05	.125
$\frac{1}{8}$	.125	.120	.116	.25	.06	.141
$\frac{9}{64}$	.141	.134	.130	.28	.06	.156
$\frac{5}{32}$	.156	.150	.146	.31	.07	.172
$\frac{3}{16}$	.188	.176	.172	.38	.09	.203
$\frac{7}{32}$	.219	.207	.202	.44	.10	.234
$\frac{1}{4}$	.250	.225	.220	.50	.11	.266
$\frac{5}{16}$	.312	.280	.275	.62	.14	.312
$\frac{3}{8}$	.375	.335	.329	.75	.16	.375
$\frac{7}{16}$	.438	.406	.400	.88	.20	.438
$\frac{1}{2}$	.500	.473	.467	1.00	.23	.500
$\frac{5}{8}$	.625	.598	.590	1.25	.30	.625
$\frac{3}{4}$	.750	.723	.715	1.50	.36	.750

[a] ANSI/ASME B18.8.1–1994.

31 Metric Equivalents

Length	
U.S. to Metric	**Metric to U.S.**
1 inch = 2.540 centimeters 1 foot = .305 meter 1 yard = .914 meter 1 mile = 1.609 kilometers	1 millimeter = .039 inch 1 centimeter = .394 inch 1 meter = 3.281 feet or 1.094 yards 1 kilometer = .621 mile
Area	
$1\ inch^2$ = 6.451 $centimeter^2$ $1\ foot^2$ = .093 $meter^2$ $1\ yard^2$ = .836 $meter^2$ $1\ acre^2$ = 4,046.873 $meter^2$	$1\ millimeter^2$ = .00155 $inch^2$ $1\ centimeter^2$ = .155 $inch^2$ $1\ meter^2$ = 10.764 $foot^2$ or 1.196 $yard^2$ $1\ kilometer^2$ = .386 $mile^2$ or 247.04 $acre^2$
Volume	
$1\ inch^3$ = 16.387 $centimeter^3$ $1\ foot^3$ = .028 $meter^3$ $1\ yard^3$ = .764 $meter^3$ 1 quart = 0.946 liter 1 gallon = .003785 $meter^3$	$1\ centimeter^3$ = .061 $inch^3$ $1\ meter^3$ = 35.314 $foot^3$ or 1.308 $yard^3$ 1 liter = .2642 gallons 1 liter = 1.057 quarts $1\ meter^3$ = 264.02 gallons
Weight	
1 ounce = 28.349 grams 1 pound = .454 kilogram 1 ton = .907 metric ton	1 gram = .035 ounce 1 kilogram = 2.205 pounds 1 metric ton = 1.102 tons
Velocity	
1 foot/second = .305 meter/second 1 mile/hour = .447 meter/second	1 meter/second = 3.281 feet/second 1 kilometer/hour = .621 mile/second
Acceleration	
$1\ inch/second^2$ = .0254 $meter/second^2$ $1\ foot/second^2$ = .305 $meter/second^2$	$1\ meter/second^2$ = 3.278 $feet/second^2$
Force	
N (newton) = basic unit of force, $kg\text{-}m/s^2$. A mass of one kilogram (1 kg) exerts a gravitational force of 9.8 N (theoretically 9.80665 N) at mean sea level.	

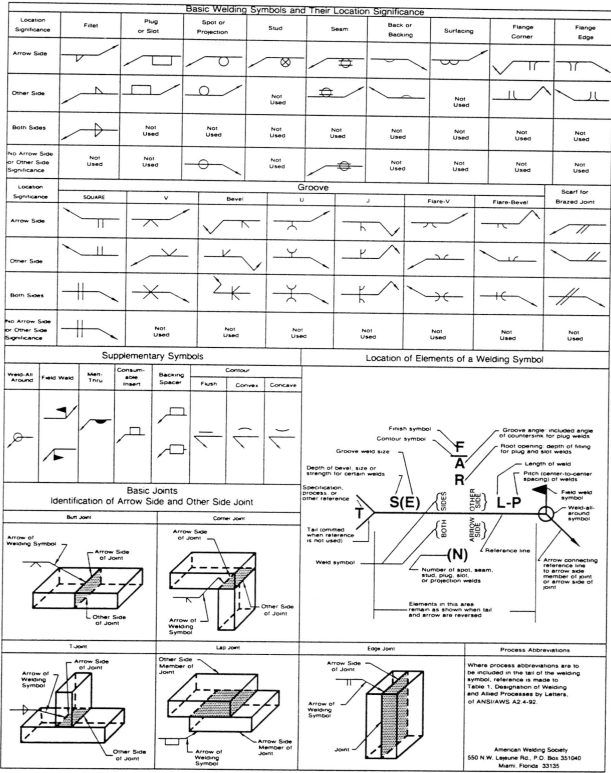

Basic Welding Symbols and Their Location Significance

Groove

Supplementary Symbols

Location of Elements of a Welding Symbol

Basic Joints
Identification of Arrow Side and Other Side Joint

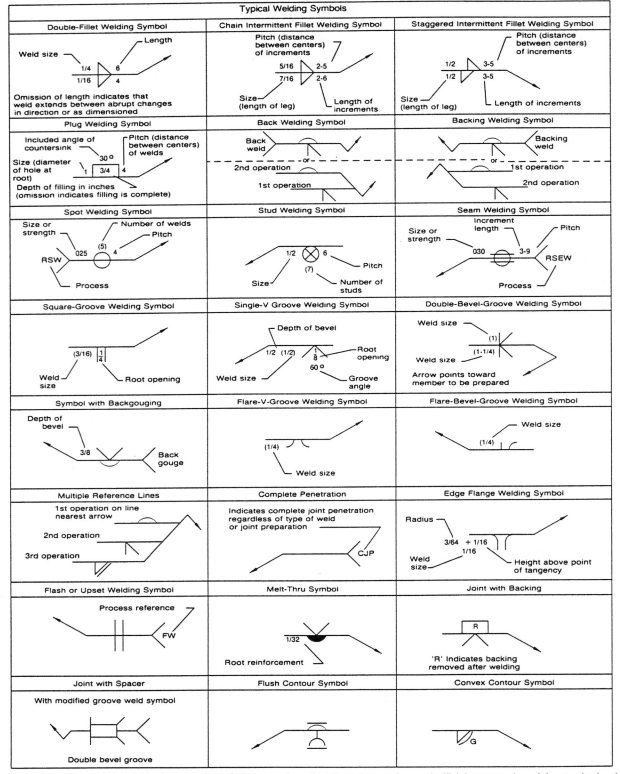

Typical Welding Symbols

* It should be understood that these charts are intended only as shop akis. The only complete and official presentation of the standard welding symbols is in A2.4.

32 Welding Symbols and Processes—
American Welding Society Standard (continued)

MASTER CHART OF WELDING AND ALLIED PROCESSES

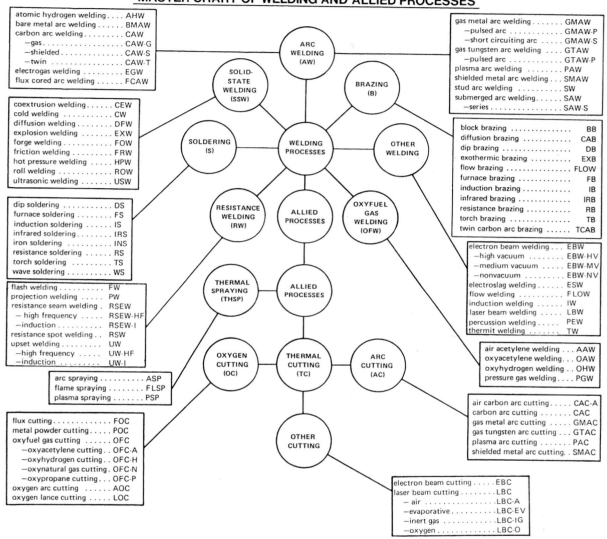

atomic hydrogen welding.... AHW
bare metal arc welding...... BMAW
carbon arc welding......... CAW
 —gas................... CAW-G
 —shielded.............. CAW-S
 —twin .... CAW-T
electrogas welding .... EGW
flux cored arc welding ...... FCAW

coextrusion welding...... CEW
cold welding .... CW
diffusion welding .... DFW
explosion welding ...... EXW
forge welding .... FOW
friction welding .... FRW
hot pressure welding ..... HPW
roll welding .... ROW
ultrasonic welding .... USW

dip soldering .... DS
furnace soldering .... FS
induction soldering ...... IS
infrared soldering........ IRS
iron soldering .... INS
resistance soldering ...... RS
torch soldering .... TS
wave soldering .... WS

flash welding .... FW
projection welding ..... PW
resistance seam welding . RSEW
 — high frequency ..... RSEW-HF
 —induction .... RSEW-I
resistance spot welding .. RSW
upset welding .... UW
 —high frequency ..... UW-HF
 —induction .... UW-I

arc spraying .... ASP
flame spraying .... FLSP
plasma spraying .... PSP

flux cutting .... FOC
metal powder cutting..... POC
oxyfuel gas cutting ...... OFC
 —oxyacetylene cutting.. OFC-A
 —oxyhydrogen cutting.. OFC-H
 —oxynatural gas cutting. OFC-N
 —oxypropane cutting... OFC-P
oxygen arc cutting ...... AOC
oxygen lance cutting ..... LOC

gas metal arc welding ...... GMAW
 —pulsed arc .... GMAW-P
 —short circuiting arc ..... GMAW-S
gas tungsten arc welding .... GTAW
 —pulsed arc .... GTAW-P
plasma arc welding .... PAW
shielded metal arc welding ... SMAW
stud arc welding .... SW
submerged arc welding...... SAW
 —series .... SAW-S

block brazing .... BB
diffusion brazing .... CAB
dip brazing .... DB
exothermic brazing .... EXB
flow brazing .... FLOW
furnace brazing .... FB
induction brazing .... IB
infrared brazing .... IRB
resistance brazing .... RB
torch brazing .... TB
twin carbon arc brazing ...... TCAB

electron beam welding ... EBW
 —high vacuum .... EBW-HV
 —medium vacuum ..... EBW-MV
 —nonvacuum .... EBW-NV
electroslag welding ...... ESW
flow welding .... FLOW
induction welding ...... IW
laser beam welding ..... LBW
percussion welding ..... PEW
thermit welding .... TW

air acetylene welding ... AAW
oxyacetylene welding... OAW
oxyhydrogen welding .. OHW
pressure gas welding .... PGW

air carbon arc cutting..... CAC-A
carbon arc cutting .... CAC
gas metal arc cutting ..... GMAC
gas tungsten arc cutting ... GTAC
plasma arc cutting .... PAC
shielded metal arc cutting.. SMAC

electron beam cutting ..... EBC
laser beam cutting .... LBC
 — air .... LBC-A
 —evaporative .... LBC-EV
 —inert gas .... LBC-IG
 —oxygen .... LBC-O

Circles (center chart): ARC WELDING (AW), SOLID-STATE WELDING (SSW), BRAZING (B), SOLDERING (S), WELDING PROCESSES, OTHER WELDING, RESISTANCE WELDING (RW), ALLIED PROCESSES, OXYFUEL GAS WELDING (OFW), THERMAL SPRAYING (THSP), ALLIED PROCESSES, OXYGEN CUTTING (OC), THERMAL CUTTING (TC), ARC CUTTING (AC), OTHER CUTTING

[a] ANSI/AWS A3.0–94.

33 Topographic Symbols

Symbol	Name	Symbol	Name
══════	Highway	── ─ ─ ──	National or State Line
++++++++	Railroad	── ─ ──	County Line
Highway Bridge		── ── ──	Township or District Line
Railroad Bridge		─ ─ ─ ─ ─	City or Village Line
Drawbridges		△	Triangulation Statio
Suspension Bridge		BM ✕ 1232	Bench Mark and Elevation
Dam		⊙	Any Location Station (WITH EXPLANATORY NOTE)
T T T T T T	Telegraph or Telephone Line		Streams in General
▪──▪──▪	Power-Transmission Line		Lake or Pond
	Buildings in General		Falls and Rapids
◉	Capital		Contours
◉	County Seat		Hachures
○	Other Towns		Sand and Sand Dunes
×-×-×-×-×	Barbed Wire Fence		Marsh
o-o-o-o-o	Smooth Wire Fence		Woodland of Any Kind
	Hedge		Orchard
○ ° ° ○	Oil or Gas Wells		Grassland in General
	Windmill		Cultivated Fields
	Tanks	☼	Commercial or Municipal Field
═══════	Canal or Ditch		Airplane Landing Field Marked or Emergency
──≪──	Canal Lock		Mooring Mast
──←──	Canal Lock (POINT UPSTREAM)	☆	Airway Light Beacon (arrows indicate course lights)
╌╌╌╌ ─ ─ ─ ─	Acqueduct or Water Pipe	★	Auxiliary Airway Light Beacon, Flashing

34 Piping Symbols—American National Standard

	FLANGED	SCREWED	BELL & SPIGOT	WELDED	SOLDERED
1. Joint					
2. Elbow—90°					
3. Elbow—45°					
4. Elbow—Turned Up					
5. Elbow—Turned Down					
6. Elbow—Long Radius					
7. Reducing Elbow					
8. Tee					
9. Tee—Outlet Up					
10. Tee—Outlet Down					
11. Side Outlet Tee—Outlet Up					
12. Cross					
13. Reducer—Concentric					
14. Reducer—Eccentric					
15. Lateral					
16. Gate Valve—Elev.					
17. Globe Valve—Elev.					
18. Check Valve					
19. Stop Cock					
20. Safety Valve					
21. Expansion Joint					
22. Union					
23. Sleeve					
24. Bushing					

[a] ANSI/ASME Y32.2.3–1949 (R1994).

35 Heating, Ventilating, and Ductwork Symbols[a]— American National Standard

High Pressure Steam	Soil, Waste or Leader (Above Grade)
Medium Pressure Return	Cold Water
Fuel Oil Flow —FOF—	Hot Water
Compressed Air —A—	Hot Water Return
Refrigerant Discharge —RD—	Fire Line —F———F—
Refrigerant Suction —RS—	Gas —G———G—
Brine Supply —B—	Sprinklers—Main Supplies —S—

Wall Radiator, Plan	Volume Damper *Elev.*
Wall Radiator on Ceiling, Plan	
Unit Heater (Propeller), Plan	Deflecting Damper
Unit Heater (Centrifugal Fan), Plan	
Thermostatic Trap	Turning Vanes
Thermostatic Float	
Thermometer	
Thermostat (T)	Automatic Dampers M
Duct Plan (1st Figure, Width; 2nd Depth) 20X12	
Inclined Drop in Respect to Air Flow D	
Supply Duct Section S 12X20	Canvas Connections
Exhaust Duct Section E 12X20	
Recirculation Duct Section R 12X20	
Fresh Air Duct Section A/F 12X20	Fan and Motor with Belt Guard
Supply Outlet	
Exhaust Inlet	
Volume Damper *Plan*	Intake Louvres and Screen

[a] ANSI/ASME Y32.2.3–1949 (R1994) and ANSI Y32.2.4–1949 (R1993).

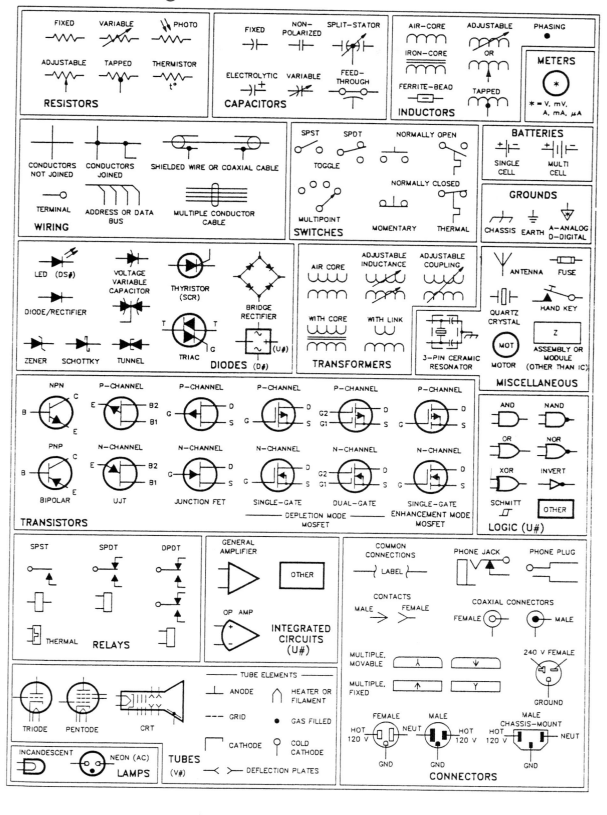

37 Form and Proportion of Geometric Tolerancing Symbols[a]

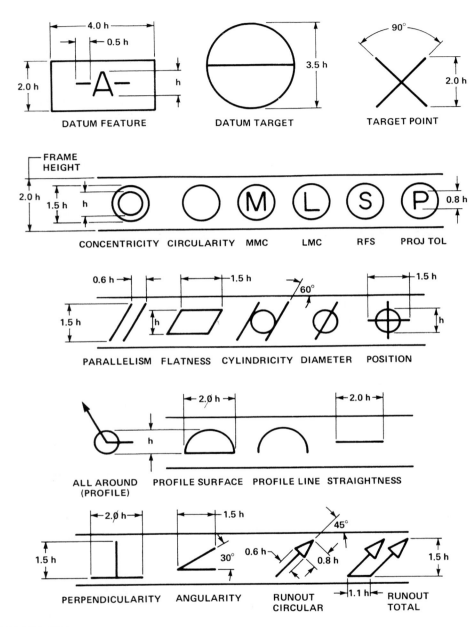

[a] ANSI/ASME Y14.5M–1994.

All dimensions are in inches except those in last two columns.

Nominal Pipe Size	D Outside Diameter of Pipe	Threads per Inch	L1[c] Normal Engagement by Hand Between External and Internal Threads	L2[c] Length of Effective Thread	Nominal Wall Thickness										Length of Pipe, Feet, per Square Foot External Surface[f]	Length of Standard Weight Pipe, Feet, Containing 1 cu. ft.[f]
					Sched. 10	Sched. 20[d]	Sched. 30[d]	Sched. 40[d]	Sched. 60[d]	Sched. 80[d]	Sched. 100	Sched. 120	Sched. 140	Sched. 160		
1/8	.405	27	.1615	.2639	…	…	…	.068	…	.095	…	…	…	…	9.431	2,533.8
1/4	.540	18	.2278	.4018	…	…	…	.088	…	.119	…	…	…	…	7.073	1,383.8
3/8	.675	18	.240	.4078	…	…	…	.091	…	.126	…	…	…	…	5.658	754.36
1/2	.840	14	.320	.5337	…	…	…	.109	…	.147	…	…	…	.188	4.547	473.91
3/4	1.050	14	.339	.5457	…	…	…	.113	…	.154	…	…	…	.219	3.637	270.03
1	1.315	11.5	.400	.6828	…	…	…	.133	…	.179	…	…	…	.250	2.904	166.62
1 1/4	1.660	11.5	.420	.7068	…	…	…	.140	…	.191	…	…	…	.250	2.301	96.275
1 1/2	1.900	11.5	.420	.7235	…	…	…	.145	…	.200	…	…	…	.281	2.010	70.733
2	2.375	11.5	.436	.7565	…	…	…	.154	…	.218	…	…	…	.344	1.608	42.913
2 1/2	2.875	8	.682	1.1375	…	…	…	.203	…	.276	…	…	…	.375	1.328	30.077
3	3.500	8	.766	1.2000	…	…	…	.216	…	.300	…	…	…	.438	1.091	19.479
3 1/2	4.000	8	.821	1.2500	…	…	…	.226	…	.318	…	…	…	…	.954	14.565
4	4.500	8	.844	1.3000	…	…	…	.237	…	.337	…	.438	…	.531	.848	11.312
5	5.563	8	.937	1.4063	…	…	…	.258	…	.375	…	.500	…	.625	.686	7.199
6	6.625	8	.958	1.5125	…	…	…	.280	…	.432	…	.562	…	.719	.576	4.984
8	8.625	8	1.063	1.7125	…	.250	.277	.322	.406	.500	.594	.719	.812	.906	.443	2.878
10	10.750	8	1.210	1.9250	…	.250	.307	.365	.500	.594	.719	.844	1.000	1.125	.355	1.826
12	12.750	8	1.360	2.1250	…	.250	.330	.406	.562	.688	.844	1.000	1.125	1.312	.299	1.273
14 OD	14.000	8	1.562	2.2500	.250	.312	.375	.438	.594	.750	.938	1.094	1.250	1.406	.273	1.065
16 OD	16.000	8	1.812	2.4500	.250	.312	.375	.500	.656	.844	1.031	1.219	1.438	1.594	.239	.815
18 OD	18.000	8	2.000	2.6500	.250	.312	.438	.562	.750	.938	1.156	1.375	1.562	1.781	.212	.644
20 OD	20.000	8	2.125	2.8500	.250	.375	.500	.594	.812	1.031	1.281	1.500	1.750	1.969	.191	.518
24 OD	24.000	8	2.375	3.2500	.250	.375	.562	.688	.969	1.219	1.531	1.812	2.062	2.344	.159	.358

[a] ANSI/ASME B36.10M–1995.
[b] ANSI/ASME B1.20.1–1983 (R1992).
[c] Refer to §11.18 and Fig. 11.18
[d] Boldface figures correspond to "standard" pipe.
[e] Boldface figures correspond to "extra strong" pipe
[f] Calculated values for Schedule 40 pipe.

Size, inches	Thickness, inches	Outside Diameter, inches	16 ft Laying Length Avg. per Foot[b] Weight (lb)	16 ft Laying Length Per Length Based on
Class 50: 50 psi Pressure—115 ft Head				
3	.32	3.96	12.4	195
4	.35	4.80	16.5	265
6	.38	6.90	25.9	415
8	.41	9.05	37.0	590
10	.44	11.10	49.1	785
12	.48	13.20	63.7	1,020
14	.48	15.30	74.6	1,195
16	.54	17.40	95.2	1,525
18	.54	19.50	107.6	1,720
20	.57	21.60	125.9	2,015
24	.63	25.80	166.0	2,655
30	.79	32.00	257.6	4,120
36	.87	38.30	340.9	5,455
42	.97	44.50	442.0	7,070
48	1.06	50.80	551.6	8,825
Class 100: 100 psi Pressure—231 ft Head				
3	.32	3.96	12.4	195
4	.35	4.80	16.5	265
6	.38	6.90	25.9	415
8	.41	9.05	37.0	590
10	.44	11.10	49.1	785
12	.48	13.20	63.7	1,020
14	.51	15.30	78.8	1,260
16	.54	17.40	95.2	1,525
18	.58	19.50	114.8	1,835
20	.62	21.60	135.9	2,175
24	.68	25.80	178.1	2,850
30	.79	32.00	257.6	4,120
36	.87	38.30	340.9	5,455
42	.97	44.50	442.0	7,070
48	1.06	50.80	551.6	8,825
Class 150: 150 psi Pressure—346 ft Head				
3	.32	3.96	12.4	195
4	.35	4.80	16.5	265
6	.38	6.90	25.9	415
8	.41	9.05	37.0	590
10	.44	11.10	49.1	785
12	.48	13.20	63.7	1,020
14	.51	15.30	78.8	1,260
16	.54	17.40	95.2	1,525
18	.58	19.50	114.8	1,835
20	.62	21.60	135.9	2,175
24	.73	25.80	190.1	3,040
30	.85	32.00	275.4	4,405
36	.94	38.30	365.9	5,855
42	1.05	44.50	475.3	7,605
48	1.14	50.80	589.6	9,435
Class 200: 200 psi Pressure—462 ft Head				
3	.32	3.96	12.4	195
4	.35	4.80	16.5	265
6	.38	6.90	25.9	415

Size, inches	Thickness, inches	Outside Diameter, inches	16 ft Laying Length Avg. per Foot[b] Weight (lb)	16 ft Laying Length Per Length Based on
Class 200: 200 psi Pressure—462 ft Head				
8	.41	9.05	37.0	590
10	.44	11.10	49.1	785
12	.48	13.20	63.7	1,020
14	.55	15.30	84.4	1,350
16	.58	17.40	101.6	1,625
18	.63	19.50	123.7	1,980
20	.67	21.60	145.9	2,335
24	.79	25.80	205.6	3,290
30	.92	32.00	297.8	4,765
36	1.02	38.30	397.1	6,355
42	1.13	44.50	512.3	8,195
48	1.23	50.80	637.2	10,195
Class 250: 250 psi Pressure—577 ft Head				
3	.32	3.96	12.4	195
4	.35	4.80	16.5	265
6	.38	6.90	25.9	415
8	.41	9.05	37.0	590
10	.44	11.10	49.1	785
12	.52	13.20	68.5	1,095
14	.59	15.30	90.6	1,450
16	.63	17.40	110.4	1,765
18	.68	19.50	133.4	2,135
20	.72	21.60	156.7	2,505
24	.79	25.80	205.6	3,290
30	.99	32.00	318.4	5,095
36	1.10	38.30	425.5	6,810
42	1.22	44.50	549.5	8,790
48	1.33	50.80	684.5	10,950
Class 300: 300 psi Pressure—693 ft Head				
3	.32	3.96	12.4	195
4	.35	4.80	16.5	265
6	.38	6.90	25.9	415
8	.41	9.05	37.0	590
10	.48	11.10	53.1	850
12	.52	13.20	68.5	1,095
14	.59	15.30	90.6	1,450
16	.68	17.40	118.2	1,890
18	.73	19.50	142.3	2,275
20	.78	21.60	168.5	2,695
24	.85	25.80	219.8	3,515
Class 350: 350 psi Pressure—808 ft Head				
3	.32	3.96	12.4	195
4	.35	4.80	16.5	265
6	.38	6.90	25.9	415
8	.41	9.05	37.0	590
10	.52	11.10	57.4	920
12	.56	13.20	73.8	1,180
14	.64	15.30	97.5	1,605
16	.68	17.40	118.2	1,945
18	.79	19.50	152.9	2,520
20	.84	21.60	180.2	2,970
24	.92	25.80	236.3	3,895

[a] Average weight per foot based on calculated weight of pipe before rounding.

40 Cast-Iron Pipe Screwed Fittings,[a] 125 lb—American National Standard

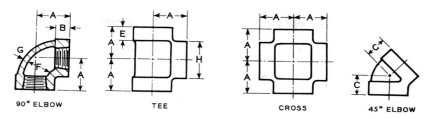

90° ELBOW   TEE   CROSS   45° ELBOW

DIMENSIONS OF 90° AND 45° ELBOWS, TEES, AND CROSSES (STRAIGHT SIZES)

All dimensions given in inches.

Fittings having right- and left-hand threads shall have four or more ribs or the letter "L" cast on the band at end with left-hand thread.

Nominal Pipe Size	Center to End, Elbows, Tees, and Crosses A	Center to End, 45° Elbows C	Length of Thread, Min. B	Width of Band. Min. E	Inside Diameter of Fitting F		Metal Thickness G	Diameter of Band, Min. H
					Max.	Min.		
$\frac{1}{4}$	.81	.73	.32	.38	.58	.54	.11	.93
$\frac{3}{8}$	.95	.80	.36	.44	.72	.67	.12	1.12
$\frac{1}{2}$	1.12	.88	.43	.50	.90	.84	.13	1.34
$\frac{3}{4}$	1.31	.98	.50	.56	1.11	1.05	.15	1.63
1	1.50	1.12	.58	.62	1.38	1.31	.17	1.95
$1\frac{1}{4}$	1.75	1.29	.67	.69	1.73	1.66	.18	2.39
$1\frac{1}{2}$	1.94	1.43	.70	.75	1.97	1.90	.20	2.68
2	2.25	1.68	.75	.84	2.44	2.37	.22	3.28
$2\frac{1}{2}$	2.70	1.95	.92	.94	2.97	2.87	.24	3.86
3	3.08	2.17	.98	1.00	3.60	3.50	.26	4.62
$3\frac{1}{2}$	3.42	2.39	1.03	1.06	4.10	4.00	.28	5.20
4	3.79	2.61	1.08	1.12	4.60	4.50	.31	5.79
5	4.50	3.05	1.18	1.18	5.66	5.56	.38	7.05
6	5.13	3.46	1.28	1.28	6.72	6.62	.43	8.28
8	6.56	4.28	1.47	1.47	8.72	8.62	.55	10.63
10	8.08[b]	5.16	1.68	1.68	10.85	10.75	.69	13.12
12	9.50[b]	5.97	1.88	1.88	12.85	12.75	.80	15.47

[a] From ANSI/ASME B16.4–1992.
[b] This applies to elbows and tees only.

41 Cast-Iron Pipe Screwed Fittings,ᵃ 250 lb—American National Standard

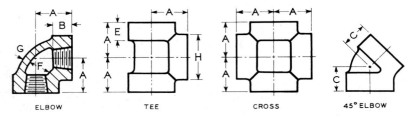

ELBOW   TEE   CROSS   45° ELBOW

DIMENSIONS OF 90° AND 45° ELBOWS, TEES, AND CROSSES (STRAIGHT SIZES)

All dimensions given in inches.

The 250-lb standard for screwed fittings covers only the straight sizes of 90° and 45° elbows, tees, and crosses.

Nominal Pipe Size	Center to End, Elbows, Tees, and Crosses A	Center to End, 45° Elbows C	Length of Thread, Min. B	Width of Band. Min. E	Inside Diameter of Fitting F		Metal Thickness G	Diameter of Band, Min. H
					Max.	Min.		
$\frac{1}{4}$	.94	.81	.43	.49	.58	.54	.18	1.17
$\frac{3}{8}$	1.06	.88	.47	.55	.72	.67	.18	1.36
$\frac{1}{2}$	1.25	1.00	.57	.60	.90	.84	.20	1.59
$\frac{3}{4}$	1.44	1.13	.64	.68	1.11	1.05	.23	1.88
1	1.63	1.31	.75	.76	1.38	1.31	.28	2.24
$1\frac{1}{4}$	1.94	1.50	.84	.88	1.73	1.66	.33	2.73
$1\frac{1}{2}$	2.13	1.69	.87	.97	1.97	1.90	.35	3.07
2	2.50	2.00	1.00	1.12	2.44	2.37	.39	3.74
$2\frac{1}{2}$	2.94	2.25	1.17	1.30	2.97	2.87	.43	4.60
3	3.38	2.50	1.23	1.40	3.60	3.50	.48	5.36
$3\frac{1}{2}$	3.75	2.63	1.28	1.49	4.10	4.00	.52	5.98
4	4.13	2.81	1.33	1.57	4.60	4.50	.56	6.61
5	4.88	3.19	1.43	1.74	5.66	5.56	.66	7.92
6	5.63	3.50	1.53	1.91	6.72	6.62	.74	9.24
8	7.00	4.31	1.72	2.24	8.72	8.62	.90	11.73
10	8.63	5.19	1.93	2.58	10.85	10.75	1.08	14.37
12	10.00	6.00	2.13	2.91	12.85	12.75	1.24	16.84

ᵃ From ANSI/ASME B16.4–1992.

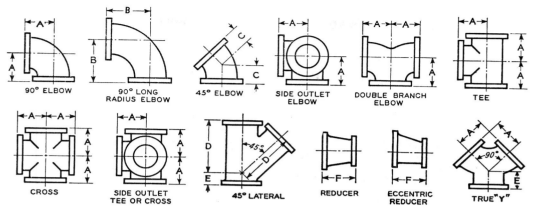

DIMENSIONS OF ELBOWS, DOUBLE BRANCH ELBOWS, TEES, CROSSES, LATERALS, TRUE Y'S (STRAIGHT SIZES), AND REDUCERS

All dimensions in inches.

Nominal Pipe Size	Inside Diameter of Fittings	Center to Face 90° Elbow, Tees, Crosses True "Y" and Double Branch Elbow A	Center to Face, 90° Long Radius Elbow B	Center to Face 45° Elbow C	Center to Face Lateral D	Short Center to Face True "Y" and Lateral E	Face to Face Reducer F	Diameter of Flange	Thickness of Flange, Min.	Wall Thickness
1	1.00	3.50	5.00	1.75	5.75	1.75	...	4.25	.44	.31
1¼	1.25	3.75	5.50	2.00	6.25	1.75	...	4.62	.50	.31
1½	1.50	4.00	6.00	2.25	7.00	2.00	...	5.00	.56	.31
2	2.00	4.50	6.50	2.50	8.00	2.50	5.0	6.00	.62	.31
2½	2.50	5.00	7.00	3.00	9.50	2.50	5.5	7.00	.69	.31
3	3.00	5.50	7.75	3.00	10.00	3.00	6.0	7.50	.75	.38
3½	3.50	6.00	8.50	3.50	11.50	3.00	6.5	8.50	.81	.44
4	4.00	6.50	9.00	4.00	12.00	3.00	7.0	9.00	.94	.50
5	5.00	7.50	10.25	4.50	13.50	3.50	8.0	10.00	.94	.50
6	6.00	8.00	11.50	5.00	14.50	3.50	9.0	11.00	1.00	.56
8	8.00	9.00	14.00	5.50	17.50	4.50	11.0	13.50	1.12	.62
10	10.00	11.00	16.50	6.50	20.50	5.00	12.0	16.00	1.19	.75
12	12.00	12.00	19.00	7.50	24.50	5.50	14.0	19.00	1.25	.81
14 OD	14.00	14.00	21.50	7.50	27.00	6.00	16.0	21.00	1.38	.88
16 OD	16.00	15.00	24.00	8.00	30.00	6.50	18.0	23.50	1.44	1.00
18 OD	18.00	16.50	26.50	8.50	32.00	7.00	19.0	25.00	1.56	1.06
20 OD	20.00	18.00	29.00	9.50	35.00	8.00	20.0	27.50	1.69	1.12
24 OD	24.00	22.00	34.00	11.00	40.50	9.00	24.0	32.00	1.88	1.25
30 OD	30.00	25.00	41.50	15.00	49.00	10.00	30.0	38.75	2.12	1.44
36 OD	36.00	28.00	49.00	18.00	...	...	36.0	46.00	2.38	1.62
42 OD	42.00	31.00	56.50	21.00	...	...	42.0	53.00	2.62	1.81
48 OD	48.00	34.00	64.00	24.00	...	...	48.0	59.50	2.75	2.00

[a] ANSI/ASME B16.1–1989.

43 Cast-Iron Pipe Flanges, Drilling for Bolts and Their Lengths,[a] 125 lb— American National Standard

Nominal Pipe Size	Diameter of Flange	Thickness of Flange, Min.	Diameter of Bolt Circle	Number of Bolts	Diameter of Bolts	Diameter of Bolt Holes	Length of Bolts
1	4.25	.44	3.12	4	.50	.62	1.75
$1\frac{1}{4}$	4.62	.50	3.50	4	.50	.62	2.00
$1\frac{1}{2}$	5.00	.56	3.88	4	.50	.62	2.00
2	6.00	.62	4.75	4	.62	.75	2.25
$2\frac{1}{2}$	7.00	.69	5.50	4	.62	.75	2.50
3	7.50	.75	6.00	4	.62	.75	2.50
$3\frac{1}{2}$	8.50	.81	7.00	8	.62	.75	2.75
4	9.00	.94	7.50	8	.62	.75	3.00
5	10.00	.94	8.50	8	.75	.88	3.00
6	11.00	1.00	9.50	8	.75	.88	3.25
8	13.50	1.12	11.75	8	.75	.88	3.50
10	16.00	1.19	14.25	12	.88	1.00	3.75
12	19.00	1.25	17.00	12	.88	1.00	3.75
14 OD	21.00	1.38	18.75	12	1.00	1.12	4.25
16 OD	23.50	1.44	21.25	16	1.00	1.12	4.50
18 OD	25.00	1.56	22.75	16	1.12	1.25	4.75
20 OD	27.50	1.69	25.00	20	1.12	1.25	5.00
24 OD	32.00	1.88	29.50	20	1.25	1.38	5.50
30 OD	38.75	2.12	36.00	28	1.25	1.38	6.25
36 OD	46.00	2.38	42.75	32	1.50	1.62	7.00
42 OD	53.00	2.62	49.50	36	1.50	1.62	7.50
48 OD	59.50	2.75	56.00	44	1.50	1.62	7.75

[a] ANSI B16.1–1989.

44 Shaft Center Sizes

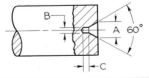

Shaft Diameter D	A	B	C	Shaft Diameter D	A	B	C
$\frac{3}{16}$ to $\frac{7}{32}$	$\frac{5}{64}$	$\frac{3}{64}$	$\frac{1}{16}$	$1\frac{1}{8}$ to $1\frac{15}{32}$	$\frac{5}{16}$	$\frac{5}{32}$	$\frac{5}{32}$
$\frac{1}{4}$ to $\frac{11}{32}$	$\frac{3}{32}$	$\frac{3}{64}$	$\frac{1}{16}$	$1\frac{1}{2}$ to $1\frac{31}{32}$	$\frac{3}{8}$	$\frac{3}{32}$	$\frac{5}{32}$
$\frac{3}{8}$ to $\frac{17}{32}$	$\frac{1}{8}$	$\frac{1}{16}$	$\frac{5}{64}$	2 to $2\frac{31}{32}$	$\frac{7}{16}$	$\frac{7}{32}$	$\frac{3}{16}$
$\frac{9}{16}$ to $\frac{25}{32}$	$\frac{3}{16}$	$\frac{5}{64}$	$\frac{3}{32}$	3 to $3\frac{31}{32}$	$\frac{1}{2}$	$\frac{7}{32}$	$\frac{7}{32}$
$\frac{13}{16}$ to $1\frac{3}{32}$	$\frac{1}{4}$	$\frac{3}{32}$	$\frac{3}{32}$	4 and over	$\frac{9}{16}$	$\frac{7}{32}$	$\frac{7}{32}$

45 Cast-Iron Pipe Flanges and Fittings,[a] 250 lb— American National Standard

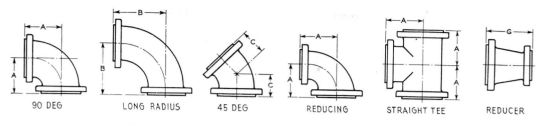

90 DEG   LONG RADIUS   45 DEG   REDUCING   STRAIGHT TEE   REDUCER

DIMENSIONS OF ELBOWS, TEES, AND REDUCERS
All dimensions are given in inches.

Nominal Pipe Size	Inside Diameter of Fitting, Min.	Wall Thickness of Body	Diameter of Flange	Thickness of Flange Min.	Diameter of Raised Face	Center-to-Face Elbow and Tee A	Center-to-Face Long Radius Elbow B	Center-to-Face 45° Elbow C	Face-to-Face Reducer G
1	1.00	.44	4.88	.69	2.69	4.00	5.00	2.00	...
$1\frac{1}{4}$	1.25	.44	5.25	.75	3.06	4.25	5.50	2.50	...
$1\frac{1}{2}$	1.50	.44	6.12	.81	3.56	4.50	6.00	2.75	...
2	2.00	.44	6.50	.88	4.19	5.00	6.50	3.00	5.00
$2\frac{1}{2}$	2.50	.50	7.50	1.00	4.94	5.50	7.00	3.50	5.50
3	3.00	.56	8.25	1.12	5.69	6.00	7.75	3.50	6.00
$3\frac{1}{2}$	3.50	.56	9.00	1.19	6.31	6.50	8.50	4.00	6.50
4	4.00	.62	10.00	1.25	6.94	7.00	9.00	4.50	7.00
5	5.00	.69	11.00	1.38	8.31	8.00	10.25	5.00	8.00
6	6.00	.75	12.50	1.44	9.69	8.50	11.50	5.50	9.00
8	8.00	.81	15.00	1.62	11.94	10.00	14.00	6.00	11.00
10	10.00	.94	17.50	1.88	14.06	11.50	16.50	7.00	12.00
12	12.00	1.00	20.50	2.00	16.44	13.00	19.00	8.00	14.00
14 OD	13.25	1.12	23.00	2.12	18.94	15.00	21.50	8.50	16.00
16 OD	15.25	1.25	25.50	2.25	21.06	16.50	24.00	9.50	18.00
18 OD	17.00	1.38	28.00	2.38	23.31	18.00	26.50	10.00	19.00
20 OD	19.00	1.50	30.50	2.50	25.56	19.50	29.00	10.50	20.00
24 OD	23.00	1.62	36.00	2.75	30.31	22.50	34.00	12.00	24.00
30 OD	29.00	2.00	43.00	3.00	37.19	27.50	41.50	15.00	30.00

[a] ANSI B16.1–1989.

46   Cast-Iron Pipe Flanges, Drilling for Bolts and Their Lengths,[a] 250 lb— American National Standard

Nominal Pipe Size	Diameter of Flange	Thickness of Flange, Min.	Diameter of Raised Face	Diameter of Bolt Circle	Diameter of Bolt Holes	Number of Bolts	Size of Bolts	Length of Bolts	Length of Bolt Studs with Two Nuts
1	4.88	.69	2.69	3.50	.75	4	.62	2.50	...
$1\frac{1}{4}$	5.25	.75	3.06	3.88	.75	4	.62	2.50	...
$1\frac{1}{2}$	6.12	.81	3.56	4.50	.88	4	.75	2.75	...
2	6.50	.88	4.19	5.00	.75	8	.62	2.75	...
$2\frac{1}{2}$	7.50	1.00	4.94	5.88	.88	8	.75	3.25	...
3	8.25	1.12	6.69	6.62	.88	8	.75	3.50	...
$3\frac{1}{2}$	9.00	1.19	6.31	7.25	.88	8	.75	3.50	...
4	10.00	1.25	6.94	7.88	.88	8	.75	3.75	...
5	11.00	1.38	8.31	9.25	.88	8	.75	4.00	...
6	12.50	1.44	9.69	10.62	.88	12	.75	4.00	...
8	15.00	1.62	11.94	13.00	1.00	12	.88	4.50	...
10	17.50	1.88	14.06	5.25	1.12	16	1.00	5.25	...
12	20.50	2.00	16.44	17.75	1.25	16	1.12	5.50	...
14 OD	23.00	2.12	18.94	20.25	1.25	20	1.12	6.00	...
16 OD	25.50	2.25	21.06	22.50	1.38	20	1.25	6.25	...
18 OD	28.00	2.38	23.31	24.75	1.38	24	1.25	6.50	...
20 OD	30.50	2.50	25.56	27.00	1.38	24	1.25	6.75	...
24 OD	36.00	2.75	30.31	32.00	1.62	24	1.50	7.50	9.50
30 OD	43.00	3.00	37.19	39.25	2.00	28	1.75	8.50	10.50

[a] ANSI B16.1–1989.

YOU SHOULD CAREFULLY READ THE FOLLOWING TERMS AND CONDITIONS BEFORE OPENING THIS CD-ROM PACKAGE. USING THIS CD-ROM PACKAGE INDICATES YOUR ACCEPTANCE OF THESE TERMS AND CONDITIONS. IF YOU DO NOT AGREE WITH THEM, YOU SHOULD PROMPTLY RETURN THE PACKAGE UNOPENED.

Prentice-Hall, Inc. provides this program and licenses its use. You assume responsibility for the selection of the program to achieve your intended results, and for the installation, use, and results obtained from the program. This license extends only to use of the program in the United States or countries in which the program is marketed by duly authorized distributors.

LICENSE

You may:

a.   use the program;
b.   copy the program into any machine-readable form without limit;
c.   modify the program and/or merge it into another program in support of your use of the program.

LIMITED WARRANTY

THE PROGRAM IS PROVIDED "AS IS" WITHOUT WARRANTY OF ANY KIND, EITHER EXPRESSED OR IMPLIED, INCLUDING, BUT NOT LIMITED TO, THE IMPLIED WARRANTIES OF MERCHANTABILITY AND FITNESS FOR A PARTICULAR PURPOSE. THE ENTIRE RISK AS TO THE QUALITY AND PERFORMANCE OF THE PROGRAM IS WITH YOU. SHOULD THE PROGRAM PROVE DEFECTIVE, YOU (AND NOT PRENTICE-HALL, INC. OR ANY AUTHORIZED DISTRIBUTOR) ASSUME THE ENTIRE COST OF ALL NECESSARY SERVICING, REPAIR, OR CORRECTION.

SOME STATES DO NOT ALLOW FOR THE EXCLUSION OF IMPLIES WARRANTIES, SO THE ABOVE EXCLUSION MAY NOT APPLY TO YOU. THIS WARRANTY GIVES YOU SPECIFIC LEGAL RIGHTS AND YOU MAY ALSO HAVE OTHER RIGHTS THAT VARY FROM STATE TO STATE.

Prentice-Hall, Inc. does not warrant that the functions contained in the program will meet your requirements or that the operation of the program will be uninterrupted or error-free.

However, Prentice-Hall, Inc. warrants the CD-ROM(s) on which the program is furnished to be free from defects in material and workmanship under normal use for a period of ninety (90) days from the date of delivery to you as evidenced by a copy of your receipt.

LIMITATION OF REMEDIES

Prentice-Hall's entire liability and your exclusive remedy shall be:

1.   the replacement of any CD-ROM not meeting Prentice-Hall's "Limited Warranty" and that is returned to Prentice-Hall, or

2.   if Prentice-Hall is unable to deliver a replacement CD-ROM that is free of defects in materials or workmanship, you may terminate this Agreement by returning the program.

IN NO EVENT WILL PRENTICE-HALL BE LIABLE TO YOU FOR ANY DAMAGES, INCLUDING ANY LOST PROFITS, LOST SAVINGS, OR OTHER INCIDENTAL OR CONSEQUENTIAL DAMAGES ARISING OUT OF THE USE OR INABILITY TO USE SUCH PROGRAM EVEN IF PRENTICE-HALL, OR AN AUTHORIZED DISTRIBUTOR HAS BEEN ADVISED OF THE POSSIBILITY OF SUCH DAMAGES, OR FOR ANY CLAIM BY ANY OTHER PARTY.

SOME STATES DO NOT ALLOW THE LIMITATION OR EXCLUSION OF LIABILITY FOR INCIDENTAL OR CONSEQUENTIAL DAMAGES, SO THE ABOVE LIMITATION MAY NOT APPLY TO YOU.

GENERAL

You may not sublicense, assign, or transfer the license or the program except as expressly provided in this agreement. Any attempt otherwise to sublicense, assign or transfer any of the rights, duties, or obligations hereunder is void.

This Agreement will be governed by the laws of the State of New York.

Should you have any questions concerning this Agreement, you may contact Prentice-Hall, Inc. by writing to:

Prentice Hall
College Division
Upper Saddle River, NJ 07458

YOU ACKNOWLEDGE THAT YOU HAVE READ THIS AGREEMENT, UNDERSTAND IT, AND AGREE TO BE BOUND BY ITS TERMS AND CONDITIONS. YOU FURTHER AGREE THAT IT IS THE COMPLETE AND EXCLUSIVE STATEMENT OF THE AGREEMENT BETWEEN US THAT SUPERSEDES ANY PROPOSAL OR PRIOR AGREEMENT, ORAL OR WRITTEN, AND ANY OTHER COMMUNICATIONS BETWEEN US RELATING TO THE SUBJECT MATTER OF THIS AGREEMENT.

ISBN: 0-13-031724-1

Worksheet 2.1
CAD system evaluation checklist

Item	Y/N	Size/Type	Comments	Cost
Central Processor				
memory (MB)				
word size (16/32 bit)				
cache				
speed (MHz/Mips)				
bus type (eisa, vesa, pci)				
expansion/upgrade				
Operating System				
32 bit				
multitasking				
software availability				
Data Input Devices				
mouse				
trackball				
digitizer				
light pen				
thumb wheel				
Display				
monochrome				
color				
screen size				
resolution				
Video Card				
memory				
software support				
dual display support				
Storage, Hard Drive				
type				
access time				
capacity				
expansion				
removable				
Maintenance				
hardware				
software				

Make copies of these pages to use for additional practice.

Worksheet 2.1 cont.
CAD system evaluation checklist

Item		Y/N	Size/Type	Comments	Cost
Storage, Floppy					
	type				
	access time				
	capacity				
	removable				
CD-ROM					
	type				
	speed				
	capacity				
	read/write				
Backup System					
	type				
	capacity				
	speed				
	automation				
Output Devices					
	type				
	provided with system?				
	medium				
	cost per sheet				
	speed				
	resolution/accuracy				
	color				
Notes					

Make copies of these pages to use for additional practice.

Worksheet 3.1
Practice sketching freehand lines

Use the spaces at right to practice sketching freehand lines of each type shown below. The first construction line is done for you.

CONSTRUCTION LINE

VISIBLE LINE

HIDDEN LINE

—15—
DIMENSION LINE

EXTENSION LINE

CENTER LINE

PHANTOM LINE

CUTTING-PLANE LINES

Construction lines	Construction lines
Visible lines	Visible lines
Hidden lines	Hidden lines
Dimension lines	Dimension lines
Extension lines	Extension lines
Center lines	Center lines
Phantom lines	Phantom lines
Cutting plane lines	Cutting plane lines

Sketching inclined lines

For inclined lines, shift position with relation to the paper or turn the paper slightly. Use the same movements as for horizontal or vertical lines.

Complete the series of inclined lines below.

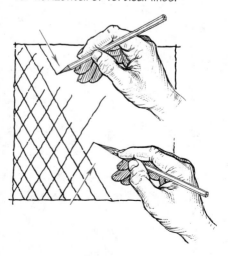

Make copies of these pages to use for additional practice.

Worksheet 3.2
Sketching circles and ellipses

Use the construction lines provided below to begin sketching circles and ellipses. Practice this technique and the others you have learned for creating circles and ellipses on your own unlined paper.

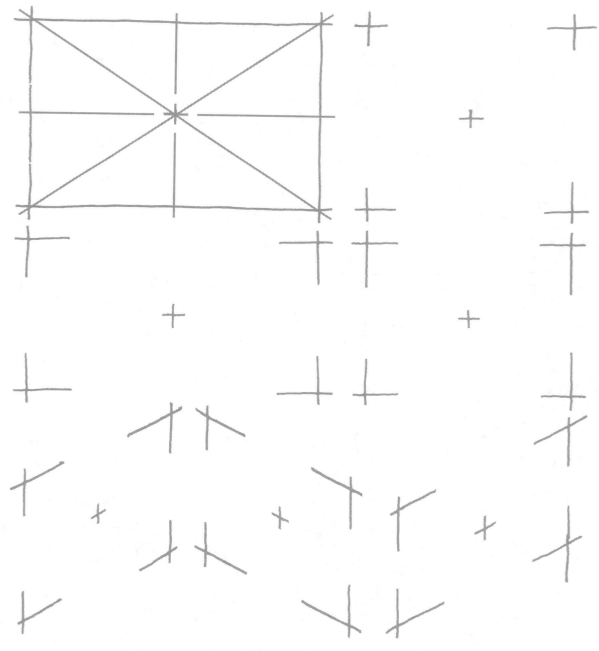

Make copies of these pages to use for additional practice.

Worksheet 3.3
Practice line and curve technique

Sketch the figures below on the grid provided.

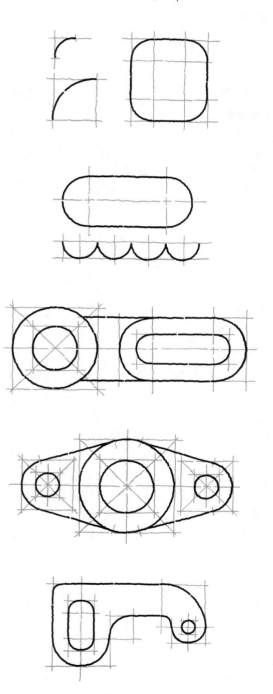

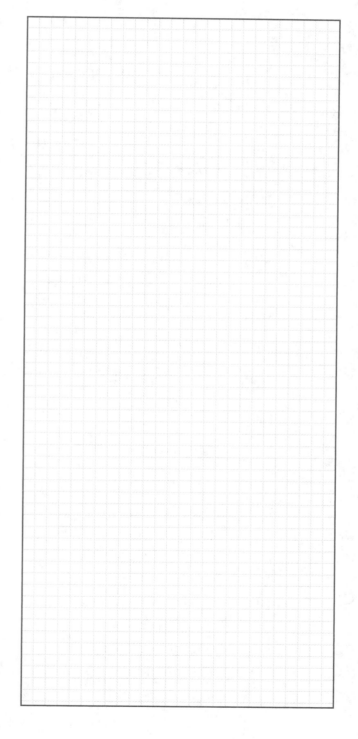

Make copies of these pages to use for additional practice.

Worksheet 3.4
Applying the squares method

Transfer the drawing of the car shown at right to the larger grid below using the squares method.

Match how the lines enter and exit each grid square.

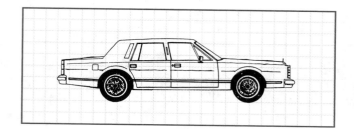

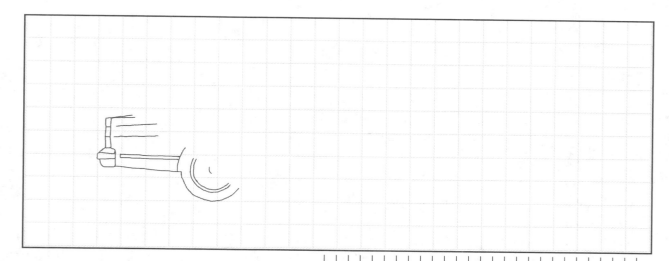

Cut a small picture out of a magazine or newspaper and fix it in the box at right.

Using the tickmarks as guides, draw a one-eighth inch grid over the picture

Use the same process you used above to transfer and enlarge the picture to the grid below.

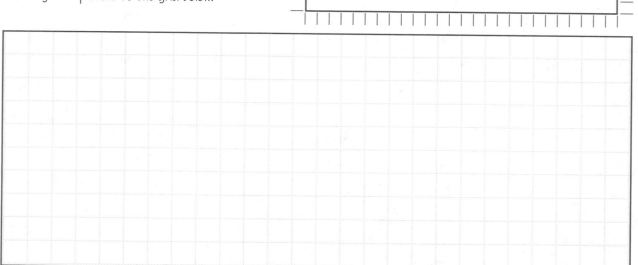

Make copies of these pages to use for additional practice.

Worksheet 3.5
Vertical lettering practice

Repeat each letter on the blue grid provided, paying careful attention to the proportions and stroke direction shown in the example. Then repeat the letter once more in the white space.

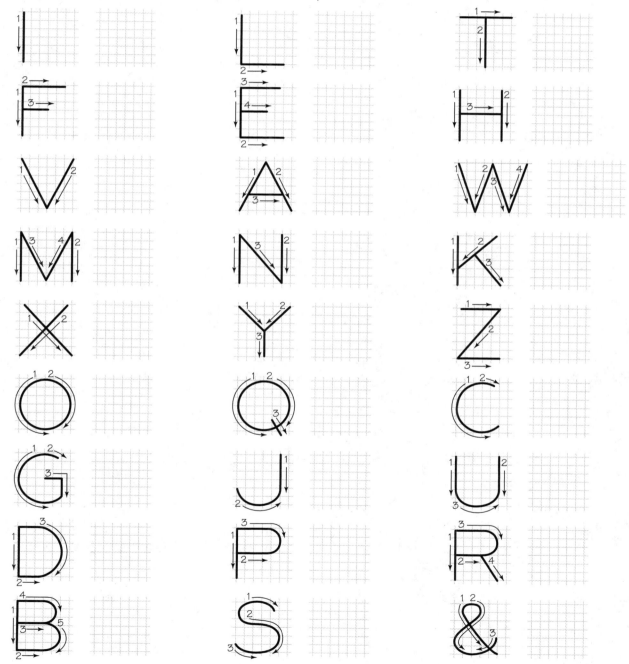

Make copies of these pages to use for additional practice.

Worksheet 3.6
1/8 Inch lettering practice

Use the guidelines provided to letter the drawing notes and
dimensions in the appropriate spaces. Use 1/8" uppercase
standard engineering lettering.

SEE DETAIL A	SAE 1115 - 4 REQUIRED
2 X .50 - 12 UNC - 2B ▼.688	ALL FILLETS AND ROUNDS R .125
.062 X .062 SLOT	3 HOLES EQUALLY SPACED
2 X ∅19.0 - 19.5	$\frac{1}{8}$ AMER. STD. PIPE TAP FOR BOWEN GREASE CUP
BOTTOM TAP	
∅.626 ⊔∅1.062	2 X #10 - 32 UNF - 2B ▼.438
MEDIUM KNURL	M14 X 2
F & R R3 UNLESS OTHERWISE SPECIFIED	
ALL MEASUREMENTS IN MILLIMETERS	JAW PLATE SCREW

Make copies of these pages to use for additional practice.

Worksheet 4.1
Sketching tangent arcs

Use sketching techniques to block in tangent arcs as shown in the example drawing. Remember a radial line (one through the center of the circle) is perpendicular to the line at the point of tangency.

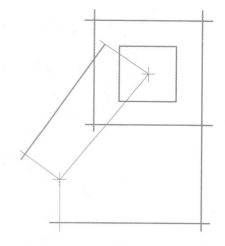

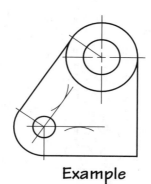

Example

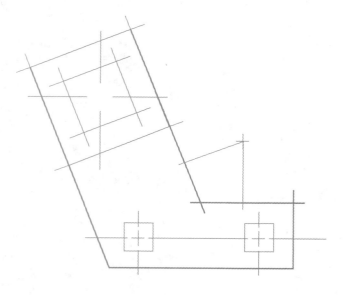

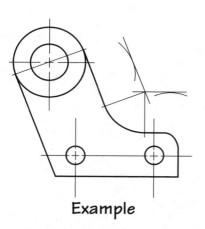

Example

Make copies of these pages to use for additional practice.

Worksheet 5.1
The glass box

Directions

The pattern at right shows the viewing planes for the six regular views unfolded. Cut on the solid lines and fold on the dashed fold lines to make a box. The front and rear viewing planes are labeled for you. Label the top, right-side, left-side, and bottom views.

1. Label the viewing planes with the principal dimensions that will show in each view. (For example, the front view will show the height and width of an object placed inside the box.)

2. Can you think of a different way to cut and fold the box so that the top and right-side views would align?

Rear

Front

Make copies of these pages to use for additional practice.

Worksheet 5.2
Transferring Depth Dimensions

Two views of a triangular plane are shown projected onto the glass box below. Vertex B has already been projected into the side view for you. Finish projecting vertices A & C and draw the side view of the triangular plane. Cut out and fold up the "glass box" to help you visualize how to transfer the vertices to the side view. Cut out a triangular piece of paper the true size of the triangle and orient it inside the "glass box."

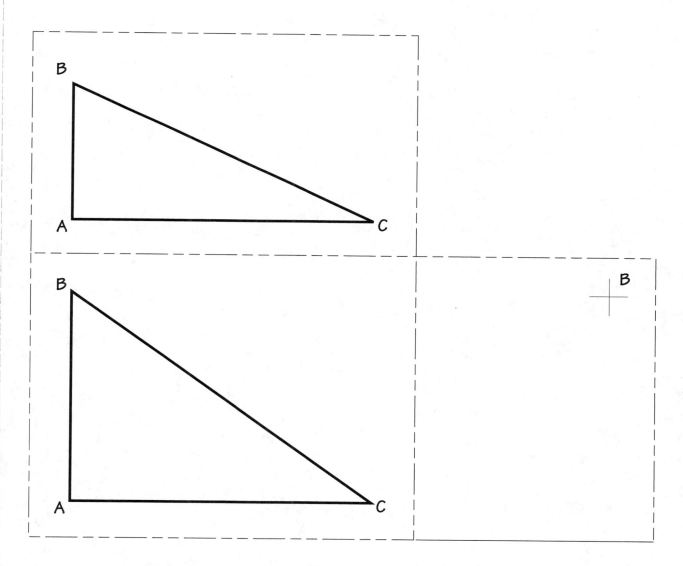

Make copies of these pages to use for additional practice.

Worksheet 5.3
Blocking a multiview drawing

Construction lines are provided below to help you create the orthographic views of the part shown. Show all features in each view. Darken the final drawing lines.

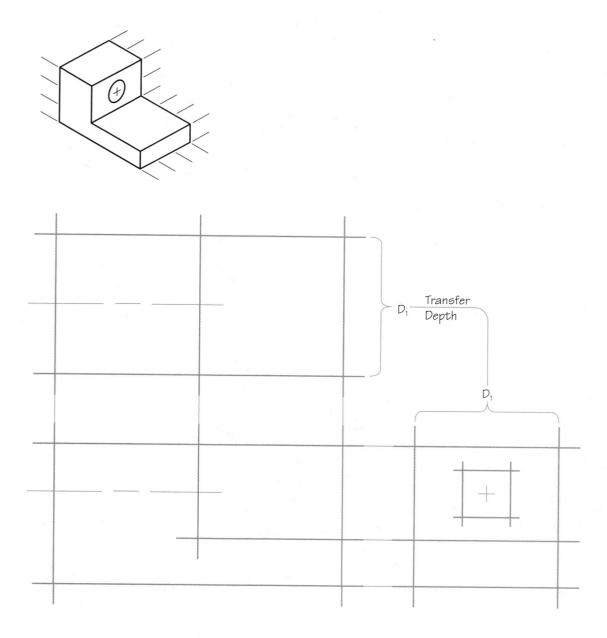

D$_1$ — Transfer Depth

D$_1$

Make copies of these pages to use for additional practice.

Worksheet 5.4
Projecting inclined surfaces

Construction lines have been drawn for you to help you sketch the orthographic views of the block shown. Surface B shows on edge in the top view. Surface A shows on edge in the front view. Finish projecting the points to the side view and complete the drawing.

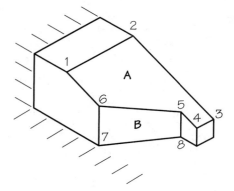

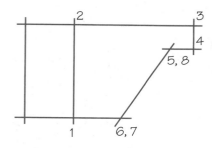

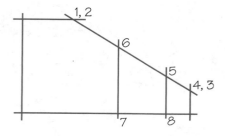

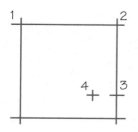

Make copies of these pages to use for additional practice.

Worksheet 5.5
Practice with hidden lines

Add hidden lines to the drawing of the part below. Use good line quality and proper hidden line technique.

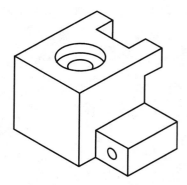

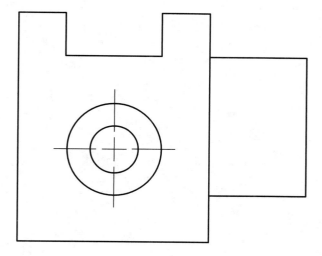

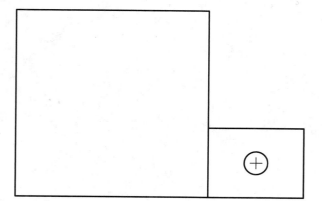

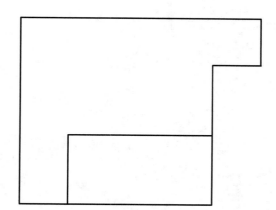

Make copies of these pages to use for additional practice.

Worksheet 5.6
Practice with centerlines

Lines are missing from the views shown below. Add the missing
lines, including centerlines. Use good centerline practices.

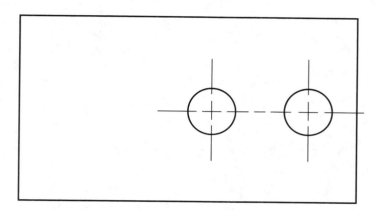

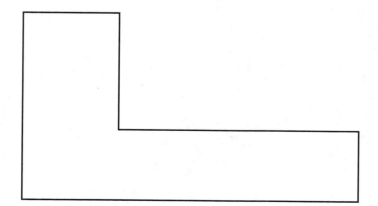

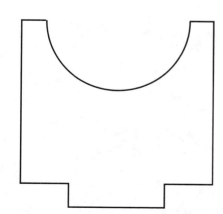

Make copies of these pages to use for additional practice.

Worksheet 6.1
Practice with hidden lines

Sketch an isometric pictorial of the object shown in the
orthographic views. A block with the correct overall
dimensions has been provided to get you started.

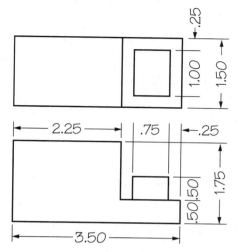

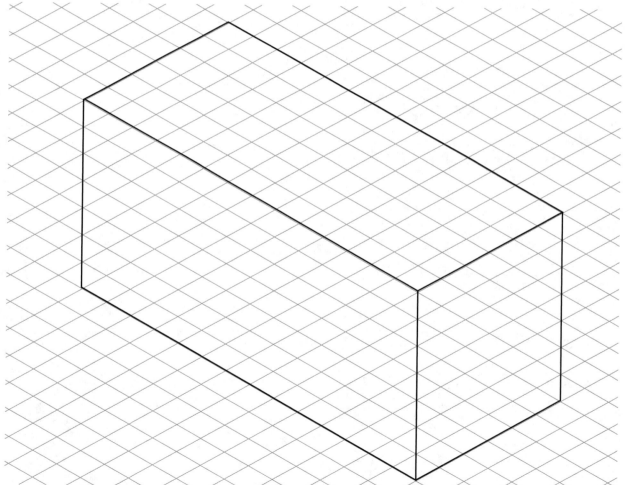

Make copies of these pages to use for additional practice.

Worksheet 6.2

Sketching curved shapes
in isometric

Finish the isometric sketch of the object shown in
the orthographic views.

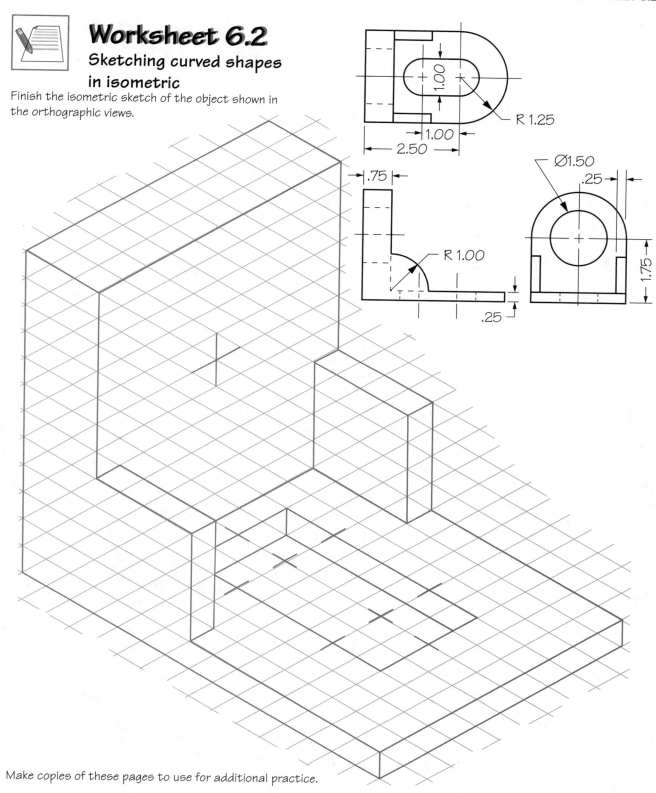

.75

R 1.00

1.00

1.00

2.50

R 1.25

Ø1.50

.25

1.75

.25

Make copies of these pages to use for additional practice.

Worksheet 7.1
Sketching a full section

Sketch the right side view as a full section. The cutting plane line has been shown, although in this case it is not necessary. Use the grid to help create thin hatch lines at an angle of 45° by drawing through the diagonals. Show hatching on the solid portions of the object cut through by the cutting plane.

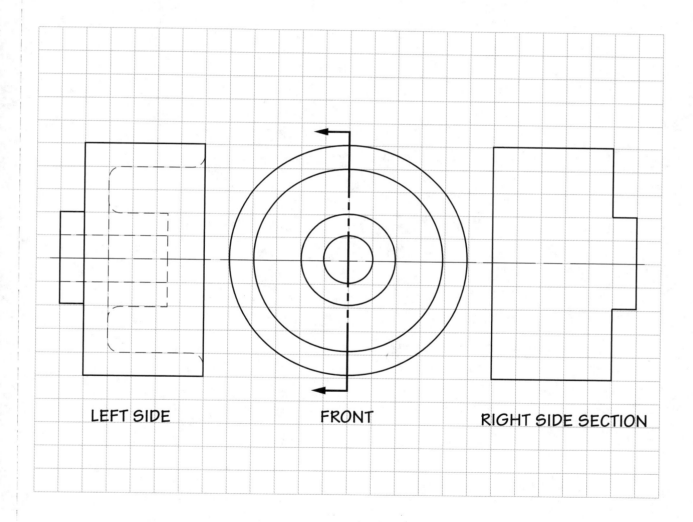

LEFT SIDE   FRONT   RIGHT SIDE SECTION

Make copies of these pages to use for additional practice.

Worksheet 7.2
Sketching a half-section

Add the right side view as a half section. Make sure to use a centerline to divide the sectioned and unsectioned halves. Notice the way the cutting plane is offset to pass through the hole. Keep in mind the practices for sectioning ribs and webs.

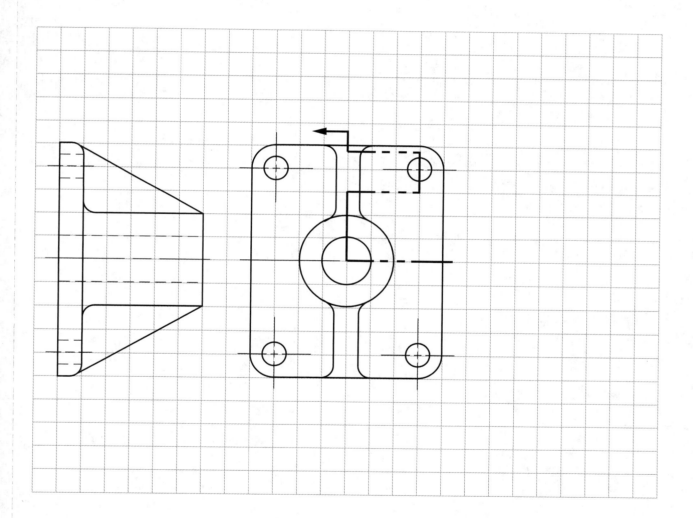

Make copies of these pages to use for additional practice.

Worksheet 7.3
Creating an aligned section

Sketch an aligned section for the right side view. Notice that in the left side view the holes are shown revolved onto the vertical centerline.

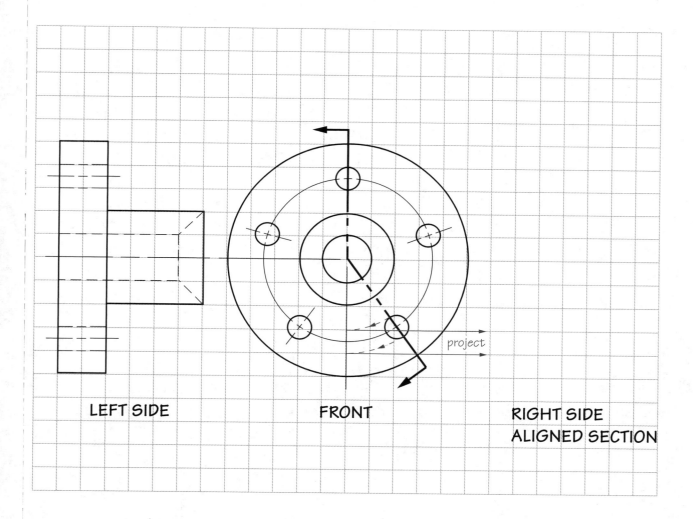

LEFT SIDE   FRONT   RIGHT SIDE
                                                     ALIGNED SECTION

project

Make copies of these pages to use for additional practice.

Worksheet 8.1
The auxiliary view glass box

Directions:

Cut out the paper representation of a glass box with an auxiliary viewing plane shown below.

Neatly letter the names "horizontal plane," "frontal plane," and "auxiliary plane" on the appropriate planes. Label the viewing planes with the principal dimensions that will show in each view. (For example, the front view will show the height and width of an object placed inside the box.)

Project an auxiliary view of the object shown pictorially at the upper right from the two views shown.

Fold on the folding lines and use a small piece of tape to join the top and auxiliary viewing planes. Then use the box to help you answer the questions below.

1. Why must the depth dimension be the same in the top view as in the auxiliary view?

2. Why can you draw projection lines between the front view and the auxiliary view?

3. Could this box have been cut and folded different way so that you could project from the top view?

4. How many other auxiliary viewing planes could you construct that would show the depth dimension?

5. Would this particular box show a true size auxiliary view for all objects?

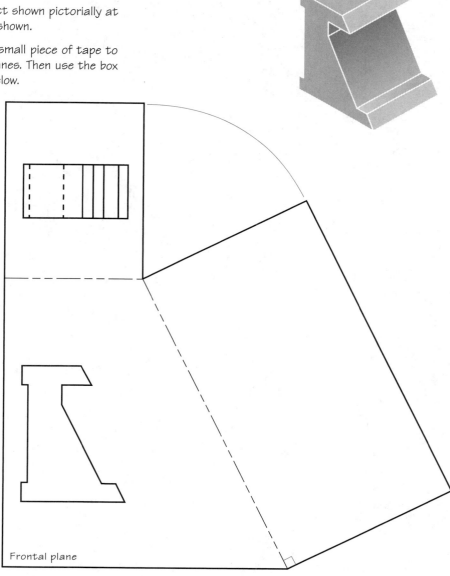

Frontal plane

Make copies of these pages to use for additional practice.

Worksheet 8.2
Developing a prism

The development of a prism explained in Step by
Step 8.4 is shown below. Cut on the solid lines
and fold on the fold lines to create the prism.
Tabs are provided to help you glue or tape the
development together.

A

D

B

1

C

4

B

3

4

A

2

1

D

1

Make copies of these pages to use for additional practice.

Worksheet 8.3
Developing a cylinder

The development of a cylinder explained in Section 8.28 is shown below. Cut on the solid lines and fold on the fold lines to create the prism. Tabs are provided to help you glue or tape the development together.

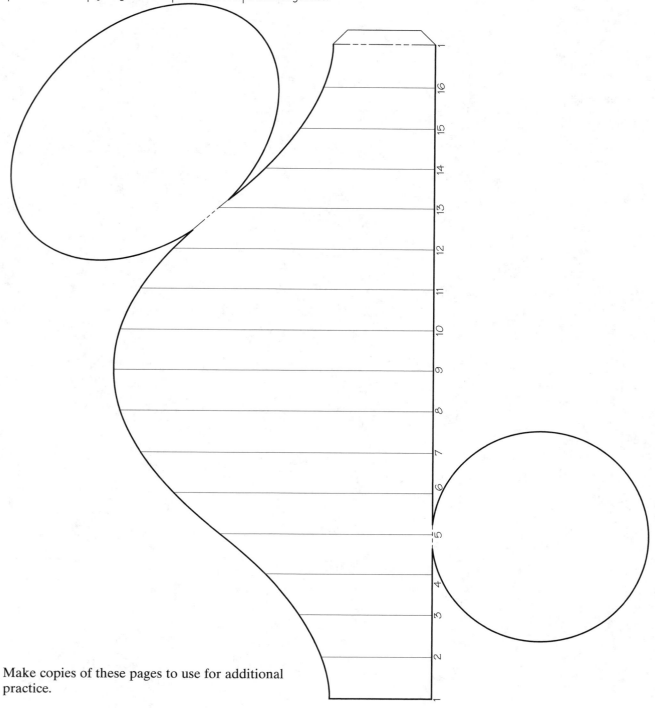

Make copies of these pages to use for additional practice.

Worksheet 9.1
Dimensioning

Add dimensions to the drawing views shown below using good technique, choice, and placement of dimension. Use the grid to help size the dimension features. Determine the dimension values by measuring the views or from the 1/8" grid spacing. The drawing is full scale. Use 2 place decimal inch measurements.

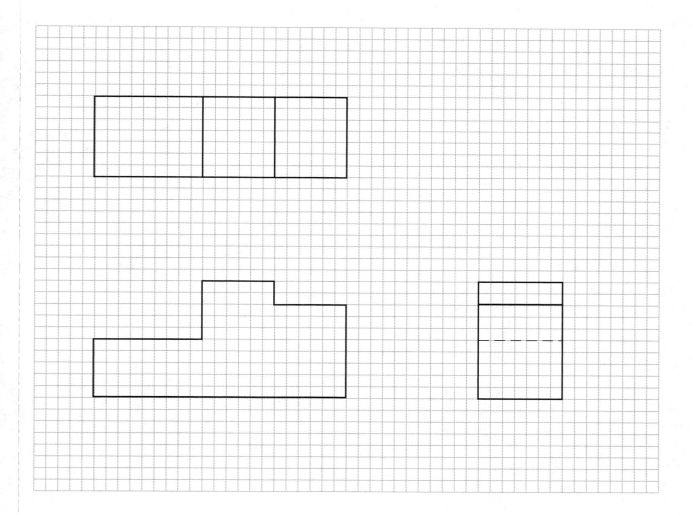

Make copies of these pages to use for additional practice.

Worksheet 9.2
Dimensioning

Some lines are missing. Add the missing lines, then measure the object and dimension using two place decimal inches. It is shown full size.

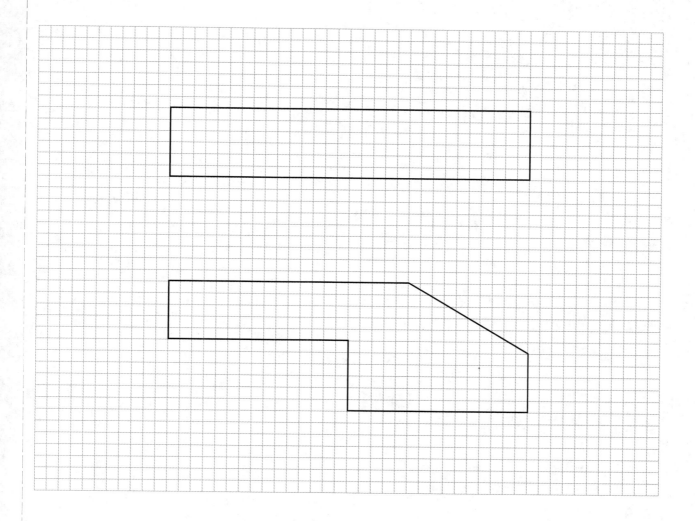

Make copies of these pages to use for additional practice.

Worksheet 9.3
Dimensioning

Measure the object in millimeters and dimension to the
nearest whole millimeter. It is shown on a 5 mm grid.

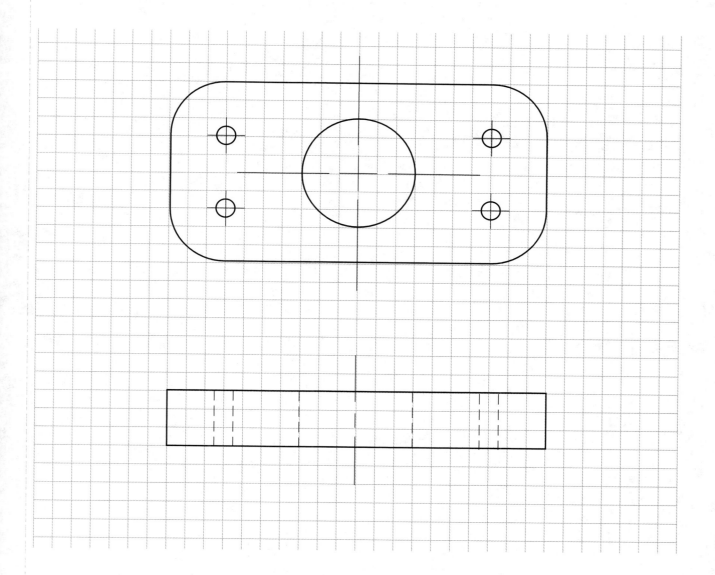

Make copies of these pages to use for additional practice.

Lettering Guides
1/8 inch and 1/4 inch lettering

Remove this sheet and use it as a guide under plain white paper. Save it for reuse throughout this book.

Grid Sheet
1/8 inch grid

Remove this sheet and place it under a sheet of plain white paper. Save it for reuse throughout this book.

Grid Sheet
5 millimeter grid

Remove this sheet and place it under a sheet of plain white paper. Save it for reuse throughout this book.

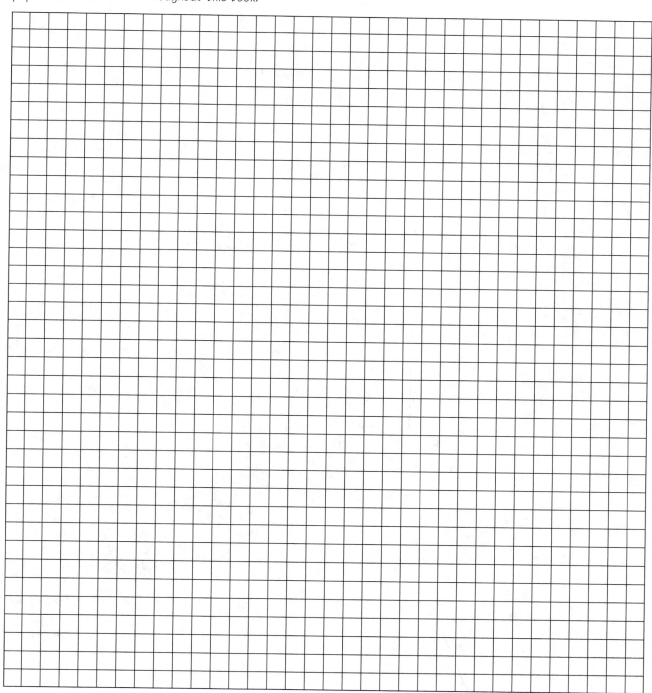

Grid Sheet
1/8 inch isometric grid

Remove this sheet and place it under a sheet of plain white paper. Save it for reuse throughout this book.

Grid Sheet
1/4 inch isometric grid

Remove this sheet and place it under a sheet of plain white paper. Save it for reuse through out this book.

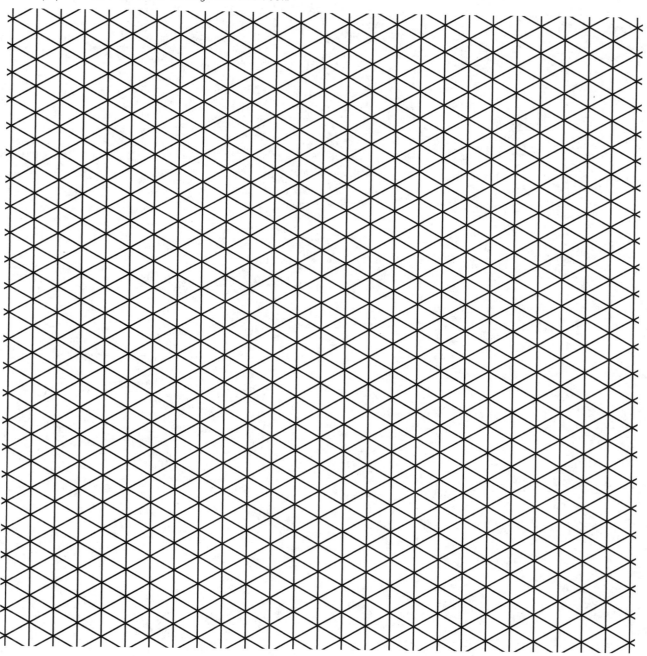

Grid Sheet
5 millimeter isometric grid

Remove this sheet and place it under a sheet of plain white paper. Save it for reuse throughout this book.

Decimal and Millimeter Equivalents

4ths	8ths	16ths	32nds	64ths	To 4 Places	To 3 Places	To 2 Places	Milli-meters	4ths	8ths	16ths	32nds	64ths	To 4 Places	To 3 Places	To 2 Places	Milli-meters
				1/64	.0156	.016	.02	.397					33/64	.5156	.516	.52	13.097
			1/32		.0312	.031	.03	.794				17/32		.5312	.531	.53	13.494
				3/64	.0469	.047	.05	1.191					35/64	.5469	.547	.55	13.891
		1/16			.0625	.062	.06	1.588			9/16			.5625	.562	.56	14.288
				5/64	.0781	.078	.08	1.984					37/64	.5781	.578	.58	14.684
			3/32		.0938	.094	.09	2.381				19/32		.5938	.594	.59	15.081
				7/64	.1094	.109	.11	2.778					39/64	.6094	.609	.61	15.478
	1/8				.1250	.125	.12	3.175		5/8				.6250	.625	.62	15.875
				9/64	.1406	.141	.14	3.572					41/64	.6406	.641	.64	16.272
			5/32		.1562	.156	.16	3.969				21/32		.6562	.656	.66	16.669
				11/64	.1719	.172	.17	4.366					43/64	.6719	.672	.67	17.066
		3/16			.1875	.188	.19	4.762			11/16			.6875	.688	.69	17.462
				13/64	.2031	.203	.20	5.159					45/64	.7031	.703	.70	17.859
			7/32		.2188	.219	.22	5.556				23/32		.7188	.719	.72	18.256
				15/64	.2344	.234	.23	5.953					47/64	.7344	.734	.73	18.653
1/4					.2500	.250	.25	6.350	3/4					.7500	.750	.75	19.050
				17/64	.2656	.266	.27	6.747					49/64	.7656	.766	.77	19.447
			9/32		.2812	.281	.28	7.144				25/32		.7812	.781	.78	19.844
				19/64	.2969	.297	.30	7.541					51/64	.7969	.797	.80	20.241
		5/16			.3125	.312	.31	7.938			13/16			.8125	.812	.81	20.638
				21/64	.3281	.328	.33	8.334					53/64	.8281	.828	.83	21.034
			11/32		.3438	.344	.34	8.731				27/32		.8438	.844	.84	21.431
				23/64	.3594	.359	.36	9.128					55/64	.8594	.859	.86	21.828
	3/8				.3750	.375	.38	9.525		7/8				.8750	.875	.88	22.225
				25/64	.3906	.391	.39	9.922					57/64	.8906	.891	.89	22.622
			13/32		.4062	.406	.41	10.319				29/32		.9062	.906	.91	23.019
				27/64	.4219	.422	.42	10.716					59/64	.9219	.922	.92	23.416
		7/16			.4375	.438	.44	11.112			15/16			.9375	.938	.94	23.812
				29/64	.4531	.453	.45	11.509					61/64	.9531	.953	.95	24.209
			15/32		.4688	.469	.47	11.906				31/32		.9688	.969	.97	24.606
				31/64	.4844	.484	.48	12.303					63/64	.9844	.984	.98	25.003
					.5000	.500	.50	12.700						1.0000	1.000	1.00	25.400

Metric measurements may be set off directly on drawings with the metric scale. Decimal measurements may be set off directly on drawings with the engineers' scale, or the decimal scale.

Symbols for Instructors Corrections

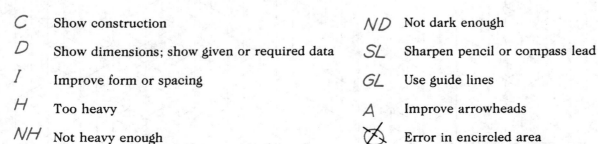

C	Show construction	ND	Not dark enough
D	Show dimensions; show given or required data	SL	Sharpen pencil or compass lead
I	Improve form or spacing	GL	Use guide lines
H	Too heavy	A	Improve arrowheads
NH	Not heavy enough	⊘	Error in encircled area